Energy, the Environment, and Sustainability

Energy, the Environment, and Sustainability

Efstathios E. Michaelides

CRC Press
Taylor & Francis Group
Boca Raton London New York

CRC Press is an imprint of the
Taylor & Francis Group, an **informa** business

eResource material is available for this title at https://www.crcpress.com/9781138038448.

CRC Press
Taylor & Francis Group
6000 Broken Sound Parkway NW, Suite 300
Boca Raton, FL 33487-2742

© 2018 by Taylor & Francis Group, LLC
CRC Press is an imprint of Taylor & Francis Group, an Informa business

---

Library of Congress Cataloging-in-Publication Data

Names: Michaelides, Efstathios, author.
Title: Energy, the environment, and sustainability / Efstathios E. Michaelides.
Description: Boca Raton, FL : CRC Press/Taylor & Francis Group, 2017. | "A CRC title,
part of the Taylor & Francis imprint, a member of the Taylor & Francis Group,
the academic division of T&F Informa plc." | Includes bibliographical references and index.
Identifiers: LCCN 2017055489| ISBN 9781138038448 (pbk. : acid-free paper) |
ISBN 9781138489172 (hardback : acid-free paper) | ISBN 9781315177359 (ebook)
Subjects: LCSH: Renewable energy sources.
Classification: LCC TJ808 .M5285 2017 | DDC 333.79--dc23
LC record available at https://lccn.loc.gov/2017055489

Visit the Taylor & Francis Web site at
http://www.taylorandfrancis.com

and the CRC Press Web site at
http://www.crcpress.com

To Emmanuel, Dimitri, and Eleni, whose generation will have to invent and develop sustainable methods for the production of the energy it needs.

Contents

Foreword

Our society is faced with an *energy challenge*. The consumption of continually increasing quantities of energy is essential for developed industrial societies to maintain their economic prosperity and affluence. In addition, as highly populated developing countries become more industrialized and affluent, they need more energy. As a result, the global energy demand is rapidly growing and has reached high levels that cannot be sustained in the future. The supply of fossil fuels, which currently provide more than 85% of the total global energy consumption, is limited. If our society continues with the current unsustainable energy mix, the fossil fuels will be depleted, very likely before the end of the twenty-first century. In addition, the widespread combustion of fossil fuels produces carbon dioxide, which is one of the causes of global warming, as well as of other adverse environmental effects (e.g., acid rain, higher ozone concentration in urban areas, particulates and aerosols) that are detrimental to air quality. The limited supply of fossil fuels and their effects on the environment point to the only sustainable long-term solution of the energy challenge: a significant increase of alternative energy sources for meeting the energy needs of the industrial and postindustrial states. This transformation will include conservation of the primary energy resources, higher efficiency for the use of energy, and storage to meet the energy demand when renewables are not in sufficient supply.

This book is aimed at two categories of readers:

1. Students of science and engineering who take elective courses on one of the subjects of *energy production, alternative energy, renewable energy, sustainability*, etc. These students will review and expand their knowledge in the traditional disciplines of thermodynamics, fluid dynamics, and heat transfer. If "repetition is the mother of learning," students in science and engineering will learn a great deal from the material of this book.

2. The educated reader, who has a basic knowledge of algebra, physics, and chemistry. The book assumes minimum prior knowledge on behalf of the reader and imparts all the required engineering concepts and material. The many worked examples in the book will help these readers to understand the engineering systems and perform quantitative analyses to draw their own independent conclusions on energy, the environment, and our sustainable options for the future.

The book is designed to provide the readers the means to understand the scientific principles of energy conversion and the operation of the technical systems that are employed for the harnessing of all the currently known energy sources. The readers of the book will become familiar with the scientific principles for the harnessing of primary energy sources and will be able to apply this knowledge to conduct feasibility studies and choose systems that make the best use of our energy resources.

Energy supply and the associated environmental issues have always had economic, social, and political effects and were strongly debated. The current debates on the global energy challenge and climate change are not exceptions. Citizens of all countries are bombarded with opinions, claims, and counterclaims on whether or not there is global warming; on the merits and demerits of solar and wind power; on the feasibility of biomass as an alternative global energy source; on the safety of nuclear reactors; and on count-less "inventions" that will magically solve our energy problem. Several of these opinions, inventions, and claims that are patently false are repeated without much thinking, due diligence, and the needed scientific analysis. The claims have created many "myths" that are often repeated and have spread confusion among the general public. A unique aspect of this book is the inclusion of sections (at the end of each chapter) that debunk many misconceptions related to energy production and the environment. The key to this is the quantitative analysis of the data that are pertinent to each one of the misconceptions. All the myths that are mentioned in this book emanate from statements of prominent "politi-cians," environmental "activists," and energy "experts." When the pertinent data are quan-titatively analyzed within the time-honored scientific framework, conclusions are drawn, and the "reality" about our sustainable options becomes apparent. I hope that exposing such myths using the scientific data and quantitative tools that abound in the book will help readers draw their own independent and educated conclusions and will develop their own ideas on how to tackle the global energy challenge.* The readers will also become familiar with the environmental, political, social, and economic issues that surround the use of energy sources, and they will be able to participate as well-informed citizens in the ongoing debates on energy, the environment, and sustainability.

A second unique feature of this book is the inclusion of chapters and the high emphasis on *nuclear energy, energy storage, energy conservation and efficiency*, and *decision-making meth-ods*. When our society decides to seriously tackle the global climate change problem, all four will play important roles in the energy supply of nations. With the intermittent wind and periodically variable insolation, large-scale energy storage is vital for the reliable and uninterrupted supply of electricity. And without energy conservation projects and more efficient engines for energy conversion processes, sustainability cannot be attained for the growing population of the planet.

A number of individuals have helped in the writing of this book: First among them are the students who took my courses on power generation, alternative energy, and sustain-ability, which I have offered in the last 37 years at four universities (Delaware, Tulane, University of Texas at San Antonio, and Texas Christian). I have learned from them more than they have learned from me. I am very thankful to my several colleagues at these universities as well as to other colleagues I regularly meet during conferences for many fruitful and often animated discussions on energy, the environment, and the great chal-lenge our society is facing. I am also very much indebted to my own family, not only for their constant support, but also for lending a hand whenever this was needed. My wife Laura has been a constant source of inspiration and help. Our three children, Emmanuel, Dimitri, and Eleni, were always there and ready to help. I owe to all my sincere gratitude.

Efstathios E. (Stathis) Michaelides
Fort Worth, Texas

---

* While working on these sections, I came across the excellent monograph of Professor Vaclav Smil that exam-ines and debunks 10 energy myths.

Author

Efstathios E. Michaelides is the Tex Moncrief chair of Engineering at Texas Christian University. Prior to this appointment, he was chair of the Department of Mechanical Engineering of the University of Texas at San Antonio, where he also held the Robert F. McDermott chair in engineering and was the founder and director of the National Science Foundation-supported Center on Simulation, Visualization and Real Time Computing. In the past, he was the founding chair of the Department of Mechanical and Energy Engineering at the University of North Texas (2006–2007); the Leo S. Weil professor of mechanical engineering at Tulane University (1998–2007); director of the South-Central Center of the National Institute for Global Environmental Change (2002–2007); associate dean for Graduate Studies and Research in the School of Engineering at Tulane University (1992–2003); and head of the Mechanical Engineering Department at Tulane (1990–1992). Between 1980 and 1989, he was on the faculty of the University of Delaware, where he also served as acting chair of the Mechanical Engineering Department (1985–1987).

Professor Michaelides has served in leadership positions in several national and international scientific organizations. He has published more than 140 journal papers and has contributed more than 230 papers and presentations in national and international conferences. He has also published four other books: *Particles, Bubbles and Drops—Their Motion and Heat Transfer, Alternative Energy Sources, Heat and Mass Transfer in Particulate Suspensions, and Nanofluidics—Thermodynamic and Transport Properties.*

Professor Michaelides was awarded the student chapter American Society of Mechanical Engineers (ASME)/Phi Beta Tau excellence in teaching award (1991 and 2001); the Lee H. Johnson award for teaching excellence (1995); a Senior Fulbright Fellowship (1997); the ASME Freeman Scholar award (2002); the Outstanding Researcher award at Tulane University (2003); the ASME Outstanding Service award (2007); the ASME Fluids Engineering award (2014); and the ASME-FED 90th Anniversary medal (2016). He holds a bachelors degree (honors) from Oxford University and masters and doctorate degrees from Brown University. He was also awarded an honorary MA degree from Oxford University (1983).

Commonly Used Abbreviations

atm	atmospheric
AWM	annual worth method
CAES	compressed air energy storage
CAFE	corporate average fuel economy
CCS	carbon capture and sequestration
CFC	chlorofluorocarbon
COP	coefficient of performance
CPV	concentrated photovoltaic
DEC	direct energy conversion
emf	electromotive force
EPBT	energy payback time
EROI	energy return on investment
EU	European Union
FBR	fluidized bed reactor
F–T	Fischer–Tropsch (process)
GCM	global circulation model
GDP	gross domestic product
GHG	greenhouse gas
GIB	grid-independent building
GSHP	ground source heat pump
HDI	human development index
HFC	hydrofluorocarbon
HFO	hydrofluoroolefin
HHV	high heating value
HTFC	high-temperature fuel cell
HVAC	heat ventilation and air-conditioning
IAEA	International Atomic Energy Agency
ICE	internal combustion engine
IEA	International Energy Agency
IPEEC	International Partnership for Energy Efficiency Cooperation
LED	light-emitting diode
LEED	Leadership in Energy and Environmental Design
LHV	low heating value
LNG	liquefied natural gas
LOCA	loss of coolant accident
MSW	municipal solid waste
NAECA	National Appliance Energy Conservation Act
NDC	nationally determined contributions
NPS	new policies scenario (EIA term)
NPV	net present value
NREL	National Renewable Energy Laboratory
OECD	Organization for Economic Cooperation and Development
OPEC	Organization of Petroleum Exporting Countries
ORC	organic Rankine cycle

PCM	phase-change material
PHS	pumped hydro system
ppb	parts per billion (by volume)
ppm	parts per million (by volume)
PV	photovoltaic
PV	present value (in economics, Chapter 9)
rpm	revolutions per minute
SEER	seasonal energy efficiency ratio
SI	Systeme Internationale—the International System of units
SOFC	solid oxide fuel cell
TPES	total primary energy supply
UCG	underground coal gasification
USGS	US Geological Survey
UN	United Nations
UV	ultraviolet
WTO	World Trade Organization
ZNEB	zero-net energy building

1

Fundamental Concepts

The beginnings of human civilization are defined by the discovery and utilization of energy for space heating and lighting. When prehistoric humans mastered the use of fire for domestic comfort and cooking, human civilization began. Later, the use of animal power (horses and oxen) significantly contributed to the development of ancient and medieval agricultural societies and helped nourish an increasing human population. More recently, the industrial revolution and the invention of energy-voracious engines enabled the development of today's *knowledge-based society*, where citizens live in greater comfort, increasingly use fast modes of transportation, and are continuously in contact with modern electronics. Throughout the centuries, the human society has evolved by using energy in increasingly larger quantities. Today, the consumption of vast amounts of energy—in thermal, chemical, mechanical, and electrical forms—is absolutely necessary for the functioning of the contemporary society, the prosperity of the nations, and the survival of our civilization.

While we commonly use the word *energy* for several related concepts, we use energy per se in different forms as source of lighting, heating, communications, and motive power. Airplanes and automobiles use liquid hydrocarbon fuels. Computers and wireless network systems use electricity; the latter is produced from power plants that convert other primary resources (coal, natural gas, nuclear, solar, wind, geothermal, hydroelectric, etc.). Contemporary households use natural gas for cooking, and heating and electricity for lighting and entertainment. Elaborate and technologically very advanced networks for the transportation and supply of *energy* to the final consumer have been developed in the last three centuries. Electricity is fed into communities by the transmission lines of the electric grid at very high voltage. Natural gas, by a complex system of pipelines, which often transcend national boundaries. Tanker ships crisscross the oceans daily to transport crude oil to refineries, which then supply us with gasoline and diesel via an intricate system of pipelines, train cars, and trucks. In today's world, energy enters all aspects of human life, economics, and politics. Actually, the use of energy defines the lives of contemporary humans.

The impact of energy supply and the energy trade is of paramount importance to all nations. The geopolitical activities of modern nations have been significantly influenced by their need for a reliable, uninterrupted, and secure energy supply. Several wars, after 1950, have been fought for the control and security of energy supplies—primarily petroleum—and many treaties and international agreements have been cemented with energy resources as the primary issue.

This first chapter is a quantitative introduction to the broad concept of energy: It explains the fundamental concepts related to energy transformations and establishes the analytical basis for the rest of the book. Relevant elementary concepts of the discipline of thermodynamics are presented to explain the fundamental principles for the conversion of energy from one form into another. Because a great deal of confusion to the nonexpert stems from the use of a myriad of units for energy, an extensive section is devoted to the explanation and relation of the different units that are still used. The definitions and relationships of common figures of merit we use daily, such as the efficiency, the coefficient of performance, and seasonal energy efficiency ratio (SEER), are given in a simple and comprehensible way. Finally, a succinct description is given of the thermodynamic cycles that produce electric power and provide refrigeration and air-conditioning.

1.1 Work, Energy, Heat

Mechanical work (or simply work), heat, and energy are related concepts measured by the same units, but they are fundamentally different. Electricity is a form of work. In everyday life, we use work for transportation, for lighting, for refrigeration or air-conditioning, and for communications. We use heat for domestic comfort; for the preparation of meals; and for a variety of industrial processes, where higher temperatures are desirable. Work is defined as the product of a force and the distance its point of application moves:

$$dW = \vec{F} \cdot d\vec{x} \Rightarrow W = \int \vec{F} \cdot d\vec{x}. \tag{1.1}$$

When work is performed on a system, its energy increases. When we lift a weight with a force equal or higher than its weight—the gravitational force—we increase its energy. The opposite happens when the weight falls. Heat, on the other hand, is not related to the macroscopic motion or any forces acting on a system. Heat is transferred from one object to another when there is a temperature difference.

Work and heat are forms of energy that may be converted to one another and are measured by the same units. Energy is a very broad concept that characterizes the potential of materials and systems to produce useful work or to supply heat. Energy (sometimes referred to as internal energy or enthalpy) is a property of the materials and systems. Systems possess energy, in different forms and quantities: a hot cup of tea contains more energy than a cold one; a fast moving vehicle possesses kinetic energy or motion energy; a ton of coal contains a great deal of energy, in chemical form; similarly, a ton of natural uranium contains a significant amount of nuclear energy. The chemical energy of the coal and the nuclear energy of the uranium may be transformed to heat in burners and nuclear reactors.

Forms of energy, which are commonly met in engineering practice and everyday applications, are as follows [1,2]:

1. The potential energy of materials at higher elevations
2. The kinetic energy of materials in motion; these include the wind and sea currents
3. The chemical energy of substances, such as coal, biomass, and hydrocarbons
4. The elastic energy of a solid material under stress or torsion
5. The electric energy of electric charges
6. The magnetic energy of magnets
7. The electromagnetic energy of solar radiation (insolation), microwaves, and all other electromagnetic waves
8. The energy of ocean waves and tides
9. The nuclear energy of radionuclides
10. The surface energy of liquids

Modern humans use the energy content of several substances to produce mechanical work, electricity, and heat and to accomplish several tasks and activities that are necessary for the smooth functioning of our society. Different forms of energy are converted to other forms in order to produce the form of energy that is most convenient or suitable for these human activities: aviation fuel is converted in a jet engine to transport a group of passengers from London to Shanghai; electric and magnetic charges facilitate the magnetic storage of data in a computer; and natural gas is converted into heat and helps produce a birthday cake. These conversions of energy from one form to another are subjected to the laws of thermodynamics, which are explained in simple form in the following sections.

Related to the concept of energy is power, which is the rate of work per unit time. From Equation 1.1, one may define the power $\dot{W}$ by dividing with a time duration, as follows:

$$\dot{W} = \vec{F} \cdot \frac{d\vec{x}}{dt} \Rightarrow \dot{W} = \vec{F} \cdot \vec{V}, \tag{1.2}$$

where $\vec{V}$ is the velocity the application point force is moving. It is common thermodynamic convention that the overdot (.) denotes the time rate of the respective variable. As the last part of the equation shows, power may be considered as the product of a force and a velocity.

1.2 Units and Unit Conversions

A novice in the subject of energy may be easily confused by the plethora of commonly used units, which have historically emerged from a variety of sources. The *Système International* (*SI*) is the most commonly used system of units in the world (the United States is one of the few exceptions of countries that have not formally adopted the system) and the joule (J) is the unit of energy in this system. The joule is defined as the work done when a constant force of 1 newton (1 N) moves its point of application by 1 meter (1 m). In electricity applications, 1 J is performed when a charge of 1 coulomb (1 C) moves through an electric potential difference of 1 volt (1 V). In the realm of energy studies, 1 coulomb volt is equivalent to 1 newton meter.

Since the advent of the industrial revolution, several other units have been defined and used in engineering applications. Units that are still in use and may be encountered in the twenty-first century are as follows [1]:

1. In the centimeter–gram–second (cgs) system of units, which is primarily used in some areas of physics and chemistry, the erg is the energy unit. One erg is equal to 10^{-7} J.

2. The British system of units was developed during the industrial revolution for the use with large heat engines. The British thermal unit, which is abbreviated as Btu (and less often as BTU) is the principal energy unit. One British thermal unit is defined as the amount of energy needed to increase the temperature of 1 lb of water from 59.5°F to 60.5°F. One Btu is equal to 1,055 J, and is often approximated as 1 kJ.

3. A larger energy unit in the British system is the therm (1 thm), and it is equal to 100,000 Btu. One therm is equal to 1.055×10^8 J.

4. The calorie (cal) is an older unit that is still used in thermal and food applications and is equal to 4.184 J. One cal is the amount of energy (heat) required to raise the temperature of 1 g of water from 14.5°C to 15.5°C.

5. When it refers to food and nutrition applications, what is commonly called calorie, "1 cal," is actually 1 kcal, equal to 4.184 kJ.

6. An extremely large amount of energy is the quad (1 Q). This unit is commonly used to measure the amount of energy consumed by regions or nations or by the entire population of the earth. One quad is equal to 10^{15} Btu or, approximately, 10^{18} J. The total energy consumed in the United States in 2017 was approximately 100 Q, and that in the entire world was approximately 540 Q.

7. Very low energies are associated with atomic and subatomic applications, e.g., the energy of one electron or one neutron. For nuclear applications, the electron volt (1 eV) has been defined as the potential energy gained by an electron when it moves through an electric potential difference of 1 V. One electron volt is equal to 1.6×10^{-19} J. The megaelectron volt (MeV), which is often used in nuclear physics, is 1.6×10^{-13} J.

A series of energy units often used in the fossil fuel industry is related to the chemical energy content of fossil fuels, such as coal, crude oil, and natural gas. Because the chemical composition and energy content of actual fuels, e.g., sweet East Texas crude, West Texas Intermediary, and Saudi Arabian crude, largely depends on the location of the mining, these units of energy have been defined according to international convention. Frequently used units of this kind and their SI equivalents are as follows [1]:

1. One ton of coal equivalent (1 tce) is equal to 2.931×10^{10} J.

2. One barrel of oil (1 bbl) is equal to 6.119×10^{9} J.

3. One ton of oil equivalent (1 toe) is equal to 10^{10} cal or 4.187×10^{10} J.

4. One cubic foot of natural gas (1 standard cubic foot [scf]), measured at standard pressure and temperature (1 atm and 25°C), is equal to 1.072×10^{6} J. One standard cubic foot is often approximated as 10^{6} J.

The unit of power (energy per unit time) in the SI is the watt (W), defined as 1 J/s. When an engine consumes or produces 1 J of work per second, its power is 1 W. Oftentimes, for larger engines, the kilowatt (1 kW = 1000 W) and the megawatt (1 MW = 1,000,000 W) are used. Another commonly used power unit in the transportation industry is the horsepower (hp), which is equal to 745.7 W.

Multiples of all the SI units, including the joule and the watt, have been defined by international convention for multiples in the range of 10^{-24} to 10^{24}. For the unit joule, they are denoted as follows:

- 1 yoctojoule (yJ) = 10^{-24} J
- 1 zeptojoule (zJ) = 10^{-21} J
- 1 attojoule (aJ) = 10^{-18} J

- 1 femtojoule (fJ) = 10^{-15} J
- 1 picojoule (pJ) = 10^{-12} J
- 1 nanojoule (nJ) = 10^{-9} J
- 1 microjoule (μJ) = 10^{-6} J
- 1 millijoule (mJ) = 10^{-3} J
- 1 kilojoule (kJ) = 10^{3} J
- 1 megajoule (MJ) = 10^{6} J
- 1 gigajoule (GJ) = 10^{9} J
- 1 terajoule (TJ) = 10^{12} J
- 1 petajoule (PJ) = 10^{15} J
- 1 exajoule (EJ) = 10^{18} J
- 1 zettajoule (ZJ) = 10^{21} J
- 1 yottajoule (YJ) = 10^{24} J

Thus, 1 Q is equal to 1.055 exajoules (1 Q = 1.055 EJ). These prefixes are not specific to energy units and are applied to all other units of SI. It must be emphasized that the SI units and their multiples follow a very precise notation, including the use of lowercase and capital letters. Errors in notation between lowercase letters and capital letters may result in calculation errors of several orders of magnitude: While mJ/mJ = 1 by definition as an identity, MJ/mJ = 10^9, and PJ/pJ = 10^{27}. Everyone has to be very careful and precise with the notation of units!

Table 1.1 shows the conversion factors for several energy units that are commonly used [1]. One calorie = 4.18 J, 1 tce is equal to 2.78×10^7 Btu, etc. The diagonal terms of this table are equal to 1 by definition, and each term is equal to the reciprocal of its diagonally mirror term, that is, $c_{ij} = 1/c_{ji}$.

Example 1.1

1. Convert to joules: 3.45 Btu; 42 bbl; 3.8 Q.
2. Convert to British thermal units: 1 MJ; 34 scf; 5.6 kWh.

Solution: From Table 1.1

1. 3.45 Btu = $3.45 \times 1{,}055$ J = 3,639.75 J
   42 bbl = $42 \times 6.12 \times 10^9$ J = 257.04×10^9 J
   3.8 Q = $3.8 \times 1.06 \times 10^{18}$ J = 4.03×10^{18} J
2. 1 MJ = 1,000,000 J = $1{,}000{,}000 \times 9.48 \times 10^{-4}$ Btu = 948 Btu
   34 scf = $34 \times 1{,}016$ Btu = 34,544 Btu
   5.6 kWh = $5.6 \times 3{,}412$ Btu = 19,107 Btu

Other properties that are extensively used in this book, their SI units, and several useful conversion factors are included in Table 1.2.

TABLE 1.1

Conversion Factors for Energy Units

	Electron Volt (eV)	Calorie (cal)	erg	Joule (J)	British Thermal Unit (Btu)	Gallon of gasoline	Barrel of Oil (bbl)	Ton of Oil Equivalent (toe)	Kilowatt Hour (kWh)	Foot Pound (ft lb)	therm	ton-TNT	Quad (Q)	Ton of Coal Equivalent (tce)	Cubic Foot of Natural Gas, scf
Electron volt, eV	1	3.82E-20	1.60E-12	1.60E-19	1.52E-22	1.21E-27	2.61E-29	3.83E-30	4.44E-26	1.18E-19	3.82E-29	3.82E-29	1.52E-37	5.46E-30	9.89E-26
Calorie, cal	2.62E+19	1	4.19E+07	4.184	3.97E-03	3.17E-08	6.84E-10	1.0E-10	1.16E-06	3.09	3.97E-08	1.00E-09	3.97E-18	1.43E-10	3.90E-06
erg	6.25E+11	2.39E-08	1	1.00E-07	9.48E-11	7.58E-16	1.63E-17	2.39E-18	2.78E-14	7.37E-08	9.48E-16	2.39E-17	9.48E-26	3.41E-18	9.33E-14
Joule, J	6.25E+18	0.24	1.00E+07	1	9.48E-04	7.58E-09	1.63E-10	2.39E-11	2.78E-07	0.74	9.48E-09	2.39E-10	9.48E-19	3.41E-11	9.33E-07
British thermal unit, Btu	6.59E+21	252	1.06E+10	1,055	1	8.00E-06	1.72E-07	2.52E-08	2.93E-04	7.78E+02	1.00E-05	2.52E-07	1.00E-15	3.60E-08	9.84E-04
Gallon of gasoline	8.25E+26	3.15E+07	1.32E+15	1.32E+08	1.25E+05	1	2.16E-02	3.15E-03	3.66E+01	9.73E+07	1.25	3.15E-02	1.25E-10	4.50E-03	123.1
Barrel of oil, bbl	3.82E+28	1.46E+09	6.12E+16	6.12E+09	5.80E+06	46.4	1	0.1462	1.70E+03	4.51E+09	5.80E+01	1.46	5.80E-09	2.09E-01	5,708
Ton of oil equivalent, toe	2.61E+29	1.0E+10	4.187E+17	4.187E+10	3.97E+07	317.8	6.84	1	1.16E+04	3.09E+10	3.97E+02	10	3.97E-08	1.43	3.91E+04
Kilowatt hour, kWh	2.25E+25	8.60E+05	3.6E+13	3.60E+06	3,412	2.73E-02	5.88E-04	8.62E-05	1	2.66E+06	3.41E-02	8.61E-04	3.41E-12	1.23E-04	3.358
Foot pound, ft lb	8.48E+18	0.32	1.36E+07	1.36	1.29E-03	1.03E-08	2.22E-10	3.24E-11	3.77E-07	1	1.29E-08	3.24E-10	1.29E-18	4.63E-11	1.27E-06
therm	6.59E+26	2.52E+07	1.06E+15	1.06E+08	1.00E+05	0.8	1.72E-02	2.52E-03	29.31	7.78E+07	1	2.52E-02	1.00E-10	3.60E-03	98.42
ton-TNT	2.62E+28	1.00E+09	4.18E+16	4.18E+09	3.97E+06	31.73	0.68	0.1	1162	3.09E+09	39.66	1	3.97E-09	1.43E-01	3,899
Quad, Q	6.59E+36	2.5E+17	1.06E+25	1.06E+18	1E+15	8.00E+09	1.72E+08	2.52E+07	2.93E+11	7.78E+17	1.00E+10	2.52E+08	1	3.60E+07	9.84E+11
Ton of coal equivalent, tce	1.83E+29	7.00E+09	2.93E+17	2.93E+10	2.78E+07	2.22E+02	4.79E+00	0.6993	8.14E+03	2.16E+10	2.78E+02	7.00	2.78E-08	1	2.73E+04
Cubic foot of natural gas, scf	1.01E+25	2.56E+05	1.07E+13	1.07E+06	1,016	8.12E-03	1.75E-04	2.56E-05	0.2978	7.88E+05	1.02E-02	2.56E-04	1.02E-12	3.66E-05	1

TABLE 1.2

Frequently Used Units and Conversion Factors

Property	Symbol	Units
Area	A	1 m² = 10.73 ft²
		1 ha = 10,000 m² (hectare)
		1 km² = 1,000,000 m²
Density	ρ	1 kg/m³ = 0.06243 lb/ft³
Energy flux, heat flux, insolation	$\dot{E}/A$, $\dot{Q}/A$, S	1 W/m² = 0.3172 Btu/(hr·ft²)
Force	F	1 N = 0.2248 lb$_f$
Heat transfer coefficient	h, U	1 W/(m²·K) = 0.1761 Btu/(hr·ft·°F)
Length	L	1 m = 3.2808 ft
		1 mi = 1,6093 m
		1 ft = 12 in
Mass	m	1 kg = 2.2046 lb
		1 ton = 1000 kg
Mass flow rate	$\dot{m}$	1 kg/s = 2.2046 lb/s
		1 kg/s = 7,037 lb/hr
Power, heat rate	$\dot{W}, \dot{Q}$	1 W = 3.415 Btu/hr
		1 W = 0.7376 (lb$_f$·ft)/s
		1 hp = 745.7 W
		1 t (refrigeration) = 12,000 Btu/hr = 3,514 W
Pressure	P	1 Pa = 1 N/m² = 0.000145 psi (lbf/in²)
		1 atm = 101,325 Pa
		1 atm = 14.7 psi
Specific heat capacity	c or c_p	1 J/(kg·K) = 0.0002388 Btu/(lb × °F)
Thermal conductivity	k	1 W/(m·K) = 0.5777 Btu/(hr × ft × °F)
Velocity	V	1 m/s = 3.2808 ft/s
		1 mph = 0.44703 m/s
		1 knot = 0.5144 m/s
Volume	V	1 m³ = 1000 L = 35.31 ft³
		1 gal (US) = 3.785 L
		1 gal (UK) = 4.546 L
		1 barrel (oil, or 1 bbl) = 42 US gal = 0.159 m³ = 159 L
		1 barrel (whiskey) = 31.5 US gal
Volumetric flow rate	$\dot{V}$	1 m³/s = 1000 L/s = 35.31 ft³/s
		1 m³/min = 0.01666 m³/s
		1 gal/min (1 gpm) = 0.0631 L/s

1.3 Elements of Thermodynamics: Principles of Energy Conversion

The development of thermodynamics was the intellectual response to the technological breakthroughs that brought heat engines and marked the beginnings of the Industrial Revolution in the middle of the eighteenth century. Following the invention of the steam engine that demonstrated the conversion of chemical energy through heat to mechanical work—motive power—the scientific community of the nineteenth century developed the subject of thermodynamics in order to explain the conversion processes from the caloric (heat and rate of heat) to motive power (work and power), to optimize the energy conversion processes and thermal engines, and to better design engines for the production of mechanical and electrical power.

Even though the technology of the heat engines was relatively well developed, in the beginning of the nineteenth century, scientists did not have a clear understanding of how chemical energy and heat were converted to mechanical power. The technologists James Watt, William Murdoch, George Stephenson, and Robert Fulton, among others, had invented and produced steam engines that powered industrial machinery, trains, and boats. However, the concepts of work, heat, and energy were understood neither by the inventors nor by the scientists of the early nineteenth century. Work was associated to motion (motive power). Heat was considered to be a weightless fluid, "the caloric," which was exchanged between hotter and colder objects.* When this weightless fluid passed from one body to another, it cooled the first and warmed the second. In the first half of the nineteenth century, several scientists and practicing engineers, chief of whom were Benjamin Thompson (Count Rumford), Robert Meyer, James Prescott Joule, Sadi Carnot, Rudolf Clausius, and William Thomson (Lord Kelvin), put together the analytical and experimental framework that demonstrated the equivalency of heat and work and established the scientific principles that govern the conversion of one energy form into another. This analytical framework, which is accepted as sound and correct in our days, is summarized by the first and second laws of thermodynamics [2,3].

The concept of the *thermodynamic system* (or simply *system*) is central to the understanding of the theory of thermodynamics. The system is enclosed by a *boundary*, and outside the boundary are the *surroundings*. The latter is the part of the universe that is affected by changes in the given thermodynamic system. Thermodynamic systems are defined as *closed* or *open*. The two types of systems are depicted in Figure 1.1. Closed systems contain a fixed amount of molecules (and mass), while open systems have inlets and exits through which mass is allowed to flow. If the sums of the mass flow rates at the inlets and the exits are equal, the open system does not receive or discharge a net amount of mass, and it is said to be at *steady state*. When the mass inside the open system varies with time, the system is *unsteady* or *transient*. One may notice in Figure 1.1 that the typical quantities of interest in a closed system are heat and work, Q and W, respectively, which are measured in joules, while the typical quantities of interest in the open systems are the time rates of heat and work and the rate of heat and the power, $\dot{Q}$ and $\dot{W}$, respectively, which are measured in watts, as well as the mass flow rates $\dot{m}_i$ and $\dot{m}_e$ that enter and exit the system. The horsepower (hp), a unit of power in the British system of units, is sometimes used with open systems. For the rate of heat $\dot{Q}$ (the heat power), we sometimes use British thermal units per second or British thermal units per hour, with 1 Btu/s = 1,055 W, and 1 Btu/hr = 0.295 W.

From the practical standpoint, the vast majority of energy conversion machinery is open systems: Pumps, boilers, turbines, compressors, nozzles, and all the various kinds of heat exchangers are all open systems. Most of them operate and are modeled as open systems at steady state. Cylinders fitted with pistons that are typical of internal combustion engines used in automobiles are modeled as closed systems.

---

* Several terms have remained in our vocabulary as reminders of that weightless fluid: Heat transfer is often described as *heat flow* or *heat flux*, and the potential of a substance to store thermal energy (internal energy or enthalpy) is called *heat capacity*, in analogy to the capacity of a vessel to hold a liquid. We also refer to *heat reservoirs* for the supply or rejection of heat.

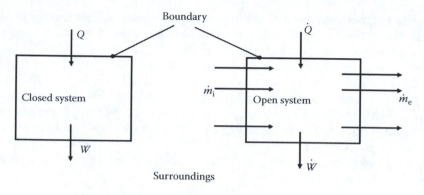

FIGURE 1.1
Schematic diagrams of closed and open thermodynamic systems.

1.3.1 First Law of Thermodynamics

After several experiments that did not meet with the approval of his pedantic academic colleagues, Joule (who was a brewer) devised an experiment [4], in the middle of the nineteenth century, to not only demonstrate that mechanical work and heat are inter-related quantities, but also derive a very accurate expression for the mechanical equiva-lent of heat: the number of mechanical work units, which are now known as "joules," that correspond to one calorie. Joule's experimental apparatus is schematically shown in Figure 1.2. A closed and well-insulated vessel contained a constant and known mass of water. A paddle wheel in the vessel was used to stir the water and a thermometer mea-sured the temperature of the water. The paddle wheel was driven by a mechanism, which was powered by a falling weight. As the weight fell slowly to the ground, it moved the wheel, which stirred the water. Thus, the potential energy of the weight was converted to the kinetic energy of the stirred water, and the latter was dissipated by friction. Joule observed that the only effect of the falling weight was an increase in the temperature of the water. The experiment implied that the potential energy of the weight was converted into kinetic energy in the fluid, and the latter, into the thermal energy (now known as internal energy) of the water in the system, an effect that was previously thought to be

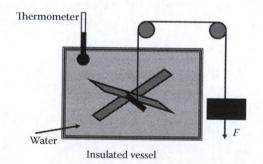

FIGURE 1.2
Joule's experiment.

achieved only by the "flow" of the caloric/heat. Joule proved that the same effect may be also accomplished by the dissipation of mechanical energy or work, and therefore, mechanical work and heat are equivalent variables. Using this experimental apparatus, he also obtained a surprisingly accurate figure for the equivalency of work to heat (1 cal = 4.159 J), which is very close to the now accepted number: 1 cal = 4.184 J [5]. Joule's meticulous experimental work convinced the scientific establishment of the nineteenth century that heat, work, and energy are equivalent; they are not destroyed, and they may be converted from one form to another, a principle that is now known as the *first law of thermodynamics*.

Two important relationships of the first law, which help understand quantitatively the energy–heat–work relationship, are the following:

1. For closed systems that contain a fixed mass m and undergo a process from state 1 to state 2, we have the relation

$$Q - W = E_2 - E_1 = \Delta E = m\Delta e. \tag{1.3}$$

   Q is the heat supplied to the system during the process; W is the work produced by the system; E is the total energy; and e is the total specific energy (or energy per unit mass) of the system. The differences ΔE and Δe denote the change in the total energy and the change in the total energy per unit mass of the system from the beginning to the end of the process. These differences may be calculated from other measurable properties, such as temperature and pressure.

2. For open systems with a single inlet and a single outlet at steady state,

$$\dot{Q} - \dot{W} = \dot{m}(h_2 - h_1) = \dot{m}\Delta h. \tag{1.4}$$

   The overdot (.) denotes the rate of the respective variable. Thus, $\dot{Q}$ is the rate of heat entering the open system; $\dot{W}$ is the rate of work (power) produced by the open system; and $\dot{m}$ is the mass flow rate that enters and leaves the system. The property h is the specific enthalpy of the system, measured in joules per kilogram, and is closely related to the specific energy e of the system. As with the difference in specific energy, the specific enthalpy difference between the exit and the entrance of the system Δh may be calculated from temperature and pressure differences.

It must be emphasized that in the last two equations as well as the entire subject of thermodynamics, the following sign conventions for the heat and work apply: Work and power produced by the system are positive, while work and power supplied to the system are negative. Heat supplied to the system is positive, while heat leaving the system is negative. This scientific convention applies to all thermodynamic expressions and is schematically depicted in Figure 1.3.

Example 1.2

A steam turbine receives 120 kg/s of steam and exhausts to a condenser. The heat loss in the turbine is 2,580 kW and the enthalpy difference Δh across the turbine is −625 kJ/kg of steam. Determine the power produced by the turbine.

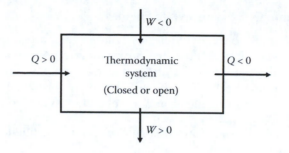

FIGURE 1.3
Thermodynamic convention for the algebraic signs of work and heat.

Solution: The steam turbine is an open system operating at steady state. Therefore, Equation 1.4 applies to this situation. The equation may be rearranged to yield the power $\dot{W}$ as follows:

$$\dot{Q} - \dot{m}\Delta h = \dot{W}.$$

Substituting $\Delta h = -625$ kJ/kg, $\dot{m} = 120$ kg/s, and $\dot{Q} = -2{,}580$ kW ($-2{,}580$ kJ/s), the power produced by the turbine is $\dot{W} = 72{,}420$ kW.

Note: The rate of heat $\dot{Q}$ has a negative sign because it represents heat that leaves the system (heat loss).

Example 1.3

The condenser of an air-conditioning system is cooled by a constant airflow. The condenser dissipates a rate of heat to the air equal to 56,500 W and does not produce or receive any mechanical power. The specific enthalpy increase of the air after passing through the condenser is 12 kJ/kg. Determine the mass flow rate of air needed for this operation.

Solution: The thermodynamic system in this case is the amount of air that is heated by the condenser. This is an open system at steady flow, and hence, Equation 1.4 applies. Heat enters the system (air), and therefore, it is positive. The rate of heat is $\dot{Q} = 56{,}500$ W = 56.5 kW. Also, $\dot{W} = 0$ because the condenser does not receive or supply mechanical power. Rearranging Equation 1.4 for the mass flow rate, we have the expression

$$\dot{m} = \frac{\dot{Q}}{\Delta h} \text{ and, hence, } \dot{m} = 4.73 \text{ kg/s.}$$

Example 1.4

A cylinder fitted with a piston contains 0.023 kg of helium gas. The gas is compressed from 1 to 15 atm by receiving a quantity of work equal to 320 J. If the change in the specific internal energy of the helium is 4,780 J/kg, determine the amount of heat transferred and state whether the heat enters or leaves the cylinder.

Solution: The cylinder–piston assembly contains a fixed amount of mass and is a closed system. Therefore, Equation 1.3 applies to this problem, which may be written as follows to yield the quantity of heat:

$Q = W + m\Delta e$, with $W = -320$ J (negative sign because the work enters the system), $m =$ 0.023 kg, and $\Delta e = 4{,}780$ J/kg. Therefore, the amount of heat involved in this process is $Q = -210.06$ J. The negative sign for the heat implies that heat leaves the cylinder and is dissipated in the surroundings.

1.3.2 Thermodynamic Cycles and Cyclic Engines

The fluid that drives a cyclic engine undergoes a sequence of processes, where the last process ends at the same state at which the first process started. Most of the everyday engines that convert heat into mechanical work (and vice versa) operate in cycles. For example, the refrigerant fluid in our refrigerator at home undergoes the following sequence of processes that constitute the refrigeration cycle [2] and are depicted in Figure 1.4:

1. Vapor compression in the compressor: Work W is supplied to the system (the refrigerant fluid).
2. Vapor condensation in the condenser to produce liquid: Heat Q_{out} is transferred from the system to the surroundings (usually through the warmer coils at the back of the refrigerator).
3. Liquid expansion in the expansion valve from a high to a lower pressure: Neither work nor heat is transferred during this process. The enthalpy of the refrigerant remains constant during the expansion process.
4. Evaporation in the evaporator to produce the vapor that is fed back to the compressor to start the first process: Heat Q_{in} is removed from the interior of the refrigerator and enters the refrigerant. The removal of the heat during this process produces the "refrigeration effect."

While undergoing the preceding four basic processes, the refrigerant fluid receives a net amount of mechanical work $W_{net} = W$ and receives a net amount of heat $Q_{net} = Q_{in} - Q_{out}$. If one considers the entire sequence of the four processes, where the final state of the refrigerant fluid is the same as the initial state, the energy of the refrigerant fluid is unchanged at the end of the cycle. Therefore, the first law of thermodynamics may be written as follows for the entire cycle:

$$Q_{in} - Q_{out} - W_{net} = 0 \quad \text{or} \quad Q_{net} = W_{net}. \tag{1.5}$$

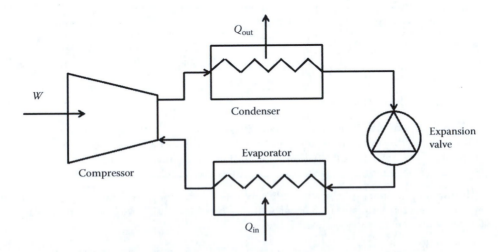

FIGURE 1.4
Components and energy exchanges in a refrigeration cycle.

The compression work is the only quantity of work supplied to this cycle, and hence, it represents the net work of the cycle. Because the work is supplied to the cycle, it has a negative numerical sign. Similarly, the net heat is negative, which implies that more heat leaves the cycle than enters. The net mechanical work obtained from a cycle always equals the net heat supplied to the cycle.

Similarly, the air in the car engines is considered to undergo a cycle [2]:

1. During the air intake, atmospheric air fills the engine cylinders through the intake valves.
2. With all the valves closed, the piston compresses the air in the cylinder. Work W_{in} is supplied to the cylinder.
3. The fuel injection and combustion processes (either spark ignition with gasoline or compression ignition with diesel fuel) supply heat Q_{in} to the air, the working fluid of the engine.
4. The expansion process produces mechanical work W_{out}, which provides the motive power to the vehicle through the transmission.
5. After the expansion process, the hot combustion gas in the cylinder is exhausted via the exhaust valves. This process is equivalent to the rejection of heat Q_{out} to the environment.
6. The exhaust valves close, the intake valves open, and air fills the cylinder as in the first process to repeat the cycle.

Again, when one applies the first law of thermodynamics to the sequence of the processes that constitute this air cycle, the net mechanical work produced by the engine is equal to the net heat received:

$$Q_{in} - Q_{out} - (W_{out} - W_{in}) = 0 \quad \text{or} \quad Q_{net} = W_{net} \tag{1.6}$$

The two equations also apply for the mechanical power and the rates of heat:

$$\dot{Q}_{in} - \dot{Q}_{out} - (\dot{W}_{out} - \dot{W}_{in}) = 0 \quad \text{or} \quad \dot{Q}_{net} = \dot{W}_{net}. \tag{1.7}$$

It must be noted that in the last expressions, the signs of work and heat follow the thermodynamic convention of Figure 1.3.

The same expression for the relationship between the net mechanical work produced and the heat supplied may be derived for all the other cyclic engines, including electric power producing plants regardless of the fuel used (nuclear, natural gas, coal), jet engines, locomotives, and heat pumps.

1.3.3 Second Law of Thermodynamics

Natural processes proceed in only one direction: if left unsupported, an apple always falls down and will not rise; a billiard ball finally stops at a position on the table; water flows from a higher to a lower elevation; a higher pressure at a location drives the wind to a lower pressure; a shattered glass does not spontaneously reconstitute itself; droplets of

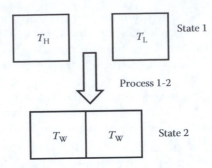

FIGURE 1.5
The spontaneous heat transfer process 1-2 that brought the two bodies to thermal equilibrium may not be reversed without the expense of mechanical work. The total entropy of the two blocks at state 2 is greater than the total entropy at state 1.

perfume evaporate and diffuse in a room; and heat is always spontaneously transferred from a hotter to a colder body.

Natural processes follow a unique direction and when they are completed, it is not possible to *spontaneously* reverse them, that is, without spending mechanical work for this reversal.

In the case of heat transfer, if we bring together two objects, one at high temperature T_H and the other at lower temperature T_L and allow them enough time, they will come to thermal equilibrium, and finally, they will both attain a common temperature T_W, which is between the two original temperatures $T_L < T_W < T_H$. This heat transfer process is schematically depicted in Figure 1.5. During this process, from state 1 to state 2, the total energy, which was originally contained in the two objects, is conserved: the sum of the energies of the two objects at state 1 is equal to the sum of the energies of the same objects at state 2.

The opposite process that brings the two objects from state 2 to state 1 is impossible to spontaneously occur. If we wish to restore the two bodies to their original temperatures T_H and T_L, we will find out that this cannot be done without the use of a refrigeration device, which consumes mechanical work. Despite the fact that the energy of state 2 is equal to that of state 1, process 2 to 1 is impossible to achieve without the addition of work, even though the reverse process 2-1 does not violate the first law of thermodynamics. Similarly, if we wish to restore the apple that has fallen from the tree, we must also perform work by lifting it back to its original level.

The origins of the second law of thermodynamics are traced to the scientific work of Sadi Carnot, who observed that heat supplied at higher temperatures has a higher "motive power" than heat supplied at lower temperatures and that there is a cascading of the "quality" of heat supplied to thermal engines. The cascading of the quality of heat is a consequence of the directionality that governs all natural processes [6,7]. Rudolf Clausius defined the property *entropy* to explain the directionality of all natural processes (sometimes called irreversible processes) and stipulated that in all isolated systems, entropy always increases. Thus, all insulated thermodynamic systems that undergo a natural, unconstrained process proceed in a direction where their entropy increases. Since at the time of this formulation, the universe as a whole was considered to be a huge isolated system, Clausius postulated the two laws of thermodynamics in his famous maxim [8]

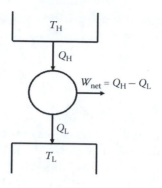

FIGURE 1.6

The net mechanical work produced by the cyclic engine, which receives heat from one reservoir and rejects heat to a second reservoir, is equal to the net heat received.

Die Energie der Welt ist konstant; die Entropie der Welt strebt einem Maximum zu ("The energy of the World is constant. The entropy of the World tends to a maximum").*

The most important implication of the second law of thermodynamics on the energy conversion processes is that *mechanical work may not be produced spontaneously by a cyclic engine, when this engine receives heat from a single heat reservoir.*[†] As a consequence of the second law, a cyclic engine—including most power plants, gas turbines, jet engines, and car engines—must communicate with at least two heat reservoirs. The engines are supplied with heat from one of these reservoirs and must reject heat to another. A schematic diagram of the operation of a cyclic engine that produces mechanical work is shown in Figure 1.6. During a single cycle, the engine receives heat Q_H, rejects heat Q_L, and produces net mechanical work: $W_{net} = Q_H - Q_L$. Typically, the high temperature heat Q_H is produced in the boiler or burner during combustion, and the exhaust heat Q_L is rejected to the atmosphere (gas turbines, jet, and car engine exhausts) or the hydrosphere (large fossil or nuclear power plants) and is called *the waste heat.* The waste heat is at low temperature; it is of very low "quality," and is impossible to use for the production of more power. It is really wasted!

The consequence of the second law of thermodynamics for energy conversion engines is that even though heat may be readily converted to work or power, it is only a fraction of the heat that is actually converted to work in all the cyclic thermal engines. This fraction is equal to the thermal efficiency the engine and is defined as the ratio of the net mechanical work produced W_{net}, and the heat input, to the cycle Q_H:

$$\eta = \frac{W_{net}}{Q_H}. \tag{1.8}$$

---

* It is debatable whether Clausius meant by the word *welt* the *world* as *the Earth* or *the Universe.* Modern physics tells us that neither the Earth nor the universe is an isolated system. While Clausius's statement is still correct in the length scales of the laboratory, it is not advisable to apply them to the length scales of the universe, without thinking of the nature of the isolated thermodynamic system where they may apply.

† This statement is often attributed to Lord Kelvin and Max Planck and called the *Kelvin–Planck statement of the second law.*

The second law and the entropy increase principle dictate that the thermal efficiency of any cyclic engine is less than the *Carnot efficiency* η_C, which is defined as

$$\eta_C = 1 - \frac{T_L}{T_H} \quad \text{and} \quad \eta \leq \eta_C \text{ or } \frac{W_{net}}{Q_{in}} \leq 1 - \frac{T_L}{T_H}. \tag{1.9}$$

The last equations are fundamental consequences of the second law of thermodynamics with a significant implication for the heat engines: no cyclic heat engine may be developed that would convert the entire quantity of the available heat into mechanical work. A significant fraction of the available heat Q_H must be always rejected, and the work produced W_{net} is limited by Equation 1.9. This is often referred to as the *Carnot limitation* of the heat engines.

The temperatures in the last equation are absolute temperatures, measured in kelvins (K) in the SI or degrees Rankine (°R) in the British system of units. Table 1.3 gives the conversion from commonly used temperatures scales to absolute scales. It must also be noted that as with Equation 1.6, Equations 1.8 and 1.9 also apply when the net power and the rate of heat input are substituted for the work and the heat input, respectively.

Example 1.5

The reactor of a large nuclear power plant produces steam at 290°C. The power plant rejects waste heat to a river at 30°C. What is the maximum thermal efficiency this power plant may achieve?

Solution: The two temperatures that are given are the high and low temperatures of the cycle T_H and T_L, respectively. When converted to absolute temperatures, they are, respectively, 563.15 K and 303.15 K. From Equation 1.9, the Carnot efficiency is calculated to be 46.17%.

This is the maximum thermal efficiency this nuclear power plant may attain.

Typical thermal efficiencies of cyclic engines are significantly lower than the calculated Carnot efficiency. An actual nuclear power plant that operates within the same temperature range would have thermal efficiency in the range of 30–35%.

Example 1.6

A coal power plant produces 400 MW of net electric power. The heat input to this power plant is 1,068 MW. Calculate (a) the thermal efficiency of the power plant, (b) the rate of waste heat that is rejected to the environment, and (c) the electric energy produced during a day, in joules and kilowatt hours.

TABLE 1.3

Temperature Conversions

K = 273.15 + °C
°C = K − 273.15
°R = °F + 459.67
K = °R × 5/9
°C = (°F − 32) × 5/9
°F = °C × 9/5 + 32

Solution: Equation 1.8 applies to the rates of work (power) and heat.

a. Therefore, the thermal efficiency of the power plant is equal to $400/1{,}068 = 0.3745 = 37.45\%$.
b. From Equation 1.7, the rate of waste heat is $\dot{Q}_{out} = \dot{Q}_{in} - \dot{W}_{net} = 1{,}068 - 400 = 668$ MW. This is a significant rate of heat that is typically dissipated in the environment and contributes to *thermal pollution*.
c. One megawatt is equivalent to 10^6 J/s, and 1 day has $60 \times 60 \times 24$ s. Hence, the power plant produces $400 \times 10^6 \times 60 \times 60 \times 24 = 34.65 \times 10^{12}$ J. This is equal to 9,600,000 kWh.

Typical thermal efficiencies of modern fossil fuel thermal power cycles are close to 40%, and typical efficiencies of power cycles used with nuclear reactors are close to 33%. Following the calculations of the last example, a typical coal power plant, which produces 400 MW of electric power at an efficiency of 40%, uses 1000 MW rate of heat and rejects 600 MW waste heat to the environment, usually to a river or a lake. For a typical nuclear power plant that produces close to 1000 MW of electric power at 33% thermal efficiency, the rate of heat produced in the reactor is 3,000 MW, and the rate of the waste heat rejected to the environment is 2,000 MW. These by all means represent very high quantities of heat power!

1.3.4 Perpetual Motion Engines

Perpetual motion engines are fictitious engines that violate one of or both the first and second laws of thermodynamics and are impossible to construct and operate. A *perpetual motion engine of the first kind* is a cyclic engine that works continuously; it produces more work than it receives, and hence, it violates the first law of thermodynamics. Such a thermal engine would have thermal efficiency more than 100%. A *perpetual motion engine of the second kind* violates the second law of thermodynamics. The thermal efficiency of such an engine may be less than 100%, which implies that the operation of the engine does not violate the first law, but if the efficiency of the cyclic engine is claimed to be higher than the Carnot efficiency ($\eta > \eta_C$), the engine is impossible to construct and operate.

Any claim for the operation of engines that violate the laws of thermodynamics is a false claim. The construction of perpetual motion engines are not simply limited by the current technology and may become feasible in the future. Such engines are impossible to create.

1.4 Thermal Efficiency and Other Figures of Merit

The thermal efficiency of a power cycle is a figure of merit, which indicates how good the performance of the power plant that operates with this cycle is. Analogous figures of merit apply to refrigeration and air-conditioning cycles. Similarly, all the power-related machinery components—pumps, heat exchangers, turbines, condensers, and solar cells—are associated with figures of merit, the efficiencies of the machines, which characterize their performance. Efficiency improvements always result in energy savings and, in most cases, less expensive operations. This section gives an overview of these figures of merit and provides more information on how to use them for the improvement of energy conversion processes.

1.4.1 Power Plants

Thermal power plants with higher efficiencies produce more power per unit heat input than power plants with lower efficiencies. Higher efficiency power plants of the same type (e.g., nuclear, coal, natural gas) operate more economically, produce cheaper power, and are in general *more efficient*. The definition of the thermal efficiency of an electric power plant is a dimensionless number and represents a benefit/cost ratio: W_{net} is the net electricity produced (e.g., in kWh) which is sold to the customers to generate revenue; Q_{in} is the heat input, which is produced from the combustion of a fuel (fossil or nuclear). The amount of heat input Q_{in} is proportional to the mass of fuel consumed, and the latter represents the variable cost for the production of electricity. The thermal efficiency $\eta = W_{net}/Q_{in}$, is a measure of the benefit/cost ratio for the power plant. Increasing the thermal efficiency of a given power plant improves the benefit/cost ratio; it always results in higher profit, and, in most cases, lessens the environmental impacts of the plant. For this reason, engineers strive to increase the thermal efficiency of power plants, by improving the cycles and the performance of the components of these plants.

Thermal efficiency comparisons of different types of power plants—e.g., nuclear and coal or coal and hydroelectric—do not have the revenue/cost meaning because the cost of the different fuels is different. In the case of most renewable energy sources, the cost of the "fuel" (the incident or input energy) is zero, and the efficiency does not represent a benefit/cost ratio. For photovoltaic (PV) cells, where the solar energy input is obtained at no cost, the ratio $\eta = W_{net}/Q_{in}$ does not represent a benefit/cost ratio. The same applies to wind turbines that receive free energy input, the wind. In the case of renewable energy, one must carefully think what the efficiency of the engines represents, what the different efficiencies mean, and avoid unjustified comparisons.

Example 1.7

A contractor has to install 25 kW of PV cells to supplement the electric power used in a building where the insolation is 1.06 kW/m². He or she has two choices of PV cells: (a) type A has an efficiency of 18% and costs $1,890 per square meter installed and (b) type B has 11% efficiency and costs $980 per square meter installed. All other things being equal, which type of PV cells should he or she install?

Solution: If PV cells of type A are used, the contractor has to install 25/1.06/0.18 = 131.03 m² of PV panels at a cost of 131.03 × 1,890 = $247,640. If PV cells of type B are used, the contractor has to install 25/1.06/0.11 = 214.4 m² of PV panels at a cost of 214.4 × 980 = $210,120.

Since everything else is the same, the contractor will choose the less costly option and install type B cells, even though their conversion efficiency is significantly lower.

This example illustrates that when one uses the efficiencies of different energy-producing devices or power plants for cost or other types of comparisons, one must be cognizant of how this ratio is defined, what it means, and whether or not the comparison pertains to the same variables of operation. A solar power plant with 10% thermal efficiency has zero heat input cost. This plant is not necessarily worse or less cost effective than a fossil fuel power plant with 40% thermal efficiency, which uses a costly fossil fuel. Comparisons based on the thermal efficiencies of power plants are meaningful only when they are performed for the same types of plants that use the same energy source. A more pertinent figure of merit for renewables, and a better basis for comparison with fossil and nuclear power plants, is the total cost of the electric energy produced, e.g., in dollars per kilowatt hour or dollars per megawatt hour. These numbers include all the cost variables that come into the production of electric power.

Example 1.8

It is proposed to make improvements to increase the efficiency of the power plant of Example 1.6 to 42% while keeping the same heat input. Calculate (a) the power the improved plant will produce and (b) the rate of the waste heat.

Solution: The power plant still has heat power input of 1,068 MW.

- a. With an efficiency of 42%, the power produced is $1,068 \times 0.42 = 448.56$ MW (a 12.12% power increase compared to Example 1.6).
- b. From $\dot{Q}_{out} = \dot{Q}_{in} - \dot{W}_{net}$, the rate of waste heat rejection is $1068 - 448.56 = 619.44$ MW (a 7.3% decrease).

Other things being equal, the net power increase will also increase the revenue of the power plant by 12.12%, a welcome outcome for the operator of the plant. The waste heat reduction would also be a welcome consequence.

It is observed in this example that an increase in the thermal efficiency of a power plant has two effects: (a) it increases the power (and revenue produced) for the same amount of fuel and (b) it decreases the waste heat that is rejected to the environment. Efficiency improvements are always beneficial.

1.4.2 Refrigeration and Heat Pump Cycles

Refrigeration* cycles operate by receiving external mechanical work W_{net} to drive a quantity of heat Q_L from a low-temperature reservoir (the interior of the refrigerator). The cycle discharges a larger quantity of heat Q_H to the environment, which is at higher temperature. Figure 1.7 is a schematic diagram of a refrigeration cycle operating with two heat reservoirs. Upon inspection, it is apparent that the refrigeration cycle is the reverse of the power cycle, depicted in Figure 1.6. In the case of the refrigeration cycle, the "benefit" is the heat extracted from the cold space Q_L. The "cost" of the cycle is the work W_{net} required to drive out this quantity of heat, typically supplied by electricity to the refrigerator. An appropriate figure of merit for refrigeration cycles is the ratio Q_L/W_{net}. When the heat and work are measured in the same units (e.g., J, kJ, kWh, and Btu), this ratio is called the coefficient of performance (COP):

$$\text{COP}_{ref} = \frac{Q_L}{W_{net}} \tag{1.10}$$

A similar figure of merit for refrigeration cycles, which is used in commercial refrigerator and air-conditioning systems, is the SEER. This is essentially the same benefit/cost as the COP, Q_L/W_{net}, with the work measured in watt hours (Wh), and the heat, measured in

---

* Air-conditioning essentially uses a refrigeration cycle. For brevity, the term *refrigeration* will be used to include air-conditioning too.

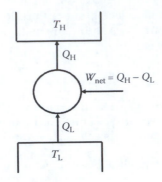

FIGURE 1.7
Schematic diagram for the refrigeration and heat pump cycles.

British thermal units. Using the conversion of British thermal unit to watt hours, one may derive the following relationship between SEER and COP:

$$\text{SEER} = 3.412 \times \text{COP}_{\text{ref}} \text{ or COP}_{\text{ref}} = \frac{\text{SEER}}{3.412} \tag{1.11}$$

An analysis of the refrigeration cycle, based on the second law of thermodynamics, proves that the coefficient of performance of the cycle must be less than the temperature ratio: $T_L/(T_H - T_L)$, with the temperatures in absolute values kelvins or degrees Rankine.

$$\text{COP}_{\text{ref}} = \frac{Q_L}{W_{\text{net}}} \leq \frac{T_L}{T_H - T_L}. \tag{1.12}$$

Refrigeration cycles have coefficients of performance in the range of 1.5–6. The higher COP values correspond to large, industrial, refrigeration plants. The lower values are more typical of small refrigerators used in households, where low cost rather than higher efficiency is of importance.

Heat pump cycles operate in the same way as the refrigeration cycles of Figure 1.7. The function of the heat pump is to extract heat Q_L from the colder environment, which is at the ambient temperature T_L. By the addition of mechanical work/power W_{net}, the heat pump rejects a higher quantity of heat Q_H to the interior of a building. Thus, the building is heated by the quantity Q_H and is kept at higher temperature than the ambient. While they are little known, heat pumps are very economical and are ubiquitous in large buildings—hotels, hospitals, schools, government buildings, and universities. The heat pump and the air-conditioning systems of a building may use the same machinery. The reversal of the airflow through the equipment by automated valves provides heat in the cold months and cooling during the hot months of the year. In the case of a heat pump, the benefit is the heat supplied to the building Q_H, and the cost is the amount of mechanical work from electricity required for the operation of the machinery. Hence, the figure of merit for the heat pumps is a coefficient of performance defined as

$$\text{COP}_{\text{hp}} = \frac{Q_H}{W_{\text{net}}}. \tag{1.13}$$

The algebraic sign of Q_H is negative because Q_H leaves the cycle. The absolute value of Q_H is used for the calculation of the COP_{hp} in Equation 1.13). For heat pumps, the second law of thermodynamics dictates that this ratio is always less than the temperature ratio: $T_H/(T_H - T_L)$, where the temperatures are in absolute values, kelvins or degrees Rankine. When the same system is used as heat pump and as air-conditioner, it follows from the last two equations that $COP_{hp} = COP_{ref} + 1$. Hence, the COP of heat pump systems is higher than that of refrigerators and air-conditioners. Actual heat pump cycles have coefficients of performance in the range of 2.5–7, depending on the ambient temperature. The SEER ratio is also used with commercial heat pump systems, with the relationship between the SEER of a heat pump and the COP_{hp} being the same as in Equation 1.11.

While the second law of thermodynamics dictates that the efficiencies of the power cycles are always less than 1 (or less than 100%), this does not apply for the coefficients of performance of both heat pumps and refrigerators. Typical COPs are significantly higher than 1 (or 100%) and sometimes as high as 6 or 7 (600% or 700%). A given quantity of mechanical work may cause the transfer of significantly higher quantities of heat from the interior of a refrigerator or to the interior of a building. This has led some authors to characterize mechanical work and electricity as "energy of higher quality" and heat as "lower-quality energy." For all applications, it is always helpful to recall that while not all the heat may be converted to mechanical work, a unit of mechanical work may facilitate the transfer of multiple units of heat when suitable engines are used.

Example 1.9

A large industrial refrigeration plant has a SEER of 12. The annual cost of electricity for this plant is $38,500. It is proposed to make improvements to the refrigeration cycle that will increase the SEER to 20.6. Determine the annual monetary savings.

Solution: It will be assumed that the operation of the refrigeration plant does not change and that the same annual quantity of heat Q_L is transferred before and after the improvements. From the definition of the SEER (and of the COP), this improvement in the value of SEER will reduce the work/electricity needed W_{net} (in Wh) to $W_{net} \times 12/20.6 = 0.5825 \times W_{net}$. The corresponding electricity cost for this work is $22,427. This represents an annual cost reduction of $16,073 or 41.75%.

Improvements to the refrigeration cycle are usually carried out with capital expenditures. For the owners of the refrigeration plant, a sound business decision is to undertake this investment if it is justified by the annual savings in the cost of electricity. The net present value method (Chapter 9) may be used for such an investment decision. In general, when the price of electricity is high, the savings are higher, and more energy saving investments are justified. This conclusion applies to regions and countries with different electricity prices. For example, the United States enjoys significantly lower electricity prices than most other industrialized countries [9]. Because of this, it is not very cost effective to make energy efficiency investments in the United States. As a result, in general, the efficiency of similar industrial and residential processes and machinery in the United States are relatively lower than countries of the European Union, Japan, Australia, and Switzerland.

1.4.3 Component Efficiencies

In addition to the cycle thermal efficiencies, engineers have defined the *component efficiencies*, to characterize the operation of separate components that are parts of cycles, such as turbines, compressors, heat exchangers, fans, and pumps. The component efficiencies are defined as ratios of the actual work W_{act} produced or consumed by the component and the corresponding ideal work W_{id}, which is produced or consumed in idealized processes, the *isentropic processes*, where the entropy remains the same. All the component efficiencies are defined using the absolute values of work and heat—not the thermodynamic convention of Figure 1.3—in a way that their numerical values are between 0 and 1 (0–100%). The definitions of component efficiencies for commonly used machinery are as follows:

1. For turbines:

$$\eta_T = \frac{W_{act}}{W_{id}};$$

$\qquad$ (1.14)

2. For pumps:

$$\eta_P = \frac{W_{id}}{W_{act}};$$

$\qquad$ (1.15)

3. For compressors and fans:

$$\eta_C = \frac{W_{id}}{W_{act}};$$

$\qquad$ (1.16)

4. For solar cells:

$$\eta_{sc} = \frac{\dot{W}_{act}}{S}.$$

$\qquad$ (1.17)

where S is the incident solar radiation (insolation) and the power $\dot{W}_{act}$ is expressed in watts per square meter. The latter varies temporally and spatially; therefore, η_{sc} is typically a function of time and location. Because the local insolation S is not associated with any cost, the efficiency of a solar cell does not represent a benefit/cost ratio.

Typical efficiencies of industrial steam turbines are in the range of 75–85%, and those of gas turbines, 80–90%. Large industrial compressor efficiencies are in the range of 70–85%, while the efficiencies of compressors that are used for domestic appliances (e.g., small refrigerators) may be as low as 30%. Pump efficiencies are in the range of 65–80%. For larger, industrial units, component efficiency charts are supplied by the manufacturers of the equipment. For commonly used solar cells, typical efficiencies are in the range of 10–20%. In calculations, an engineer first calculates the ideal isentropic work W_{id} from the theory of thermodynamics using the first and second laws and then calculates the actual work W_{act} using the pertinent expression from Equations 1.14 through 1.16.

The figure of merit used for all types of heat exchangers (including condensers, evaporators, regenerators, car radiators, burners, and boilers) is sometimes called *effectiveness* or *efficacy*. This is defined as the heat that is actually transferred from one medium to another (e.g., refrigerant to ambient air) divided by an ideal quantity of heat, which is the heat that would have been transferred if the heat exchanger had an infinitely large area. Typical heat exchanger effectiveness values are in the range of 60–90%. Heat exchangers with higher effectiveness have larger surface area and are more costly.

Example 1.10

A solar panel has an active area of 3.5 m² and an efficiency that is constant at 16%. What is the power the cell produces when the local insolation is 1, 0.8, and 0.5 kW/m²?

Solution: From Equation 1.17, the power produced per unit area of the solar panel is: 0.16, 0.128, and 0.08 kW/m² respectively. Since the area of the solar panel is 3.5 m², the corresponding power produced by the solar cells is 0.56, 0.448, and 0.28 kW respectively.

1.5 Practical Cycles for Power Production and Refrigeration

The vast majority of thermal electric power plants in the world make use of two main types of cycles: vapor cycles and gas cycles. A succinct description of the essential components of the two types of cycles will follow. Practical methods and processes for the improvement of the thermal efficiency of the cycles are discussed in Chapter 8. More detailed descriptions of the practical cycles may be found in textbooks of engineering thermodynamics, such as those by Moran and Shapiro [2] and Kestin [3].

1.5.1 Vapor Power Cycles: The Rankine Cycle

The *Rankine cycle* is the most commonly used vapor power cycle. In most practical systems, water is the working fluid for this cycle, and the produced vapor is steam. There are a few systems that utilize an organic fluid, typically a hydrocarbon or ammonia, which produce the vapor of the organic fluid. In such cases, the cycle is referred to as *organic Rankine cycle*. Water–steam Rankine cycles and their variations are the principally used cycles of all coal and nuclear power plants. A schematic diagram of this thermodynamic cycle with its basic components is in Figure 1.8.

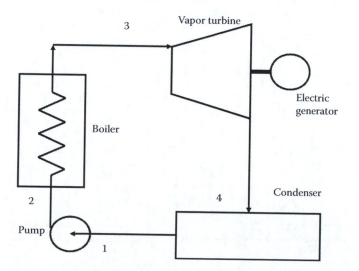

FIGURE 1.8
Principal components of a Rankine vapor power cycle.

Liquid water at state 1 is pressurized in the pump from where it exits at high pressure at state 2. The pressurized water enters the boiler, where it absorbs heat, and exits as superheated steam at high temperature, state 3. The high-temperature, high-pressure steam enters the vapor turbine, where it expands to a much lower pressure—usually subatmospheric pressure—and its temperature drops to almost ambient temperature and exhausts at state 4 in the condenser. In the condenser (the cooling system of the power plant), the spent steam condenses to water and is fed back to the pump as liquid water to repeat the cycle.

The thermodynamic processes that comprise the basic Rankine cycle are shown in Figure 1.9a and b on the temperature–entropy and pressure–volume diagrams, respectively:

1. Process 1-2 is the pressurization of the liquid water effluent from the condenser. The pump carries out this process, which is almost isentropic. Typical inlet conditions at state 1 for water are 6–10 kPa (6–10% of atmospheric pressure), and outlet pressures at state 2 vary from a few megapascals to 30 MPa for supercritical cycles. The pump is driven by a small fraction, typically 1%, of the total power produced in the cycle. The temperature of the pressurized water at state 2 is almost the same as that at state 1, that is, $T_2 \approx T_1$.

2. The boiler (also called steam generator, burner, etc.) is the component where fuel combustion (coal, petroleum, natural gas) or nuclear reactions produce a large amount of heat (Q_H in Figure 1.6) that is transferred to the pressurized water. Process 2-3 produces steam at high pressure and high temperature. The boiler is the high-temperature reservoir for the cycle. Heating process 2-3 occurs with very low-pressure loss and is considered to be isobaric (constant pressure).

3. From the boiler, the steam enters a single turbine or several steam turbines in large power plants. The expansion of the steam in the turbine provides the motive power for the rotation of the turbine, which drives an electricity generator via a shaft. The pressure and temperature of the steam are reduced in the turbine and the steam is exhausted in the condenser at very low pressure, in the range of 6–10 kPa, and at low temperatures, in the range of 35–50°C. The steam expansion process in the turbine, process 3-4, is almost isentropic. The net work produced by the cycle W_{net} in Figure 1.6 is equal to the difference of the work produced by the

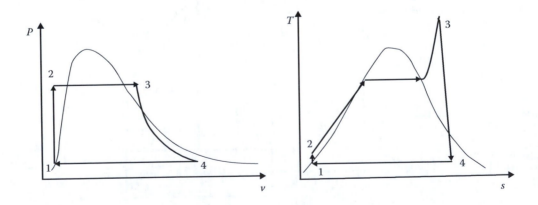

FIGURE 1.9
Basic Rankine cycle depicted on (a) pressure–volume and (b) temperature–entropy coordinates.

turbine and the work consumed by the pump. This is the electric work the plant produces.

4. The condenser receives the exhaust steam from the turbine and cooling water from the cooling system of the power plant. The steam transfers latent heat to the cooling water and condenses to become liquid water. The condenser is essentially a heat exchanger, where steam from the cycle is condensed by warming up large quantities of cooling water. Heat is rejected in this process of the cycle to the cooling water and, finally, to the environment. This is the waste heat of the power plant Q_L in Figure 1.6. Condensation process 4-1 occurs at constant pressure (isobaric process).

If the mass flow rate of water in the cycle is denoted by $\dot{m}$, then the amount of power consumed by the pump is $\dot{W}_P = \dot{m}(h_2 - h_1)$; the rate of heat produced by the boiler and transferred to the water/steam is $\dot{Q}_H = \dot{m}(h_3 - h_2)$; the amount of power produced by the turbine is $\dot{W}_T = \dot{m}(h_3 - h_4)$; and the rate of heat rejected by the condenser via the cooling water to the environment is $\dot{Q}_L = \dot{m}(h_4 - h_1)$. The net electric power, which is produced by this plant in the generator, is equal to the difference of the power produced by the turbine and the power consumed by the pump: $\dot{W}_{net} = \dot{m}\left[(h_3 - h_4) - (h_2 - h_1)\right]$. The thermal efficiency of the power plant is

$$\eta_t = \frac{(h_3 - h_4) - (h_2 - h_1)}{h_3 - h_2}. \tag{1.18}$$

The enthalpy and the other properties of the water and steam at the states 1, 2, 3, and 4 are obtained from *steam tables* that are standard appendices of books on engineering thermodynamics [2,3].

1.5.2 Gas Cycles: The Brayton Cycle

Gas cycles typically use air as the working fluid. Several advanced gas cycles for nuclear reactors have used carbon dioxide or helium. The arrangement of the basic components in a simple gas cycle is shown in Figure 1.10. The pressure–volume and temperature–entropy diagrams of this cycle are depicted in Figure 1.11a and b, respectively. The series of the main processes that produce electric power are

1. Air at ambient temperature and pressure, state 1, enters a compressor where its pressure and temperature rise to state 2. Typical pressures at the exit of the compression process are 8–30 atm, and oftentimes, intercoolers are used.

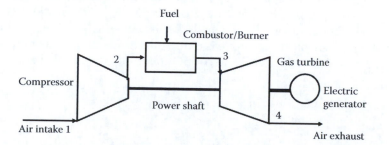

FIGURE 1.10
Principal components of a Brayton gas power cycle.

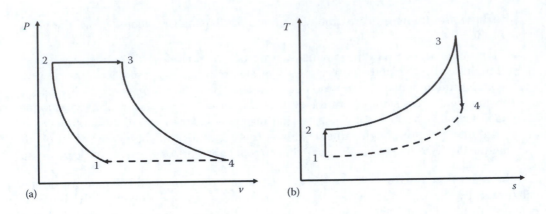

FIGURE 1.11
Basic Brayton cycle depicted on (a) pressure–volume and (b) temperature–entropy coordinates. Virtual process 4-1 is shown by the dashed curves.

The compressor consumes a significant amount of power, on the order of 35% of the total power produced by the turbine. In order to avoid additional losses, the compressor is usually mechanically coupled to the turbine by a shaft and is driven directly by the turbine.

2. The compressed air enters the combustor or burner where fuel (pulverized coal, natural gas, or liquid hydrocarbons) is injected and burns by combining with the oxygen in the air. The combustion produces gases at very high temperatures in the range of 1,000–2,000°C.

3. The high-temperature gaseous combustion products are introduced to the gas turbine where they expand to atmospheric pressure and then discharged to the environment at state 4. An electric generator coupled to the turbine produces the electric power, which is approximately equal to the difference of the power produced by the turbine and the power consumed by the compressor.

The gas cycle is not a cycle per se because the turbine exhaust is at a different state and has a different composition (contains the combustion products) from the compressor input. This series of processes is treated as a cycle, because the input to the compressor is always air at ambient temperature, and the rest of the atmosphere may be thought of as an enormous reservoir of mass and heat, which receives the hotter output of the turbine; cools it; purifies it to the ambient temperature and composition; and finally, allows it to be fed to the compressor at the same, ambient pressure, temperature, and composition. The fictitious cooling process that occurs in the atmosphere is denoted by the broken lines 4-1 in Figure 1.11a and b.

The thermal efficiency of the gas cycle is defined in the same way as in the vapor cycles and may be expressed as follows in terms of the states depicted in these two diagrams:

$$\eta_t = \frac{(h_3 - h_4) - (h_2 - h_1)}{h_3 - h_2}. \tag{1.19}$$

As with the vapor cycles, the enthalpy and other properties of the air at the states 1, 2, 3, and 4 may be obtained from *air tables* that are standard appendices in engineering thermodynamics books.

1.5.3 Refrigeration, Heat Pump, and Air-Conditioning Cycles

If the operation of the power generation cycle is reversed, the cycle absorbs heat Q_L from a colder heat reservoir and dissipates an amount of heat Q_H to a hotter reservoir, while also consuming work W. Such a reversed cycle may be used as a *refrigeration cycle*, a *heat pump cycle*, or an *air-conditioning* cycle. The working fluid in the refrigeration cycles is one of the common refrigerants, which are compounds of carbon hydrogen and fluorine. The thermodynamic diagram in the (T, s) coordinates of a typical refrigeration cycle, which is a slightly modified and reversed Rankine cycle, is shown in Figure 1.12:

1. The compressor admits the vapor refrigerant and raises its pressure from states 1 to 2. This device is driven by a motor, which consumes electric work. The refrigerant exits the compressor as superheated vapor at higher temperature.

2. The superheated refrigerant is directed to the condenser, where it undergoes condensation process 2-3. During this process, the refrigerant dissipates heat to a heat sink at a higher temperature T_H. In a heat pump, the heat sink is the interior of a building. In a refrigerator, the heat sink is the room (kitchen) where the refrigerator is located, and the heat transfer occurs through coils at the back of the refrigerator, which always feel warmer to the touch. In an air-conditioning cycle, the heat sink is the outside air or the ground.

3. Process 3-4 is caused by an expansion valve and produces a mixture of liquid and vapor at significantly lower temperature ($T_1 = T_4$) and at the original pressure of the cycle P_1. The expansion valve is insulated, and this process is considered isenthalpic ($h_3 = h_4$).

4. The refrigerant finally passes through the evaporator and undergoes process 4-1, where it evaporates. Because the temperature of the refrigerant at this stage is very low, it absorbs heat from the heat source at the low temperature T_L and leaves the evaporator at original state 1. In an air-conditioning cycle, the lower-temperature heat source is the air in the interior of the building. In a refrigeration cycle, it is the interior of the refrigerator, which is kept at lower temperature. In a heat pump cycle, the lower-temperature heat source is the outside ambient air or the ground.

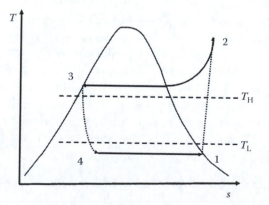

FIGURE 1.12
Thermodynamic diagram of the operation of a refrigeration cycle. The same cycle is used for heat pumps and air-conditioning units.

It must be noted that the main difference between a reversed Rankine cycle and a refrigeration cycle is that process 3-4 of the refrigeration cycle does not produce work. The actual work that may be produced from the expansion of the liquid refrigerant at state 3 is too low to justify the additional cost of an expander/turbine. The simple expansion valve used for process 3-4 is by far cheaper and fulfills the task to lower the temperature of the refrigerant.

1.6 Exergy: Availability

The concept of the Carnot efficiency is useful in defining a measure of the maximum efficiency of a cyclic heat engine, when it converts a given quantity of heat Q to work W. However, quantities of heat at high temperatures are not found in nature as energy resources. Heat is normally artificially released in burners and boilers from the combustion of fossil fuels or from nuclear reactions. Instead of heat, *natural resources* or *energy resources*, such as fossil fuels, nuclear fuels, geothermal fluids, incident solar radiation, and wind power, exist in nature. We harness these natural resources to produce heat and work.

One may ask the question "What is the maximum work, one may obtain from a given quantity of an energy resource?" Clearly, the Carnot efficiency does not offer an answer to this question, because the Carnot efficiency applies to heat engines and involves the production of heat and the conversion of this heat to work. The conversions of an energy resource to heat (e.g., by combustion) and finally to useful work always involve thermodynamic irreversibilities that limit the final quantity of work produced. The direct conversion of the energy resource to work entails lesser irreversibilities and produces more work. The concept of *exergy* or *availability*, which gives a quantitative measure of the maximum amount of work one may obtain from an energy resource, is a very useful tool in energy studies [2,10].

When energy conversion processes are considered, one makes the following observations [10,11]:

1. All the energy conversion processes occur in the environment—the atmosphere and the hydrosphere—which take the place of reservoirs of mass, heat, and work. The environment may absorb or provide very large quantities of mass, heat, and work without any appreciable changes in its properties.

2. Energy conversion processes are possible, because there are natural substances, the *energy resources*, which are not in thermodynamic equilibrium with the environment. For example, coal is not in equilibrium with the atmosphere. Coal will come to equilibrium with the atmosphere when it combines with oxygen to form carbon dioxide, one of the atmospheric constituents.

3. The maximum work is extracted from an *energy resource*, when, after several processes, the energy resource is brought in thermodynamic equilibrium with the environment. This is often called the *dead state* for the constituent materials of the resource.

The maximum work from an *energy resource* will depend on the following:

1. The type of the resource (chemical, nuclear, solar radiation, wind, etc.)
2. The thermodynamic properties of the environment, primarily its temperature and pressure

3. The interaction between the resource and the environment

4. The energy dissipation or irreversibilities during the conversion processes

The maximum work that may be obtained from an energy resource is not described by a single formula, such as the one of the Carnot efficiency. One should perform an analysis for the specific resource using the first and second laws of thermodynamics, which is often called *exergy analysis* and leads to the thermodynamic concept of *exergy*. The following are three examples of exergy analysis for three types of natural energy resources [10].

1.6.1 Geothermal Energy Resources

Naturally occurring geothermal reservoirs supply power plants with geothermal fluid, which is in the one of the following three conditions:

1. Superheated steam

2. Mixture of steam and water

3. Geopressured liquid water at high temperature

The geothermal power plant is an open thermodynamic system that receives a mass flow rate of the geothermal resource $\dot{m}$, at high temperature T, and pressure P. The specific enthalpy h and specific entropy s of the geothermal resource is calculated from the pair of properties (T, P). The plant operates in the environment, which is at temperature T_0 and pressure P_0. After all the energy conversion processes, the power plant discharges the spent geothermal fluid as an effluent at temperature T_e and pressure P_e. The specific enthalpy and specific entropy of the effluent are denoted as h_e and s_e, respectively. One may write the two laws of thermodynamics for the operation of this open system as [10]

$$\dot{Q} - \dot{W} = \dot{m}(h_e - h) \tag{1.20}$$

and

$$\dot{m}(s_e - s) - \frac{\dot{Q}}{T_0} = \dot{\Theta} \quad \text{with } \dot{\Theta} > 0. \tag{1.21}$$

Substituting the rate of heat from the second expression to the first, we obtain the following expression for the power produced by the geothermal unit:

$$\dot{W} = \dot{m}\left[(h - T_0 s) - (h_e - T_0 s_e)\right] - T_0 \dot{\Theta}. \tag{1.22}$$

The geothermal power plant will deliver maximum power, when the following conditions are satisfied:

1. The rate of entropy generation $\dot{\Theta}$ is minimized to a value close to zero.

2. The state of the effluents e is the same as the state of the environment, that is, the effluents from the power plant are at temperature T_0 and pressure P_0, as implied by observation 3, of the previous section. At this state—the so-called dead state—the specific enthalpy and entropy of the effluents is denoted as h_0 and s_0.

TABLE 1.4

Specific Work Produced[a] by Several Types of Geothermal Resources

Steam at 280°C, 15 atm	955
Saturated steam at 220°C	929
50% water and 50% steam at 220°C	552
Saturated liquid water at 200°C	158
Saturated liquid water at 160°C	97
Geopressured liquid water at 140°C, 25 atm	73

[a] In kilojoules per kilogram.

Hence, the maximum rate of work, or power, one may obtain from the geothermal resource is

$$\dot{W}_{max} = \dot{m}\left[(h - T_0 s) - (h_0 - T_0 s_0)\right] = \dot{m}(e - e_0). \tag{1.23}$$

The quantity $e = h - T_0 s$ is called the *exergy* of the geothermal resource. The exergy of an energy resource, which is utilized by an open thermodynamic system, is a function of the thermodynamic state of the resource as well as of the state of the environment, where the energy conversion process occurs.

Table 1.4 shows values of the specific work, the quantity in the square brackets of Equation 1.23 for several states of the geothermal fluid. The properties of the geothermal fluid are approximated with the properties of water. The environmental temperature T_0 is 298 K (25°C), and the environmental pressure P_0 is 1 atm. It is observed in this table that resources with high amount of vapor (steam) provide much higher amount of work than liquid water resources. Also, the amount of specific work significantly diminishes when the temperature of the resource is relatively low. For vapor–liquid systems, such as steam–water, the numerical value obtained for the exergy at the dead state is very close to zero, that is, $e_0 \approx 0$, which implies

$$\dot{W}_{max} \approx \dot{m}e. \tag{1.24}$$

1.6.2 Fossil Fuel Resources

Fossil fuels provide heat and work because they are not in chemical equilibrium with the atmosphere [10–12]. An equilibrium state is reached when the fossil fuels combine with the oxygen of the atmosphere and produce carbon dioxide and water, two constituents of the environment. Let us consider the conversion process of a chemical substance—e.g., a hydrocarbon, such as octane (C_8H_{18}), which is the main constituent of gasoline—to mechanical work. The fuel is supplied to an engine at a molar flow rate $\dot{n}$, measured in kilomoles per second or moles per second.

Because the octane and all the other fossil fuels undergo chemical reactions to reach equilibrium with the environment, the molar flow rates and properties of the reactants, denoted by the subscript R, and of the products, denoted by the subscript P, must enter the derivation of the maximum power the engine may produce. One may write the laws of thermodynamics for this open chemical system as follows:

$$\dot{Q} - \dot{W} = \sum_P \dot{n}_P h_P - \sum_R \dot{n}_R h_R \tag{1.25}$$

and

$$\sum_P \dot{n}_P s_P - \sum_R \dot{n}_R s_R - \frac{\dot{Q}}{T_0} = \dot{\Theta} > 0.$$

(1.26)

The elimination of the rate of heat from the last two expressions yields the following expression for the power produced by an engine that utilizes this fossil fuel:

$$\dot{W} = \left[\sum_R \dot{n}_R (h_R - T_0 s_R) - \sum_P \dot{n}_P (h_P - T_0 s_P) \right] - T_0 \dot{\Theta}.$$

(1.27)

As with the geothermal fluid, the maximum power is produced by the engine when

1. The internal irreversibilities are minimized so that the rate of entropy production is zero
2. The products of the chemical reaction reach thermodynamic equilibrium with the atmosphere, at temperature T_0 and pressure P_0

Under these conditions, the terms in the square bracket may be written as the product of the *Gibbs free energy* of the reaction ΔG^0 and the molar flow rate of the fuel $\dot{n}_F$:

$$\dot{W}_{max} = -\dot{n}_F \Delta G^0.$$

(1.28)

From Equation 1.28, one may deduce that the maximum amount of specific work per kilomole of the fossil fuel is

$$\tilde{w}_{max} = -\Delta G^0.$$

(1.29)

The units of ΔG^0 are kilojoules per kilomole or British thermal units per pound mole.

The Gibbs free energy of the reaction is a thermodynamic property for all reactions and may be found in standard thermodynamic and chemical tables. The superscript 0 denotes that this function is evaluated at the atmospheric pressure and temperature.*

1.6.3 Radiation: The Sun as Energy Resource

Contrary to popular thinking, not all the solar radiation energy may be converted to mechanical work and power. The energy flux (in W/m^2) emitted by a source at temperature T_H is equal to σT_H^4, where σ is the Stefan–Boltzmann constant. By combining the first and second laws of thermodynamics and substituting the appropriate expressions for the specific energy and entropy of radiation, one may obtain the following expression [10] for the exergy flux from a radiation source, at temperature T_H, when the exergy is utilized in an environment at temperature T_0.

$$E = \frac{\sigma}{3} \left(3T_H^4 - 4T_H^3 T_0 + T_0^4 \right).$$

(1.30)

---

* Fuel cells, which convert the chemical energy of the fossil fuels directly to electric power, approach the thermodynamic limit ΔG^0 closer than thermal engines, which first convert the chemical energy in the fuels to heat.

The units of this expression are power per unit area (e.g., W/m²). In the case of solar energy—insolation—the temperature of the source of radiation, the sun, is approximately 5,800 K, and the temperature of the receiving object, the Earth, is approximately 300 K. When these numbers are substituted in the last equation, it is concluded that approximately 93% of the incident solar energy flux may be converted to work.

It must be noted that unlike energy, the exergy is not conserved during a process. When energy resources undergo conversion processes (e.g., the burning of coal and the refinement of petroleum), their exergy of the process products always decreases. The optimization of energy conversion processes may always be formulated as dissipation minimization, irreversibility minimization, and exergy conservation [1,13,14].

1.7 Myths and Reality about Energy Conversion

Myth 1: The machinery we employ for the production of electricity is old and very inefficient. If we invest in better power plants, we can get three times more electric power than what we get now.

Reality: Typical thermal efficiencies for the conversion of fossil and nuclear fuels to electric power are in the range of 30–40%, and this implies that for every 100 units of heat input, we obtain 30–40 units of electricity. The efficiency of the thermal power plants may somehow improve, but it is subjected to the Carnot limitations, which would place the upper limit of thermal efficiency of the existing power plants in the range of 50–70%. Better and more efficient power plants may produce additional, small amounts of power, but will never produce three times more. If current power plants produced three times more power with the same heat input, then they would produce 90–120 units of work for every 100 units of heat input. Such power plants are impossible to construct because they would be perpetual motion engines of the first kind or the second kind.

Myth 2: The oceans are massive and contain an enormous amount of energy. We can use some of this energy to satisfy the energy needs of the planet. Actually, if we lower the ocean temperature by 1°C, we will produce enough electricity for 10 years for the entire planet. In addition, we will neutralize the threat of global warming.

Reality: It is correct that the oceans have an enormous amount of energy. The volume of the oceans is approximately 1.3×10^{18} m³, and the mass of the water in the oceans is slightly more than 1.3×10^{21} kg. If we were able to lower the average temperature of the ocean water by 1°C, we would obtain $1.3 \times 10^{21} \times 1 \times 4.184 = 5.4 \times 10^{21}$ kJ. Since the global primary energy consumption in 2016 was approximately 540 Q (~540×10^{15} kJ), the quantity of energy from lowering the temperature of the oceans would be enough to supply the energy needs of the planet for approximately 10,000 years! However, in order to develop a thermal engine to extract all this energy from the ocean, we must have a heat reservoir at a lower temperature than the ocean, as in Figure 1.6, and such a natural heat reservoir does not exist. A cyclic engine, which takes energy from the oceans and delivers work, would be a perpetual motion engine of the second kind, and unfortunately, it is impossible to create one.

Myth 3: After so many years of research, the efficiency of PV devices remains under 20%, while the efficiency of new coal power plants is well over 40%. This proves that solar energy will never be as affordable as energy from fossil fuels.

Reality: This is an example of a gratuitous comparison of efficiencies of different types of power plants. The coal power plant consumes a fossil fuel and its efficiency (energy produced divided by the heat input) represents a benefit/cost ratio. The energy input to the PVs is free of cost, and its efficiency (energy produced divided by the incident solar energy) does not represent a benefit/cost ratio, because solar energy is free. A more meaningful comparison of the two types of electric energy generation would be the cost of energy produced, in dollars per kilowatt hour. If the industry manages to produce cheaper PV cells, and the unit cost of electricity from the PV cells becomes very low (e.g., $0.02/kWh), the lower efficiency of the PV cells would be entirely irrelevant.

PROBLEMS

1. Convert the following to joules: (a) 5 tce, (b) 15 Btu, (c) 15 MeV, (d) 45 scf of gas, and (e) 3000 kWh.

2. Convert the following to British thermal units: (a) 10 MJ, (b) 45 kWh, (c) 215 MWh, (d) 5 tce, and (e) 115 bbl of crude oil.

3. Convert the following to kelvins: (a) 45°C, (b) 1,085°F, and (c) 679°R.

4. A nuclear power plant produces 1000 MW of power and has thermal efficiency of 33%. What is the rate of heat input of this plant and what is the rate of waste heat from the plant?

5. A coal power plant produces 400 MW of power and has thermal efficiency of 38%. How much heat input, in kilojoules, does this power plant need to operate every day? How many tons of coal equivalent (tce) correspond to this quantity of heat?

6. What is the maximum efficiency a steam cycle may achieve if the highest temperature of the cycle is 600°C and the ambient temperature is 30°C?

7. A modern gas turbine engine achieves thermal efficiency of 46% and produces 80 MW of power. If the gas turbine operates for 50% of the time during a year, what is the annual heat input of this cycle and how many cubic foot of gas (scf) does it consume during a year?

8. The average retail price of electricity for residential use in the southwest part of the United States is $0.095/kWh. A typical large household in the region consumes 2600 kWh/month. The corresponding numbers for Germany are €0.34/kWh and 1100 kWh/month. What would be the annual savings in US dollars if, by using higher efficiency appliances, the typical household in each country reduced its energy needs by 15%? (Note: You will need to use the current euro to dollar conversion factor.)

9. The interior of a refrigerator is kept at 38°F, while the ambient temperature is 98°F. What is the maximum value of the COP for this refrigerator? Is it possible for this refrigerator to have a SEER of 27? How about a SEER of 119?

10. The solar radiation in Fort Worth, Texas, during a day of the year is given by the expression $S = 860 \times \sin[(t - 7)\pi/13]$ W/m^2, where t is in hours and 7:00 < t < 20:00. A solar panel in Fort Worth has an effective area of 2.5 m^2 and efficiency that may be considered constant at 15%. Determine how much electric power the solar panel produces at each hour from 7:00 am until 8:00 pm (20:00) in kilowatts and the total energy it produces during the day in kilowatt hours.

11. A pump is to be used to lift water to a high-level reservoir. It is calculated that the power needed for this task is 12.5 kW. If the pump efficiency is 77%, what is the electric power that needs to be supplied to the motor that drives this pump?

12. Determine from thermodynamic tables what the heat (ΔH^0) and the maximum work (ΔG^0) per kilogram the following chemical substances may produce: (a) hydrogen, (b) methane; (c) methanol; (d) ethanol.

References

1. Michaelides, E.E., *Alternative Energy Sources*, Springer, Berlin, 2012.
2. Moran, M.J., Shapiro, H.N., *Fundamentals of Engineering Thermodynamics*, 6th edition, Wiley, New York, 2008.
3. Kestin, J., *A Course in Thermodynamics*, vol. 1, Hemisphere, Washington, DC, 1978.
4. Joule, J.P., On the existence of an equivalent relation between heat and the ordinary forms of mechanical power, *Philosophical Magazine*, **3**, 27, 205–207, 1845.
5. Joule, J.P., On the mechanical equivalent of heat, *Philos. Trans. Royal Soc. London*, **140**, 61–82, 1850.
6. Carnot, S., *Reflexions sur la Puissance Motrice du Feu, et sur les Machines Propres à Developer Ceite Puissance*, Bachelier, Paris, 1824.
7. Clapeyron, E., Memoir sur la puissance motrice du feu, *J. Ec. Polytech.*, **19**, 1–89, 1832.
8. Clausius, R., Über verschiedene fur die Anwendung bequeme Formen der Hauptgleichungen der mechanischen Wärmetheorie, *Abh. Über die mech. Wärm.*, **11**, 1–44, 1867.
9. International Energy Agency, *Key World Statistics*, International Energy Agency, Paris, 2016.
10. Michaelides, E.E., Exergy and the conversion of energy, *Int. J. Mech. Eng. Educ.*, **12**, 65, 1984.
11. Kestin, J., Availability: The concept and associated terminology, *Energy*, **5**, 679–692, 1980.
12. Baehr, H. D., Schmidt, E. F., Definition and calculation of fuel exergy (in German), *Brennstoff-Warme-Kraft*, **16**, 12, 589–596, 1963.
13. Bejan, A., Tsatsaronis, G., Moran, M., *Thermal Design and Optimization*, John Wiley & Sons, New York, 1996.
14. Tsatsaronis, G., Pisa, J., Gallego, J., *Thermodynamic Analysis and Improvement of Energy Systems*, Proceedings of the International Symposium, Pergamon Press, Beijing, 195–200, 1989.

2

Energy Demand and Supply

Energy is used in different forms: airplanes and automobiles use liquid hydrocarbon fuels; electric power plants convert the energy of coal, natural gas, nuclear, and flowing water into electricity; households use electricity and natural gas for domestic comfort with heating and air-conditioning, communications, entertainment, and the preparation of meals; factories use natural gas, petroleum, and coal for processing materials and manufacturing; and commercial buildings use natural gas, petroleum products, and electricity for heating and air-conditioning.

Because most functions in our society are based on the use of specific forms of energy, elaborate networks for the supply of the needed energy forms have been developed in the last two centuries. Electricity is fed into our communities by the transmission lines of the electric grid at very high voltage. Natural gas enters our dwellings by a complex system of underground pipelines, which often transcend national boundaries. Large tanker trucks supply the gasoline stations with petroleum products. Tanker ships crisscross the oceans daily to bring petroleum to the refineries that, in turn, supply the tanker trucks and the gasoline stations. The complex and ubiquitous energy network feeds the national economies and sustains modern human society.

We classify the different forms of energy in three general categories:

1. *Primary* energy encompasses the forms of energy that are directly consumed as they are found in nature, without any processing. The various forms of coal, petroleum (crude oil), natural gas, hydraulic energy, wind energy and solar energy are among the primary forms. The total primary energy source (TPES) is a measure of the level of energy consumption in a country. In most cases, it is the TPES that is commonly referred to as "energy used" or "energy consumed" in national and regional statistics.

2. *Secondary* energy forms are used by the consumers in a refined, processed form. The liquid petroleum products derived from crude oil, such as gasoline, diesel, and kerosene, are secondary energy forms. Ethanol fuel and biodiesel derived from biomass, thermal energy from solar collectors, and district heating from geothermal fluids are also considered secondary energy forms.

3. *Tertiary* energy forms involve one or more transformations of energy from one form to another. Electric energy, in any way it is produced, is a tertiary form of energy. Nuclear energy, natural gas, coal, wind power, and most of the other renewable energy sources, which are categorized as primary energy sources, contribute to the supply of tertiary energy when they are used to produce electricity.

The processing and transformations that produce secondary and tertiary energy forms always entail energy dissipation/losses—oftentimes significant for tertiary energy. The energy dissipation during such transformations is governed by the first and second laws of thermodynamics. The efficiency of the transformation process is an indication of how well the original energy of the primary resource is transformed to the energy of the secondary form. Fossil fuels (primary energy sources) are converted to heat (a secondary energy form) with 80–90% efficiencies. This implies that only 80–90% of the original chemical energy stored in the fossil fuels is transformed to useful heat. Similarly, 75–85% of the falling water energy is converted to electricity in a hydroelectric power plant. Converting the energy of the fossil fuels to electricity through a series of processes in thermal power plants has efficiencies in the range of 25–40%, which means that only 25–40% of the chemical energy of the fossil fuels is converted to electrical energy. The rest of the energy is dissipated as waste heat to the environment. According to the first law of thermodynamics, energy is always conserved. However, during all energy transformations, a fraction of the originally available energy (larger for thermal power plants, smaller for hydroelectric plants) is dissipated to waste heat that is not needed or otherwise used by the human society.

2.1 Demand for Energy: Whither Does It Go?

At the beginning of the twenty-first century, it is difficult to imagine our society without sufficient energy to function. The previous two centuries have been marked with unprecedented industrial productivity that was initiated in northern Europe and was globally exported. Material affluence; rapid communications; consummate economic growth; fast population growth, matched by adequate food production; and exceptional progress in every aspect of the human society characterize today's human civilization. The main reason for the development of this civilization and the unparalleled prosperity the modern humans enjoy is the invention, development, and optimization of a multitude of electrical and mechanical engines that have substituted human and animal labors, provided assistance to contemporary humans in all facets of their lives, and enriched their communications.

The engines we now use are by far more powerful and faster than the humans and animals they have substituted. They continuously work, produce more, and move faster. The engines, which operate only by consuming energy forms, have liberated human workers from several menial occupations and enabled them to pursue more fulfilling careers. During the Middle Ages, it was necessary for 98% of the European population to work in the fields and agricultural occupations in order to supply enough food for the subsistence of their communities. With the invention and operation of modern agricultural machinery, in the early twenty-first century, a single farmer in Kansas provides enough food for 128 additional persons. Modern humans have liberated themselves from toiling the earth to produce their food and may now pursue other professions and more pleasant activities (including writing books on energy!) that fulfill their lives and help them in the pursuit of their personal goals and happiness.

Using a variety of modern tools and machinery, a furniture factory worker in the People's Republic of China produces by far more chairs and tables per day than his/her counterparts seven centuries ago. The vastly increased supply of these products has made furniture affordable for many and enabled most of us to have more comfortable homes and places of work. Heating, ventilation, and air-conditioning has made modern dwellings be much better and infinitely more hygienic than dwellings of the workers in the Middle Ages. New and faster means of transportation—airplanes, cars, ships, and bullet trains—bring people, families, and nations together more often than in previous centuries. Modern computers, a multitude of new communication devices, and an electronic web that covers the wide world disseminate sounds, images, and information throughout the globe instantly.

2.1.1 Economic Development, Quality of Life, and Human Development

The prosperity and affluence of a nation are directly related to the amount of energy the nation consumes per capita. Citizens in the more affluent nations, as a group, buy more energy-intensive consumer goods; they use more frequently private means of transportation; they travel often and further; and they have higher levels of energy-intensive comfort—e.g., heating and air-conditioning—in their homes. All these amenities require higher amounts of energy, and this makes the energy consumption per capita in affluent nations to be significantly higher than the energy consumption in the less affluent nations. Figure 2.1, which depicts the gross domestic product (GDP) per capita of several nations vs. the total primary energy consumption per capita, shows the relationship between societal affluence and energy use in several nations. The total energy consumption is in tons of oil equivalent per person. The GDP values in the data of this figure have been adjusted to take into account the different purchasing power of the US dollar in the respective countries [1].

It is apparent from this figure that there is a direct relationship between the perceived affluence of a nation and the average energy its citizens consume. Nations in the Organisation for Economic Co-operation and Development (OECD) group,* which encompasses most of the industrialized and more affluent nations, have significantly higher GDP per capita and consume significantly more energy per capita than developing nations. The OECD countries appear at the top of the figure. At the other end of the figure are developing countries where societies are in general less affluent and the energy consumption is significantly less. In the middle of the graph, there are nations in intermediate stages of economic development. The energy consumption in these countries is commensurate with their respective GDP. Among these, the economies of China, India, and Brazil have significantly improved in the period of 2000–2017, and this has been reflected in their energy consumption. Citizens of these and similarly developing countries have become more affluent in the last two decades, and as a result, the total national energy consumption as well as the energy consumption per capita have significantly increased.

Figure 2.2 depicts an interesting consequence of affluence and energy consumption for all nations. This figure shows the average life expectancy versus their energy consumption per capita. The outlier countries in this figure are principally countries at war or with high incidence of cases of the acquired immune deficiency syndrome epidemic. It is apparent in Figure 2.2 that a significant and positive correlation exists between energy consumption and the average longevity of the citizens of a nation. It is important to note that energy consumption is not the *cause* of affluence and longevity in these nations, but the *effect* of affluence: A more affluent society consumes more goods and services, travels more, and typically has better public or private healthcare than a less affluent society. Because of this, the average citizen of an affluent nation consumes more energy and lives longer as the last two figures demonstrate.

---

* The OECD group of countries comprises most of the European Union countries, Switzerland, Canada, the United States, Chile, Israel, Mexico, Turkey, South Korea, Australia, and New Zealand. The countries in this group are sometimes referred to as the "developed" countries.

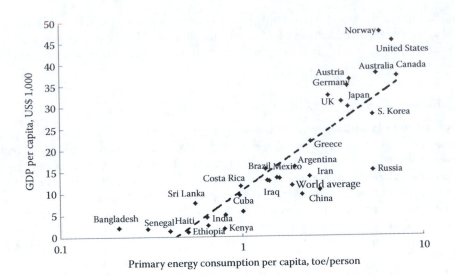

FIGURE 2.1

Relationship between TPES consumption per capita and GDP adjusted for purchasing power. (Data from International Energy Agency, *Key World Statistics*, IEA-Chirat, Paris, 2016.)

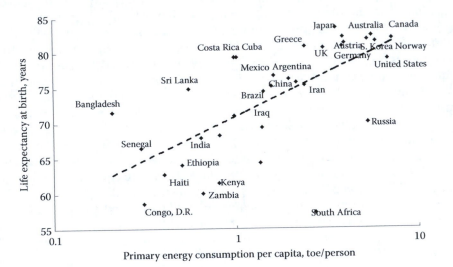

FIGURE 2.2

Relationship between TPES consumption per capita and average life expectancy at birth in several countries. (Data from International Energy Agency, *Key World Statistics*, IEA-Chirat, Paris, 2016; United Nations Development Programme, *United Nations Human Development Report 2015*, United Nations Development Programme, New York, 2015.)

One may also establish a cause-and-effect relationship between longevity and energy use by looking at the uses of energy since the beginnings of humans. Prehistoric humans had to perform all the tasks by themselves and lived in unsanitary caves. The fossils of their bones show a great deal of "wear and tear" of their bodies, and modern scientists estimate that their life expectancy was approximately 20 years. In the early historic period, humans domesticated animals, such as horses and oxen, to perform most heavy tasks for

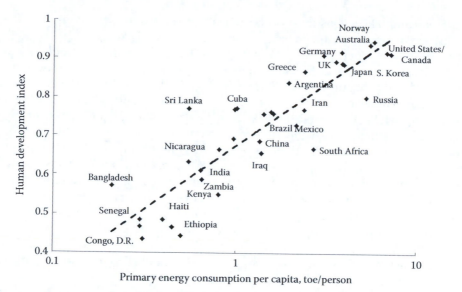

FIGURE 2.3

Relationship between TPES consumption per capita and the UNDP index in several countries. (Data from International Energy Agency, *Key World Statistics*, IEA-Chirat, Paris, 2016; United Nations Development Programme, *United Nations Human Development Report 2015*, United Nations Development Programme, New York, 2015.)

them, but still worked hard in agriculture and did not lead healthy lives. The life expectancy of these humans increased to only 30–35 years. The humans of the early twenty-first century are employing machinery to do most of their manual work including manufacturing and agriculture; their vast majority do not physically stress their bodies (unless they do this for sport) and enjoy several types of private and public healthcare, which also needs a great deal of energy. As a result, the human life expectancy has dramatically increased in the last two centuries.*

The human development index (HDI) is a quantitative measure developed by the United Nations Development Programme (UNDP) that gives a quantitative measure of the quality of life in the member nations of the United Nations (UN) [2]. The HDI, which is a number between 0 and 1, is a statistical measure that takes into account quality-of-life indicators including life expectancy at birth, public education and years of compulsory schooling, population and fertility rate, rate of inflation, and income per capita. Figure 2.3 depicts the relationship between the HDI and the total primary energy consumption per capita. The strong correlation between the average energy consumption per capita and the perceived quality of life by the UNDP is also apparent in this figure.

National and regional governments that look after the welfare of their citizens aim to increase the economic prosperity of their people through long-term strategic planning and infrastructure development. By doing so, they also increase the total consumption of energy in the country and, by extent, the electric energy, which is necessary for the functioning of an industrial society. National, regional, and local governments adopt long-term plans to secure the adequate supply of energy products and build electric power units for the production of sufficient electric power. The relationship between economic

---

* While it often said that "money does not buy happiness," Figure 2.2 and this cause–effect relationship demonstrate that, at least, money can buy a longer life.

development and TPES consumption becomes apparent when one considers the recent GDP increase of several countries that have experienced significant GDP growth in the early twenty-first century. Among these, the countries of Brazil, the People's Republic of China, India, Indonesia, and Mexico—where 40% of the earth's population lives—have experienced significant and unprecedented GDP growth and infrastructure development. Table 2.1 offers statistical information of the GDP of the country, its population, and the TPES consumption for the years 2001 and 2014. The corresponding variables of five OECD nations are also shown in this table for comparison.

It is apparent from Table 2.1 that the economic growth of the developing nations is accompanied by a significant increase in the TPES consumption. With respect to Figure 2.1, a developing nation's position on the graph does not simply move upward to higher GDP. The movement is also sideways, toward higher energy consumption per capita. An interesting observation in Table 2.1 is that the energy consumption of the OECD/developed countries actually decreased by a small fraction, despite the (also small) increase in their populations and GDP growth during the 2001–2014 period. This is primarily attributed to recent energy conservation measures and concerted national programs to reduce the TPES consumption in OECD countries.

The consumption of electricity is a significant characteristic of a modern developed nation. The nexus of the electric energy demand to affluence is illustrated by looking at three fast-growing economies—Brazil, India, and the People's Republic of China—of the last few decades. Figure 2.4 depicts the GDP per capita (adjusted for purchasing power parity) and the total electric energy consumption in the country. It is observed that the evolution of the GDP per capita, which is a measure of affluence in a country, has been accompanied by an equivalent increase in the electric energy (as well as of the total energy) demanded by its citizens.* The national demand for total energy as well as for electric energy is highly correlated with the affluence (or perceived affluence) of the citizens. The trend is particularly evident in the case of the People's Republic of China, where the rapid industrialization and rise of GDP in the first decade of the twenty-first century were accompanied by an equally rapid increase of the use of electric energy.

Regarding future trends, it must be noted that the three countries in Figure 2.4 account for more than 37% of the entire population of the earth. It is expected that the continued industrialization of these three as well as of similar countries with high populations will result in increased affluence and increased global energy demand, at least in the near future.

2.1.2 Benefits to the Human Society from Mechanization and Energy

It is not uncommon for the twenty-first-century humans to be blamed for environmental calamities, global climate change, and several societal ills, which are consequences of the historically high energy use. While part of these accusations may be truthful, the societal benefits derived by the higher energy use in all nations, developed and developing, by far outweigh any real or perceived damages. The following is a summary of the benefits of the higher energy usage for the humans of the twenty-first century:

1. Through mechanized agriculture and food transportation, the global food production is now adequate for a human population that increased sevenfold in the last two centuries. The production and transportation of adequate food supplies

---

* The observed fluctuations in the GDP per capita of Brazil are due to several devaluations of its currency in the 1990s. The long-term trends for Brazil are apparent in this figure.

TABLE 2.1

Effect of Economic Development on the Energy Demand

Country	2001 Population, Million	2001 GDP, US$ Billion	2001 TPES, Quads	2014 Population	2014 GDP, $US Billion	2014, TPES, Quads	2001–2014 Percentage of GDP Increase	2001–2014 Percentage of Energy Increase
World	6,103	42,374	400	7,249	101,463	543.5763	139.4	35.9
Brazil	172	1,040	7.3	206	3,061	12.02	194.3	63.8
PRC	1,272	4,708	45.2	1,364	16,841	121.1	257.7	167.9
India	1,032	2,707	21.1	1,295	6,902	32.7	155.0	55.0
Indonesia	209	561	6.0	254	2,501	8.9	345.8	48.3
Mexico	99	807	6.0	119.7	1,939	7.5	140.3	25.0
US	286	8,978	90.5	319	16,157	87.9	80.0	−2.9
Japan	127	3,126	20.6	127	4,437	17.5	41.9	−15.0
Germany	82.3	1,922	13.9	81	3,438	12.1	78.9	−12.9
France	61	1,394	10.6	66	2,407	9.6	72.7	−9.4
UK	59	1,293	9.3	64.6	2,441	7.12	88.8	−23.4

Source: International Energy Agency, *Key World Statistics*, IEA-Chirat, Paris, 2016; International Energy Agency, *Key World Statistics*, IEA-Chirat, Paris, 2003.

Note: PRC, People's Republic of China.

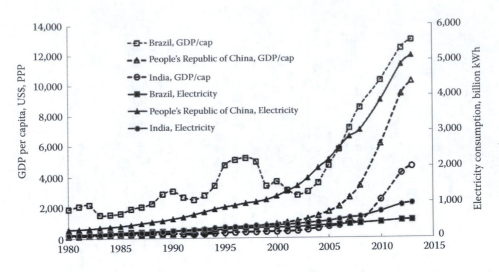

FIGURE 2.4

Electric energy–affluence relationship: the evolution of GDP per capita and electric energy consumption in Brazil, People's Republic of China, and India since 1980. The GDP is adjusted for PPP. (Data from International Energy Agency, *Key World Statistics*, IEA-Chirat, Paris, 2016; United Nations, Department of Economic and Social Affairs, Population Division, 2015 and their previous editions.)

have eradicated regional famines in the early twenty-first century.* Fertilizers, pesticides, and insecticides used for food production have, in general, a detrimental effect on the soil and the environment. However, the benefits from adequate food supply for the entire population of the earth by far outweigh this modest environmental damage. Regenerative agriculture uses very low quantities of chemicals for the soil and is friendlier to the environment. All types of agriculture use a great deal of energy for field machinery, chemicals, and mechanized transportation.

2. The mechanized agriculture practices removed the need for most humans to work in agriculture and food production for their sustenance and enabled them to pursue other occupations and more intellectual careers. The world now has proportionately a much higher number of artists, scientists, office workers, university professors, and students than any other epoch in human history.

3. Modern healthcare, both private and public, which includes medical tests and equipment to ensure longer life and a better quality of life.

4. Public health measures have significantly improved; human dwellings and places of work are more sanitary; and the life expectancy for humans more than doubled.

5. Climate control for buildings (heating, ventilation, and air-conditioning) has made life, leisure, and work easier to appreciate and enjoy under all weather conditions.

6. Modern systems of transportation have considerably shortened the time to travel between regions, countries, and continents.

---

* The last regional famines occurred in 1985 (Ethiopia) and 1995 (North Korea). Food shortages still occur in several regions, but they are caused from regional wars and civic strife, not from a global lack of food and available transportation to alleviate the food shortage.

7. Factories and workshops produce by far more goods. Affluence (or relative affluence) is enjoyed by a much higher number of humans.

8. Communications are almost instantaneous. The human race is more connected now than at any other time in history.

9. Every human enjoys some form of entertainment through radio, television, the Internet, etc.

10. Because of the mechanized higher productivity, most workers take lengthy vacations from work, a concept that was known to only a few privileged ones two centuries ago.

2.1.3 Global Trends of the Demand for Energy

The brave new and prosperous world that was created in the last two centuries is based on the use of ever-increasing quantities of energy. Our civilization evolved and reached the current age of the locomotive, the nuclear power plant, the automobile, the airplane, the personal computer, and the wireless Internet. The human society has continuously evolved by increasingly using energy in its several forms. Figure 2.5 shows the historical global energy consumption and the population of the planet since 1800. The data in the figure are from Michaelides [4] and the UNDP [5] with the data before 1950 being estimates by the UN. The unprecedented growth of energy consumption and population during the last two centuries is apparent in Figure 2.5: between 1800 and 2014, the population of the earth increased sevenfold and the energy consumed increased by a factor of 28. This also implies that the per capita consumption of energy has also increased fourfold. The per capita annual energy consumption for the earth's population is also shown in this figure

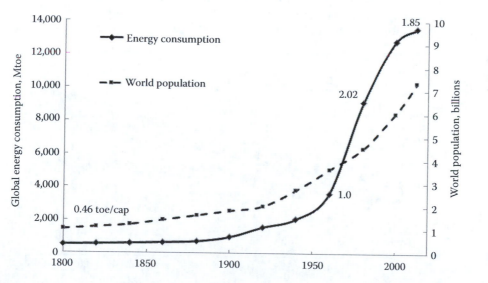

FIGURE 2.5
World population, in billions of inhabitants, and global energy consumption, in 10^6 toe, since 1800. The numbers inside the figure show the energy consumption per capita (toe/capita) in 1800, 1960, 1980, and 2014. (Data from United Nations, Department of Economic and Social Affairs, Population Division, *Human Development Report* 2015; with kind permission from Springer Science+Business Media: *Alternative Energy Sources*, 2012, Michaelides, E. E.)

for the years 1800, 1960, 1980, and 2014. Two persisting global trends are apparent in Figure 2.5 and corroborated by the data of Table 2.1:

1. The increasing population of the planet demands higher quantities of energy.
2. Not only the earth's population is increasing, but also the energy consumption per capita is increasing as individuals and entire nations pursue more affluent lifestyles.

The combination of the two trends has resulted in the 28-time energy demand growth. In the early twenty-first century, the very existence of the human society is based on the use and consumption of energy. An adequate supply of energy is necessary for the basic functions of the society, the prosperity of the nations, and the survival of our civilization.

The significant increase in the energy demand/consumption, which is characteristic of rapidly developing economies, was experienced in the past by countries that are currently considered developed countries. The OECD group of nations consumes a significantly larger proportion of the global TPES than the fraction of its population. Figure 2.6 depicts the evolution of the global energy demand by region, in quads, from 1970 to 2014 in several large geographical regions of the planet [1,4]. The People's Republic of China and Middle Eastern countries are depicted separately from the rest of Asia. It is observed in this figure that since 1970, the global energy consumption more than doubled; actually, it increased by 220%. The relative share of the OECD countries has fallen from 62% of the total to 39.2%. However, the absolute amount of energy consumed by the OECD group still increased from 101 Q to 147.5 Q, a 47% rise. Figure 2.6 also shows that while the energy consumed by the OECD countries significantly increased, their average rate of growth was less than the average rate of growth of the global energy consumption. During this time, the People's Republic of China and the rest of Asia almost doubled their share of the total energy consumption, while the share of Latin American and African nations also significantly increased. As the economies of the countries in the developing regions of the earth grew and populations became more affluent; the consumption of energy increased according to

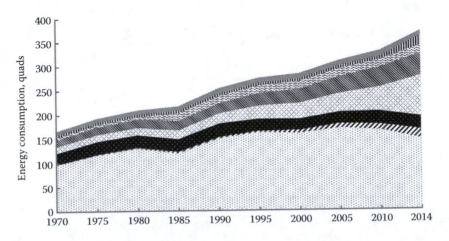

FIGURE 2.6
Evolution of the global energy demand by region and groups of countries. The areas (from bottom to top) represent the OECD nations, Middle East, non-OECD Europe, People's Republic of China, the rest of Asia, Latin America, Africa, and bunker fuels (fuels internationally used for transportation). (Based on data from International Energy Agency, *Key World Statistics*, IEA-Chirat, Paris, 2016.)

the trends of Figure 2.1. These trends are expected to continue in the near future, and the global energy demand is expected to increase further [1].

Figure 2.7 depicts the recent evolution of the energy forms consumed, also in quads. The data in this figure show that during this period, when the total global energy consumption increased by 220%, the consumption of electricity increased by 322%, and that of petroleum products, by 238%. Interestingly enough, the contribution of the "other forms" that include renewable sources experienced the highest growth (590%), but the actual amount of the "other energy forms" is still very low as the thin sliver that represents this energy form in the figure signifies. It must be noted that Figure 2.7 represents the final energy forms that are globally consumed, not the total primary energy used by the consumers. Because several of the energy-demanding processes have efficiencies less than 100%—e.g., the production of electricity—more primary energy is needed and used for the production of the final energy forms we consume.

It must also be noted that great deal of the energy globally consumed is rejected as waste heat or waste by-products. For example, the efficiency of automobiles and trucks, which use petroleum products, is close to 25%, and this means that only 25% of the energy in the diesel fuel or gasoline—both are petroleum products—is used to produce the motive power of these vehicles. The remaining 75% is rejected to the environment through the exhaust pipes of the vehicles as waste heat. Clearly, increasing the efficiency of all energy-demanding processes will reduce the global energy consumption in all its forms. Regarding the economic sectors where these energy forms are demanded, the final energy quantities shown in the last two figures were consumed by the following sectors of the global economy [6]:

- 50% industrial
- 20% transportation
- 18% residential
- 12% commercial

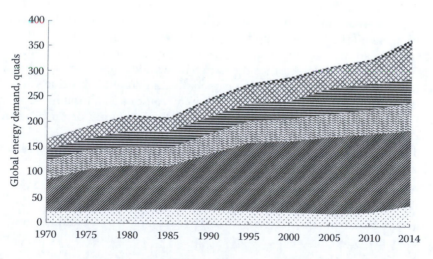

FIGURE 2.7

Evolution of the global energy consumption by final energy form. The areas in the graph (from bottom to top) represent coal, petroleum, natural gas, wood and waste, electricity, and other energy forms. (Based on data from International Energy Agency, *Key World Statistics*, IEA-Chirat, Paris, 2016; with kind permission from Springer Science+Business Media: *Alternative Energy Sources*, 2012, Michaelides, E. E.)

A rather somber conclusion about total energy consumption trends the OECD countries is that despite the energy conservation and improved energy efficiency measures that were adopted since the mid-1970s, the total energy consumed in OECD nations actually increased by 47%. The energy consumption in the developing nations more than doubled, despite national and regional higher efficiency and conservation policies. The global energy consumption—and the environmental impacts associated with it—continue to grow at an accelerated pace in the first two decades of the twenty-first century. Past and recent trends indicate that the global energy consumption will continue to increase in the foreseeable future [1,3,6]. The highest growth of energy demand will occur in Asia, Africa, and Latin America, regions where most of the earth's population lives and where most of the economic growth is expected. Increasing affluence and population growth will be the driving forces for the growth of global energy demand at least until the middle of the twenty-first century.

The economic and societal impacts of a continuous and uninterrupted energy supply are of paramount importance to all modern nations. Adequate energy supply is a matter of "national security" for most nations, and energy security is relentlessly pursued through political and military interventions. For this reason, the geopolitical and military activities of modern nations are significantly influenced by their need for uninterrupted and secure energy supply to the OECD nations. Many of the recent wars, after 1950, have been fought for the control and security of petroleum supplies, and several alliances, treaties, and international agreements have been cemented with energy resources as the primary consideration.

2.2 Energy Supply: Whence Does It Come?

The energy we use daily is produced from the primary energy sources, which are not uniformly distributed on the planet. The Arabian peninsula contains a great deal of oil; the United States and the United Kingdom have large quantities of coal; Mali and Canada have large quantities of uranium; Sri Lanka and Sudan enjoy more solar energy per unit area than Germany and Iceland. Since the natural laws dictate that energy may neither be created nor destroyed, all the energy that is globally consumed is produced from existing energy sources. The aggregate energy demand in all nations as a total must be met by the global energy supply, which is mainly derived from the following primary sources:

1. Fossil fuels, which are the various forms of coal—anthracite, lignite, bituminous coal, and peat—petroleum, and natural gas
2. Nuclear fuels, uranium and thorium, which are contained in minerals such as uranite
3. Renewable energy forms, such as hydroelectric, solar, wind, biomass, geothermal, and wave energies

The first two forms of energy are minerals that were formed several millennia ago. Their formation processes took place over very long periods (geological periods). Our society at present consumes vast quantities of these minerals, but the reproduction rate of these minerals is very low. Because of this, fossil and nuclear fuels are expected to be depleted at some point in the future. On the contrary, renewable energy forms are inexhaustible. Humankind may reasonably expect that renewable energy forms will continue to supply energy to humans in the foreseeable and far future.

The secondary and tertiary forms of energy, which society uses at high rates, must be produced from TPESs. The principal primary energy sources that have satisfied the global energy demand since 1970 are shown in Figure 2.8 [1,4]. The most important of the TPESs are the following:

1. The various forms of coal (anthracite, bituminous, lignite, peat)
2. Crude oil/petroleum
3. Natural gas
4. Nuclear energy
5. Hydroelectric energy or water energy
6. Biomass and waste, which primarily comprise wood from trees, grass, and bushes
7. Other renewable forms, such as solar, wind, and geothermal

The last three primary energy sources are renewable and are considered inexhaustible. A comparison of Figures 2.7 and 2.8 shows that the primary energy supply is significantly higher than the energy consumed: in 2014, the total energy consumed in its final form was 374.0 Q, while the TPES that produced this amount of consumed energy was 543.6 Q. The difference is due to the energy dissipation when primary energy forms are transformed to secondary and tertiary forms. The energy dissipation upon processing and transformations are a consequence of the second law of thermodynamics and may be reduced with processes and equipment that have higher efficiency.

A glance at the data of Figure 2.8 proves that the energy supply increased by 237% between 1970 and 2014, following the increased global energy demand/consumption.

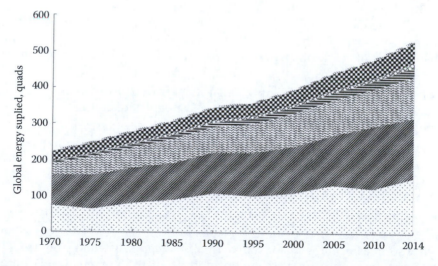

FIGURE 2.8
Evolution of the global primary energy supply, by energy form. The areas in the graph (from bottom to top) represent coal and peat, petroleum, natural gas, nuclear, hydroelectric, wood and waste, and other energy forms. (Based on data from International Energy Agency, *Key World Statistics*, IEA-Chirat, Paris, 2016.)

TABLE 2.2

Evolution of the World Primary Energy Supply between 1973 and 2014

	1973, Total of 242 Q	2014, Total of 543 Q
Coal and coal products	24.5%, 59.3 Q	28.6%, 155.5 Q
Crude oil	46.1%, 111.6 Q	31.3%, 170.2 Q
Natural gas	16.0%, 38.7 Q	21.2%, 115.2 Q
Nuclear	0.9%, 2.2 Q	4.8%, 26.1 Q
Hydraulic	1.8%, 4.4 Q	2.4%, 13.0
Wood, biomass, and wastes	10.6%, 25.7 Q	10.3%, 56.0
Other renewables	0.1%, 0.2 Q	1.4%, 7.6

Source: International Energy Agency, *Key World Statistics*, IEA-Chirat, Paris, 2016; with kind permission from Springer Science+Business Media: *Alternative Energy Sources*, 2012, Michaelides, E. E.
Note: Values are in percentages of the total and absolute values in quads.

It is of interest to know the contributions of the several primary energy sources to the total global energy demand during the period covered in Figure 2.7. Table 2.2, which was produced from the same data, shows two snapshots of the total global energy supply in the years 1973 and 2014 and elucidates the changes that occurred in the supply of the several primary energy forms.

Several trends about the use of primary energy sources in the recent past are apparent in Table 2.2:

1. The total primary energy consumption more than doubled in the last 40 years.

2. The coal consumption almost tripled, despite several national and international programs to curb the production of CO_2.

3. While the relative amount of crude oil consumption significantly decreased, from 46.1% to 31.3% of the total, the absolute amount of crude oil consumed actually increased from 111.6 to 170.2 Q. Despite all the national campaigns for the reduction of crude oil consumption, the increased mileage of cars, and national measures for national energy independence in OECD countries, the consumption of crude oil is higher than ever and will likely increase in the near future.

4. Most of the global energy supply, 81.1%, comes from fossil fuels (coal, oil, and natural gas) which are in finite supply, are fast depleted, and will be exhausted at some point in the future.

5. Both the relative and the actual amount of nuclear energy consumed have significantly increased. However, most of this increase occurred during the 1970s and early 1980s. Only a handful of nuclear reactors have been built the OECD countries since 1986, when the Chernobyl accident occurred.

6. The production of hydroelectric energy, in kilowatt hours, almost tripled.

7. Only 14.1% of the primary energy consumption is derived from renewable energy sources, and the major source in this category is wood from trees. Wood consumption cannot increase much more than the current levels because it will lead to deforestation. This implies that for a sustainable energy future, a great deal more energy must be produced from solar, wind, geothermal, and hydropower.

8. Even though the use of solar, wind, and geothermal energies has increased by a factor of 38 in this period, these renewable energy sources contribute only 1.4% of the total primary energy supply. The entire world has to make significant technological progress and great strides to achieve a sustainable energy future that is based on renewable energy sources.

One of the reasons why approximately 85% of the total primary energy is supplied by fossil and nuclear fuels is the high energy density (e.g., in MJ/kg) of these fuels and the convenience/ease of their utilization. A small amount of each fuel is capable to provide a great deal of thermal energy or electricity. This implies that by using nuclear and fossil fuels, a great deal of energy may be produced in relatively small facilities. In contrast, renewable energy sources, such as solar, wind, hydraulic, and geothermal energies, have significantly lower energy density (in MJ/kg or MJ/m² for wind and solar) and entail much larger facilities to produce an equivalent amount of energy.

An important conclusion from the data of Table 2.2 and Figure 2.8 is that the global primary energy is principally supplied by exhaustible energy sources—coal, oil, natural gas, and nuclear. These fossil fuels are depleted at fast rates; they will become scarce; and they will be exhausted at some point in the future. Because the entire human civilization is based on the consumption of energy, the continuation of our civilization necessitates that we must ensure the adequate supply of energy from renewables before the other energy sources are depleted. Technologies for the utilization of solar, wind, geothermal, and hydrogen fusion energies* must become widely available in the not-so-distant future to satisfy the global energy demand and to ensure the continuation of the contemporary civilization.

All the countries utilize and promote domestic energy sources to satisfy their energy demand to the extent possible. Whatever a country cannot produce, it imports from others by transportation systems that have been developed in the last two centuries: coal, petroleum and petroleum products, uranium, and natural gas are traded and transported daily in tankers, trainloads, and pipelines. The flow of primary energy forms in the main regions of the world is shown in Table 2.3, together with the population of the regions and the average GDP adjusted for the purchasing power parity (PPP) of the US dollar [1].

Table 2.3 shows that energy imports are primarily directed to the OECD countries and the People's Republic of China. It is also of interest to note that the OECD countries have the highest energy consumption per capita, 0.165 Q per million inhabitants, while the average for the entire world is 0.075 Q per million inhabitants. If the entire world had the same level of affluence and consumed primary energy at the same rate as the OECD countries in 2014, the global TPES would have been 1,196 Q or 2.20 times higher than it actually was during that year. Given the trend for all countries to approach the OECD standards of living and higher energy consumption per capita, it is likely that the future global energy consumption may tend toward this very high level. This will entail a great deal of strain in the energy resources and the environment. Serious national and international efforts in energy conservation and energy efficiency as well as international mandates for conservation and the curtailment of energy consumption, globally, may alleviate this future threat to the society and the environment.

---

* Hydrogen is not a naturally occurring energy source and must be artificially produced. The amount of hydrogen on earth (as water in the oceans) is very high, and the energy released by fusion is large enough to make this energy source virtually inexhaustible (Chapter 5).

TABLE 2.3

World Population; GDP; and Primary Energy Production, Imports, and Total Supply

Region	Population, Million	GDP Billion, $US, (PPP)	Production, Q	Imports, Q	TPES, Q
OECD	1,267	46,238	164.4	52.5	209
Middle East	224	4,946	71.7	−41.7	28.6
Rest of Europe and Eurasia	343	5,534	72.2	−26.4	44.6
People's Republic of China	1,372	17,214	102.9	21.2	121.7
Rest of Asia	2,408	15,871	59.4	12.6	69.1
Latin America	480	6,528	32.4	−6.0	30.6
Africa	1,156	5,131	44.8	−13.4	30.6
Total world	7,249	101,463	547.8	n/a	543.6[a]

Source: International Energy Agency, *Key World Statistics*, IEA-Chirat, Paris, 2016.

Note: Negative imports are exports. Data are for 2014.

[a] The TPES for the entire world includes international aviation fuel, international marine fuel, traded electricity, and traded heat.

2.2.1 Energy Prices, Economics, and Politics

Price is the variable that connects the supply of and demand for a commodity, which is traded in a free market. Consumers buy less of a commodity if its price increases, and hence, its demand is inversely related to its price. The demand for energy commodities drops only by a small amount when the price increases, because they are necessary for the functioning of the society. Our experience shows that consumers will buy petroleum at high prices to run their cars, trucks, and trains. Commodities that are necessities and will be bought at high prices exhibit a sharper, more vertical demand curve, which is called *inelastic*. Supply curves are positively correlated with price, because suppliers produce more of the commodity when the commodity price rises. This is particularly pertinent to petroleum and natural gas: when their market price rises, more wells are drilled and more extraction rigs are built to increase the supply of the two energy commodities. At the intersection of the supply and demand curves, the supply and demand curves of a given commodity are considered to be *at equilibrium*, and the intersection point of the two curves determines the equilibrium price of the commodity. Figure 2.9 depicts typical demand and

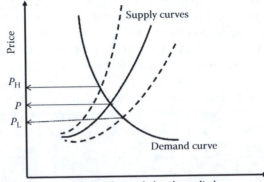

FIGURE 2.9

Typical demand and supply curves of a commodity such as petroleum.

supply curves for an inelastic commodity, such as petroleum. The equilibrium price P is shown at the intersection of the actual demand and supply curves, which are represented by the solid curves.

Let us consider a political event that takes place and affects one of the oil-producing regions—a war, an embargo, or political unrest in the Middle East. This event disrupts the petroleum production in the region and decreases the produced quantity. Other suppliers or other methods may enter the market to supply the demand, but this takes time and the supply curve temporarily shifts to the left as denoted by the upper dashed line in Figure 2.9. Given the demand, a new equilibrium is reached with a higher price P_H. It must be noted that because of the higher price, the quantity demanded and produced has slightly decreased at this equilibrium state. On the contrary, if the suppliers see high future profits and decide to increase the quantity of the commodity produced, the supply curve shifts to the right and is represented by the lower dashed curve in Figure 2.9. The new equilibrium is reached at the lower price P_L. At this equilibrium, the consumption of the commodity is higher than that at the original equilibrium position, because many consumers respond to the lower price and use more of this commodity.

The influence of political events on the supply, the demand, and the price of petroleum has been demonstrated in the aftermath of the recent events that are commonly known as "Arab Spring": In the first three months of 2011, political upheavals in Tunisia, Egypt, Libya, Syria, and several other nations caused a temporary disruption to the petroleum supplies from North Africa and a few parts of the Middle East. The immediate result was a jump in the oil price from approximately $80/bbl in January 2011 to $118/bbl in April 2011, a 47.5% increase. When a few months later, the political situation in these countries became more stable, the price of petroleum dropped to approximately $95/bbl in June 2011. At the same time, and as a result of the temporary higher petroleum price, more petroleum production rigs were developed and were drilled in 2011 and the next three years in the United States, Canada, and other parts of the world. The additional production shifted the global supply curve to the right. The increased supply caused the significant drop of the petroleum price in 2015. The price of petroleum reached $34 in December 2015 and remained in the price range of $35–$50 in the entire year of 2016 and most of 2017.

The prices of petroleum and other hydrocarbons fluctuate daily, responding to demand and supply conditions. Absolute crude oil prices are historically much more volatile than prices of other commodities for two reasons:

1. They are significantly influenced by international politics and events.
2. A group of relatively few countries, the Organization of the Petroleum Exporting Countries (OPEC), which effectively controls a high fraction of the petroleum production for export, supplies a high fraction of the petroleum demand to the rest of the world.

The OPEC is an economic cartel composed of Algeria, Angola, Ecuador, Iran, Iraq, Kuwait, Libya, Nigeria, Qatar, Saudi Arabia, the United Arab Emirates, and Venezuela. Put together, the OPEC members possessed 81.2% of the total global proven petroleum reserves in 2015 [7]. The cartel has quarterly meetings, usually in Vienna, Austria, where the representatives of the member countries attempt to regulate the supply of crude oil by issuing quotas among the member countries. In 2015, the OPEC accounted for the export of 23,569,500 bbl/day of crude oil [7]. This quantity is equivalent to 62.8 Q for the year and represents 56.6% of all the oil exports in the world.

Since 1973, the OPEC has been effective in controlling the global price of petroleum by periodically adjusting the quantity of petroleum produced by its member countries. By regulating the crude oil supply by its member nations, the actions of OPEC manipulate the supply curve for petroleum. By the supply–demand mechanism of Figure 2.9, the equilibrium position of the market and the price of crude oil are significantly affected.* Effectively, the manipulation of the petroleum supply by OPEC regulates the global price of this commodity and, by extent, the prices of gasoline, diesel, and most liquid fuels.

The prices of the other energy forms also fluctuate, following the petroleum prices, but with lower variability. Political events in the Middle East—the major supply region of crude oil to the OECD countries and the Far East—have significant influence on the production and prices, not only of petroleum, but also of all liquid hydrocarbons. Between 1960 and 2015, the price of crude oil and natural gas temporarily jumped significantly (sometimes by more than 100%) within a short time in the aftermath of the following political events:

1. The Arab–Israeli war of 1973 and the following oil embargo by the Arab nations
2. The aftermath of the Iranian revolution in 1979
3. The Iraqi invasion of Kuwait and the Persian Gulf War of 1990–1991
4. The production reduction of the mid-1990s by the OPEC
5. The 2003 invasion and prolonged war in Iraq
6. The Arab Spring in 2011

The prices of all energy commodities are characterized as *real* and *nominal* prices. The nominal prices are those quoted in the everyday transactions and frequently appear on the gasoline station billboards. For example, the *nominal price* of 1 L of gasoline in Oxford, England, was £0.21 in 1976 and £1.65 in 2015. The *real price* of gasoline and other petroleum products are more stable and account for the inflation and the different purchasing power of a currency over the years. The real price is often quoted in constant currency of a particular year, such as constant 1990 US dollars or constant 2000 British pounds.† The official inflation rate in a country, which is often given by the respective national Treasury Department, determines the annual rate of inflation of the currency for the corrections to be made.

The fluctuations of the yearly averaged crude oil prices from the end of the Second World War to 2015 are shown in Figure 2.10, which depicts the nominal and real crude oil prices, the former measured in constant 2010 US dollars [4,7].‡ Several of the important sociopolitical events in the Middle East that globally influenced the price of petroleum are also noted in the figure. A glance at the price fluctuations proves the following:

1. The effect of major international events in the price of crude oil is significant but, typically, of short duration. Prices are restored close to their previous level once the political or social "crisis" is over.

---

* Another reason for the higher petroleum prices in the early twenty-first century is the increased demand in the People's Republic of China, India, Brazil, and other developing countries.
† The conversion to Euro in many European nations, which occurred between 1995 and 2005, makes this computation more cumbersome. Most of the energy resources are quoted in constant US dollars or British pounds.
‡ The price of crude oil (petroleum) is a well-publicized economic parameter. This price is followed daily and appears in most news outlets. The price is usually quoted per barrel of oil, dollars per barrel of oil (1 bbl = 42.0 gal = 159 L).

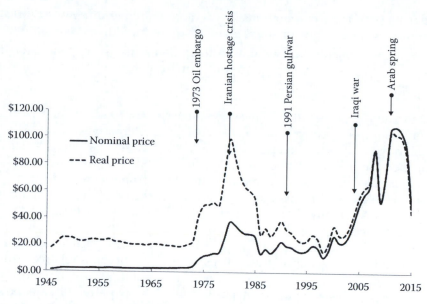

FIGURE 2.10

Nominal price (*solid line*) and real price in 2010 (US$) (*dashed line*) of petroleum. Significant political events that influenced the supply and price of petroleum are also shown.

2. Despite the significant fluctuations in its nominal price, the real price of crude oil, measured in constant 2010 US dollars, has been historically in the range $20–$40 per barrel. The real price of petroleum was in this range in 46 of the 70 years covered in the graph. However, the period 2004–2015 appears to be an exception, with the price of petroleum being significantly higher than this range. The Iraqi war, the Arab Spring, and the increasing demand from several developing nations are the main reasons for the significant jump of the real petroleum prices during this period.

The prices of the other fuels and other energy commodities have a high, positive correlation to the price of petroleum. The prices of the other fuels have higher inertia and do not exhibit the wild fluctuations of petroleum prices. In extended periods of high energy prices, consumers may substitute one energy form with other, cheaper energy forms. This *substitution effect* has had significant and beneficial effect for the development of alternative energy sources, energy efficiency, and conservation. When energy prices increase, consumers, and the society as a whole, strive to satisfy their energy needs with lesser energy consumption or with other sources of energy, such as wind, solar, and geothermal. Some of the effects in the period 2004–2014 that have been responses to the higher petroleum prices include the following:

1. Substitution of low-mileage cars to higher-mileage cars, including hybrids.
2. Substitution to cars that are powered by natural gas, which is cheaper than liquid fuels.
3. Substitution to electric cars in the regions where electricity is priced cheaper or where subsidies are offered by national policies.
4. Use less energy at home for heating and air-conditioning.

5. Electricity-producing corporations invest more in higher-efficiency power plants as well as in power plants that use alternative energy sources.

6. National governments initiate and promote policy measures that favor the use of natural gas, alternative energy sources, more energy conservation, and higher efficiency.

When petroleum and the other conventional energy sources become expensive, alternative energy sources become more viable economic alternatives to satisfy the energy demand of the society. The experience of the period 2004–2014 demonstrates how the significant increase in energy prices resulted in more conservation measures in OECD countries (Table 2.1 and Figure 2.6 corroborate this trend) and increased use of alternative energy sources. As shown in the insets of Figure 2.5, the globally averaged energy consumption per capita actually decreased in the second decade of the twenty-first century as a result of higher energy prices.

On the contrary, when there is an "oil glut," all energy prices drop and energy conservation, alternative energy, and higher efficiency projects become relatively more costly and less important to the general population. The period 1984–2002 is an example of such a low-energy prices era: during this period of very low petroleum and other energy prices, there was very little investment in solar and wind energy projects. Conservation and higher efficiency efforts were relaxed or abandoned, and the automobile companies throughout the world thrived with the production of large, energy-guzzling vehicles with very low mileage, such as the seven-seat family caravan, the large SUV, and the Hummer.

2.3 Energy for Transportation

The internal combustion engine (ICE or IC engine) and the jet engine are two of the most notable inventions of engineering. Used in automobiles, boats, trucks, and trains, the ICE revolutionized the way people and goods are transported. In the last two centuries, people enjoy much faster and comfortable transportation than any time in human history. Cars and busses are much faster, cheaper, convenient, cleaner, and hygienic than animal-drawn carriages and provide a great deal of freedom of movement to individuals. The automobile transformed the urban communities and allowed individuals and families to take more frequent trips. In the early twenty-first century, the mechanized and highly efficient agricultural production around the world in combination with the faster global food transportation system has alleviated regional famines. Mobile industrial equipment—bulldozers, forklifts, elevators, and cranes—increased human productivity and have played a major role in the development of the modern urban centers.

The jet engine, a product of the twentieth century, has significantly shortened the time of travel between continents. In the early twenty-first century, it is possible for individuals to travel from New York to London in less than 6 hours, or from Frankfurt to Tokyo in about 10 hours. Global travel, which was very slow, dangerous, expensive, and an activity enjoyed by a privileged few in the past, has become an ordinary activity for many. The ICE and the jet engine have significantly contributed to human gratification, global awareness, and social equality.

The ICE was invented in the nineteenth century and was immediately fitted in ships and smaller boats that became significantly faster. A few light trucks and automobiles appeared late in the nineteenth century, but were primarily the property of the wealthier part of the population. The T-model, designed by Henry Ford, was the first vehicle to be massively manufactured in an assembly line. It was designed to be affordable by the general population and transformed the way families travel and the entire transportation industry. The assembly line and the massive production of automobiles were soon adopted by other car manufacturers in several countries. By the 1970s, the automobile, from a luxury item, became a necessity for individuals and families in OECD countries. In the United States, which has the highest number of automobiles per 1000 persons, the number of registered vehicles rose from a mere 8,000 in 1,900 to 34 million in 1940 to 74 million in 1960 [8]. The economic affluence that followed the Second World War in the United States had a significant impact on automobile purchases and use. Almost all American families purchased one or two cars; the urban population soon expanded in the suburbs; and automobiles became a necessity for commuting to work and shopping. At the same time, more trucks and buses became necessary to transport goods and people in the ever-expanding urban environments. Figure 2.11 depicts the number of the registered vehicles and aircraft in the United States in the period 1960–2013 [9]. It is observed that the number of vehicles skyrocketed from 74 million in 1960 to 256 million in 2013. This amounts to approximately 800 vehicles per 1000 people or 0.8 vehicles per capita in the United States. The same trends have been observed in other OECD countries with a time lag: In Japan, the number of private automobiles rose from less than 1 million in 1950 to more than 60 million in 2013 [10]. In Germany, there were 48.8 million registered vehicles in 2014 (0.588 per capita); in Australia 17.3 million (0.731 per capita); and in the United Kingdom, 32.9 million (0.519 per capita) [11]. Figure 2.11 also shows that the number of commercial and private aircraft, which only use liquid petroleum products, has also skyrocketed in the last 50 years in the United States. As with the numbers of automobiles, the other OECD countries experienced the same trends with the numbers of registered aircraft.

A similar, dramatic increase has been observed in the number of passengers and the tonnage of goods transported. While OECD countries have first experienced the increase in air transport, in the early twenty-first century, developing countries are participating in the growth of air transport of goods and passengers with significant numbers. The global international and domestic traffic of passenger miles increased from 3.9 trillion in 2005

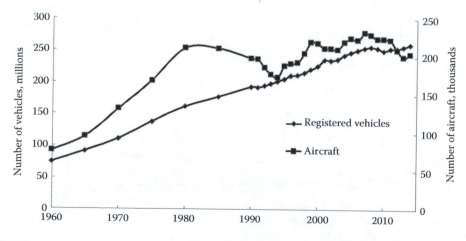

FIGURE 2.11
Numbers of registered vehicles and aircraft in the United States.

to 6.6 trillion in 2015. At the same time, the global airborne freight of goods transported increased from 154 billion ton-kilometers to 198 billion ton-kilometers [12].

Approximately one-third of the total primary energy produced in the world is used for transportation [1], and most of this energy is derived from petroleum. Land vehicle engines and aircraft engines predominantly use fuels composed of liquid hydrocarbon mixtures—gasoline, diesel, kerosene—all products of petroleum distillation. Although petroleum products have been used for other activities in the past—e.g., for industrial and residential heating, for the production of chemicals, and for electricity production— regional and global transportations consume an increasingly higher fraction of petroleum [1,4]. The increased number and increased usages of land vehicles, ships, and aircraft have caused a significant increase in the global petroleum consumption. Figure 2.12, which covers the same time period as Figure 2.11, shows the corresponding increasing demand for petroleum products in the United States [13]. A couple of significant trends may be observed in this figure: The use of petroleum products has significantly increased in the transportation sector, while it has remained the same or declined in the other economic sectors. As a result, the fraction of petroleum products consumed by the transportation sector gradually rose from approximately 50% in 1960 to more than 71% in 2015. The growth in the demand for petroleum has occurred despite the significant price increase of petroleum relative to other fuels since 1973 and despite several national efforts to curb the petroleum demand and develop alternative transportation fuels.

The significant increase in the demand for petroleum products, which was observed in the past in OECD countries, is also evident in the trends of the developing countries. In the period 2012–2016, approximately 25% of all the automobiles globally produced were sold in the People's Republic of China, where the demand for petroleum products increased from 41×10^6 toe in 1973 to 506×10^6 toe in 2014, a 12-fold increase [1]. The number of private automobiles in India, Indonesia, and Brazil is continuously increasing at a fast rate and generates a similar increased demand for petroleum products. At the same time, the transportation of passengers and goods by sea and by air is increasing globally. All these recent developments and trends have resulted in the 30.2% increase in petroleum demand

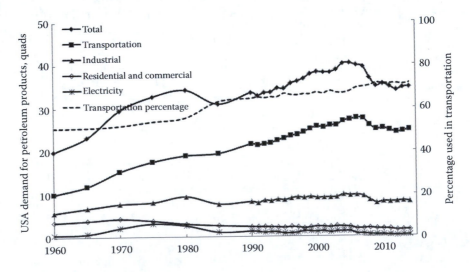

FIGURE 2.12
US demand for petroleum products between 1960 and 2013 (*left axis*). Percentage of petroleum used in the transportation sector (*right axis*).

globally between 2000 and 2014 (which is apparent in Figure 2.8) despite the higher global petroleum prices, the consumption reduction in OECD countries, and efforts to decrease the petroleum imports in all the petroleum importing nations. All indications are that the current trend of increasing petroleum demand will continue at least in the near future [14,15].

The only transportation sector with decreasing relative demand for petroleum products is rail transport. Several of the national railroad systems throughout the globe have converted their propulsion systems from ICEs to electric motors. Between 1950 and 2015, most of the gasoline and diesel locomotives in OECD countries have been replaced by electric locomotives, which are more powerful and operate without local emissions.* This trend continues globally with the development of the modern electric railway systems that are much faster and operate in urban environments without causing additional local pollution. The continuing evolution of the railway propulsion systems to electric power has shifted some of the global demand for petroleum products to higher electric power demand, which must be satisfied by other primary energy sources.

2.4 Production of Electricity

More than 40% of the TPES globally is used for the production of electricity, another energy form that was developed on a large scale in the early twentieth century, and has become indispensable in contemporary human society. Electricity is a transformative technology that has entered homes, businesses, and industrial facilities to conveniently (with a simple switch) deliver lighting, entertainment, communications, heating, ventilation, air-conditioning, and power for machinery. The production and consumption of electric energy has been considered as an indicator of development for nations. In the twenty-first century, the uninterrupted and reliable production of adequate electric power, which mostly occurs in centralized power plants, has become one of the functions of regional and national governments.

Electric energy is a *tertiary* form of energy because it is predominantly produced in power plants after several transformations of the primary forms of energy: In a coal power plant, the chemical energy of the coal is converted to thermal energy (heat) in a boiler; the heat is transferred to a working fluid (steam or air); it is then converted to mechanical energy in a turbine; finally, the mechanical energy of the turbine is transferred by a power shaft to a generator (alternator) where it is converted to electric energy. The electric power produced in the power plant is finally transported to the points of consumption by the electric power grid that connects the central power plant with the consumer. The several energy conversion processes in an electric power plant are always accompanied by energy dissipation in the form of waste heat, which is removed in the cooling system of the plant and is finally dissipated in the environment (thermal pollution). Typical thermal efficiencies of thermal power plants—coal, natural gas, and nuclear—are in the range of 30–40%. This implies that for every unit of electric energy produced by the power plants, 3.3–2.5 units of a primary energy source is actually used as heat input and the rest, 2.3–1.5 units of energy, becomes waste heat and dissipates in the environment.

---

* This does not imply that the electric railway transport is not associated with pollutant emissions and carbon footprint. The emissions occur at the power plants that produce the electricity for the railways.

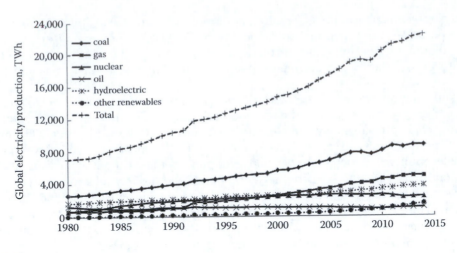

FIGURE 2.13
Global electricity production by primary energy source.

Coal, gas (natural and synthetic), nuclear, hydroelectric, petroleum, wind, geothermal, and solar are the main primary energy sources that contribute to the production of electricity. Figure 2.13 depicts the contribution of these primary energy sources to the global production of electricity in the period 1980–2014 [16,17]. A glance at this figure leads to the following conclusions:

1. The total global electric energy production has constantly increased since 1980. The average annual rate of increase in the period covered by the graph is 4.93%. At this rate, the electric energy production doubles every 14.5 years.

2. Coal is the major primary energy source that supplies electric energy globally. In 2014, coal power plants produced 39% of the total electric energy. It is important to note that both the total quantity and the relative fraction of coal for the production of electric energy are increasing, despite international agreements and appeals of environmental groups for the reduction of carbon dioxide emissions.

3. Petroleum supplies a continuously decreasing amount of the globally produced electric energy, reflecting the higher price of this fossil fuel relative to others.

4. The contribution of nuclear energy to the production of electricity has remained almost constant since 1990, reflecting the absence of building and commissioning new nuclear power plants during the period 1990–2014.

5. The amount of electricity produced by hydroelectric energy has steadily increased. In 2014, hydroelectric energy supplied 16.7% of the global electric energy. Hydroelectric energy is by far the highest contributor among the renewable energy sources in global electricity production.

6. There has been a very significant increase of the contribution of the "other renewable sources" to the global electric energy demand since 2005. The average annual increase in this category is 13.1%. This has been primarily driven by the increased number of solar photovoltaic (PV) panels and wind turbines that have become ubiquitous in the landscapes of many countries. However, and despite the dramatic relative growth, the renewable energy sources as a group contributed less than 7% of the total electricity production in 2014 [1].

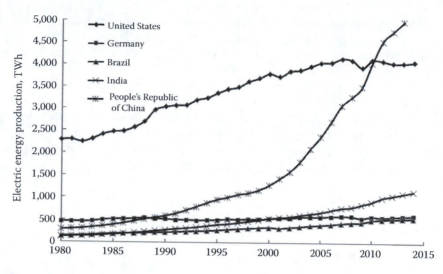

FIGURE 2.14

Trends of electricity production in the United States, Germany, Brazil, India, and People's Republic of China in the period of 1980–2014.

As with the other energy forms, the growing electric power consumption is primarily due to the continuing industrialization and increased economic development of several nations. The development and increased industrialization of national economies causes the implementation of higher numbers of heavy industrial equipment and the consumption of increasing amounts of electric energy. At the same time, their citizens become more affluent; they purchase more automatic household equipment and use more electric energy. The use of air-conditioning, which consumes a great deal of electric energy, is an example of this trend: While air-conditioning was a novelty before the 1960s even in OECD nations, it has now become almost a necessity in private dwellings, offices, and industrial establishments, in both OECD and developing nations that experience high temperatures during the summers. A second factor for the globally increased electric energy consumption shown in Figure 2.13 is the steadily increasing human population. More people, naturally, consume more electric energy.

The effect of these two factors are shown in Figure 2.14, which depicts the electric energy demand trends in two OECD countries, the United States and Germany,* and three developing countries, Brazil, India, and the People's Republic of China in the period 1980–2014 [1,16,17]. It is observed that the electric energy production in Germany has remained almost constant. At the same period, the population of Germany has been almost constant, and the electricity demand for air-conditioning is extremely low. The electricity consumption in the United States has increased at an average annual rate of 1.7%, and this is primarily due to the increasing population and the more widespread use of air-conditioning in this country. The highest electric energy growth, with an annual average rate 9.2%, is observed in the People's Republic of China, which has a very high but almost constant population, and has experienced a rapid rate of industrialization and economic development during this period. Because of the sheer size of its population, the People's Republic of China has surpassed the United States in electricity production. Brazil and India have also experienced very high increases in electric energy demand with average annual growth rates

---

* Data for East and West Germany were added in the period 1980–1991.

4.3% and 7.0%, respectively. Both growth rates are primarily due to the increased industrialization in the two countries and secondarily to their population growth.

The high growth of the global consumption of electricity is also associated with significant emissions of pollutants and gases from electric power plants that contribute to the global climate change. Several countries—including several EU countries, the United States, India, and the People's Republic of China—have constructed electric power plants that burn low-grade coal, lignite, and peat. Apart from carbon dioxide, the combustion of these fuels emits harmful particulates and chemicals that are precursors to smog formation. As a result, areas in the vicinity of the low-grade coal electric power plants experience higher occurrences of smog and environmental degradation.

2.5 Future TPES Demand

The global TPES consumption has been continuously increasing after the industrial revolution, and this increase is expected to continue at least in the near future. The spread of affluence in all the countries and the continuing growth of the world's population are expected to accelerate the global energy consumption. On the other hand, the rise of the prices of fuels is expected to promote conservation measures that would mitigate the growth of TPES consumption. Higher fuel prices would also promote fuel substitution that would increase the contribution of the renewable sources.

The globally averaged compounded annual rate of the TPES growth between 1970 and 2014 was 1.91%. During this period, the TPES growth rate in OECD countries was a mere 0.85%, and the growth was significantly higher in the developing countries. Given that the developing countries have most of the world's population and currently experience higher GDP growth rates, it is expected that the global energy consumption will continue to increase, albeit at a lower annual rate. The *International Energy Agency* (IEA), which is an agency of the UN, predicts that if nothing changes in the global regulatory environment, the TPES will increase from the 543.6 Q of 2014 to 567 Q in 2020, to 655 Q in 2030 and to 734 Q in 2040 [1,18].

International agreements and regulations designed to curtail the global greenhouse gas emissions are the main reasons that the TPES consumption growth rate may decrease in the future. As it will be seen in more detail in Chapter 3, the combustion of fossil fuels produces large quantities of carbon dioxide, which are expected to increase the temperature of the atmosphere and result in global environmental change. Several countries have already adopted limited measures for the reduction of fossil fuel emissions and the curtailment of carbon dioxide emissions. It is likely, actually desirable, that all nations will follow with similar regulations and that the world will have meaningful and binding restrictions for the production of carbon dioxide. If this occurs, the consumption of all fossil fuels—coal, liquid, and gaseous hydrocarbons—will be curtailed, and the total TPES will also be reduced.

The Paris Agreement of December 2015 calls for the reduction of the fossil fuel consumption and of TPES. However, this is a general agreement that lays down broad initiatives and does not specify details for adoption as national policies for the reduction of fossil fuel combustion. In addition, it is not enforceable. This regulatory vacuum contributes a great deal of uncertainty to future forecasts of TPES consumption.

The IEA has prepared two scenarios for future TPES consumption that extend to the year 2040 [1]. The first scenario, the New Policies Scenario (NPS), assumes that there will not be any major regulatory intervention on TPES consumption by the international community, and only national policies under consideration in 2015 will take place. The second scenario, 450S, is based on the assumption that a new binding international treaty will aim to reduce the future carbon dioxide emissions and to stabilize the atmospheric concentration of this gas at 450 parts per million. The effect of this action will be a significant reduction of the fossil fuel combustion, which will result in a corresponding mitigation of the TPES growth. Such an international treaty would also result in the substitution of a fraction of fossil fuel consumption by alternative energy, conservation, and higher efficiency measures.

Tables 2.4 and 2.5 show the expected consumption of TPES in the year 2040 under the two IAE scenarios. The first table depicts the expected energy demand in 2040 in the several major regions of the globe and the second table gives the supply of primary energy forms that are predicted to satisfy this demand. The People's Republic of China appears

TABLE 2.4

Expected TPES Consumption in 2040 in Quads by Region

	NPS in 2040		450S in 2040	
	Total TPES	**Percentage**	**Total TPES**	**Percentage**
OECD countries	202	27.6	185	28.9
People's Republic of China	163	22.2	139	21.7
Rest of Asia	129	17.6	111	17.4
Rest of Europe and Eurasia	70	9.5	63	9.9
Middle East	51	6.9	40	6.2
Africa	58	7.8	52	8.1
Latin America	38	5.1	32	5.0
International transportation fuel	24	3.3	18	2.8
Total	734		639	

Source: International Energy Agency, *Key World Statistics*, IEA-Chirat, Paris, 2016; International Energy Agency, *Key World Statistics—2015*, IEA-Chirat, Paris, 2015.

TABLE 2.5

Expected TPES in 2040 in Quads by Primary Energy Form

	NPS in 2040		450S in 2040	
	Total TPES	**Percentage**	**Total TPES**	**Percentage**
Coal	175	23.8	103	16.1
Petroleum	206	28.1	139	21.7
Gas	171	23.2	151	23.6
Nuclear	60	8.1	71	11.2
Hydroelectric	16	2.2	20	3.1
Other renewables	107	14.6	155	24.2
Total	734		639	

Source: International Energy Agency, *Key World Statistics*, IEA-Chirat, Paris, 2016; International Energy Agency, *Key World Statistics—2015*, IEA-Chirat, Paris, 2015.

separately, because of the magnitude of its population and its significance in the world economy.

It must be noted in the last table that from approximately 63.6 Q in 2014 (last two entries in Table 2.2), the other renewable sources are to increase to 107 Q under the NPS scenario and to 155 Q under the 450S scenario. Currently, this energy category primarily reflects the use of wood, which is becoming a depleting resource in many regions of the globe. The increase to 107 or 155 Q by 2040 cannot be achieved merely by burning more wood and must come from other renewable energy sources, primarily solar, wind, and geothermal.

An analysis of the numbers in the last two tables leads to the following conclusions:

1. Despite all the good intentions and the conservation measures adopted since the 1970s, the TPES consumption is continuously increasing and is not expected to decrease in the near future. Under the NPS scenario, the expected compound annual growth rate of TPES consumption is 1.2%, and under the more restrictive 450S scenario, this annual growth rate is 0.63%. An encouraging fact is that in 2009, these two numbers were calculated as 1.75% and 0.9%, respectively, for the period of 2007–2030 [4]. At least the rate of growth of the TPES is decreasing and is expected to decrease in the near future.

2. Given that the 2014 TPES consumption in OECD countries is approximately 209 Q, it is apparent that the expected TPES consumption in these countries will remain almost constant between now and 2040, at 202 Q under the NPS scenario, or will decrease to 185 Q under the more restrictive 450S scenario.

3. A comparison of Tables 2.3 and 2.4 reveals that the TPES is expected to increase in non-OECD countries between 2014 and 2040 under both scenarios.

4. Comparing the NPS scenario and the more restrictive 450S scenario, all the regions are expected to contribute to the reduction of the TPES. The relative TPES reductions are 8.8% for the OECD countries, 14.6% for the People's Republic of China, 13.8% for the rest of Asia, 9.1% for the rest of Europe and Eurasia, 21.9% for the Middle East, 10.3% for Africa, and 15.8% for Latin America. The highest fraction of the relative TPES savings is expected to come from developing nations from a combination of lower rate of economic growth and lower rate of population growth.*

5. Fossil fuels, coal, petroleum, and gas, are still expected to provide a high fraction, 62%, of the energy consumption in 2040 even under the 450S scenario.

6. If the 450S scenario is adopted by an international agreement, the contributions of nuclear energy and renewable energy (primarily wind and solar) are expected to increase significantly. The contribution of hydroelectric energy is not expected to increase as much, primarily because the most powerful hydroelectric resources have already been utilized or are expected to have been utilized by 2040 under the NPS scenario.

---

* Such assumptions do not appear to be politically acceptable to most of the developing nations and may become reasons for not reaching a definitive international agreement for the CO_2 stabilization in the future. The developing countries have repeatedly denounced international accords that do not include significant TPES reductions in OECD countries, which currently enjoy higher TPES per capita.

2.6 Energy Resources and Reserves

Fossil fuels—coal, natural gas, and petroleum—as well as nuclear fuels—uranium and thorium—are mineral resources, which have been formed eons ago and are currently extracted from the earth's crust. The current rate of extraction of these fuels, and especially that of the fossil fuels, by far exceeds the replenishment rate of these natural resources. Because the mass of the earth's crust is finite, it is reasonable to deduce that the total amount of these natural resources is finite too. With the continuous and increasing mining of such finite resources, it is also reasonable to deduce that the fossil fuel and nuclear energy resources will be depleted at some point in the near or more distant future. Herein lies the *energy challenge* contemporary humans are facing: unless the human society develops sufficient renewable energy sources before the depletion of the currently used primary sources, we will experience severe energy shortages that will jeopardize our civilization.

Of the total amount of the existing fossil fuel resources, a small fraction is located in known regions and may be economically extracted with existing technology. This fraction, which may be economically extracted under the present conditions, is referred to as *proved* or *proven reserves*. Another part of the fossil fuel resources is known to exist, but may not be economically recovered at present. These are the *potential reserves*. If the prices of fossil fuels rise or if a technological breakthrough that lowers the cost of extraction is achieved, the potential reserves become proved reserves and are available to be extracted and consumed. The potential reserves are often classified as *probable reserves* and *possible reserves*. The probable reserves are more likely to be recovered than the possible reserves. The Society of Petroleum Engineers suggests a probabilistic analysis for the reserves: after the analysis of all the geotechnical and engineering data, the proved reserves have 90% probability to be recovered; the probable reserves 50% probability to be recovered; and the possible reserves have 10% probability to be recovered.

In addition to the reserves, there are also larger quantities of the fuel resources, located in the vicinity of the proven and the potential reserves. Such resources have not been fully explored and quantified at present. From past experience with mining, it may be reasonably assumed that when they are explored and quantified, they will be economically recovered with the currently known technology. These are the *inferred resources*. Finally, there are additional resources that might exist in the crust of the Earth, but they have not been discovered yet and may not be economically extracted or may not be extracted with the current technology. These resources may become available at a future date, and they are the *undiscovered* or *hypothetical resources*.

The actual quantity of all the resources is estimated with analytical, probabilistic methods. For this reason, all the numbers that reflect the quantity of the resources must be treated as probabilistic estimates with high degree of uncertainty. The total mass of the resources by far exceeds the mass of the reserves. Petroleum engineers have suggested that the total volume/mass of a fossil fuel resource may be schematically depicted in a triangle as in Figure 2.15 [19]. The "conventional" resources represent the proven and a large part of the potential reserves. These resources, often referred to as *conventional resources*, may be economically extracted at present and with currently available technology. The total volume (or mass) of the conventional resources is rather small, estimated at 10–20% of the total, and is represented by the upper part of the triangle that has the smaller area.

The bottom of the triangle with the larger area represents the quantity of the resources that exist in the crust of the earth, but may not be extracted at present, either because they are not economical to extract or because the necessary technology has not been developed.

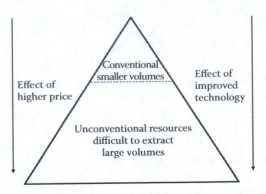

FIGURE 2.15
Energy resource triangle.

These are the *unconventional resources* and their estimated quantity is 5–10 times, higher than that of the conventional resources. For petroleum, the unconventional resources include shale oil and viscous or heavy oil in low-permeability reservoirs. Increased prices and improved technology make possible the development and extraction of the unconventional resources. The effect of both is to lower the dividing line in the resource triangle of Figure 2.15.

It becomes apparent that *reserves* and *resources* are strongly tied to the technology and economics of mineral extraction. Their quantities are not fixed by a scientific law, but vary according to the current price and the available technology. Reports of their quantities fluctuate significantly. When the price of a fuel rises, and technology improves, potential reserves become proven reserves. Inferred or hypothetical resources eventually become proven resources with additional exploration and technological improvements. For example, during the 1970s, the oilfields in the North Sea were classified as potential reserves or inferred resources [20]. With the technological advances in offshore exploration, drilling, and oil production as well as with the significant increase in crude oil prices during the 1970s, these resources became economical to be extracted; they were promoted to proven reserves; and by 2000, they have been producing petroleum. Similarly, the technological advances of the 1990s that allowed deep sea drilling operations have resulted in the promotion of several offshore oilfields in the Gulf of Mexico from resources to proven reserves. Natural gas resources have followed similar trends: the increased price of the natural gas in combination with horizontal drilling and hydraulic fracturing methods has triggered the economical extraction of gas from gas shale formations, which were not considered as reserves until the 1990s.

Because of the strong dependence of the reserves and resources on the current technology and energy prices, and because the quantities of the fossil fuel resources suffer from high uncertainty, reports on reserves and resources fluctuate significantly. Several reports of the past have been proven to be unreliable. In addition to the scientific uncertainty of their estimates, corporations and national governments have intentionally fudged the numbers and underestimated or overestimated their reserves and resources for political and commercial purposes [21].

2.6.1 Finite Life of a Resource

Because the amount of all mineral resources is finite, if we continue to indiscriminately extract them and use them for the production of energy, these resources will be exhausted at some point in the near or far future. Therefore, the question for a society, which over-whelmingly relies on fossil fuel resources for the production of its energy, is not *if the resources will be exhausted*, but *when the resources will be exhausted*. Sooner or later the major sources of today's energy will be depleted, and the human society will need other, viable, reliable, and more sustainable alternatives.

One of the rather fortunate factors for the human society is that, since 1950, with the advances of technology and the intensified fossil fuel exploration in all the continents, new mineral energy resources have been discovered, and many of these resources have been mined and exploited. Among these are the near-surface coal deposits in Wyoming, United States; the oil deposits of Alaska; and the oil in the North Sea. The discovery and development of these natural resources has contributed to the lower nominal and real prices of energy resources during the 1980s and 1990s and to the dramatic rise in the production and consumption of fossil fuels since the 1970s, as it is apparent in Table 2.2.

If continued to be unchecked and unmitigated, the increasing rate of fossil fuel con-sumption will accelerate the depletion of the existing energy resources. If other alternative energy sources are not available—or they are too expensive to be used—when the deple-tion of a major energy resource happens, there will be dire consequences for our society. Finite resources that are consumed at accelerated rates deplete at continuously faster rates and will be suddenly exhausted, leaving a very short time for the society to find substitu-tions for them [20].

Let us consider a hypothetical resource with finite total amount, equal to 400,000 units. This resource was consumed at an annual rate of 10 units from the beginning of the industrial revolution to 1850, at a rate of 20 units from 1850 to 1900, and at a rate of 30 units from 1900 to 1950. From that point in time onward, the consumption of the hypothetical resource doubles every 20 years. The global petroleum consumption has followed similar growth trends. Figure 2.16 schematically depicts the depletion sce-nario of this resource. The calculations and the figure show that because of the continu-ingly increasing consumption rate of the resource, this finite resource will be exhausted shortly after 2070. Although the resource is hypothetical, the following conclusions may

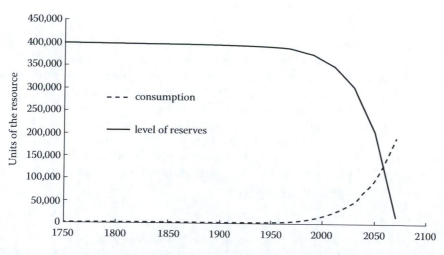

FIGURE 2.16
Life cycle of a hypothetical resource, whose consumption doubles every 20 years since 1950.

be drawn from Figure 2.16 for resources that experience similar consumption growth trends:

1. The total consumption of the resource in its first 200 years is minimal and hardly makes a dent on the available quantities of the resource.

2. The most significant drop in the proven reserves of this resource occurs in the last 20 years of its consumption.

3. In 2050, almost 20 years before the depletion of this resource, one may observe that there is still as much of this resource available for consumption (207,500 units) as it was consumed in the past 200 years. Twenty years before the complete exhaustion of the resource, the consumers may be lulled into a false sense of security by statements such as ". . . we have as much of this resource left as we have consumed in the last three centuries." When the potential of an energy resource is to be evaluated, it is the future consumption of the resource that matters and not the past history of utilization of the resource.

Before our society is lulled into a false sense of security with our fossil fuel (especially petroleum and natural gas) reserves, we will have to consider alternatives for their eventual substitution as primary fuels.

The *lifetime* of a resource is often used as a measure of the time in the future when the resource will be depleted. This lifetime depends on the total amount of the reserves Q_∞, as well as the annual consumption/production rate of the resource P_A. In the simple case when the consumption of the resource is constant, the lifetime of the reserves T, is simply equal to the ratio Q_∞/P_A. For resources whose consumption grows at an annual compound rate r, the lifetime of the resource, in years, may be calculated from the following equation:

$$P_A + P_A(1+r) + P_A(1+r)^2 + \ldots + P_A(1+r)^T = Q_\infty \Rightarrow$$

$$T = \frac{\ln\left[r\dfrac{Q_\infty}{P_A}+1\right]}{\ln(1+r)} - 1. \tag{2.1}$$

Predictions for the lifetime of several fossil fuel resources have been made using this and similar expressions. It must be noted, however, that the total amount of the reserves of an energy resource Q_∞ is highly uncertain, and this uncertainty significantly affects the accuracy of T. A glance in the petroleum data and predictions since the 1950s proves that predictions for T have been grossly inaccurate [15,20].* The main reason for this is that the total reserves of petroleum Q_∞ have been adjusted upward several times, because higher prices and technological breakthroughs promoted inferred and hypothetical resources to potential and proven reserves [15]. One must be reminded, however, that because of the finite amount of hydrocarbons on the planet, this type of promotion to new proven reserves cannot be sustained indefinitely.

---

* During the energy crisis of the 1970s, there was an almost unanimous belief, which was promulgated in the popular press by "experts," that the lifetime of the petroleum reserves was 30–35 years and that, hence, petroleum would have been almost exhausted by 2010. Similarly, "experts" in the late 1990s advocated that petroleum would be exhausted by 2020.

2.7 Sustainable Energy Supply and Limitations

The reliable supply of sufficient energy is essential for the global society. Securing energy supplies for the future has become one of the principal considerations of the governments of all nations. Energy sufficiency through domestic production and imports has played an important role in regional and global political and military conflicts since the 1950s [4,20,22]. The energy sources have become prime commodities, on which the economic growth and national prosperity are based, and the control of vital energy supplies—petroleum, in particular—has been the reason for several regional military conflicts.

Because with the current consumption trends, the fossil fuels are expected to be exhausted sometime in the twenty-first century, alternative energy sources, which include all the renewables as well as nuclear, are becoming more important for the sustainable economic development of the nations and the sustenance of our civilization. In the not too distant future, alternative energy sources will have to replace a high fraction of the fossil fuel consumption and to provide sufficient energy for the human society to maintain its current or desired standards of living.

In addition, societal concerns about global climate change may impose regulatory limitations on the consumption of fossil fuels in the near future. For this reason, it is very important to know the following: what are the alternative energy sources that may be harnessed with the current technology as well as with expected short-term technological breakthroughs; what are the engineering systems that will successfully harness this energy in the future; how may the alternative energy sources satisfy the energy demand of the human society in the future without causing significant environmental problems; and what is the potential of these sources to supply energy on a scale that would meaningfully benefit the human society. The rest of this book attempts to answer these and similar questions about the current and future global energy supply.

When completing the extensive use of renewable sources, the society must be cognizant of several limitations in their use. Resource availability, technological know-how, cost factors, and sociopolitical limitations are the most apparent factors that may limit the available quantity and use of renewable energy.

Resource availability is an obvious limitation for all energy forms: e.g., if there are no available geothermal resources within a region, then electricity cannot be produced by geothermal energy in that region. Limitations are also encountered with resources that appear to be inexhaustible. An example is the production of electricity with PVs from solar energy. While the solar radiation that reaches the earth's surface is enough to provide more than 11,000 times the total primary energy used by the earthlings, PV cells are currently manufactured with rare materials (gallium, selenium and indium), which may not be available in the large quantities needed for the extensive development of solar energy. The availability of these scarce material resources imposes a limitation on how much solar energy may be converted to electricity by PV cells. Unless technological advances achieve the production of PV cells with different, more abundant materials, the availability of such rare materials will soon become a constraint on the utilization of solar energy.

Availability of technology limitations has been with us since the invention of the first locomotives. Fundamental research and engineering development efforts constantly expand the frontiers of science and develop devices for more efficient and environmentally benign energy production. Limitations of pertinent technology range from complete to partial use of an energy source. The harnessing of fusion energy is an example of complete use, where the lack of fundamental understanding has prevented the development of fusion technology to engineering systems that may have the capability to supply abundant energy everywhere in the world. Partial use technology limitations at present include the production of PV cells from more common materials; the production of more efficient PV cells; the production of efficient, reliable, and durable fuel cells; the development of gas turbines with thermal efficiencies closer to the Carnot efficiency; and the production of convenient, high-capacity, and economical energy storage methods.

Economic and financial limitations include the cost of the several energy production methods and the ability of state and private corporations to finance projects that will produce energy. In a market-oriented society, informed consumers will choose the least-expensive alternative to satisfy their energy needs. If renewable energy is to be more widely utilized, the cost of it should be competitive with the cost of energy from fossil fuels, and this may happen in two ways:

1. The price of renewable energy drops, because of technological advances that develop more efficient energy production systems, as it happened since the 1990s with wind turbines and solar cells.

2. The price increases of fossil fuels, as it happened in the period of 2002–2014.

When price parity materializes, the availability of capital (financial limitation) to build the required infrastructure for harnessing and using the renewable energy will become of paramount importance for the continued utilization of renewable energy sources.

Sociopolitical limitations include the perceptions of the population about forms of energy and the political–regulatory framework that promotes or bans certain forms of energy. The development of nuclear energy in the United States and Japan are prime examples of such limitations: the 1979 accident at the Three-Mile Island nuclear power plant and the public perception of nuclear energy in the early 1980s resulted in the cessation of the design and construction of new conventional nuclear power plants in the United States after 1982 and the complete ban of the development of breeder reactors. The 2011 accident in Fukushima had a similar effect for the nuclear industry in Japan. More recently, since 2010, the public perception of nuclear energy in the United States has changed—primarily because of global climate change fears—leading to the development and beginning of construction of four new conventional nuclear reactors in the country. Political decisions, which usually follow public perceptions, generate national and international regulations/rules that stimulate the development of energy sources and sanction others. Renewable energy has always enjoyed favorable public opinion—primarily because of its cleanliness and environmental benefits—and has been assisted by national policies that provide incentives for its development. A significant boost to renewable energy investment and production will be the adoption of a binding international treaty to limit the global CO_2 emissions.

PROBLEMS

1. Convert the following to British thermal units: 5.3 TWh; 2,000 bbl; 25 toe.

2. Convert the following to megajoules: 38 TWh; 25 Mbbl; 15 tce.

3. What was the energy demand in the country you currently live in 2015 and which primary sources supplied the demand?

4. How much was the electricity production in the country you currently live in 2015 and which primary energy resources produced it?

5. Based on the data of Figure 2.4 (or other more recent sources) derive a correlation between the GDP per capita and electricity consumption for the People's Republic of China. The GDP per capita of this country is expected to increase by an average rate of 8% in the next 10 years. Estimate the corresponding increase in the rate of electricity consumption for the People's Republic of China at the end of the 10-year period.

6. How many kilowatt hours of energy were globally produced in 2014 from renewable sources? How do these compare to the energy from coal?

7. What is the projected demand for petroleum, in barrels of oil, in 2040 according to the NPS and the 450S scenarios? Based on your knowledge of petroleum product consumption, suggest four systems or methods that will result in decreasing future petroleum demand.

8. "Based on the average oil demand of the last hundred years, we have enough petroleum reserves to last us for another three centuries." Comment in an essay of 300–350 words.

9. The world total coal reserves (adjusted for the heating value of the various types of coal) are estimated to be approximately 1.4×10^{12} tce. The current annual coal consumption rate is 6.62×10^9 tce/year, and the consumption of coal increases by 0.85% per year. Estimate the lifetime of this resource.

10. The world's petroleum reserves are approximately 1.7×10^{12} bbl. The global petroleum consumption in 2015 was 33.6×10^9 bbl/year, and the consumption growth rate was 0.34×10^9 bbl/yr^2. Estimate the lifetime of the global petroleum reserves. Also, estimate the lifetime of the global petroleum reserves if the reserves increased to 3.4×10^{12} bbl.

11. The global natural gas reserves are approximately $5,605 \times 10^{12}$ scf. The global natural gas consumption in 2015 was 124.6×10^{12} scf/year, and the growth rate was 1.93 scf/yr^2. Estimate the lifetime of the global natural gas reserves. Estimate the lifetime of the global natural gas reserves if the reserves increased to $11,210 \times 10^{12}$ scf.

References

1. International Energy Agency, *Key World Statistics*, IEA-Chirat, Paris, 2016. (For some figures, editions of this source in the period of 2007–2015 were consulted.)
2. United Nations Development Programme, *United Nations Human Development Report 2015*, United Nations Development Programme, New York, 2015.
3. International Energy Agency, *Key World Statistics*, IEA-Chirat, Paris, 2003.
4. Michaelides, E.E., *Alternative Energy Sources*, Springer, Berlin, 2012.
5. United Nations, Department of Economic and Social Affairs, Population Division, Population Trends, 2015.
6. US Department of Energy, *International Energy Outlook 2016*, US Department of Energy, Washington, DC, 2016.
7. OPEC (Organization of the Petroleum Exporting Countries), *Annual Statistical Bulletin*, OPEC, Vienna, 2016.
8. US Department of Transportation, *State Vehicle Registrations by State 1900 to 1995*, Washington, DC, 1997.

9. US Department of Transportation Bureau of Transportation Statistics, *National Transportation Statistics*, US Department of Transportation Bureau of Transportation Statistics, Washington, DC, 2015.

10. Statistics Japan—Prefecture Comparisons, *Automobiles Registered*, http://stats-japan.com/t/kiji /10786 (website), last visited 2014.

11. World Bank Data, *Motor Vehicles (per 1,000 People)*, https://data.worldbank.org/restricted -data, 2015.

12. ICAO (International Civil Aviation Organization), *Annual Report of the ICAO Council: 2015*, ICAO, Montreal, 2015.

13. US Department of Transportation Bureau of Transportation Statistics, *Energy Used by the Transportation Sector*, US Department of Transportation Bureau of Transportation Statistics, Washington, DC, 2016.

14. US-EIA (US Energy Information Administration), *Petroleum and Other Liquids—Annual Projections to 2040*, US-EIA, Washington, DC, 2015.

15. Michaelides, E.E., A new model for the lifetime of fossil fuel resources, *Natural Resources Research*, **26**, 161–175, 2016.

16. TSP (The Shift Project), *The Shift Project Data Portal—Electricity Statistics*, www.tsp-data-portal.org /all-datasets?field_themes_tid=2, 2015.

17. US-EIA, *International Energy Statistics—Electricity*, US-EIA, Washington, DC, 2015.

18. International Energy Agency, *Key World Statistics—2015*, IEA-Chirat, Paris, 2015.

19. Masters, J.A., Deep basin gas trap, western Canada, *AAPG Bull.*, **63**(2), 152–181, 1979.

20. Foley, G., *The Energy Question*, Penguin Books, Middlesex, 1976.

21. Laherrere, J.H., World oil reserves—Which number to believe? *OPEC Bulletin*, 9–13, February 1995.

22. Grathwohl, M., *World Energy Supply*, De Gruyter, Berlin, 1982.

3

Environmental Effects of Energy Production and Utilization

The exponentially increasing energy consumption since the beginning of the industrial revolution has produced significant changes in the environment of the planet. Chief among these is the increase in the average concentration of *greenhouse gases* (GHGs) in the atmosphere, including carbon dioxide (CO_2), methane (CH_4), and nitrogen oxides (NOX). GHGs play the role of a blanket: they reflect part of the infrared radiation emitted by the earth and cause the warming of the troposphere. Climatologists predict that the increase in the GHGs will cause an increase in the average temperature of the troposphere accompanied by regional and global climatic changes. Other significant environmental effects of energy consumption are the several ecological problems caused by acid rain, which has threatened in the past the ecosystems of several lakes and rivers; ozone depletion in the stratosphere (the ozone hole); lead contamination in the atmosphere; nuclear waste, which is produced by the approximately 450 nuclear reactors in continuous operation worldwide; and the waste heat rejection by all thermal power plants, which necessitates the consumption of vast amounts of freshwater.

The environmental threats of energy consumption may be neutralized by public policy, either national policy or concerted international treaties and protocols. Despite the efforts of the environmental community, there is not yet a comprehensive, binding global agreement for the mitigation of the GHG emissions that cause global climate change (GCC), the principal environmental threat in the twenty-first century. The problem of nuclear waste is being addressed at several national and regional levels, and it appears that viable solutions for the long-term storage of radionuclides will become available in the near future. National policies and international collaborations have almost resolved the acid rain and the lead contamination problems, while the ozone depletion problem appears to be in check.

This chapter starts with a short section on the environment and ecosystems; continues with the success stories of environmental threat mitigation from acid rain, lead contamination, ozone depletion, and NOX emissions; examines the problems and possible consequences of GHG emissions and global warming, which have significant societal and political implications; and concludes with an exposition and the up-to-date progress for the solution of the problems of nuclear waste, thermal pollution, and freshwater use.

3.1 Energy, Ecology, and the Environment

At the beginning of the exposition of the environmental and ecological effects of energy consumption, it is advisable to define the most important concepts and terms that will be used. The *environment* is everything that surrounds the humans and where our economic activity occurs. The lithosphere (solid earth), the atmosphere (air), and the hydrosphere (sea, lakes, rivers, etc.) are the three distinct components of our environment. Climatic processes and events interact in different ways with the environment. For example, a hurricane is formed in the atmosphere and encompasses water that comes from the hydrosphere. When the hurricane washes over land, it dumps on the ground large quantities of rainwater. The latter causes local flooding, erodes the soil, and carries parts of the lithosphere into the sea or the lakes. These types of interactions produce *environmental changes*, most of which are undesirable.

Ecology is the study of the relationships of living organisms with one another and the relationship of organisms to their environment. The subject incorporates parts from the scientific disciplines of biological sciences, physics, physiology, and chemistry. The *ecosystem* is a rather loose concept that refers to a subdivision of the landscape or a geographic region that is relatively homogeneous. An ecosystem is made up of organisms, environmental factors, and physical or ecological processes. The concept of the ecosystem comprises living organisms, species and populations; soil and water; climate and other physical factors; and physicochemical processes, such as nutrient cycles, energy flow, the carbon cycle, water flow, freezing, and thawing.

Although the two are related and are often confused, there is a clear distinction between environmental and ecological changes as well as between environmental and ecological concerns: The ecological concerns always involve effects on ecosystems. For example, a tropical storm will wash a great deal of soil into the sea and will change the coastline of an entire region. If we are concerned only with the physical process of the soil erosion, the suspension of sediment in the water, and the subsequent deposition of soil at the bottom of the sea—three purely physical processes—then we have an *environmental concern*. If we are concerned with the effect of the soil erosion on the agricultural crops, the loss of habitat of living organisms, or with the effect of increased concentration of pesticides that follows the soil erosion in the aquatic life, then we have an *ecological concern*.

Because ecosystems are closely related to their environment, every environmental change is accompanied by ecological consequences. The observed increase in carbon dioxide concentration in the atmosphere and the expected global and regional climate changes are related environmental changes. Their consequences in the ecosystems include altered patterns of crop production and migration or disappearance of several species. Similarly, the discharge of pollutants, such as dioxin and lead, is an environmental event that has ecological consequences.* When one considers the effects of industrial pollutants on subsurface organisms, the effects of the leaching of the pollutants in nearby aquifers, streams, or lakes and its ultimate effects on animals and humans that drink the water, we have the ecological effects of an environmental problem.

Another example on the distinction of environmental and ecological effects and concerns is the accident in the Chernobyl nuclear power plant, which occurred in 1986: The steam explosions in the nuclear reactor released into the environment a great deal of radionuclides, which were accumulated in the vicinity of the power plant or were transported to other regions by atmospheric currents. As a result of runoff from the rainfall, a great deal of radioactive cesium and strontium is now physically buried in the sediment at the bottom of rivers and lakes or in the subsurface of the land. These are environmental changes. The ecological effects that are consequences of such environmental changes include the mutations in the cells of living species that absorbed radionuclides via the food chain; the decimation of herds of reindeer in Lapland, which consumed grass contaminated with

---

* In most countries, it is illegal to discharge pollutants in the environment.

radionuclides; the forests with trees that have uptaken radionuclides from the soil; and the significant increase in childhood leukemia and cancer incidents in the local human populations that were affected by the release of the radionuclides.

3.2 Recent Successes in Environmental Stewardship

It is worth describing three environmental threats of the past that were successfully mitigated by a concerted effort of scientists, engineers, and regulators who adopted a process that included the following:

1. The scientific formulation of the problem
2. The identification of the major sources of the problem
3. National and international collaborations that resulted in effective regulations for the mitigation of the environmental threat

The three environmental problems that posed significant threats in the past are known as acid rain, lead contamination, and stratospheric ozone depletion (ozone hole).

3.2.1 Formation of Sulfur Dioxide and Nitrogen Oxides

Fossil fuels, coal, natural gas, and crude oil, contain small amounts of sulfur, typically in the range of 0–2%. When fossil fuels burn—e.g., in the boiler of a coal power plant or in the engine of a car—gaseous sulfur dioxide (SO_2) is formed and released in the atmosphere. Since all combustion processes use the ambient air, which contain 79% nitrogen (N_2), and take place at very high temperatures, a small amount of nitrogen combines with oxygen and forms the three nitrogen oxides, N_2O, NO, and NO_2, which are denoted as NOX. SO_2 and NOX are air pollutants and precursors to the formation of harmful acids. When their concentrations in the atmosphere increase, they cause significant respiratory problems to humans that may possibly lead to death.

These gases are produced in large quantities in *stationary sources*, such as coal power plants, gas turbines, and oil-fired power plants, and *mobile sources*, primarily cars and trucks that use the refined products of crude oil. Even though they are produced in significantly lower amounts than carbon dioxide, the main product of the combustion process, these chemicals constitute an environmental threat, because they are significantly more toxic than CO_2.

Example 3.1

A 400 MW power plant has thermal efficiency 37% and burns anthracite, which has a heat content 29,000 kJ/kg and contains 1.2% sulfur by weight. This anthracite contains 94% carbon by weight. It is estimated that during the combustion process, NO_2 is also formed at the rate 0.02% of the formation rate of CO_2. Determine the amounts of CO_2, SO_2, and NO_2 produced by this power plant per day, per week, and per year.

Solution: At first, we need to calculate the amount of heat used by this power plant and the amount of coal/anthracite needed to produce this heat. With a thermal efficiency of

37%, the power plant uses $400/0.37 = 1,081$ MW, or $1,081,000$ kW of heat. This corresponds to $1,081,000 \times 60 \times 60 \times 24 = 93.4 \times 10^9$ kJ per day. And this quantity of heat is supplied by the combustion of $93.4 \times 10^9/29,000$ kg $= 3.22 \times 10^6$ kg of anthracite (or 3,220 t).* At 94% carbon content, the power plant burns $0.94 \times 3.22 \times 10^6 = 3.03 \times 10^6$ kg carbon per day.

From the chemical reaction $C + O_2 \rightarrow CO_2$, 1 kg of C produces $44/12 = 3.67$ kg of CO_2, and, hence, this power plant produces 11.1×10^6 kg of CO_2 daily. This is the equivalent of 27.75 t of CO_2/(MW·day). The weekly amount of CO_2 produced is 77.77×10^6 kg, and the annual amount is 4.06×10^9 kg of CO_2 (4.06 million t!).

Since the amounts of NO_2 produced are 0.02% (0.0002) of the amount of CO_2 produced, the power plant produces $11.1 \times 10^6 \times 0.0002 = 2.22 \times 10^3$ kg, 15.56×10^3 kg, and 811×10^3 kg (811 t) of NO_2 daily, weekly and annually, respectively.

For the calculation of the SO_2 released, it is noted that this power plant burns daily $0.012 \times 3.22 \times 10^6 = 38.64 \times 10^3$ kg of sulfur, which has an atomic weight 32 kg/kmol. From the reaction $S + O_2 \rightarrow SO_2$, 32 kg of sulfur produces 64 kg of SO_2, and hence, the 38.65×10^3 kg of sulfur burned daily produces 77.3×10^3 kg of sulfur dioxide daily, 541.1×10^3 kg of SO_2 weekly, and 28.21×10^6 kg (28,210 t!) of SO_2 annually.

It is apparent from example 3.1 that very large quantities of the pollutant gases SO_2 and NOX are produced from a single coal power plant. Because the pollutants are formed and released at a single site, the power plant, it is easier for engineers to take measures for their reduction. The worldwide effort for the significant reduction in SO_2 and NOX emissions primarily targeted the stationary sources of such pollutants, the power plants, large industrial complexes, and petroleum refineries.

A moment's reflection on the results of example 3.1 proves that this rather typical coal power plant produces 3.5×10^9 kWh of electricity annually accompanied by more than 4 million t of CO_2. In 2014, the entire world produced $23,816 \times 10^9$ kWh, of which 40.8% were produced from coal combustion, with an estimated release of 11,271 million t of CO_2. The total anthropogenic CO_2 production exceeded 32,300 million t. Herein lies the problem with the global CO_2 emissions: too much of this gas is produced from the several anthropogenic activities, and all of it is released in the atmosphere. Because avoiding or minimizing an undesirable environmental impact is by far easier and less costly than neutralizing the effects of the impact after it has taken place, the efforts of the global community should better concentrate on restricting the emissions of this gas by inventing new methods to satisfy its energy needs.

3.2.2 Acid Rain

Acid rain or acid precipitation is the return to the terrestrial aquatic environment of the oxides of carbon, nitrogen, and sulfur in the form of an acid. The last section and example 3.1 describe the formation of these oxides. When released in the atmosphere, the two oxides combine with water vapor or water droplets to form mild acids. For example, the hyposulfuric acid and the carbonic acid, two weak acids, are formed in the atmosphere by the following reactions:

$$SO_2 + H_2O \rightarrow H_2SO_3 \quad \text{and} \quad CO_2 + H_2O \rightarrow H_2CO_3. \tag{3.1}$$

---

* It must be noted that this is typical and represents a very large amount of coal: approximately 100 railroad cars are needed *daily* to haul this amount of anthracite from the mine to the power plant.

Part of the SO_2 in the atmosphere may also combine with ozone first and then with water vapor to form the much stronger sulfuric acid:

$$SO_2 + O_3 \rightarrow SO_3 + O_3 \quad \text{and} \quad SO_3 + H_2O \rightarrow H_2SO_4. \tag{3.2}$$

In a similar way, the several NOX compounds that are formed during the combustion processes may also combine with water vapor in the atmosphere to form the weaker nitrous acid (HNO_2) and the much stronger nitric acid (HNO_3). All the NOX gases are contributors to some form of acid precipitation.

The acids in the atmosphere are formed within small droplets or on the side of very fine particles, which are called aerosol particles (aerosols). The sizes of aerosols are in the submicron range—their diameters are less than 1 μm—and this implies that the aerosols settle extremely slowly and may remain airborne in the atmosphere for weeks or months following the air currents. This enables acid pollution to be transported over distances of hundreds of kilometers. However, during rain and snow, the aerosols combine with the larger raindrops or snowflakes, precipitate on the ground, and are thus removed from the atmosphere. The rain or snow runoff, which contains higher concentration of the acids, eventually feeds the water of rivers and lakes and makes them acidic. For this reason, the precipitation of such aerosols has been called *acid rain*, *acid snow*, or, in general, *acid precipitation*.

The pH of pure water is 7. Acids have lower pH and bases have higher pH: the pH of vinegar is 2.5 and the pH of (the much stronger) sulfuric acid is less than 1. The normal pH level of natural freshwater is in the range of 6.8–7.4. Most fish and several other aquatic organisms do not thrive outside this natural range of pH. The addition of the acids in lakes and rivers by the acid precipitation process significantly lowers the pH in rivers and lakes, with the result that several species languish or disappear. Some of the more dramatic acid precipitation observations of the 1970s in North America and Europe are [1] as follows:

1. A storm in Scotland in 1974 dropped rain with pH 2.4, which mixed with local waters.

2. The pH of rain in Kane, Pennsylvania, on September 19, 1978, was 2.32. This is lower than the pH of common vinegar!

3. For the entire year of 1975, rains in Norway and Sweden recorded pH less than 4.6.

4. During the 1970s, the pH of 80% of drizzles in Holland was less than 3.5 and sometimes as low as 2.5.

It is apparent that when acid precipitation of such low pH mixes with the water of the rivers and lakes, the pH of the water drops significantly below the 6.8 value, which is considered "safe" for the aquatic species. This has many adverse effects on the ecosystems of rivers and lakes, because several animal species do not survive at low pH and disappear. Other species also disappear from the ecosystem because of lack of nutrients. The low pH resulting from acid deposition decimated the fish population in several rivers and lakes in the 1970s and 1980s and became a big environmental threat. In addition, the high acidity in the precipitation rendered the soil of several regions more acidic with a significantly adverse effect on crop yields and partial deforestation. A few of the environmental and ecological effects of acid precipitation that were observed in the 1970s are [1] as follows:

1. As the water of the streams becomes more acidic, a shift to acid-tolerant plants, such as green algae, occurs. The appearance (color) of several lakes, especially in the northern latitude, was changed by the green algae.
2. Acid-sensitive species, such as snails, clams, and amphipods, disappear. Organisms that feed on them also disappear.
3. The high pH causes the higher concentrations of Al^{3+} and other metal ions. These ions damage the gills of fish and enhance the precipitation of dissolved organic matter in the water, which is a source of food for fish. With decreased food supply and injured gills, fish become emaciated or die.

These causes had catastrophic effects on the aquatic populations of most rivers and lakes in northern Europe and North America during the 1970s. For example, salmon in several Norwegian rivers did not reproduce for years and became almost extinct. The fish disappeared entirely from 190 lakes in the Adirondacks region of the New York State and Canada and from more than 2,000 lakes in southern Norway. Acid rain became the preeminent global environmental threat of the 1960s and 1970s.

What accentuated the environmental problems of acid deposition and created political problems is that in most cases, the production of SO_2 and of the other oxides actually occurred in other neighboring states. The oxides and the acid-laden aerosols were carried by the predominant air currents over the state and national boundaries and affected neighboring states and nations. Given that most of the economic activity of the world is produced in the northern latitudes between the 30th and the 60th parallels, where the predominant winds are southeasterlies—directed from southeast to northwest—the pollutants produced in countries to the south were deposited as acid rain in countries to the north. Thus, the acid oxides produced in Ohio and Michigan, United States, formed acidic aerosols that were carried north and affected the lakes in the Ontario Province of Canada. Similarly, acid oxides produced in the industrial Ruhr region of Germany affected the aquatic environment of Holland and Scandinavia. Environmental pollution does not respect national boundaries, and when it occurs, it becomes an international issue.

A concerted international effort to reduce acid precipitation and mitigate its effects started in the early 1970s and continued in the 1980s and 1990s with considerable success: Despite the protests of the coal and electricity production industries, one after another, national governments enacted regulations to limit the emissions of SO_2. In the United States, a goal was set to reduce SO_2 emissions from 28 million t per year in 1970 to less than 9 million t per year by 2010. The Environmental Protection Agency (EPA) of the United States incorporated this program in an amendment to the *Clean Air Act** and developed a market-based initiative to achieve the reduction in SO_2 emissions at their sources. This amendment sets annual upper limits (caps) for the emissions of SO_2 at electric power plants and refineries. The EPA then issues permits, which are called *annual allowances*. The *allowances* decrease every year to be consistent with the overall goals for the national reduction of the emissions. Corporations that expect to exceed their reduction targets in a given year may trade their *allowances* to others that do not meet these reduction targets. This is the so-called cap and

---

* The Clean Air Act in the United States was enacted in 1963 and was amended in 1970 for SO_2 and in 1990 for the NOX pollution.

trade program. The program directs all the entities that produce environmental pollution to gradually decrease their polluting emissions. The program also creates a market incentive for these entities to exceed their own goals and trade the differences of their annual allowances to others for a profit [2]. Cap and trade policies encourage significant efforts that result in meaningful progress for environmental pollution reduction.

On the other side of the border, Canada took similar measures [3]. The European Union (EU) countries also took measures for the reduction in the SO_2 and NOX emissions: the Cooperative Program for Monitoring and Evaluation of the Long-range Transmission of Air Pollutants in Europe, which started in 1980, achieved a 75% decrease in the SO_2 emissions between 1980 and 2005 [4]. The more recent *Gothenburg Protocol* is more ambitious and covers more types of pollutants.

Acid precipitation reduction programs have been immensely successful in both Europe and North America, where the 2010 reduction goals were met before 2007. As a result, in the beginning of the twenty-first century, acid deposition has dropped by more than two-thirds from its peak in the 1970s and is not considered to be the environmental and ecological threat it was in the past. Because of the resolute international action, the ecosystems in most of the affected lakes, rivers, and forests have recovered. Figure 3.1, with data from EPA [2,3], Environment Canada [4], Vestreng et al. [5], European Environmental Agency [6], and Smith et al. [7], shows the dramatic drop of the emissions of SO_2 and NOX in the United States and Europe as a result of the implementation of the acid rain reduction measures. It is remarkable that in approximately 30 years after the enactment of national and international regulations, the SO_2 production in the two continents decreased by 85% and that acid deposition is not considered a threatening environmental problem. Most of the environment and the aquatic ecosystems in these regions were returned to their previous, almost pristine, states.

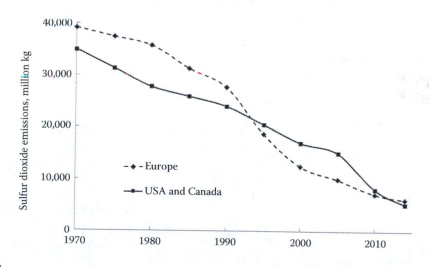

FIGURE 3.1

Emission reductions of SO_2 in Europe and in United States–Canada. (Data from EPA, *SO_2 and NO_x Emissions, Compliance, and Market Analyses: Progress Report—2012*, EPA, Washington, DC, 2012; EPA, *National Trends in Sulfur Dioxide Levels*, https://www.epa.gov/air-trends/sulfur-dioxide-trends, EPA, Washington, DC, 2014. Environment Canada, *Canada–United States Air Quality Agreement Progress Report 2012*, Environment Canada website, Environment Canada, Gatineau, 2012; Vestreng, V. et al., *Atmos. Chem. Phys.*, 7, 3663–3681, 2007; European Environmental Agency, *Sulfur Dioxide (SO₂) Emissions*, APE 001, European Environmental Agency, Copenhagen, 2014; Smith, S. J. et al., *Atmos. Chem. Phys.*, 11, 1101–1116, 2011.)

The engineering solutions for compliance with the reduced SO_2 emission standards varied. In some cases, it was a reduction in the amount of coal used for the production of electricity. In others, it was the implementation of new technologies for the *in situ* removal of SO_2. In the United States and the EU, the principal technical approach was the reduction of the SO_2 emissions at the sources using the *flue gas desulphurization* method. The method, which is schematically depicted in Figure 3.2, removes the SO_2 from the stack gases by *scrubbers* before the SO_2 is discharged to the atmosphere. The gaseous products of combustion enter the scrubber, where they are "showered" with a basic water solution, typically a limestone–water mixture that contains the basis calcium hydroxide $Ca(OH)_2$. Most of the gaseous effluents are not affected, but the SO_2 is absorbed by the basic limestone solution to first form the hyposulfuric acid (H_2SO_3) and then calcium sulfite ($CaSO_3$):

$$SO_2 + H_2O \rightarrow H_2SO_3 \quad \text{and} \quad H_2SO_3 + Ca(OH)_2 \rightarrow 2H_2O + CaSO_3. \qquad (3.3)$$

$CaSO_3$ is a solid that precipitates and is removed as sludge. Since it is not a pollutant, it is usually disposed underground at the site of the production.

Example 3.2

The 400 MW power plant of example 3.1 has a thermal efficiency of 37% and burns anthracite, which has a heat content 29,000 kJ/kg and contains 1.2% sulfur by weight. This anthracite contains 94% carbon by weight (similar to example 3.1). It is desired to remove 99.5% of the sulfur dioxide using limestone solution. Determine how much $Ca(OH)_2$ is needed and how much $CaSO_3$ is produced per day, per week, and per year.

Solution: From example 3.1, this power plant burns $0.012 \times 3.22 \times 10^6 = 38.65 \times 10^3$ kg of sulfur daily and produces 77.3×10^3 kg of sulfur dioxide; 99.5% of this quantity, which will be removed, is 76.9×10^3 kg. From Equation 3.3, 64 kg of SO_2 is removed using 74 kg of $Ca(OH)_2$ to produce 120 kg $CaSO_3$. Therefore, the needed daily quantity of $Ca(OH)_2$ is $76.9 \times 10^3 \times 74/64 = 88.9 \times 10^3$ kg, and the daily production of $CaSO_3$ is $76.9 \times 10^3 \times 120/64 = 144.2 \times 10^3$ kg. The corresponding quantities of the two chemicals for a week are 622.3×10^3 and $1,009.3 \times 10^3$ kg; and for a year, $32,449 \times 10^3$ and $52,628 \times 10^3$ kg, respectively.

The last number implies that a landfill must be available that will store approximately 53 thousand t of calcium sulfide per year.

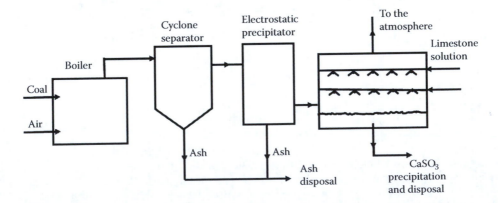

FIGURE 3.2
Flue–gas desulfurization process. Up to 99.9% of SO_2 is removed in the scrubber with the addition of limestone solution.

Crude oil also contains a small fraction of sulfur, which is removed during refining. The sulfur removal is usually accomplished by a catalytic reduction/oxidation process—*redox process*—often called the *Claus process*. At first, the sulfur compounds in the crude oil mixture are converted to hydrogen sulfide (H_2S) in bubble columns where hydrogen gas is introduced as bubbles. Subsequently, the hydrogen sulfide is partly oxidized in the presence of a catalyst to produce water and elemental sulfur:

$$RS + H_2 \rightarrow H_2S + R \quad \text{and} \quad H_2S + \frac{1}{2}O_2 \rightarrow H_2O + S. \qquad (3.4)$$

R in the two reactions is an organic radical, which is finally converted to a hydrocarbon. The sulfur produced is in the solid phase. It is separated from the rest of the petroleum products and is used to produce industrial sulfuric acid (H_2SO_4), which is sold in the market. The low-sulfur liquid products of petroleum (gasoline and diesel) fetch higher prices in the fuels market than high-sulfur products.

Other methods that have been used worldwide for the significant reduction of SO_2 emissions are the following:

1. Using fluidized bed reactors (FBRs) for the combustion of coal in new power plants: FBRs mix limestone particles with coal during the combustion process to remove SO_2 by converting it directly to solid $CaSO_3$.* The latter is removed with the solid materials of the ash.
2. Blending high-sulfur coal with low-sulfur coal.
3. Switching coal as fuel to natural gas or a mixture of coal and natural gas.
4. Retiring old electricity generation units and replacing them with newer units, which operate with FBRs or with SO_2 scrubbers.
5. Purchasing or transferring emissions allowances from other units.
6. Increasing the demand-side management and conservation efforts to reduce the electric power consumption (or reduce the growth rate of power consumption).
7. Power purchases from other power generators that use low-sulfur coal or other fuels.

The regulatory framework that resulted in the significant reduction of SO_2 also had an impact on the reduction of the emissions of NOX in both the United States and the EU. In the United States, it started with the 1990 amendment to the Clean Air Act, which included NOX as pollutants targeted to be reduced. A great deal of the NOX emission reductions is achieved by catalysts (including vehicle catalysts) and, at the power plants, by FBRs and newer, better-designed burners. Figure 3.3, with data from EPA [2,3,8], shows the dramatic reductions in the average concentrations of SO_2 and NO_2, two pollutant gases on their own, in the continental United States. It is also of interest that the recent slope of the two curves is negative, which implies that the significant emission reduction of the two pollutant gases will continue at least in the near future.

In the case of acid rain, it is of significance that not only the long-term goals of the programs were achieved ahead of the deadlines, but also the costs of the abatement

---

* In an FBR, SO_2 comes in direct contact with particles of $Ca(OH)_2$ and reacts as $SO_2 + Ca(OH)_2 \rightarrow CaSO_3 + H_2O$. The solid $CaSO_3$ is removed with the ash.

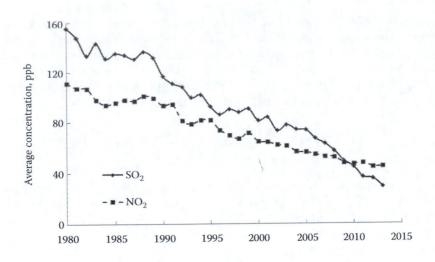

FIGURE 3.3
Reductions in the average atmospheric concentrations of SO_2 and NO_2 in the United States as a result of the Clean Air Act, in parts per billion. (Data from EPA, *SO_2 and NO_x Emissions, Compliance, and Market Analyses: Progress Report—2012*, EPA, Washington, DC, 2012; EPA, *National Trends in Sulfur Dioxide Levels*, https://www .epa.gov/air-trends/sulfur-dioxide-trends, EPA, Washington, DC, 2014. EPA, *National Trends in Nitrogen Oxide Levels*, EPA website, EPA, Washington, DC, 2014.)

implementations to the businesses and the consumers were lower than the original predictions. It is estimated that, in the United States, the total cost of the SO_2 emissions reduction implementation strategies was in the range of $1–$2 billion. This is only one-fourth of the original estimates by the coal industry and the electricity generation corporations. A large fraction of the sulfur abatement costs was recovered by the increased efficiency of the new equipment in the power plants. It is also important that the sulfur abatement did not cause any disruptions of the electric power production and no inconvenience to the consumers. All in all, with the concerted and coordinated efforts of the international scientific community, acid precipitation has been significantly reduced in the early twenty-first century, and its detrimental environmental effects have been mitigated at a very small cost to the electricity generation industry and the consumers.

3.2.3 Lead Abatement

Gasoline and diesel fuel are mixtures of liquid hydrocarbons derived from crude oil after refinement. While the diesel–air mixture in the diesel engines is designed to self-ignite at the end of the compression process, when a high temperature is reached, the gasoline–air mixture is designed to ignite during the ignition stage by a spark. Because very high temperatures are reached at several stages of the compression process, a few of the hydrocarbons in the gasoline mixture reach their ignition point, autoignite, and release heat prematurely. This causes the "knocking" problem in gasoline engines, where the autoignition has triggered premature engine detonation, severe vibrations, low cycle efficiency, and engine damage.

Autoignition in gasoline engines may be prevented by chemical additives, the most common of which in the past was tetraethyl lead ($Pb(C_2H_5)_4$). When added to the gasoline, tetraethyl lead prevents premature combustion, knocking, and engine damage. The use of this chemical was widely adopted by the refining and automobile industries as an antiknock additive to the gasoline in the early part of the twentieth century. However, $Pb(C_2H_5)_4$ burns with the fuel, and its combustion releases lead oxides to the atmosphere, primarily PbO and Pb_2O, as well as atomic Pb. Humans and animals will breath or ingest these chemical compounds. A second pathway for the human absorption of lead is via the lead-based pigments and the sweetener lead acetate, which is also known as "sugar of lead." Pigments with lead base have been used for centuries to provide bright colors of yellow ($PbCrO_4$), red (Pb_3O_4), and white ($PbCO_3$). Lead-based pigments were used by the paint industries, worldwide, for the manufacturing of paints for artists and for the walls of buildings in the 1980s.

Recent toxicological studies have proven that chemicals containing lead are harmful to the health of the humans, because lead compounds affect the synapses in brain cells, especially those of children. Prolonged exposure to lead has been proven to cause mental retardation and brain disorders to humans, including dementia.* The ingestion of large quantities of lead will cause death (lead poisoning).

Lead-based chemicals used as engine antiknock additives are also incompatible with several types of catalytic converters in automobiles and significantly reduce the useful life of the catalysts. During the 1970s, regulations were enacted in several countries to phase out the use of $Pb(C_2H_5)_4$ from gasoline additives. In addition, lead-based chemicals were phased out from paints and other commonly used materials as well as from all the equipment used in the food industry. In the United States and the countries of the EU, the sale of leaded fuel for automobiles has been completely banned since the early 1990s. Limited amounts of leaded gasoline are still allowed for marine engines, racing cars, and a few types of agricultural equipment. Most of the other countries have followed with similar regulations. Tetraethyl lead has been replaced by other additives, typically made by aromatic hydrocarbons. The few countries where leaded gasoline is allowed have adopted emerging environmental regulations to significantly restrict its use. It is anticipated that by 2020, the use of leaded gasoline will be globally banned, with very few exceptions for special engines.

The restrictions and eventual ban on leaded gasoline, as well as leaded paint, resulted in millions of tons of lead not being released in the environment. The average concentration of lead in the air significantly decreased in the last 20–25 years in all the OECD countries. Figure 3.4 depicts this dramatic decrease of the average lead concentration in the air in the United States [9]. It is observed that between 1980 and 2013, the ambient lead concentration decreased by a factor of 12. The lower amount of ambient lead caused the lowering of lead in the human bloodstream, especially in children. This is expected to become the means to better public health, lesser neurological disorders, and improvements of the quality of life. As with the acid rain, the elimination of these large quantities of lead from the environment was achieved with new technology and minimal cost and inconvenience for the consumers and industry.

---

* Several historians attribute the apparent early dementia and other mental disorders of the Roman emperors and senators to chronic lead poisoning: the sweetener lead acetate was widely used as a wine additive in the Roman Empire, and unscrupulous merchants from the East Mediterranean mixed the imported and expensive pepper and spices with cheaper lead dust.

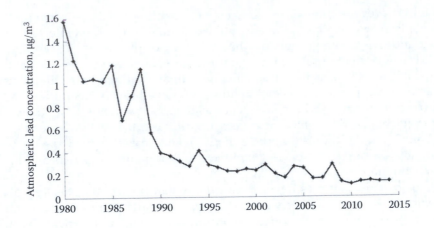

FIGURE 3.4
Average atmospheric lead concentration in the United States.

3.2.4 Ozone Depletion: The "Ozone Hole"

The ozone gas (O_3) is composed of three oxygen atoms and is very reactive. Ozone gas in the atmosphere is both a pollutant and toxic to humans as well as very important for the preservation of life on earth. Because of its very high reactivity, ozone in the biosphere is very toxic and harmful when it is in contact with plants and animals. High ozone concentrations in the lower atmosphere can cause throat and lung irritation, asthma, and emphysema to humans. On the other hand, ozone as a gaseous component of the stratosphere—the part of the atmosphere between 10 and 50 km—is necessary for the absorption of the high-energy ultraviolet (UV) sunlight and for the preservation of life. Without ozone in the stratosphere, the UV radiation of the sun would sterilize the Earth's surface and kill all living organisms. Ozone in the upper atmosphere absorbs most of this radiation. It effectively provides a "sunscreen" that eliminates all the most energetic, UV-c radiation, about 90% of the UV-b radiation, and 50% of the less harmful UV-a radiation. Excessive UV-B and UV-A radiations, which are almost eliminated by ozone in the stratosphere, cause severe sunburns that may develop to skin cancers and eye damage. In short, ozone in the lower atmosphere is detrimental to human health, but its presence in adequate concentrations in the stratosphere is necessary for the preservation of human life.

Figure 3.5 depicts the concentration profile of ozone in the atmosphere. It is seen that only traces of the gas are present at the surface of the earth, where the average concentration is less than 1 part per million (ppm). At that low altitude, ozone is formed from anthropogenic activities, primarily from combustion products and NOX gases. The concentration of ozone significantly increases in the stratosphere where it reaches a maximum close to 8 ppm and then gradually decreases at higher altitudes.

It was observed in the spring of 1985 that the total stratospheric ozone concentration above Antarctica had significantly fallen, because of the presence of inorganic chlorine [10]. The observations confirmed an analytical study of 1974 [11], which concluded that the continuously increasing concentration of chlorofluorocarbon (CFC) refrigerants—compounds relatively inert that may persist in the atmosphere between 40 and 150 years—produces significant amounts of chemically active chlorine atoms that combine with atmospheric ozone in the troposphere and cause its destruction. CFCs were used as refrigerants, fluids that run the refrigeration and air-conditioning cycles. The reduced total concentration of

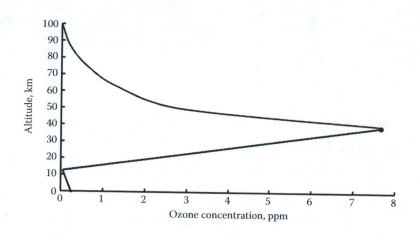

FIGURE 3.5

Profile of the average ozone concentration in the atmosphere in parts per million.

ozone in a vertical column of the atmosphere was dubbed "ozone hole." Actually, it is not a hole per se but a very large area above the Antarctica, where the total concentration of ozone is significantly less than the globally averaged concentration. As a result of the ozone hole, higher fractions of harmful UV radiation penetrated the upper atmosphere and reached the surface of Antarctica.

The appearance and growth of the ozone hole in the Antarctica is alarming, because its extension to other, more densely populated areas of the planet has the potential to cause disastrous health effects to humans. An initial international conference was held in Vienna (1985), where it was decided that the CFCs that contribute to the ozone destruction should be gradually banned. The decision was ratified in 1987 in Montreal, with the goal that by 1996, there would be zero production of ozone-depleting CFCs. This is the so-called Montreal Protocol. The resolute international effort for the elimination of the production of the harmful CFCs was successful: new refrigerants (e.g., R-134a and R-510) were produced between 1987 and 1996 and substituted the CFCs in refrigerators and air-conditioners. Within a few years, the atmospheric concentration of the CFCs gradually decreased and the ozone concentration above Antarctica was stabilized and, actually, started to gradually increase in 2001. As a result, the ozone hole in the Antarctica is slowly recovering and climatological projections predict that the ozone layer in that continent will return to the 1980 levels sometime between 2050 and 2070 [12].*

The abatement of lead, the mitigation of acid rain, and the recovery of the ozone layer are environmental success stories. A threat to the environment and the ecosystems was correctly identified based on scientific data. The international scientific community analyzed the data, drew its conclusions about the threats, and proposed sound solutions to the political and regulatory branches of governments. With concerted international effort, the cause of the threat was eliminated; healing processes for the environment started; and the results of these processes have proven to be successful. The technical community played a significant role in the three processes of environmental remediation: sulfur scrubbers were developed for power plants; lead-free antiknock substances were invented and massively produced; and refrigerants that do not affect the atmospheric ozone substituted the

---

* A problem with the substitution of the CFC refrigerants is that the new refrigerants are very potent GHGs. The Kigali agreement of 2016 (Section 3.3.8) mandates their substitution with hydrofluoroolefins that do not harm the ozone layer and are less potent GHGs.

ozone-destroying CFCs. Technologists and regulators found common ground with industry to alleviate three severe environmental threats in a relatively short time and rather inexpensively.

3.3 Global Climate Change

GCC or global warming is the predominant environmental problem of the twenty-first century that has repercussions in the social, economic, and political arenas. GCC and its expected effects that range from sea-level rise to the desertification of vast and presently prosperous agricultural regions are ubiquitous in scientific journals and the mass media of information. This issue, which has started with the scientific calculations and predictions of the effects of increasing CO_2 and other GHGs in the atmosphere, has become politicized in several countries and is a divisive national and international issue. It is rather disconcerting that several popular arguments about GCC are based on exaggerations and outright myths rather than common sense based on cause–effect relationships and sound interpretation of the available scientific data.

3.3.1 Greenhouse Effect

Greenhouse effect is a general term used for the observed higher average temperature of the biosphere as a result of the increased concentration of several atmospheric gases, the GHGs. A consequence of the first law of thermodynamics is that when a system receives more heat than it rejects, the temperature of this system rises. On the contrary, if the system rejects more heat than it receives, its temperature falls. The biosphere of the earth, which may be considered as a thermodynamic system, continuously receives heat from the sun in the form of solar radiation (insolation), and simultaneously, it emits radiation in all directions to the outer space. If the incoming radiation is higher than the outgoing radiation, the temperature of the biosphere rises. This is observed during the daytime when the temperature rises locally because of the increase in insolation. During nighttime, the earth continues to emit radiation to the outer space, while the insolation is zero and the temperature drops locally. Similarly, because days are longer in the summer and more insolation enters the biosphere during the day, the average local summer temperatures are higher than the average winter temperatures.

While the preceding paragraphs describe in very simple terms the energy exchange between the earth and outer space, several components of the planet influence and mitigate the radiation exchanges: clouds and ice sheets reflect a great deal of the insolation back to the outer space during the day, while clouds during the night reflect to the earth some of the terrestrial radiation. Figure 3.6 shows several of the features of the land, water, and atmosphere that affect and mitigate this energy exchange. The most important energy exchanges that affect the temperature of the biosphere are the following:

1. Because the earth's core is much hotter than its surface, terrestrial heat is conducted from the interior of the earth to the biosphere.
2. The earth's surface reflects a high fraction of the incoming insolation back to the outer space. Ice reflects more than 90% of the insolation.

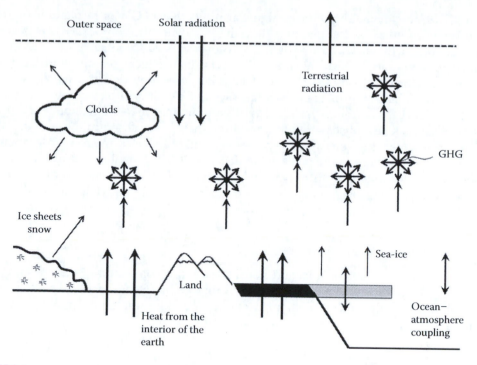

FIGURE 3.6
Sources and sinks of energy that affect the heat balance and the average temperature of the biosphere.

3. Clouds in the atmosphere reflect a high fraction of the insolation.

4. The oceans, because of the enormously high mass of water they contain, absorb, and release heat to the surrounding atmosphere and regulate the temperature.

5. Part of the terrestrial radiation is reflected back to the earth by clouds. Because of this, cloudy nights are warmer.

6. Part of the terrestrial radiation is also reflected back by GHGs. The GHGs are selective in the wavelengths of radiation they absorb and reflect. They do not reflect back to the outer space a great deal of the sun's radiation, but they reflect back to the earth a high fraction of the emitted infrared radiation. By their selective emission-absorption properties, GHGs contribute to the increased temperature of the biosphere.

The last process is very significant for the development of the greenhouse effect, because GHGs reflect a large quantity of energy back to the biosphere and, hence, contribute to its warming. Essentially, GHGs are a "blanket" that has warmed the biosphere for millennia. If GHGs were entirely absent from the atmosphere, the outgoing terrestrial radiation would have been significantly higher and the energy balance of the planet would have resulted in much lower average temperatures than the current temperatures. Climatic models conclusively show that if GHGs were entirely absent from the atmosphere, the average temperature of the biosphere would have been approximately 33°C (59°F) lower than its present value. At such average temperatures, most of the oceans and the surface waters would have been frozen and the planet would have been inhospitable to life in its current forms.

It is rather ironic that without the benign warming effect of GHGs, human life might not have evolved as it has on the planet. While the low concentration of GHGs is necessary for the life on earth, significantly higher concentrations of these gases further increase the average temperature with detrimental effects for the local ecology and the human economic activities. If this blanket of the GHGs becomes too thick, the average temperature of the biosphere will increase accordingly. The result will be several regional and global long-term effects that are harmful to life and disruptive to the economic activities of humans. Scientists have discovered that with the increased concentrations of GHGs, this blanket continuously thickens and has caused a small but significant average temperature increase in the biosphere in the last several decades. It is reasonable that if the causes for the observed modest temperature increase are left unchecked and the average biosphere temperature continues to rise, significant climatic effects will be triggered on the planet that will have catastrophic consequences for the human society.

3.3.2 Greenhouse Gas Emissions

All fossil fuels—coal, petroleum, and natural gas—are composed of carbon and other atoms, typically hydrogen. The carbon atoms form CO_2 upon combustion, as, for example. in the following complete combustion equations of carbon (the main ingredient of coal) and methane:

$$C + O_2 \rightarrow CO_2; \quad CH_4 + 2O_2 \rightarrow CO_2 + 2H_2O. \tag{3.5}$$

Humans in their everyday activities have always used small amounts of fossil fuels for their energy needs. The use of fossil fuels has dramatically increased since the industrial revolution, because all thermal engines operate with very high quantities of heat, which are mostly produced from the combustion of fossil fuels. All the current trends on energy demand indicate a continuous and accelerated increase in fossil fuel use in the near future. The result of two and a half centuries of fossil fuel combustion has been the cause of the significant increase in the average atmospheric concentration of CO_2. Figure 3.7 shows the evolution of the average concentration by volume of the CO_2 in the earth's lower atmosphere from the beginning of the industrial revolution to 2015 [13,14]. While the

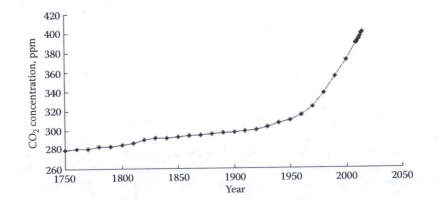

FIGURE 3.7
Globally averaged atmospheric concentration of CO_2.

concentration of this gas was almost constant for millennia, at approximately 280 ppm, it started steadily rising with the increased use of fossil fuels and reached the level 400 ppm in 2015, a 40% increase from its historical levels. It is apparent in this figure that the rate of increase of the CO_2 concentration has significantly accelerated since 1950. This is highly correlated with the increased fossil fuel consumption and, particularly, with the significant increase of coal and petroleum product combustion since the 1950s. During the period of 1950–2015, we also notice the following:

1. The significant increase in the earth's population
2. The accelerated use of coal and other fossil fuels for the production of higher quantities of electricity
3. The widespread use of the personal automobile and other modes of transportation that predominantly use liquid fossil fuels

A close look at these effects and correlations leaves no doubt that the 40% increase in the CO_2 in the atmosphere is the result of human activities. The significant increase in the atmospheric CO_2 in the last two centuries is certainly of *anthropogenic* origin.

It must be noted that CO_2 is not the sole GHG. CH_4, N_2O, and several CFC compounds also reflect the terrestrial radiation and contribute to the thickening of the earth's blanket. The atmospheric concentration of these gases has also significantly increased in the last 60 years because of human activities: the global average atmospheric concentration of CH_4 has increased from 715 parts per billion (ppb) at the preindustrial age to 1774 ppb in 2005. At the same period, the atmospheric N_2O concentration increased from 270 to 319 ppb [13]. Several of the other GHGs are significantly more potent than CO_2 in reflecting the infrared radiation back to the earth. Table 3.1 is a list of some of the high-potency GHGs [15]:

Of interest in this table is that several refrigerants, such as R-134a and R-142b, which were adopted as a result of the *Montreal Protocol* for the remediation of the ozone layer hole, are very potent GHGs. The Kigali agreement, signed in October 2016, calls for their substitution globally with other refrigerants that are less potent GHGs [16,17].

It is also apparent in Table 3.1 that the other GHGs are by far more potent than CO_2. However, their emission rates and concentrations in the atmosphere are much lower than that of CO_2 and their overall warming effect is less significant. Figure 3.8 depicts the equivalent GHG emissions in the United States during the period of 1990–2014 assuming that the effect of CO_2 is one [18]. The areas represent carbon dioxide, methane, nitrogen oxide,

TABLE 3.1

Potency of GHGs in Reflecting the Terrestrial Radiation, Relative to CO_2

Substance	Relative Potency per Unit Volume (and Unit Mole)	Relative Potency per Unit Mass
CO_2	1	1
Methane, CH_4	21	58
Nitrogen oxide, N_2O	206	206
Refrigerant-12	1,580	5,750
Refrigerant-114	1,830	4,710
Refrigerant-134a	9,570	4,530
Refrigerant-142b	10,200	4,130

Source: Fisher, D. A. et al., *Nature*, 344, 513–516, 1990.

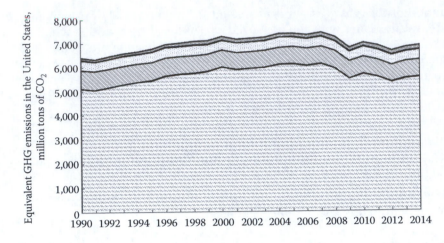

FIGURE 3.8
Equivalent GHG emissions in the United States in the period of 1990–2014. The areas from the bottom represent CO_2, CH_4, N_2O, and combined refrigerants and their atmospheric products.

and refrigerants and their atmospheric products that include several fluorine compounds. It is observed in the figure that CO_2 emissions account for the highest part of the emission with their contributions being in the range of 79–83% of the total. During the same period, the methane emissions have been in the range of 10–12%; the nitrogen oxide emissions have been in the range of 5–6%; and the refrigerants (including all the fluorine compounds formed by them in the atmosphere) have been in the range of 2–3% of the total.

The earth is a complex, highly nonlinear, dynamic system, where small changes in the composition of the atmosphere have the potential to cause significant local and global environmental and ecological effects. One of these effects has become apparent and has been confirmed by the scientists: the average annual temperature of the biosphere shows a clear, increasing long-term trend in the last several decades. This is shown in Figure 3.9, which depicts the temperature deviation of the lower atmosphere [19,20] with respect to

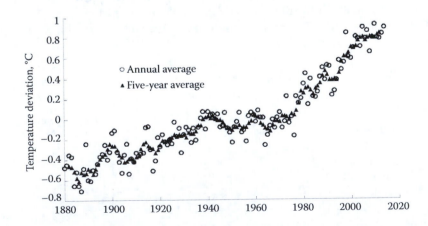

FIGURE 3.9
Global temperature deviation trends averaged annually and 5-year averaged. The reference, zero, is the average temperature of the period of 1901–2000.

the time-averaged temperature of the twentieth century (years 1901–2000). The two sets of data show the annually averaged temperature and the 5-year average temperature, which has lesser variability. Both graphs illustrate that the average temperature in the biosphere increases in the long run and that the temperature increase has accelerated since the 1960s. A glance at Figure 3.7 proves that there is a corresponding atmospheric CO_2 concentration increase since the 1960s, which suggests a cause–effect relationship.

There is a noteworthy variability in the annually averaged data and a lesser variability in the 5-year-averaged data: While the temperature deviation in 1983 was 0.39°C, the temperature deviation in 1984 was 0.22°C. The corresponding 5-year averages show a lesser degree of variability and, for this reason, are preferred for the identification of long-term trends. It may also be seen in Figure 3.9 that there are several time periods ranging from a few years to 20 years (e.g., 1940–1965) when it appears that the temperature deviation trend is negative, with the implication that the biosphere is cooling. The 10-year data from 2005 to 2014, which are depicted in more detail in Figure 3.10, portray a period of almost constant temperature deviation, and actually, data of shorter periods, e.g., 2005–2009, may even indicate a negative trend. This, in combination with the measurement uncertainty of the data, has led a few individuals—primarily nonexperts in statistics and climatology—to doubt that the temperature of the biosphere is increasing in the long-term. In order to draw conclusions on the climatic effects, one must be reminded that climate is the long-term average weather and that data over very long periods must be consulted for any valid scientific conclusions to be reached. The 5-year averaged data, also shown in Figure 3.10, indicate that there is a perceptible long-term upward temperature trend. The year-to-year variability of the temperature data is explained by the appearance or disappearance of sun spots and planetary weather phenomena, such as the El Niño or La Niña sea currents.

Other effects and phenomena that corroborate the conclusion that the temperature of the biosphere is increasing are the following [21,22]:

1. The long-term rising of the average sea water temperature
2. The severely decreasing amounts of ice coverage in the polar regions and the thinning of the glaciers
3. The reduction in the average amount of snowfall

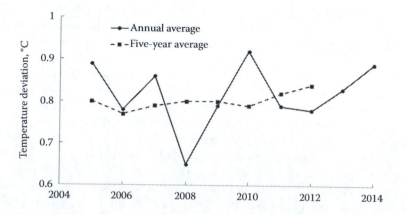

FIGURE 3.10

Average temperature deviation data within short time periods are not indicative of long-term trends. The 5-year-averaged data often give better indications of long-term trends.

4. The apparent rising of the sea level by 19 cm from 1990 to 2010

5. The unprecedentedly high atmospheric concentrations of GHGs, which are now higher than in the last 800,000 years. The cause–effect relationship between their presence and the atmospheric energy balance foretells a temperature increase.

Example 3.3

It is estimated that if all the air in the atmosphere were under standard conditions (1 atm pressure and 298 K), its volume would be approximately 4×10^{18} m^3. Estimate from Figure 3.7 the mass of CO_2 that was added since the year 1750.

Solution: The parts per million in the figure refer to volumetric quantities, which also correspond to molar quantities. In the year 1750, 1 million (10^6) m^3 of air contained the equivalent of 280 m^3 of CO_2. Under standard conditions, 1 kmol of any gas occupies 22.4 m^3. Therefore, the 1 million m^3 of air contains 280/22.4 = 12.5 kmol, or 550 kg of CO_2. Hence, the entire atmosphere contained approximately $4 \times 10^{12} \times 550 = 2.20 \times 10^{15}$ kg of CO_2.

In 2015, the concentration of 400 ppm corresponds to 400 m^3 of CO_2/million m^3 of air. This is the equivalent of 400/22.4 = 17.86 kmol of CO_2, or 786 kg of CO_2/million m^3 of air. The entire atmosphere contains $4 \times 10^{12} \times 786 = 3.14 \times 10^{15}$ kg of CO_2. The difference between the beginning of the industrial revolution and our days is 0.9610^{15} kg of CO_2, which was added in the atmosphere since the year 1750.

3.3.3 Weather and Climate

Before any discussions on the climate, it is prudent to define and distinguish between weather and climate. The *weather* is the short-term product of all the complex interactions between the sun and the earth, including the atmosphere, the hydrosphere, the continents, and their features (such as mountains, vegetation, urban environment, and ice sheets). Weather is a short-term phenomenon that results from the temporary thermal interactions between the solar radiation, the atmosphere, and the hydrosphere. The weather may be predictable over short times, e.g., a few days, but is unpredictable over long periods, e.g., months or years. The *climate* is the long-term result of the weather. It may be said that climate is the average weather, taken over a period of several decades. Unlike weather, climate may be predicted in the long run: we know with certainty that the summer, 10 years from now, will be warmer than the winter and the spring of the same year; that the average wind velocity in the spring will be higher than that of the summer; and that in the northern hemisphere, there will be lower temperatures and more snowfall in January than in July.

Agriculture, which supplies food for all humans, has been based on the predictability of the climate and the seasons for millennia. Most of the other economic activities of humans have developed over the centuries based on the fact that the regional and global climates have been and will be constant. For centuries, one could rely on the fact that the January days in Hanover, Germany, will be exceptionally cold; the month of March in London will carry a significant amount of rain; that July in Fort Worth, Texas, will be very hot and dry; and that the month of January will be warmer in Buenos Aires than in New York. Weather may bring rain on a particular October day in the Sahara desert, but one can rely on the climatic fact that during a given year or a given season, the rainfall in the Sahara will be significantly lower than the rainfall in the Amazon basin.

Weather changes frequently, but the weather phenomena are brief and do not significantly impact the ecology, the environment, and the human activities. On the contrary, climate changes, whether regional or global, significantly affect the environment, the eco-systems, and the human economic activities. For example, a long-term 2–3°C temperature increase in Nebraska and Kansas accompanied by drought will convert that region to a desert, will destroy the local economy, and will deprive the United States of its bread-basket. A 1 m rise of the sea level will have catastrophic consequences for Bangladesh, Louisiana, and the Maldives. The increase in the average temperature of the earth is a *de facto* global climate change that has the potential to impact severely all human activities and, consequently, the entire global economy.

3.3.4 Potential GCC Effects on the Climate

Several climatological and statistical studies including [19–21] concluded that the long-term temperature data of Figure 3.9 show at the 95% confidence level that the temperature of the biosphere is increasing. The cause–effect relationship between the GHGs and tem-perature has convinced the scientific community that the global average temperature will continue to increase and may even accelerate in the future. In summary, the increased thickness of the blanket of the GHGs in the atmosphere will continue to have a warming effect on earth's climate, and this warming effect will increase with the thickness (the GHG atmospheric concentration) of this blanket. What is uncertain is the exact magnitude of the temperature rise during the twenty-first and the twenty-second centuries. The pre-dictions of the several models that have been used to predict the average temperature of the biosphere for the year 2100 vary by several degrees. It is noteworthy that all the mod-els agree that there will be a significant (more than 2°C) average temperature increase if the CO_2 concentration in the atmosphere exceeded the 450 ppm level [21,23]. A separate, independent study concluded that when other effects are taken into account, the 2°C level will be surpassed in 2036, even if the atmospheric CO_2 concentration increases only to 405 ppm [24].

The vast majority of climatologists predict that the change in the composition of the atmosphere of the planet and the gradually increasing temperature of the biosphere will have significant impacts on earth's climate regionally and globally that will disrupt the human economic activities and may even become catastrophic. The most important con-sequences of GCC, which are detrimental to the environment, the ecosystems, and the human society, are [22,23] as follows:

1. *Melting of the polar ice caps:* An increase in the average atmospheric temperature will also result in an increase in the temperature of the polar regions. Global circulation models (GCMs) predict a higher temperature rise in the polar zones than in the equatorial and temporal zones. An immediate effect of the polar temperature rise is the melting of part or of the entire ice caps in the polar regions. Since ice reflects 92% of the incident sunlight, the polar ice caps reflect a high fraction of the incident solar radiation. The disappearance or simply the reduction of the polar ice caps will effectively increase the amount of insolation absorbed in the biosphere, which will further increase the temperature and will accelerate the rate of global warming.

2. *Sea level rise:* The total or partial melting of the ice caps will create enormous masses of liquid water. The thermal expansion of the oceans will also entail higher volume for the existing liquid water on earth, which will cause additional sea level rise.* Following the hydrological cycle, a very large fraction of the mass of the

---

* Because of expansion and ice melting, the average sea lever rose by 0.19 m between 1950 and 2013 [17].

additional freshwater will end up in the oceans. The level of the oceans will rise to accommodate the additional mass of water. It is estimated that if only the entire West Antarctic sheet ice melted, the average sea level would rise by approximately 5 m (16.6 ft). If all the ice on the surface of the Earth melted, the sea-level rise would be more than 60 m (200 ft). Any significant (more than 1 m) sea level rise will bring large parts of the continents under water and will threaten other coastal parts with floods. Since 76% of the population of the planet lives within 60 km from the coasts, even a moderate rise of the seal level will have catastrophic consequences on large parts of the human community and their economic activities. Although it is not expected that the entire polar ice mass will melt in the foreseeable future, the model predictions maintain that the average sea level in the year 2100 will be between 0.5 and 1.4 m above the 1990 level [25]. Even at this lower range of sea level rise, several coastal communities will be severely threatened, and a large fraction of the earth's population will need to be displaced. With a 0.5–1.4 m rise of the average sea level, not only cities such as Venice and New Orleans, which have been historically vulnerable to floods, will be underwater, but also modern and thriving economic hubs, such as New York, Shanghai, Los Angeles, Karachi, Rio de Janeiro, and London, will be severely threatened. Large coastal parts of the planet will become uninhabitable and a large fraction of the earth's population will need to be relocated.

3. *Regional climate change:* The average global temperature rise will be accompanied by regional temperature rises, some more significant than others. As a consequence, the climate of several regions will dramatically change. GCMs are not sufficiently validated to make accurate predictions on regional climate changes, and for this reason, the current regional predictions of the models have high variability and uncertainty. Two predictions that have become widely known are (a) the development of a large desert in the Midwest and the Great Plains parts of the United States caused by chronic drought and (b) significant rainfall in the Sahara, which will transform the desert to fertile land.

Such specific predictions of the GCMs may not materialize in the future. However, scientific reasoning, common sense, and all the scientific models agree that the global average temperature will increase and the regional climate will be subject to changes. This will certainly have unwelcome consequences on the agricultural and other economic activities of the local populations that depend on a constant regional climate, predictable seasons, and relatively predictable rainfall.

Regional climate change and sea level rise of any magnitude have the potential to disrupt the entire economic life of the planet. Their combined effect will necessitate the displacement of coastal populations with the severe socioeconomic problems this entails. Our current global infrastructure and level of preparedness are insufficient to cope with problems of such magnitude: In 2005, following the hurricane Katrina, the entire world watched with horror the hardship of approximately 1.2 million persons who were temporarily displaced from the coastal area of the northern Gulf of Mexico. Potential consequences of global warming may necessitate the permanent displacement of 1–2 billion. Most of the contemporary nations are not ready to respond to such dramatic consequences of global warming. With massive, uncontrolled population displacements, several of today's states will crumble under the socioeconomic pressures that will follow, and it is likely that the human bonds, which form the society and the nations, will break and may be replaced by chaos, anarchy, and destruction.

4. *Other effects of lesser impact* are [21,22] as follows:

   - Changes in the global water cycle in response to the warming will be non-uniform. The difference in precipitation between wet and dry regions and between wet and dry seasons will increase.

   - The intensity of weather phenomena will increase. Tropical storms, cyclones, and hurricanes will be stronger; on the average, winters will become colder and harsher and summers hotter.

   - The continuous uptake of CO_2 by the oceans will increase ocean acidification, with detrimental effects to aquatic ecosystems [26].

   - Increased temperatures will affect the atmospheric carbon cycle processes in a way that will aggravate the increase in CO_2 in the atmosphere.

3.3.5 Mitigating and Remedial Actions

Even though the scientific models predicted the global temperature rise since the time of Fourier in 1824, until the end of the twentieth century, there have not been accurate measurements and independent and reliable confirmations of GCC. In the early twenty-first century, the scientific data show with a high degree of certainty that the average global temperature is increasing as shown in Figure 3.9. The rate of increase is much higher than that of the previous two centuries. The United Nations (UN) Intergovernmental Panel for Climate Change (IPCC)* attributed most of the measured temperature rise to anthropogenic activities and the observed increase of the GHG concentrations [13]. The reliable and independent scientific confirmations of the GCC have alarmed the international scientific community, which has called the scientists and political leaders to action.

Since the 1990s, it has become apparent to the scientific community that a more concerted, rigorous, and inclusive global effort is necessary for the actual and meaningful reduction of GHG emissions. The following list includes some of the proposed remedial actions nations and the global community may take to, first, reduce the growth of the annual CO_2 emissions and, secondly, to reduce the actual concentration of GHGs in the atmosphere.

1. *Reduction of energy consumption:* Since most GHGs are produced by energy-related activities—primarily from fossil fuel combustion—this is the most effective of the actions the global community may take to at least reduce the growth of the annual CO_2 emissions, especially in the wealthier, developed nations. The reduction of unnecessary energy consumption is the best, least disruptive, least expensive, and most feasible alternative to mitigate GCC. It may be accomplished with energy conservation and higher engine efficiency, subjects that are discussed in Chapter 8. It must also be realized that the total energy consumption and the production of GHGs is strongly correlated to the population of the earth, as shown in Figure 2.5. The long-term stabilization—and even the reduction—of the global population is part of the solution to reduce global energy consumption, reduce the GHGs in the atmosphere, and mitigate the global GCC.

---

* The members of the IPCC shared the 2008 Nobel Peace Prize.

Example 3.4

A salesperson does 30,000 mi/year and uses a light truck that consumes 12 miles per gallon (mpg). How much CO_2 does this truck produce per year? It is suggested that the salesperson substitute the truck with a small car that with 32 mpg consumption. How many gallons of gasoline will be saved annually and what will be the reduction in the CO_2 emissions? You may assume that the gasoline is composed solely of octane (C_8H_{18}).

Solution: At 12 mpg, the light truck uses 30,000/12 = 2,500 gal of gasoline per year (9,463 L). From thermodynamic tables, one may find that the density of gasoline is 6.093 lb/gal. The 2,500 gallons correspond to 15,232 lb of gasoline or 6,924 kg or approximately 60.73 kmol of octane.

The combustion equation of octane is $C_8H_{18} + 12.5O_2 \rightarrow 8CO_2 + 9H_2O$.

Therefore, 114 kg of C_8H_{18} produce 8 × 44 = 352 kg of CO_2. The 6,924 kg of C_8H_{18} produce upon combustion 21,380 kg of CO_2 (47,036 lb). This light truck produces approximately 21.4 t of CO_2 annually.

If the truck is substituted by the smaller car, the annual consumption of gasoline is 30,000/32 = 937.5 gal of gasoline per year. The reduction in the gasoline consumption is (2,500 − 937.5) = 1,562.5 gal/year (5,915 L). The corresponding reduction in CO_2 emissions is (21,376 × 1,562.5/2,500) = 13,361 kg/year (approximately 13.4 t per year).

One notes that the substitution to a smaller car results in significant CO_2 emission reductions. This is accompanied by a significant financial advantage: At approximately $3/gal, the annual monetary savings of this substitution is $4,787.5. One may also note a few useful numbers: One gallon of gasoline weights 6.073 lb and, when used in an internal combustion (IC) engine, produces approximately 18.85 lb of CO_2. The weight ratio of CO_2 produced to the gasoline is approximately 3 to 1.

2. *Substitution of coal with nuclear fuel for the production of electricity:** This remedial action is currently technologically feasible. The nuclear reactor technology is well developed, and the OECD countries have had more than 60 years' experience with such reactors. For some of the developing countries, the lack of nuclear reactor technology and the lack of safety standards make this option a rather risky solution. A concerted international program that would involve nuclear technology transfer to developing nations—or the construction and operation of nuclear power plants by internationally approved organizations and consortia—as well as international oversight of the nuclear reactors on a global scale is a viable long-term solution for the reduction of the GHG emissions. However, before this solution becomes widely adopted, the global community will have to address the environmental problems of nuclear energy, most notably the long-term storage of nuclear waste. Example 3.5 elucidates the calculations for the CO_2 emission avoidance by using nuclear energy instead of coal.

---

* The United States would have been in compliance with the first phase of the Kyoto protocol, in the first decade of the twenty-first century, if it simply diverted 15% of its electricity production from coal to nuclear. This would have been achieved with the addition of 56 nuclear power plants [22,30].

Example 3.5

It is suggested that five older coal power plants with a total 2,000 MW capacity be substituted by two 1000 MW nuclear power plants. The average thermal efficiency of the older plants is 34.5%. Determine the annual CO_2 emission reduction.

Solution: The five coal power plants collectively use 2,000/0.345 = 5,797 MW of heat that comes from the combustion of carbon. This amount of heat corresponds to $5,797 \times 60 \times 60 \times 24 \times 365 = 182.8 \times 10^9$ MJ per year. Since the heat of combustion of carbon is approximately 32,800 kJ/kg, 5.57×10^9 kg (5.57 million tons) of carbon are burned annually by the five coal plants.

The amount of CO_2 produced by the coal is $5.57 \times 44/12 = 20.43$ million t of CO_2 per year. If the coal-to-nuclear substitution happens, this is the total annual CO_2 emissions reduction.

3. *Higher use of renewable energy sources*: The use of solar, wind, hydraulic, and geothermal energy not only for the production of electricity, but also for other processes (heating of buildings, clothes drying, etc.), has the immediate effect of avoiding coal and other fossil fuel combustions that produce GHGs. Most of the renewable energy sources are abundant on earth and are available to be used in both the developed and the developing nations. The increased use of renewable sources worldwide alleviates fossil fuel consumption. However, wind and solar energies, the most common and most widely available renewable energy forms, are not continuously available. Other renewable energy forms are at present very expensive to harness. The development of reliable and cheaper methods for renewable energy production and storage will enable us to substitute fossil fuels with the increasing use of renewable energy sources and reduce the amount of GHGs emitted in the atmosphere.

4. *Capture and sequestration of CO_2* (CCS) at the stationary production sites—primarily electric power plants—and subsequent land storage. While there is current technology and several proven and reliable methods for CO_2 capture, the processes require the compression or liquefaction of the gas, which uses very large amounts of energy. Because of this, CCS is intrinsically very much energy consuming and very costly. The cost of carbon sequestration on the price of the produced electric energy is very high. Conservative estimates for coal power plants range from 130% to 230% higher electricity prices, and those for natural gas power plants are in the range of 100–150%. Even if the global community succeeds to produce inexpensive methods for CCS, the safe, reliable, long-term storage of the produced CO_2, in liquid or supercritical form, is not feasible because of the following reasons:

   - There is not enough empty space in large geological formations, e.g., depleted oil and gas fields, to hold the vast quantities of CO_2 produced. This becomes apparent in examples 3.6 and 3.7 on the large-scale CO_2 sequestration in oil fields.

   - Supercritical CO_2 is one of the most effective and powerful solvents. Its storage in land formations has not been demonstrated to be feasible and reliable at the long timescale of 1000–5000 years required for sequestration. It would be an environmental calamity if the CO_2 that is stored underground in 2020 finds its way to the surface and starts leaking in 2025 by forming "CO_2 geysers." An experimental study by the US Geological Survey in Frio, Texas, showed [27] that there is very high probability that CO_2 will behave totally unexpectedly and will be released in the long run.

Example 3.6

It is proposed that 90% of the CO_2 produced by the 400 MW power plant of example 3.1 be sequestered and stored in an oil field, where the temperature is 80°C and the pressure is 700 bar. If the capacity of the oil field is 0.5 billion barrels, how long will take for the oil field to fill with CO_2?

Solution: At the proposed pressure and temperature, the CO_2 is supercritical with density approximately 860 kg/m³. Since the annual CO_2 output of the power plant is 4.06 × 10^9 kg of CO_2 (4.06 million t), 90% of the annual output have mass of 3.654 × 10^9 kg and occupy a volume 4.25 × 10^6 m³.

The total capacity of the oil field in Système International (SI) units is 0.5 × 10^9 × 0.159 = 79.5 × 10^6 m³. At the filling rate of 4.25 × 10^6 m³/year, it will take approximately 18.7 years for the oil field to completely fill with the sequestered CO_2.

It is worth noting in this example that the almost 19 years for the filling of the oil field is less than half the expected lifetime of the power plant, which is close to 50 years or more with good maintenance. It is necessary for this power plant to completely fill two to three similar large oil fields during its lifetime, and the infrastructure to transport the liquefied CO_2 gas (pipelines, etc.) will need to be developed.

Example 3.7

The electricity production in the United States in 2014 was 4,137 TWh, and 39% of the total was produced by coal power plants at a thermal average efficiency 37%. It has been proposed to sequester 80% of the total CO_2 production from the coal power plants and store it in oil fields. If the average oil field has capacity 100 million barrels, how many oil fields are needed for this project annually?

Solution: The electricity produced by the coal power plants in 2014 is 1,613.4 TWh. At an average thermal efficiency of 37%, this amount of electricity was produced by 4,359 TWh or 15.7 × 10^{15} kJ of heat. Since the heat of combustion of C is approximately 32,800 kJ/kg, this quantity of heat was produced from the combustion of approximately 478.5 × 10^9 kg, of C and the combustion process produced 1.75 × 10^{12} kg of CO_2.

Because the supercritical CO_2 density is approximately 860 kg/m³, the annual quantity of the CO_2 produced occupies a volume of 2.04 × 10^9 m³; 80% of this volume that is proposed to sequester is 1.63 × 10^9 m³.

The average oil field has a capacity of 100 × 10^6 barrels or 15.9 × 10^6 m³. Therefore, 1.63 × 10^9/15.9 × 10^6 = 103 oil fields would be needed to sequester the 80% of the annually produced quantity of CO_2. In addition, this sequestration effort would need the energy that would transport the CO_2 from each power plant to the destination oil field. Clearly, CO_2 sequestration at this large scale is not a feasible task. Sequestration at a much smaller scale will not make an impact on the atmospheric CO_2 emissions.

The two examples illustrate the futility of CCS for the meaningful reduction of the GHG emissions in the atmosphere: The United States alone produced 5,560 million t of CO_2 in 2014, with a high fraction of this for the generation of electricity. Even if only 40% of the CO_2 produced were to be captured, the underground sequestered CO_2 would need to occupy a volume of approximately 6.5 × 10^9 m³, the equivalent of 405 average oil fields.

Such an extensive CCS effort would require the filling of more than one such oil reservoir every day, an impractical task.

As with everything else, prevention is better than the cure, and similarly with CCS, it is best (and by far less expensive) to avoid the emissions of CO_2 than to try and remove it or to mitigate its undesirable effects.

5. *Reforestation*: This process—also called afforestation—entails the planting of trees on unused land without using the produced wood as biomass. Reforestation always removes a small quantity of the CO_2 from the atmosphere because all trees absorb CO_2 to produce carbohydrates. While reforestation has beneficial effects for the regional and global environment, its impact on the atmospheric CO_2 concentration is very weak, because of the large magnitude of the daily CO_2 emissions. As seen in example 3.1, a typical 400 MW coal power plant produces 11.1×10^6 kg (11,100 t) of CO_2 during a single day. It will take approximately 9,500 fully grown pine trees to remove this amount of CO_2, which is produced daily by a single power plant! Similar calculations show that it will take eight fully grown eucalyptus trees to remove the CO_2 emissions produced by the engine of a single sport utility vehicle (SUV) which runs for 15,000 mi. Clearly, reforestation alone cannot capture the vast amounts of CO_2 produced daily by our society. Given the scarcity of land, globally, and the need to produce food in arable areas, reforestation is not a viable option for GCC mitigation. The increasing population of the planet needs the available land for food production, not carbon storage.

6. *Seeding large ocean regions with iron and nitrogen-rich fertilizer*: This action promotes the rate of CO_2 absorption by aquatic organisms, which form more complex organic compounds. This option has not been tried on a large scale, and it is doubtful that it will result in a significant and meaningful atmospheric CO_2 reduction. In addition, the alteration of the ecological function of aquatic organisms from food sources to CO_2 sinks will cause many other undesirable effects, such as eutrophication and hypoxia (Section 3.5.2).

7. *Deep ocean sequestration*: The oceans cover more than 70% of the earth's surface; their average depth is approximately 3,800 m; the pressure at the average depth is more than 382 atmospheres; and the capacity of the oceans to absorb CO_2 is on the order of 10^{19} t of the gas [28]. At the current rate of CO_2 production, the deep ocean sequestration has the capacity to store the CO_2 produced by anthropogenic activities for approximately 300 million years! However, the collection, transportation, and pumping of the gas at the appropriate depths of more than 1000 m are very energy intensive and very costly. In addition, the injected CO_2 in the seawater forms CO_3^{2-} and HCO_3^- ions that increase the pH of the seawater and pose a severe threat to aquatic life. Several international treaties including the *London Convention on Ocean Dumping* prohibit the storage of radioactive and industrial waste in the deep ocean, and CO_2 from power plants is definitely industrial waste. Deep ocean sequestration will cause the acidification of the seas with dire consequences for the aquatic ecology. It is strongly and almost unanimously opposed by the national associations of marine scientists.

A case study by *Greenpeace International* [29] estimated that the cost of CCS in oceans, which may be as high as US$92/t, is prohibitive and concluded that spending such sums for CCS "diverts resources away from real solutions." Even if ocean acidification were not considered a problem, because the liquid or compressed CO_2 is significantly lighter than sea water—the liquid CO_2 density is approximately 860 kg/m^3, the supercritical CO_2 density is in the range of 470–800 kg/m^3, while the deep sea water density is approximately 1,030 kg/m^3—the discharge of large quantities of CO_2 in saturated regions of the deep ocean will cause "CO_2 bubbles" to rise fast through the seawater. These bubbles have the potential to reach the sea surface and release the CO_2 back into the atmosphere.

8. *Conversion to minerals*: When several metallic minerals are exposed to aqueous CO_2 (e.g., to carbonic acid or bicarbonic acid), the metals will form carbonate or bicarbonate salts that are solids and precipitate out of the solution. A glance at the magnitude of the CO_2 produced will prove that this method is not effective, simply because there are not sufficient supplies of minerals to be used. In order to sequester the 11.1×10^6 kg (11,100 t) of CO_2 produced during a single day from the typical 400 MW coal power plant of example 3.1 as $MgCO_3$, one needs 10.1×10^6 (10,100 t) of magnesia (MgO), a very large quantity to produce and transport daily. The processes for the CO_2 conversion to minerals require a great deal of energy; the processes are costly; and in the end, there are not enough minerals to make a dent on the enormous quantities of CO_2 emitted.

9. *Commercial use of CO_2*: Carbonated drinks, dry ice, urea, several chemicals, and fluids used for the secondary petroleum recovery contain CO_2. It has often been proposed that CO_2 may be sequestered in power plants and then sold commercially. However, all the commercial products that contain CO_2 are too few and cannot account for the amount of CO_2 emitted daily. The global market for CO_2 is approximately 0.14 million t/day, while the CO_2 emissions are 104 million t/day. Simply, the "CO_2 market" does not have the capacity to absorb the tremendous amount of CO_2 produced globally.

It becomes apparent that the only practical solutions to avoid atmospheric CO_2 buildup and the implied GCC are (a) the reduction of the energy used, (b) higher use of nuclear power, and (c) higher use of renewable energy sources. All three reduce or avoid the CO_2 emissions at the source of production. Avoidance or minimizing an undesirable environmental impact is by far preferable and cheaper to its neutralization after the pollution has taken place.

3.3.6 The Kyoto Protocol

Since global warming and GCC are caused by the increased anthropogenic emissions of the GHGs, it is apparent that any mitigation of the problem is centered on the reduction of the rate of the GHG emissions. Of these, CO_2 is the most abundant; the atmospheric increase in CO_2 is the main factor for the acceleration of the GCC threat; CO_2 is primarily produced by the combustion of fossil fuels; and CO_2 emissions are the most feasible to be curtailed. A significant, global reduction of CO_2 emissions will have a significant impact in the mitigation of the GCC threat.

For any concerted and potentially successful action for the reduction of the anthropogenic CO_2 emissions, one must take into account that GCC is a global, not a national or regional, environmental problem. A CO_2 molecule produced in Rome or Dallas has the same adverse effect as a molecule produced in Madras or in Beijing. For this reason, the collaborative and coordinated action of the entire global community is required to avert the adverse effects of global warming. Several international scientific panels have recommended global restrictions for the CO_2 emissions. The agreements have succeeded in creating global awareness for the GCC threat, resulted in small reductions of the CO_2 emissions in several OECD countries, but came short in lowering the global rates of CO_2 emissions.

The *Kyoto protocol*, which was created within the *UN Framework Convention on Climate Change*, was an agreement reached between several nations, both developed and developing, for the reduction of the global CO_2 emissions. The first phase of the protocol essentially called for the industrialized countries to reduce their collective GHG emissions by an average 5.2% from the 1990 levels. It also made provisions for the transfer of energy conservation technology to the developing nations and suggested restrictions for the growth of CO_2 emissions in the developed nations. Specifically, the Kyoto protocol stipulated 8% CO_2 emissions reduction for the countries of the EU, 7% reduction for the United States, 6% for Japan, and 0% for Russia. The protocol has been ratified by most (192) UN countries, with two notable exceptions: the United States* and the People's Republic of China (PRC). While most of the signatories of the first phase, especially the EU and Japan, have taken significant and meaningful steps for the reduction of their GHG emissions, in the period of 1997–2010, the United States actually increased its CO_2 emissions by 16%, and the PRC, by 130% [23,30].

The second phase of the Kyoto protocol—sometimes referred to as the *Doha Amendment to the Protocol*—calls for further emission reductions. While the EU countries have signed to the amendment, most other countries that ratified the first phase have not and have rejected any new CO_2 emission cuts. In particular, China, India, and the United States have sent strong signals that they will not ratify any treaty that will legally commit them to reduce CO_2 emissions.

Because—according to the provisions of the protocol—most developing countries did not have to reduce their own GHG emissions, this international agreement has only had symbolic and not real impact on the anthropogenic CO_2 and GHG global emissions, which continued to increase in the period of 1992–2017. In short, 25 years after its signing, the Kyoto protocol has been ineffective, and the GHG concentration in the atmosphere has continued to increase at an alarming rate as shown in Figure 3.7.

3.3.7 The Paris Agreement

The Kyoto agreement was followed by other meetings, including one in Copenhagen, Denmark, (2009) and another in Cancun, Mexico, (2011). The result of both meetings has not been significant in curbing the GHG emissions. In December 2015, most of the UN countries sent representatives to Paris, France, where a celebrated agreement was reached among several nations, including EU countries, the United States, PRC, India, and Russia. The main elements of this agreement are [31] as follows:

1. Reaffirm the goal of limiting global temperature increase below 2°C and urge efforts to limit the increase to 1.5°C.

2. Ask for binding commitments by all parties to make "nationally determined contributions" (NDCs).

---

* President Clinton of the United States signed the Kyoto protocol in 1998. However, the protocol has not been ratified by the US Senate as required by the US Constitution and does not have the legal and binding effect of a treaty.

3. Commit all signatories to regularly report on their emissions and the progress made in implementing and achieving their NDCs and to undergo international review.

4. Commit all countries to submit new NDCs every 5 years, with the clear expectation that they will represent progress relative to previous NDCs.

5. Reaffirm the binding obligations of developed countries to support the efforts of developing countries.

6. Encourage voluntary actions by developing countries.

7. Extend the current goal of mobilizing $100 billion annually in support of the agreement by 2020 through 2025, with a new, higher goal to be set for the period after 2025.

8. Extend a mechanism to address "loss and damage" resulting from climate change. This will not "involve or provide a basis for any liability or compensation."

9. Require no "double-counting" in reporting international emissions trading.

With more than 55 nations formally signing the agreement by April 2016, the Paris agreement is considered "binding," for its signatories. However, a close look at the agreement proves that it stops short from imposing limits on CO_2 emissions and does not include any sanctions for countries that increase their emissions. The commitments by the signatories are "to report," "to monitor," "to submit new NDCs," "to encourage," "to ask for commitments," etc. In the entire text of the agreement, no single nation and no single government are obligated to actually curb their CO_2 emissions or to do anything that specifically and materially addresses their growing CO_2 emissions.

When it comes to actually reducing the national and global emissions, the Paris agreement makes use of the word *should*, which implies advisement rather than legal obligation, instead of the word *shall*, which denotes an obligation in legal practice. It is characteristic of the lack of any commitment and concrete action of the Paris agreement that 2 years after the signing of the agreement, no progress has been made in actually reducing the global CO_2 emissions and that the two biggest pollutants, PRC and the United States, still do not have specific and binding national programs for the reduction of their emissions. Clearly, a great deal more progress needs to be made after the Paris agreement for the reduction of the global CO_2 emissions, and this is very difficult to be achieved without global consensus and intensive political action in all the countries.

A significant problem with the future implementation of the Paris agreement is that within the United States, the agreement is not considered binding, because it has not been ratified by the US Senate. Even after the US government signed the Paris agreement, the agreement had no legal effect in the United States because the US Senate had not ratified it. With the change of government in early 2017, the United States withdrew altogether from the Paris agreement. In the absence of ratification, and adherence to the agreement by the United States, it is rather unlikely that other nations—especially developing nations—will unilaterally proceed with significant domestic CO_2 emission curtailments. In particular, and since the United States was expected to be the major donor of the $100 billion per year fund for aid to the developing nations, funds may not be available for the developing nations to "leapfrog" from their current energy mix to renewable energy sources.

In the absence of a subsequent definitive and enforceable international agreement on the specific global CO_2 emission reductions, it is rather unlikely that the Paris agreement will succeed. The same occurred with the Kyoto protocol, which failed to achieve the stated objectives of global GHG emission reductions. An agreement on CO_2 emissions must set specific goals, must be legally binding for all the signatories, and must be enforceable if it is to have any tangible impact on the global environment. So far, the international community has failed to agree to that.

3.3.8 The Kigali Agreement on Hydrofluorocarbons

A more positive, recent development for GCC is the signing of the *Kigali agreement* (Rwanda) on hydrofluorocarbons (HFCs) in October 2016. The Kigali agreement calls for the total substitution of the more potent GHGs that are used as refrigerants (Table 3.1) with other refrigerants, which are more benign to the ozone layer and are not as potent as GHGs [16,17]. Specifically, the Kigali agreement calls for the substitution of HFCs with hydrofluoroolefins (HFOs) globally. This substitution will start in the developed countries after 2018. The developing countries will have until 2028 to acquire the technology and implement the substitution. It is estimated that the effect of the elimination of the HFCs by 2050 will reduce the potential global warming at the end of the twenty-first century by 0.5°C [18].

Unlike the Paris agreement for CO_2, the Kigali agreement is very specific, it addresses solely the HFCs, and (as with the Montreal protocol for the HFCs) it has a very high probability to be ratified and become globally adopted. In its favor are the following:

1. Ready substitutes for the HFCs are available, the HFOs.
2. The manufacturing of HFOs is inexpensive, and the engineering modifications to the refrigeration equipment design are modest and may be accomplished with low cost. Thus, it will not have a disruptive effect on the economies of the countries that adopt it.
3. As with the Montreal protocol, it is politically a more palatable and more popular treaty to sign and ratify.

On the other hand, the legal adoption of the Paris agreement—by both developed and developing nations—is more expensive for the national economies and has the potential to become disruptive to the economies of several developed nations. The significant reduction of the CO_2 emissions globally and the establishment of the $100 billion annual fund for the support of the treaty are more formidable political tasks than the substitution of HFCs by HFOs. The GCC problem is global and threatening, but it is also a very complex and unique problem for the national economies.

3.3.9 Uniqueness of the GCC Problem

The scientific data and climate models unequivocally show that the average temperature of the biosphere continuously increases and that the cause of this increase is the discharge of CO_2 and other GHGs. The vast majority of scientists have warned that the continuous emissions of GHGs in the atmosphere will lead to catastrophic consequences for the environment and will threaten the socioeconomic bonds of the nations. The potential long-term effects of the CO_2 and the other GHGs are simply too serious to be ignored. The UN, several national governments, and a myriad of international institutions have called for the adoption of regulations for the reduction of the GHG emissions. However, and despite the urging and good intentions of these institutions, global GHG emissions and especially the CO_2 emissions have been continuously increasing.

Of all the atmospheric gases that contribute to the GCC, the CO_2 emissions have received the highest attention by the scientists and the public. Strategies for the mitigation of the GHG effects are centered on the reduction of the CO_2 emissions for the following reasons:

1. The atmospheric CO_2 is the most significant contributor to the GCC problem. This is corroborated in Figure 3.8, which shows that in the United States, the CO_2 emissions account for more than four times the equivalent effect of all the other GHGs.

2. Most of the CO_2 emissions are energy related. These emissions may be curtailed by energy conservation measures or using renewable and nuclear energy.

3. A great deal of the CO_2 emissions occurs in stationary sources and in concentrated form in the electric power plants and cement production plants. Stationary sources are easier to monitor and regulate than moving ones.

4. CH_4, which is the second largest contributor to the GCC, is primarily produced by nonstationary agricultural activities that are widely distributed and almost impossible to monitor, regulate, and mitigate. The small amount of CH_4 that escapes natural gas pipelines is already regulated.

5. The contribution of the other GHGs, N_2O, and all the halocarbons (refrigerants) is not very significant and limited to a maximum of 1°C of global warming potential. The Kigali agreement will eliminate most of this threat.

The solution to the problem of CO_2 emissions reduction is very different from acid rain reduction, lead abatement, and ozone depletion, which were successfully remediated by concerted national and international activities. It is also significantly more complex than the HFC with HFO substitution, which is the subject of the 2016 Kigali agreement. The following are a few details that prove how unique, difficult, and elusive the GCC problem is:

1. Acid rain and lead represent regional rather than global threats. Regional and national guidelines and agreements have largely solved the two problems. The CO_2 emissions problem, by its global nature, requires the cooperation and adherence to the guidelines and regulations by all the nations on the planet.

2. Acid rain, lead, and ozone depletion are caused by chemicals—sulfur, lead and halogens—that appear in very small quantities, almost traces, in the fuels and in the environment. On the contrary, carbon is the main constituent of the fuels, and any kind of fossil fuel combustion produces high amounts of CO_2. In 2012, more than 35 billion t of CO_2 were emitted globally, a quantity that is by far higher than the quantities of SO_2, lead, and halogens that were discharged in the past. The sheer magnitude of the CO_2 emissions makes any remediation effort (after the emissions have taken place) almost futile.

3. The solutions to the other environmental problems are, in comparison, rather easy and inexpensive to implement. Substitute chemicals were quickly found for the halogens and tetraethyl lead in gasoline. The sulfur abatement processes were known, and they were put in place with relatively low investment cost. On the contrary, the solution to the CO_2 emissions problem is not as easy to implement. Substitute fuels do not exist at present in large quantities, and the carbon sequestration processes are very much capital intensive and not feasible to be implemented on a large scale as it becomes apparent in Example 3.7.

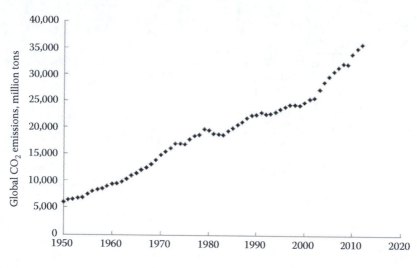

FIGURE 3.11
Annual global CO_2 emissions, in million metric tons.

4. Unlike the other three environmental problems, where the solution required new technology that was easy to obtain, the technology to alleviate the CO_2 emissions problem is not yet available to all the nations.

5. Adopting the solutions for the other three environmental problems does not threaten an economic sector or a national economy. The unilateral curtailment of CO_2 emissions by a single nation or by a small group of nations will make energy very expensive and will not solve the global GCC problem. The unilateral adoption of CO_2 curtailment measures may potentially ruin their national manufacturing sectors and have severe effects on their national economies. The CO_2 emission problem is of global scale; it presents a greater threat than the other environmental problems and must be tackled by the entire global community.

Figure 3.11 depicts the global annual CO_2 emissions from 1950 to 2012 [32]. It is apparent in this figure that the CO_2 emissions have increased almost continuously.* Of particular interest is that the CO_2 emission growth significantly accelerated from 2000 to 2012, when the first phase of the Kyoto protocol was in effect and several developed countries—most notably the EU nations, Australia, and Japan—curtailed their CO_2 emissions. Clearly, the first phase of the Kyoto protocol was not successful in reducing or even stabilizing the global CO_2 emissions.

The main reason for the continuous increase in global CO_2 emissions has been the accelerating increase in CO_2 emissions from developing nations, including the PRC, Brazil, India, Pakistan, Vietnam, and Indonesia. The annual CO_2 emissions growth in the developing countries in the first part of the twenty-first century more than offset the CO_2 emission curtailments in the EU and other countries. It is characteristic of this trend that in 2009, the PRC surpassed the United States as the country that emits the highest amount of atmospheric CO_2. Table 3.2 shows the 10 countries/economies with the highest annual amounts of CO_2 emissions [32]. The table also shows the percentage CO_2 emissions of these countries and the emissions per capita. It is apparent from the table that highly populous

---

* Approximately 50% of the annual CO_2 emissions is absorbed by plants and by oceans [33]. The remainder contributes to the atmospheric CO_2 increase.

TABLE 3.2

Total CO_2 Emissions by Country

Country	CO_2 Emissions, Million t	Percentage of Total	CO_2 Emissions per Capita, t
PRC	8,782	24.6	6.50
United States	5,143	14.4	16.36
India	2,186	6.1	1.77
Russia	1,647	4.6	11.47
Japan	1,231	3.4	9.65
Germany	734	2.1	8.97
Iran	584	1.6	7.64
Republic of Korea	575	1.6	11.49
Indonesia	536	1.5	2.17
Canada	529	1.5	15.16
All other countries	13,768	38.6	4.08
Global total	35,713	100.0	5.08

Source: Boden, T. A. et al., *Global, Regional, and National CO_2 Emissions, Trends: A Compendium of Data on Global Change*, ORNL Report, Carbon Dioxide Information Analysis Center, Oak Ridge, TN, 2012.

countries, e.g., the PRC, India, and Indonesia, emit a great deal of CO_2. However, these countries are not at the top of the list in emissions per capita. The United States and Canada are at the top of the list, followed by other industrialized countries, such as South Korea, Russia, Japan, and Germany.

A glance at the historical data of CO_2 emissions from all countries proves that the CO_2 emissions are highly correlated with the degree of industrialization. Industry, and especially the manufacturing industry, consumes a great deal of energy, which at present is primarily derived from fossil fuels. In addition, a higher degree of industrialization in a nation brings affluence to its citizens, who can afford to buy more energy-intensive consumer goods—automobiles, refrigerators, air-conditioners, television sets, etc. The widespread use of these consumer goods further accelerates the national use of energy from fossil fuels and increases the CO_2 emissions.

For all nations, developing and developed, energy from fossil fuels is the least expensive and most convenient form of energy to use. Beyond simple conservation measures, significant mandatory CO_2 emission reductions for a single nation (or a group of nations) are currently tantamount to increasing the average cost of energy for all its users. This makes deep cuts of CO_2 emissions unfavorable to an economy, because it carries the danger of making most manufactured products expensive and, perhaps, noncompetitive in the global economy. By reducing the global competitiveness of its products, with the consequence of driving the national economy to a recession or depression and possibly depriving its citizens from affluence, a significant and mandated CO_2 emissions reduction policy is a political anathema. For a single nation or a small group of nations, such a policy may have catastrophic consequences for the economies that adopt it. This leads to the conclusion that the entire community of the nations (perhaps through the UN) should decide to simultaneously impose a global CO_2 emissions reduction that will be observed by everyone and will equally affect all the economies and societies. If all nations share the burden of decreasing CO_2 emissions, the economies (and prosperity) of a small group of nations will not be threatened.

The problem in this case will become how any mandatory CO_2 emission reductions will be born in a "fair" manner among the community of nations. Among the questions that have been raised for a "fair allocation of CO_2 emissions" are as follows:

1. What should be the global limit of total CO_2 emissions?
2. Should OECD countries, which produce the highest CO_2 emissions per capita, also have the highest CO_2 emission reductions?
3. Is it fair that some nations should have higher CO_2 emissions per capita than others?
4. Should the developing countries share in the CO_2 emission reductions or should they be allowed to continue growing their economies with cheaper energy by increasing their CO_2 emissions?
5. Should the global community aim at a uniform per capita level of CO_2 emissions?
6. Should countries with fast growing populations be allocated the same per capita CO_2 emissions allowances as countries with stagnant populations?
7. How will the populations of countries such as the United States and Canada, which have historically very high emissions per capita, adapt to such a mandate?
8. Should landlocked countries in colder regions—Switzerland, Tajikistan, Mongolia, and similar countries—which are not threatened by GCC—participate in a global attempt for CO_2 emission reductions?
9. Should the global community be troubled about the most vulnerable to GCC countries—such as the Maldives, Mauritius, and small other island nations—or should they adopt a relocation policy for such sparsely populated islands?

It is apparent from the preceding questions that the GCC problem has legal, ethical, moral, as well as technological implications. The global allocation problem of CO_2 emissions becomes more complex when one considers the differences in the population numbers of the affected nations: India's population is approximately 1,250 million, while the combined population of the United States and Canada is 350 million. If the United States and Canada agree to reduce the annual per capita CO_2 emissions from the current levels to 10 t (more than 33% reduction), there will be 2.178 million t of CO_2 emission reduction annually. However, if at the same time, the citizens of India were to increase their emissions from the current 1.77 to 5 t/person, there would be 4.038 million t of CO_2 emission increase, which will more than offset the annual decrease triggered by the "sacrificial actions" of the populations of Canada and the United States. The recent experience related to the Kyoto protocol—the significant increase in the global CO_2 emissions between 2000 and 2015 despite the emission reductions by the EU, Japan, and a few other nations—proves that unless all nations in the global community agree to a formula of CO_2 emission reductions, unilateral actions or actions by a small group of nations will have an insignificant global effect.

Example 3.8

The populations of the United States, Canada, India, and Indonesia are approximately 320, 30, 1,200, and 250 million. Assume that the United States and Canada agree that no more than 8 t of CO_2 per capita should be emitted by any nation, effectively halving their CO_2 productions from their current levels, while India and Indonesia raise their CO_2 production to 5.0 t of CO_2 per capita. What will be the net effect of these changes?

Solution: From Table 3.2, the per capita CO_2 reduction in the United States and Canada would reduce their total missions by $(320 \times 10^6)(16.36 - 8.0) + (30 \times 10^6)(15.16 - 8.0) = 2,890 \times 10^6$ t of CO_2.

The corresponding increase in the CO_2 production by India and Indonesia is $1200 \times 10^6 \times (5.0 - 1.77) + 250 \times 10^6 \times (5.0 - 2.17) = 4,834 \times 10^6$. This is 1694×10^6 tons of CO_2 more than the reduction of the other two nations.

This rearrangement of the per capita CO_2 production, which appears to lead to a "higher degree of equality among nations," results in a significant (67.4%) increase in the combined emissions of the four nations.

3.3.10 Myths and Reality Related to GCC

The GCC is a unique, global, intergenerational, environmental threat with tremendous economic and social repercussions. It has attracted the attention of everyone, including scientists, politicians, economists, and the mass information media. It is rather unfortunate that the problem and its solution have not been framed in proper scientific terms, but in terms of "beliefs" of individuals and groups. The general public is frequently bombarded with "news" that encompass opinions of individuals who are unfamiliar with basic science, climate, and statistics, but freely pontificate and air out their beliefs on global warming and GCC. The tendency of the mass information media to "create a balance in every story" further complicates matters by deliberately bombarding the public with beliefs, conflicting opinions, questionable "facts," and false conclusions.

From the beginning, one has to realize that global warming and the GCC are scientific matters to be proven or disproven by the expert scientists. The beliefs of individuals, politicians, and "celebrities" do not carry any weight in such matters. Scientists resolve scientific questions, such as the GCC, by obtaining accurate data and, based on the data, by drawing logical conclusions. The beliefs of individuals are irrelevant in matters of science.* Because so many opinions and beliefs are expressed daily, several myths have been promulgated about GCC, its effects, and its possible solution. In this section, a few of the myths are exposed and a response is offered, the reality, which is based on science and the available scientific data.

Myth 1: There is no global warming; the global temperature variability is due to solar activity.

Reality: While solar activity affects the total radiation from the sun and the earth's temperature, solar activity is almost periodic with a timescale of 22 years. The data of Figure 3.9 that span more than six solar cycles show that the long-term average temperature of the biosphere has a long-term rising trend. Since the average temperature of the biosphere continues rising after six solar activity cycles, something other than solar activity (GHGs) cause this long-term temperature increase.

Myth 2: The recent 3-year temperature data show that there is no global warming and, ergo, no GCC.

Reality: While there have been short periods of a few years when the earth's temperature did not increase, or when it actually decreased, these years were followed by periods when the average temperature significantly increased and the long-term increasing trend was validated. As seen in Figures 3.9 and 3.10, the average temperature of the biosphere declined during the periods of 1914–1918 and 1945–1965, only to increase to higher levels in the periods of 1919–1926 and 1965–1981. The global climate is a long-term concept, and only long-term temperature data should be used to draw conclusions on GCC.

---

* Let us not forget that only a few centuries ago, most people "believed" that the earth is flat, supported by four pillars, and that the sun rotates around the earth. Accurate scientific data have proven that these beliefs are entirely false.

Myth 3: How can we have global warming? Here is a snowball I picked up from the streets of Washington DC today!

Reality: Here is someone who confuses weather with climate. Snow in Washington, DC, does not mean that the climate—the long-term, average weather—is cooling or that there is no increase in the *average* yearly temperature. Even if the average annual temperature in Washington, DC, increased by 5°C or 10°C, there would still be a few days in the winter with snowfall and, even, snow blizzards.

Myth 4: The average global temperature data are inaccurate. We need several more decades of data to draw accurate conclusions.

Reality: While the average global temperature data have significant uncertainty, accurate statistical analyses of the data have shown that the long-term, increasing trend is correct and significant. There is more than 95% statistical confidence that the atmospheric temperature increased by 0.55–0.67°C in the twentieth century [23]. In addition, there is the cause-and-effect relationship between GHG and the planetary average temperature, which supports the conclusion that the recent increases of GHGs (the "blanket that keeps the planet warmer) will, in the future, cause even higher temperatures and GCC.

Myth 5: Hurricanes such as Katrina, Sandy, Harvey, and Irma happen because of global warming. If we reduce the CO_2 buildup in the atmosphere, these disasters will stop.

Reality: While several GCMs predict the intensification of tropical storms—including cyclones and hurricanes—these storms are weather phenomena and do not signify climate change. Hurricanes and strong tropical storms have been recorded in the last five centuries, before any evidence of global warming was detected, and will continue to occur even if we reversed the warming trend.

Myth 6: All we need to do in the United States to remedy GCC is to levy a small tax on the coal industry.

Reality: Given the current global structure of the electricity production industry, a small tax on coal will not significantly reduce the CO_2 emissions in the United States or any other industrialized country. A significantly high tax on coal in the United States alone will not only make wind and solar energies more competitive, but will also have an adverse impact on the industry and the citizens of the country that consume a great deal of energy. In addition, a coal tax in the United States alone will hardly make a dent on the global CO_2 emissions if other countries do not take similar measures for their own CO_2 emission reductions. The GCC and CO_2 emissions are global problems. Curtailing the CO_2 emissions in a single country or a small group of countries will not mitigate the GCC problem without an enforceable global accord on the global reductions of CO_2 emissions. The failure of the *Kyoto protocol* to reduce the global CO_2 emissions—despite the reductions in the EU, Japan, and other nations—is a proof of this.

Myth 7: We should not rush to take any measures about CO_2 emissions because scientists still disagree on global warming.

Reality: The overwhelming majority of recognized experts on climate change agree that the earth's temperature is increasing and have urged the political establishment to take action. The few "scientists" that have different opinions are either unfamiliar with climate models or very closely connected to the CO_2 emissions industry and have conflicts of interest to voice unbiased opinions.

Myth 8: Stabilizing the world's climate will require high-income countries to reduce their emissions by 60–90% from the 2006 levels by 2050.

Reality: In the absence of significant technological breakthroughs with renewable energy sources, such drastic energy reductions in the near future will ruin the economies of the "high-income countries," and their citizens will be unable to maintain their "high-income" status. Citizens of the high-income countries will never condone unilateral CO_2 emissions reductions of this magnitude without commensurate concessions and reductions from "lower-income countries." Another practical barrier here is that such a drastic CO_2 reduction will adversely impact air-conditioning and automobile transportation. Life and the economic activity in several regions—e.g., the southwest United States, the Middle East, Japan, and South China—will become very difficult if not impossible without the energy-intensive air-conditioning. GCC is a global threat, and *all* countries, wealthy and poor, should contribute—by a smaller or larger measure—to its solution.

Myth 9: The consumers will benefit with cheaper electric energy when fossil fuel power plants are replaced by solar and wind power.

Reality: The main reasons fossil fuels are used for the production of electric energy are low cost and reliability for the production and distribution of electricity. An electric grid based on renewables with the same degree of reliability will supply more expensive electricity to the consumers. This has been the experience of consumers in several European countries that converted to a higher mix of renewables in the period of 1995–2015 (Figure 9.5). In addition, an electric grid with a very high fraction of electricity supplied by renewables must include a high capacity for energy storage (Chapter 7) for reliability, a costly process that also dissipates a great deal of the electric energy produced.

Myth 10: We should all plant trees to offset our carbon footprint.

Reality: There is not enough available land for this option. One fully grown pine tree (50′ tall, 12″ trunk diameter, 30′ canopy diameter) with its root system weighs approximately 2,000 lb (909 kg) and contains 800 lb (364 kg) of carbon (C). If all this carbon were absorbed from the atmosphere, 2,933 lb (1,333 kg) of CO_2 would be eliminated. As one may see in Figure 3.11, in 2012, approximately $36,000 \times 10^9$ kg of CO_2 was emitted from anthropogenic activities. Even if we offset one-half of this global carbon footprint, we would need to plant 13.5 billion trees per year. This enormous planting effort requires an area of 337,584 km² per year, which is approximately equal to the area of a country such as Finland or Germany. Reforestation at this scale is clearly unsustainable. In addition, good fertile land should be used to feed the increasing global population not to offset our carbon footprint!

Myth 11: Regenerative agriculture is the solution to global warming. Mix and bury the crops with the soil, and the carbon is captured.

Reality: Crops remove a small part of the CO_2 from the atmosphere, and when they are mixed with soil, their carbon content is buried. However, the typical agricultural crops contain a very small amount of carbon, and even if this practice becomes widespread, the amount of CO_2 sequestered will be a very small fraction of the annual CO_2 emissions. An added hazard to the practice of burying agricultural products is that in the absence of oxygen or with oxygen shortage, they undergo *anaerobic decomposition*, which produces methane CH_4. This gas is 21 times more potent GHG than CO_2 (Table 3.1) and will cause more damage than the removed CO_2. Regenerative agriculture is a very good and environmentally friendly method for food production, but is not a solution to the GCC problem.

Myth 12: We use CO_2 products in our lives every day. We can sequester all the CO_2 and sell it in the market.

Reality: While several commercial products—e.g., carbonated drinks, dry ice, and Styrofoam—contain CO_2, the annual consumption of these products is not sufficient to absorb the very large amounts of the gas emitted, which are shown in Figure 3.11. The yearly emissions of CO_2 are close to 36,000 million t, while the amount of the gas used for commercial products is estimated to be in the range of only 50–60 million t (0.15–0.16% of the total emissions). Simply, there is no market for the enormous amounts of CO_2 we produced daily.

3.4 Nuclear Waste

Nuclear power plants produce a great deal of radioactive waste that must be safely stored for centuries or even for millennia. Any uncontrolled release of radioactive chemicals is harmful to all the life in the affected region of the planet. The production of nuclear waste is a significant environmental and ecological threat. In 2015, there were 437 nuclear reactors in the world with 69 more in the construction stages; 103 of these reactors operated in the United States. Heat in a nuclear power plant is produced by the fission (breakup) of the nuclear fuel, primarily uranium-235 with small contributions from plutonium-239 and uranium-238. The fission of the radioactive materials produces a plethora of other isotopes, most of which are radioactive. The typical nuclear reactor is a closed system, where all the fuel is stacked and operates without any outside interference for 18–24 months. Throughout this period of operation, the nuclear reactor produces tons of radioactive products, which are extracted during the next refueling stage and must be stored as radioactive waste. In addition to the reactor products, nuclear waste is generated in industrial facilities that produce, refine, and reprocess the reactor fuel. Table 3.3 shows a few of the isotopes that make the nuclear waste, their half-lives, and the level of radioactivity in becquerels (Bq) (1 Bq is one disintegration per gram per second). More details on the half-lives of isotopes, the theory of nuclear reactions, and the nuclear power plants are given in Chapter 5.

It is apparent from Table 3.3 that nuclear waste continues to be radioactive and poses a health threat to the human population for millennia. Permanent nuclear waste storage facilities must be constructed that are reliable and will be capable to safely store it for thousands of years, until the final residue does not pose a public health threat. This presents a significant scientific and engineering problem, simply because of the timescale of the storage: there is not a technically proven and reliable method for the storage of the isotopes during the thousands or tens of thousands of years required for their remediation. Any accidental or intentional (e.g., by an act of terrorism) release of radioactive materials from the storage sites may render entire regions uninhabitable.

The safe and reliable storage of nuclear waste is an environmental issue of paramount importance. However, the public and the governments of several countries, including the United States, have not come to grips with the full magnitude of this problem and have not prepared permanent storage facilities for the nuclear waste that has been continuously produced since the 1950s. Nuclear waste is usually stored in temporary facilities in the vicinity of the reactors that produced it. A typical temporary storage facility is a large pool of water, where canisters filled with radioactive waste are immersed. Any heat produced by nuclear reactions in the waste is taken up by the water in the pool. These storage

TABLE 3.3

Nuclear Waste Isotopes and Their Characteristics

Isotope	Half-Life, Years	Radioactivity, Bq
Americium-231	433	11.84×10^{10}
Americium-234	7,900	0.7×10^{10}
Iodine-129	17,000,000	5.9×10^{6}
Plutonium-239	24,400	0.23×10^{10}
Plutonium-240	6,600	0.81×10^{10}
Technetium-99	210,000	6.29×10^{8}

arrangements are temporary, mostly unsecured, and pose a threat to the surrounding communities.* The permanent disposal of nuclear waste in safe and controlled sites is of paramount importance.

Before the general public or a nation accepts the reliability of sites for the long-term storage of radionuclides, the long-term structural integrity of the containers, where the nuclear waste is stored, must be addressed. In the long run (hundreds or thousands of years), metal and composite tanks degrade by corrosion and other chemical reactions. Cracks in the tanks are developed and the radioactive waste leaks. The release of heat from the decaying radionuclides only accelerates the deterioration of the containers. At present, there is no known storage material that may be reasonably expected to preserve its structural integrity and keep the nuclear waste confined for the long periods required for the remediation of the waste. Several methods for the confinement of nuclear waste, at least for the next few hundred years, have been adopted and involve the stages that follow in the next two subsections.

3.4.1 Initial Treatment of the Waste

The initial treatment of the nuclear waste aims at reducing the intensity of the radioactivity and the immobilization of the radioactive isotopes within another stable solid material, such as glass or concrete. Methods currently used for the initial treatment are [22] as follows:

1. *Concentration* of the waste, where its volume is reduced by concentrating it into a smaller volume, which may be disposed of or stored better and more economically: Flocculation (agglomeration of fine particles) with ferric hydroxide is often used to remove highly radioactive metals from the aqueous solutions of the nuclear waste. The remaining low-level radioactive materials are stabilized and immobilized by mixing with ash and cement to form concrete blocks that may be stored anywhere, because they do not pose a severe threat to the environment and the population. The removed high-level radioactive waste must be disposed of by another method, usually vitrification.

2. *Vitrification,* or *glassification,* of the high-level radioactive waste is the most common method for stabilization before storage: The nuclear waste is mixed with sugar and heated until all the water and nitrates in the waste are evaporated. The mixture is then combined with glass and heated to a higher temperature until the glass melts. This melt is poured into stainless steel containers, where it solidifies and forms a glass-like substance. The radioactive chemicals are trapped in the vitrified substance and stored in the steel cylinders. Since all vitrified materials are very stable and do not easily leak, it is considered that the treated waste may be reliably stored for several hundreds or even thousands of years.

3. *Production of Synrock,* which is a complex chemical material similar to concrete: The components of Synrock bind and immobilize the isotopes in the nuclear waste in physicochemical bonds that are similar to the bonds in vitrification.

---

* The 2011 nuclear power plant accident in Fukushima Dai-ichi, where the stored nuclear waste was exposed and contributed significantly to the pollution in the region underscores this environmental problem.

3.4.2 Long-Term Disposal

After the initial treatment, the long-term disposal of nuclear waste includes one or more of the following proposed methods:

1. *Geological disposal*, either in deep and stable formations in the crust of the earth or in the deep sea: The proposed *Yucca Mountain* repository in the United States and the *Schacht Asse* repository in Germany, which briefly operated in the 1990s, are two examples of such ground disposal sites. The nuclear waste repositories are located in stable and arid geological formations, where water leakage is not a problem.

2. *Transmutation* is the transformation of radionuclides to other materials that are not radioactive: Special nuclear reactors are needed for the transmutation processes. In the United States, research activity on the transmutation has ceased since the late 1970s, because plutonium, which is used for the production of nuclear weapons, is a by-product of the process. Relevant research work has continued in the EU, where the reactor *Myrrha* is planned to be completed in Belgium after 2017. The reactor will be used for research projects and for the transmutation of isotopes in nuclear waste.

3. *Waste reuse* usually accompanies the concentration process, which was described in the previous section: The produced high-level radioactive materials are reused in existing nuclear reactors for the production of additional power. Because a great deal of the current nuclear waste is the isotope uranium-238, this isotope is separated from the waste to be used in the breeder reactors of the future.

4. *Space disposal* is an alternative that has been advocated by a few nonexperts: Given that it costs more than $25,000 to lift 1 kg of mass into the space, and at a very high-energy expenditure, this option is extremely expensive and not practical. Additional considerations related to the adverse effects of "space debris" to satellites and the international communications network make this a prohibitive option.

5. *Disposal in another, less populated country*: The sparsely populated Saharan nations have been proposed to receive nuclear waste from other countries (and one or two governments indicated they would accept to store nuclear waste for fees). However, this solution brings the ethical dilemma of whether or not a less wealthy society should accept the potentially harmful waste of an affluent society.

A legal problem associated with the disposal of nuclear waste is *stewardship cessation*. This legal term implies the shifting of the burden for the safe maintenance and perpetual management of nuclear waste from the producer to the entity that undertakes its transportation and storage. In most national legal systems, once the nuclear waste is removed from a site, the entity that removes and stores it has *stewardship* of it and becomes responsible for its safe transport, subsequent storage, and monitoring. Commercial entities—and governments of small nations—that undertake the interstate or transnational disposal of nuclear waste may not have the technical ability to reliably store the waste for long periods of time. If leakage occurs following the disposal and vast areas of land are contaminated, who is responsible for the containment and eventual remediation of the pollution? Under most national legal systems, the original owner/producer of the waste is immune and not legally responsible for any such remediation and for any compensation to the affected populations. Several environmental organizations oppose the transfer of stewardship and advocate the *perpetual stewardship*, which implies long-term management and monitoring of nuclear wastes, by their producers.

3.5 Thermal Pollution

All thermal power plants are subject to the second law of thermodynamics and reject a great deal of low-temperature heat to the environment. As explained in Section 1.3.3, a typical 1000 MW fossil fuel plant with an overall efficiency 40% receives Q_H = 2,500 MW of heat power from the combustion of fuel, converts 1000 MW of this to electric power, and rejects the remaining Q_L = 1,500 MW to the environment. For nuclear power plants that have lower efficiencies, close to 33%, the reactor produces approximately Q_H = 3,000 MW of heat, of which 1000 MW is converted to electricity and the remaining Q_L = 2,000 MW is rejected as waste heat to the environment. The vast amounts of heat power are rejected at low temperatures, typically in the range of 30–45°C.

A common misconception, even among scientists and engineers, is that the waste heat from power plants may be somehow used for the production of more power. This is impossible: The power plants are designed to produce the maximum possible amount of power and any heat that needs to be rejected is dictated by the second law of thermodynamics. The waste heat from power plants is at very low temperatures, and it is not possible to be of further use for the production of power. Small quantities of the waste heat may be used for the heating of buildings, for aquaculture or for agriculture—primarily heating of the soil to produce a higher yield or a second harvest. Economic considerations on the transport of heat considerably limit the amount of waste heat that is utilized. A few possible uses of the waste heat from power plants and the overall potential of waste heat utilization are discussed in the chapter on energy conservation and efficiency (Section 8.4).

In 2014, the annual global electricity production from thermal power plants—including coal, natural gas, oil, nuclear, and geothermal units—was approximately 19,000 TWh, which is equivalent to 68.4×10^{15} kJ [34]. At an average efficiency of 35%, these power plants reject 127×10^{15} kJ of heat to the environment annually. Additional *thermal pollution* is produced from the transportation industry in the form of the exhaust heat from the IC engines of automobiles, ships, and airplanes.

Thermal pollution is certainly a vast amount of energy released to the environment and causes concerns among a few environmentalists. However, this wasted energy is only a very small fraction of the total energy that enters the earth's atmosphere from the sun and of the energy that is radiated from the earth itself to the space. The annual energy received from the sun is equal to 5.46×10^{24} kJ, and the annual amount of heat radiated by the earth to the outer space is of a comparable magnitude. An order of magnitude comparison—the radiation from the sun is eight orders of magnitude higher than the waste heat produced in the entire planet—proves that thermal pollution does not contribute in any sizable measure to global warming and does not pose a threat to be such a contributor in the near future. The rate of heat that is absorbed and diffused in the atmosphere by GHGs is by six to seven orders of magnitude higher than the thermal pollution caused by the anthropogenic activities.

3.5.1 Energy–Water Nexus

Freshwater availability for the production of electric power and biofuels is fast becoming an environmental issue in the twenty-first century. Freshwater is a natural resource, which is not abundant in all the regions of the planet. Even though 71% of the surface of the planet is covered by sea, only 3% of the water on the planet is freshwater. However, even the remaining 3% is not available to humans: 68.7% of the freshwater inventory is in the form of ice glaciers, primarily in the polar regions, and another 30.1% is underground water [35]. The remaining 1.2% of the freshwater on the planet comprises permafrost and ground ice, 60.0%; lakes, 20.9%; soil moisture, 3.8%; swamps and marshes, 2.6%; and rivers, 0.5%. Of these, only the lakes, the rivers, and part of the underground water are available to be used for human activities. It quickly becomes apparent that the freshwater, which is available for human activities, is only a small fraction of the freshwater on or near the surface of the earth.

Heat rejection processes in thermal power plants make a significant claim on the freshwater resources of the planet. Steam power plants reject heat primarily from their condensers, which are cooled by closed or open circuits of *cooling water*. The schematic diagram of a power plant cooling process, known as evaporative cooling, is shown in Figure 3.12. Colder water enters the condenser, warms up as it removes the waste heat from the condensing steam, and enters as warmer water a wet cooling tower. Part of the warmer water evaporates by the airflow in the cooling tower, and this process cools the remaining volume of the water. The colder water from the cooling tower is circulated back to the condenser and the cyclic process repeats.

The latent heat of evaporation h_{fg} of the water is approximately 2,400 kJ/kg. Assuming that all the waste heat $\dot{Q}_L$ is released to the environment solely by evaporation, the amount of water that evaporates $\dot{m}_{ev}$ is

$$\dot{Q}_L = \dot{m}_{ev} h_{fg}. \tag{3.6}$$

Therefore, the nuclear power plant that produces 1000 MW of power and rejects 2,000 MW of heat also causes the evaporation of approximately 833 kg/s of water or 72,000 t/day.

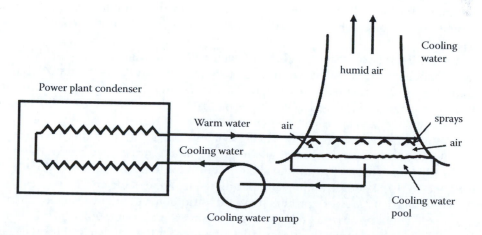

FIGURE 3.12
Evaporative cooling system of a thermal power plant.

On the average, approximately 3.6 kg (1 gal) of water evaporate for every 1 kWh of electric energy produced by a steam power plant.

An alternative heat dissipation method is the *once-through* cooling system, which eliminates the use of a cooling tower. Colder water uptake is supplied from the river or lake; circulates through the condenser, where it warms up; and is then discharged downstream. Once-through cooling systems are simpler to construct and are favored for larger steam power plants. This is the principal reason why large thermal power plants are usually built close to a natural source of freshwater. In the United States, 42 of the 103 nuclear reactors in operation are located near the banks of the Mississippi River or one of its larger tributaries. Most of the other nuclear power plants are located next to the Great Lakes and other large rivers. In France, 85% of the nuclear power plants are located near the banks of the Seine, the Loire, and the Rhone Rivers.

The once-through cooling process uses very high amounts of circulating freshwater, because the temperature rise ΔT between the colder and the warmer water is constrained by environmental regulations to be in the range of 5–12°C [36]. A heat balance in the condenser yields the following equation for the rate of waste heat rejection $\dot{Q}_L$:

$$\dot{Q}_L = \dot{m}_{cw} c_p \Delta T, \qquad (3.7)$$

where $\dot{m}_{cw}$ is the mass flow rate of the cooling water that is needed and c_p is the specific heat capacity of water (4.18 kJ/kg K). A quick computation proves that a typical nuclear power plant, which rejects approximately 2,000 MW of heat, would need close to 48,000 kg/s of circulating water when $\Delta T = 10°C$. The same power plant needs twice as much water (96,000 kg/s) if ΔT is constrained to 5°C. Because the plant effluent is warmer, a small fraction of the once-through water evaporates in the river or the lake after it is discharged.

A second energy sector with very high demand for freshwater is the production of biofuels derived from plants and agricultural products. Agricultural processes are water intensive and consume most of the freshwater supply on the planet for the production of foodstuff. The production of a single litter of ethanol from corn requires approximately 1000 L of freshwater. The large-scale production of biofuels from trees, corn, sugar cane, and switchgrass, creates higher demand for the water supplies worldwide. Naturally, the higher demand for biofuels will divert some of the water from agriculture, and the biofuels will be in direct competition with the production of foodstuff from the available water resources.

While the availability of freshwater has not been an issue in the past, it is quickly becoming a significant environmental and political issue in the twenty-first century. With the rise of the planet's population and the desired increase in the standard of living in all countries, there is a higher demand for freshwater for agricultural and domestic uses. Freshwater is becoming scarce in several parts of the planet, as for example, in the southwestern part of the United States, the Middle East, and several regions of Asia and northern Africa. High-volume water consumers, such as large steam power plants, must compete for this natural resource. In the near future, the human society will have to make hard decisions on how to better allocate this resource among competing users. Renewable energy sources for the production of electricity are very promising in this situation, either because they use direct conversion and do not produce a great deal of waste heat—e.g., photovoltaics, wind, hydroelectric, tidal—or because they produce sufficient water for the needs of their own cooling systems—e.g., geothermal.

3.5.2 Effects on the Aquatic Life

The rejection of very large amounts of heat in rivers and lakes causes the temperature of these water bodies to increase, at least locally. In extreme cases, this may cause the extinction of sensitive aquatic species. In usual warming cases, the temperature increase has the following detrimental effects on the aquatic environment and the aquatic ecology:

1. Decrease in the dissolved oxygen concentration
2. Increase in the rates of other chemical reactions
3. Changes in the behavior, rates of reproduction, and growth patterns of organisms that disturb the food chain in the ecosystem

Environmental regulations have been adopted in most countries to limit the increase in the cooling water temperature and the discharge of harmful chemicals in the biosphere. Typical temperature increase figures from these once-through cooling systems within the boundaries of the power plant are 5–12°C. The water temperature significantly falls as it mixes with the water of the river or the lake [36]. Even these rather small temperature variations may cause stress on the aquatic ecosystems.

A detrimental effect to lakes from thermal pollution is *stratification*, which is caused by withdrawing colder water from the bottom of the lake and adding warmer water to the top layer. Because the warmer water is lighter, it remains at the surface and restricts mixing by convection. This process impedes the natural mixing of the layers of the lake, the transfer of nutrients from the bottom of the lake (hypolimnion) to the top layer (epilimnion), and the transfer of oxygen to the bottom layer. In addition, water taken from the bottom layer brings to the top algae nutrients, phosphorus, and nitrogen. Several species of algae, especially those that are tolerant to high-temperatures, may increase uncontrollably at the top layer of the lake. Some of these algae are toxic to fish, and when they die, the organisms that decompose them compete for the available dissolved oxygen with the remaining fish population. The result is a (sometimes severe) decrease in the fish population at the top layer of the lake. The detrimental effects of the stratification process may be intensified if the lake receives high quantities of algae nutrients from agricultural waste or sewage discharge that stimulate the uncontrolled growth of algae. This series of processes is called *lake eutrophication*.

3.5.3 Myths and Reality Related to Water Use

Myth 1: Water has become a very scarce commodity.

Reality: What is often meant by this statement is that water is not free to be wasted on unprofitable activities. A "scarce commodity" is one with limited supply and significantly higher demand. A good that periodically falls freely from the sky, as rain, is not scarce at all. One of the characteristics of a scarce commodity is high price (e.g., the prices of gold, diamonds, rare earth metals). The price of freshwater, wherever water is sold as a commodity, is not high at all—it is typically $1–2 per cubic meter, and this cost mainly covers the cost of water transportation and distribution. The desalination of seawater produces freshwater, which may be distributed at slightly higher cost. A country or a region may produce unlimited quantities of freshwater from seawater and rainwater, but at a cost that will make several of the wasteful uses of water unprofitable.

A different type of water scarcity, *economic water scarcity* is a term occasionally used by the UN agencies for the lack of infrastructure to provide clean, potable water to the human population within a region. This is entirely different from physical water scarcity. Economic water scarcity is easily relieved with current technology by engineering infrastructure projects that purify and distribute potable water.

Myth 2: The United States will run out of water in the twenty-first century.

Reality: There is no indication that the rainfall in the United States is decreasing and that water will become scarce in this or the next century. There is sufficient rainfall to maintain the flow of the major rivers and satisfy all reasonable water uses, including agriculture. In addition to the large rivers, the United States and Canada share the five Great Lakes, which constitute a giant natural reservoir that contains 70% of all the freshwater on the surface of the earth. The United States is one of the better-endowed countries with freshwater and will not run out of it in the twenty-first century.

Myth 3: The next wars will be fought not for petroleum but for water.

Reality: A glance at the world map will prove that the relative lack of freshwater is a regional problem that is not contained by national boundaries. If a country does not have enough freshwater, it is very likely that its immediate neighbors suffer from the same problem. To prove that this is a myth, one should look at the world map and ask the question: Which two countries on this map will fight for water, and how will the winners of this war transport the water to their country?* Unlike oil, which is expensive and transportable by large tankers at low cost relative to its price, water may be produced inexpensively, but it costs to transport large quantities of it. The transportation of vast amounts of water (the plunders of war) will cost more than the water production by desalination or purification. If a country has the financial means to transport vast amounts of water by tankers, they may do so without a war: simply send the tankers to a port at the mouth of a large river—e.g., New Orleans on the Mississippi and Rotterdam on the Maas—and fill them with free freshwater to be transported anywhere.

Myth 4: Air is a better coolant for large power plants than water.

Reality: While air may also be used as a coolant for fossil fuel power plants as well as for large engines, water is a superior coolant because

- The specific heat capacity of water is four times higher than that of air (4.184 kJ/(kg·K) vs. 1.005 kJ/(kg·K)). The ratio of the volumetric heat capacity (in kJ/(m³·K)) of water is 3500 higher than that of air.

- The thermal conductivity of water is 26 times higher than that of air (0.63 W/(m·K) vs. 0.024 W/(m·K)).

- Water is almost incompressible. Large volumes of water may be pumped to cool the condenser of a power plant at the expense of small quantities of electric power. The fans that will move the equivalent mass of air require significantly more power.

- The ambient air temperature is usually higher than the ambient water temperature. This implies a higher condenser temperature and lower efficiency for the air-cooled plants.

The few power plants that are currently cooled by air (dry cooling systems) have significantly lower thermal efficiencies than the equivalent water-cooled plants.

---

* One of the most absurd notions is that a Saharan country declares war on Germany or Holland and transports the water of the Rhine or Maas River to the Sahara.

Myth 5: Water cannot be transported over long distances.

Reality: Hydraulic engineering and pump design have sufficiently progressed for this statement to be false. Water pipelines and a series of pumping stations may easily transport water for very long distances over land or sea, but there is a cost associated with this transportation. If the transportation distance exceeds 300–400 km, it is more likely that the desalination of seawater or the construction of rainwater saving projects will produce freshwater at lower cost.

Myth 6: When we convert natural resources into usable forms of energy, we degrade the environment.

Reality: This is a true statement! The production of all energy forms—even the renewables solar, wind, geothermal, and hydroelectric—entail a degree of environmental degradation and ecological disruption. However, considering the paramount necessity of energy for our economy and the functioning of our society, it is absolutely important to produce the energy we need (and to not waste this energy). While producing sufficient energy for the societal needs at present, we should be cognizant of the vital needs of future generations and try to harness our energy resources with minimum environmental and ecological impact.

3.6 Energy Sustainability and Carbon Footprint

The concept of *sustainability* has been used in various disciplines—e.g., biology, architecture economics, environmental science, engineering, and sociology—and with a variety of contexts and meanings. Several individuals and organizations have attempted to articulate a general definition of sustainability and the principles that govern sustainability with limited success. The concept has different meanings and different principles in each of the several disciplines it has been applied. In the case of energy and the environment, sustainability may be framed in the context of *intergenerational ethics* under the principle that one generation should pursue economic development without jeopardizing the living space (the environment) of the following generations. In this context, sustainability implies that regional and global economic developments that make use of primary energy forms should be pursued without causing irreparable damage to the ecology and the environment. The origins of this notion of sustainability are not in scientific principles, but in ethics.

Because the economic activity is intricately connected to energy production and use, sustainability includes all the global economic activities from the production of goods and services to the transportation and energy production. The related concept of *sustainable development* includes politics (the regulatory branch), economics, ecology, and the local and regional cultures and may be summarized as *development for the present generation without diminishing the ability of future generations to provide for their own needs*. Adhering to the simple principles of sustainable development will help avoid a Malthusian future, where the human population and cultures are extinguished due to lack of food and other resources [37].

Because it is apparent that CO_2 emissions threaten not only the ecology and the environment of humans, but also their economic activities, the *carbon footprint* has become a popular measure to assess the environmental effect of human activities. The carbon footprint is defined as the amount of CO_2 produced by all the engines and processes that are used for the completion of an economic activity. A simple example for the calculation of the carbon footprint is transportation by a car. Driving 1000 km in a small car with a mileage of 30 km/L of gasoline (at the consumption of 33.3 L or 23.3 kg of gasoline) is 73 kg of CO_2, while the carbon footprint of the same trip with a 5 km/L in an SUV is 437 kg of CO_2. Examples for the calculation of the carbon footprint are given at the end of this section. It must be noted that the carbon footprint is the most popular but not the only measure to assess environmental impact. Table 3.2 effectively represents the annual carbon footprint (tons of CO_2 emitted per capita) of several countries. Similar metrics—e.g., the methane footprint, the SO_2, and the water footprint—have been adopted for the emissions of other pollutants as well as for the usage of water.

Central to the subject of energy/environmental sustainability is the realization that significant global threats, such as global warming and pollution prevention, may be tackled only by a combination of technological advances, social awareness, and public policy. The emergence of sustainability as an environmental discipline happened in the late twentieth century as a response of the environmental community to the unchecked effects of rampant economic development and energy-related anthropogenic activities. It has been realized that many energy-related activities, including the combustion of hydrocarbons and carbon cannot be sustained indefinitely: At first, there will be a point in the future when the last carbon atom will have burned and fossil fuels will not be an option to satisfy the energy demand of the humanity. Secondly, even before this point is reached, rising temperatures and seal level may render uninhabitable several regions of the planet. Therefore, concerted remedial measures must be taken to (a) provide humans with energy from sources that can be sustained in the future and (b) reverse the current adverse environmental effects and ensure that energy-related human activities will not endanger the environment and ecosystems for future generations.

It has become apparent that the continuation of the current practices on energy consumption and the *laissez faire* or *market* energy policies are not sustainable in the long run. If continued to be unchecked at the currently growing levels, such practices will inevitably bring environmental disasters that will be followed by agricultural crop failures and long-term disruptions of our socioeconomic and cultural activities. The long-term sustainability of the human economic activities and their relationship to human society is a subject that needs careful consideration.

Sustainability as an academic subject is still undergoing evolution, its definitions are mainly subjective, and its metrics and principles are still debated among scientists of different disciplines and policy makers. The global adoption of at least the most sensible of these principles, especially in the energy field, appears to be a reasonable and realistic way for the sustained global economic expansion and prosperity in the long run. For the long-term environmental health and preservation of ecosystems, and for the long-term sustainable economic development of nations, several human economic activities must be reengineered to ensure that their effects on the environment will not impede the progress of future generations.

It must be noted that sustainability does not necessarily advocate the banning of all emissions and all environmental pollutants. Rather, it supports an economic system with gradual pollution reductions through higher efficiency. It also supports countermeasures for every pollution emission action. Thus, pollutant emissions are treated as economic phenomena that must be counteracted by other activities so that their long-term effect on the environment is neutral. For example, if an activity emits a quantity of CO_2, a countermeasure (offset) must follow that removes the same quantity of CO_2 from the atmosphere.

While promoting sustainability countermeasures and actions that will preserve the environment, one has to be aware that these actions themselves must be sustainable, that they do not use other vital resources for future generations, and that they may be implemented on a large scale. Performing quantitative analyses for the wide applicability of such countermeasures always helps to ensure that they are sustainable in the long run. Two examples of such actions that are not sustainable when applied on a large scale and in the long run are given in the following:

1. Consider the operation of the SUV with mileage of 5 km/L that was mentioned earlier, which is driven for 1000 km and its footprint is 437 kg of CO_2. The owner of the SUV may plant a tree that will absorb this amount of CO_2 from the atmosphere or may support someone else who will undertake to complete this environmental offset. Actually there are several organizations and websites that encourage and undertake such actions (for a fee or donation). If the tree grows to become a mature tree of approximately 500 kg, it would have removed enough CO_2 from the atmosphere to counteract the equivalent of two trips of 1000 km. One may drive for about 2,000 km in this SUV and all the CO_2 that is emitted would be counteracted (offset) by the planting and growth of this tree. Similar offset measures may be taken in every field to remove the environmental effects of all activities. However, a closer look at such offsetting practices proves that their large-scale application is unsustainable because the planting of trees requires large areas of land, which is an agricultural resource. If every driver on the planet were doing the same, very soon, this action would compete with agricultural land, which is better used for food. The debunking of myth 10 in Section 3.3.10 provides a simple, quantitative argument that such a practice is not sustainable in the long run: If we tried to offset 50% of our annually produced carbon footprint, we will need to plant 13.5 billion trees per year, a countermeasure that requires land of approximately the area of Finland or Germany. If this practice is continued for 10 years, the entire Europe with the exception of Russia would need to be covered with trees. Reforestation at this scale is clearly unsustainable. The reduction of the carbon footprint through conservation and high efficiency measures rather than offsetting remediation is a better path to sustainability.

2. A rather extreme notion of sustainability actions that is advocated by a few in the early twenty-first century is a "return to the fundamentals," where the society goes back to its agrarian roots; individuals produce their own food and energy by becoming farmers and herders and leave the urban centers to withdraw in simple farms, where they grow all they need including their energy. A few of these advocates have already moved to farms and rural communities. However, this is not a solution to the global environmental problems and is not a sustainable action for the current population of the planet, which exceeds 7.3 billion humans. A small, family farm that would grow food as well as fuel for the needs of a family of four persons is approximately 150 acres (60 ha).* In most of the countries, there is simply not enough arable land for each family to withdraw and own such a small farm. For example, in the United States—one of the least densely populated countries of the world—such a societal system would support only 21% of its current population, even assuming that the entire land, including the Rocky Mountains and the deserts will become arable. The current land mass of PRC would support less than 5% of its current population. The entire area of Belgium—one of the most densely populated countries in the world—would support a mere 1.9% of its

---

* Most of this land is needed for the energy needs of the modern family, not for the growing of food.

current population. Clearly, the planet earth is too small for all the humans living in the twentieth century to return to not only a simpler, agrarian, and more sustainable, but also more inefficient economic system.

A more realistic alternative for sustainability, and one that encompasses the entire current population of the planet, is the wider use of alternative energy sources, combined with increased efficiency and energy conservation. The use of alternative energy sources, including nuclear energy, is fundamental to tackling several pollution problems, most important of which is the continually increasing global CO_2 emissions. As it becomes apparent from example 3.5, the substitution of five coal-fired power plants by two 1000 MW nuclear reactors will have the net effect of removing 20.43 million t of CO_2 per year from the atmosphere annually.* Similarly, the substitution of a 60 MW gas turbine for peak power generation that operates for 20% of the year with solar power will have the effect of removing 20,400 t of CO_2 annually. Most importantly, these actions may be repeated several times and in several countries to prevent the emission of hundreds of millions of tons of CO_2. The extensive utilization of the global hydroelectric potential, tidal, and wind powers and the more widespread use of more efficient and smaller electric cars avert the further emissions of CO_2, SO_2, NOX, and other pollutants from an environment which is currently inhabited by more than 7.3 billion humans. Electric energy production from alternative sources, energy conservation measures in buildings and transportation, and higher efficiency of industrial processes are the most effective and long-term solutions to achieving global sustainable development.

Example 3.9

What are the annual carbon footprints of an SUV (10 mpg) and a hybrid car (45 mpg) that travel 18,000 mi/year?

Solution: It will be assumed that the two vehicles use gasoline, which is modeled as octane (C_8H_{18}). Since the density of octane is 6.093 lb/gal, the SUV consumes 10,967 lb or 4,985 kg of C_8H_{18}. The corresponding numbers for the hybrid car are 400 gal, 2,437 lb, or 1,108 kg of C_8H_{18}.

The combustion equation of octane is $C_8H_{18} + 12.5O_2 \rightarrow 8CO_2 + 9H_2O$, which implies that 114 kg of C_8H_{18} produces 352 kg of CO_2.

Therefore, the SUV produces 15,392 kg of CO_2 annually and the hybrid car produces 3,421 kg of CO_2 annually. These are the annual footprints of the two vehicles. It is apparent that if a commuter switches from the SUV to the hybrid car, his/her annual footprint will be reduced by 11,971 kg of CO_2.

Example 3.10

A residential building is located at a region where 62% of the electricity is produced by carbon combustion at an average efficiency 38%. The annual electricity consumption of the building is 26,000 kWh. Determine (a) the heat from carbon that corresponds to this amount of electricity; (b) the amount of carbon in kilograms, used annually for the production of this electricity; (c) the carbon footprint of the building; and (d) by how much would the carbon footprint be reduced if through energy conservation measures, the electric energy use of the building dropped to 18,000 kWh?

---

* This will create more nuclear waste, for which storage technologies exist but are not effectively used at present.

Solution:

a. Of the 26,000 kWh used, 26,000 × 0.62 = 16,120 kWh is produced from carbon combustion. This corresponds to 58.03 × 10^6 kJ of electric energy. At an efficiency of 38%, the production of this energy is corresponds to 152.7 × 10^6 kJ of heat supplied to the power plants.

b. Since the heat of combustion of carbon is approximately 32,800 kJ/kg, this heat is produced by the combustion of 4,655.4 kg of carbon.

c. From the combustion equation for carbon $C + O_2 = CO_2$, 12 kg of carbon produces 44 kg of CO_2. Therefore, the 4,655.4 kg would produce 17,070 kg of CO_2, and this is the annual CO_2 footprint of the building.

d. If the annual electricity consumption dropped to 18,000 kWh and everything else remained the same, the annual CO_2, footprint would be proportionately reduced from 17,070 to 11,818 kg CO_2.

PROBLEMS

1. A type of anthracite contains 90% carbon by weight and 0.5% sulfur. How much CO_2 and SO_2 are produced from the combustion of 1 metric ton of this anthracite?

2. How much CO_2 (in lb) is produced from the combustion of 1 Mcf (1,000,000 ft^3 at standard conditions) of propane (C_3H_8) gas? The density of propane at standard conditions is 0.1175 lb/ft^3.

3. How much CO_2 is produced from the combustion of 1 Mcf methane at standard conditions? At standard conditions, methane is an ideal gas.

4. A 400 MW electric power plant with an overall efficiency 38% uses bituminous coal, which contains 70% carbon and 2% sulfur with the rest being volatile matter and ash. The heating value of this coal is 26,500 kJ/kg. Determine (a) how much heat the plant needs annually, if it operates continuously; (b) how much of this coal does the power plant use daily and annually; and (c) how much CO_2 and SO_2 does the power plant produce annually.

5. Three coal power plants with a total power producing capacity of 1000 MW and an average thermal efficiency 36% are substituted by one nuclear power plant with 33% thermal efficiency. What is the annual amount of CO_2 that is not emitted to the atmosphere? What is the increase in the waste heat produced? The heating value of the coal that was used is 28,000 kJ/kg and the coal contains 80% carbon.

6. Three coal power plants with a total capacity of 1000 MW and an average thermal efficiency 36% are substituted by 10 smaller gas units that use methane. The new units have an average thermal efficiency of 43%. What is the annual amount of CO_2 that is not emitted to the atmosphere because of this substitution? Assume that the heating value of the coal is 30,000 kJ/kg and contains 90% carbon. The heating value of methane is 50,020 kJ/kg.

7. The coal power plant of problem 4 is fitted with a sulfur abatement system that has 99.6% efficiency. How much of the SO_2 mass is removed by the abatement system and how much is released in the atmosphere?

8. What effect would the following parameters have on the long-term temperature of the earth's atmosphere? Write a short statement to explain your reasons.

   a. A decrease in the earth's core temperature

   b. An increase in the average cloudiness

c. A decrease in the earth's reflectance

d. An increase in the amount of atmospheric methane

e. An increase in the surface temperature of Venus

f. Producing 10% of the total electric power from solar cells

9. "The melting of the polar ice caps will be a major environmental calamity because it will increase the average temperature by 10.41°C." Comment by writing a short (250–300 word) essay.

10. Three FBRs consume bituminous coal with 65% carbon, 2.3% sulfur by weight, and heating value of 24,000 kJ/kg. The FBRs supply with heat power a 600 MW coal power plant with 42% overall thermal efficiency. Determine (a) How much heat does the set of FBRs produce annually; (b) how much coal do they consume; and (c) how much $Ca(OH)_2$ must be supplied to the FBRs in order to remove the SO_2 produced?

11. A type of leaded gasoline contains 1.2% of tetraethyl lead by weight. How much lead oxide (PbO) is released in the environment with every gallon of this gasoline? Assume that all Pb is converted to PbO.

12. Calculate the amount of heat, in terajoules (10^{12} J), rejected annually from the following types of electric power plants assuming that they operate continuously.

a. A 400 MW coal plant with thermal efficiency of 40%

b. A 1000 MW nuclear power plant with thermal efficiency of 33%

c. A 35 MW geothermal power plant with 14% thermal efficiency

d. A 10 MW thermal solar plant with 18% thermal efficiency

e. An 80 MW natural gas power plant with 46% thermal efficiency

13. The cooling system of a 200 MW coal power plant with 40% thermal efficiency uses river water. It is desired that the temperature difference in the cooling water does not exceed 7°C. What is the mass flow rate of the cooling water in kilograms per second?

14. A 1000 MW nuclear power plant with a 32% thermal efficiency discharges its waste heat in a lake with 22 km^2 surface and 3 m average depth. If there is no other cooling effect for the lake, what would be the average increase in the water temperature annually? What factors would mitigate and nullify this temperature increase?

15. What is the carbon footprint of the following activities? For all, you will need to find the wattage of the pertinent appliances in your residence. In the case of electric appliances, you may assume that 70% of the electricity comes from coal power plants with an overall thermal efficiency 38%.

a. Watching television for 1 hour

b. Using a microwave oven for 10 minutes

c. Forgetting to switch off a 100 W light bulb for 12 hours

d. Driving for 2,000 mi in an SUV, which consumes 12 mpg

e. Driving for 2,000 mi in a compact car, which consumes 40 mpg

16. "The human society in its current state has failed to create a sustainable future for us and the next generations. Responsible citizens must abandon the cities and urban life to create a sustainable future for themselves and their children by

producing their own food and fuel from products they have grown for millennia and can depend on in the future." Comment by writing a short (250–300 word) essay.

17. "As the stewards of the Earth and its environment we must recognize the dangers of GCC and uniformly reduce our carbon footprint to the historical, preindustrial revolution levels." Comment by writing a short (250–300 word) essay.

18. "The postindustrial economic expansion is leading us to an environmental and economic disaster. For the survival of the human race, we must go back to the fundamentals, produce our own food, and leave in harmony with our surroundings and other human beings." Comment by writing a short (250–300 word) essay.

References

1. Gates, D.M., *Energy and Ecology*, Sinauer, Sunderland, MA, 1985.
2. EPA (Environmental Protection Agency), *SO_2 and NO_x Emissions, Compliance, and Market Analyses: Progress Report—2012*, EPA, Washington, DC, 2012.
3. EPA, *National Trends in Sulfur Dioxide Levels*, EPA website, 2014, https://www.epa.gov/air-trends/sulfur-dioxide-trends.
4. Environment Canada, *Canada–United States Air Quality Agreement Progress Report 2012*, http://ec.gc.ca/air/default.asp (website), 2012.
5. Vestreng, V., Myhre, G., Fagerli, H., Reis, S., Tarrason, H., Twenty-five years of continuous sulphur dioxide emission reduction in Europe, *Atmos. Chem. Phys.*, **7**, 3663–3681, 2007.
6. European Environmental Agency, *Sulfur Dioxide (SO_2) Emissions*, APE 001, European Environmental Agency, Copenhagen, 2014.
7. Smith, S.J., van Aardenne, J., Klimont, Z., Andres, R.J., Volke, A., Delgado Arias, S., Anthropogenic sulfur dioxide emissions: 1850–2005, *Atmos. Chem. Phys.*, **11**, 1101–1116, 2011.
8. EPA, *National Trends in Nitrogen Oxide Levels*, EPA website, 2014, https://www.epa.gov/air-trends/nitrogen-dioxide-trends.
9. EPA, *National Trends in Lead Levels*, EPA website, 2014, https://www.epa.gov/air-trends/lead-trends.
10. Farman, J.C., Gardiner, B.G., Shanklin, J.D., Large losses of total ozone in Antarctica reveal seasonal ClOx/NOx interaction, *Nature*, **315**, 207–210, 1985.
11. Molina, M.J., Rowland, F.S., Stratospheric sink for chlorofluoromethanes: Chlorine atom-catalysed destruction of ozone, *Nature*, **249**, 810–812, 1974.
12. Douglas, A.R., Newman, P.A., Solomon, S., The Antarctic ozone hole: An update. *Phys. Today*, **67**, 42–48, 2014.
13. Intergovernmental Panel for Climate Change, Climate Change 2007: Synthesis Report. Contribution of Working Groups I, II and III to the Fourth Assessment Report of the Intergovernmental Panel on Climate Change, IPCC, Geneva, Switzerland, 2007.
14. NOAA (National Oceanic and Atmospheric Administration) Global Monitoring Division, *Recent Global CO_2*, NOAA, Silver Spring, MD, May 2015.
15. Fisher, D.A., Hales, C.H., Wang, W.C., Ko, M.K.W., Sze, N.D., Model calculations of the relative effects of CFCs and their replacements on global warming, *Nature*, **344**, 513–516, 1990.
16. Seidel, S., Ye, J., Andersen, S.O., Hillbrand, A., *Not-in-Kind Alternatives to High Global Warming HFC's*, Center for Climate and Energy Solutions, Arlington, VA, October 2016.
17. EIA (Energy Information Administration), *Kigali Amendment to the Montreal Protocol: A Crucial Step in the Fight Against Catastrophic Climate Change*, Briefing to the 22nd Conference of the Parties (CoP22) to the United Nations Framework Convention on Climate Change (UNFCCC), Marrakech, November 2016.

18. EPA, *Inventory of U.S. Greenhouse Gas Emissions and Sinks: 1990–2014*, EPA 430-R-16-002, EPA, Washington, DC, 2016.
19. NOAA, *The Global Anomalies and Index Data*, NOAA, Silver Spring, MD, July 2015.
20. Hansen, J., Ruedy, R., Sato, M., Lo K., Global surface temperature change, *Rev. Geophys.*, **48**, RG4004, 2010.
21. UN (United Nations), *UN and Climate Change—Climate Summit 2014*, UN, New York, September 2014.
22. Michaelides, E.E., *Alternative Energy Sources*, Springer, Berlin, 2012.
23. IPCC, *Climate Change 2014: Impacts, Adaptation and Vulnerabilities*, IPCC, Geneva, 2014.
24. Mann, M., False hope—Earth will cross the climate danger threshold by 2036, *Scientific American*, 310, 4, April 2014.
25. Rahmstorf, S., A semi-empirical approach to projecting future sea-level rise, *Science*, **315**, 368–370, 2007.
26. Ballantyne, A.P., Alden, C.B., Miller, J.B., Tans, P.P., White, J.W.C., Increase in observed net carbon dioxide uptake by land and oceans during the last 50 years, *Nature*, **488**, 70–72, 2012.
27. Kharaka, Y.K., Cole, D.R., Hovorka, S.D., Gunter, W.D., Knauss, K.G., Freifeld, B.M. Gas–water–rock interactions in Frio formation following CO_2 injection: Implications for the storage of greenhouse gases in sedimentary basins, *Geology*, **34**, 7, 577–580, 2006.
28. Winson, T.R.S., The deep ocean disposal of carbon dioxide, *Energy Conv. Manag.*, **33**, 627–633, 1992.
29. Rochon, E., *False Hope—Why Carbon Capture and Storage Won't Save the Climate*, Greenpeace International, Amsterdam, 2008.
30. EIA, *Electric Annual Report 2013*, US Department of Energy, Washington, DC, 2015.
31. C2ES (Center for Climate and Energy Solutions), Outcomes of the U.N. climate change Conference in Paris, C2ES, Arlington, VA, December 2015.
32. Boden, T.A., Marland, G., Andres, R.J., *Global, Regional, and National CO2 Emissions, Trends: A Compendium of Data on Global Change*, ORNL Report, Carbon Dioxide Information Analysis Center, Oak Ridge, TN, 2012.
33. Sarmiento, S.L., Wofsy, S.C., *A US Carbon Cycle Plan*, US Global Change Research Program, Washington, DC, 1999.
34. International Energy Agency, *Key World Statistics*, IEA-Chirat, Paris, 2016.
35. US Geological Survey, *The World's Water*, US Geological Survey, Washington, DC, 2015.
36. Madden, N., Lewis, A., Davis, A.M., Thermal effluent from the power sector: An analysis of once-through cooling system impacts on surface water temperature, *Environ. Res. Lett.*, 8, 035006, 2013.
37. Malthus, T.R., *An Essay on the Principle of Population as it Affects the Future Improvement of Society*, Johnson, London, 1798.

4

Fossil Fuels

It is difficult to perceive today's society without the use of fossil fuels. Fossil fuel combustion supplies more than two-thirds of the electricity generated in the world, almost the entire motive power used for transportation, more than 80% of the household energy for heating and cooking, and more than 90% of the energy for industrial processes and products. At the beginning of the industrial revolution, coal alone supplied the heat for the operation of the steam engines and locomotives. The use of fossil fuels expanded early in the twentieth century, when liquid petroleum products powered the internal combustion (IC) engines of automobiles, trucks, agricultural tractors, and, later, airplanes. At about the same time, a large network of pipelines brought natural gas to cities and supplied the households with a cleaner fuel for heating and cooking than wood and coal. All three types of fossil fuels—solids, liquids, and gases—have been used for the production of electricity, which provides power in a very controlled way to railways, computer networks, telecommunications, household lights, refrigerators, air-conditioners, sensors, etc. Fossil fuels have become an essential part and a necessity for our lives, and it is difficult to imagine contemporary life without them.

Fossil fuels were essential in ushering the modern industrial age. They have been necessary for the development and operation of heavy machinery for industry and agriculture. The extensive use of fuel-devouring engines liberated large numbers of humans from the necessity to produce their own food as farmers and was the catalyst in the development of the working class and the enlargement of the middle class. These engines helped usher the modern *affluent society* [1] and *the new industrial state* [2] with all the opulence and the problems attributed to them. With abundant power, fossil fuels provided the accelerated development of technology; the economies of most nations dramatically advanced (Figure 2.1); and the lives of the populations significantly improved and extended further (Figure 2.2). In the early twenty-first century, all nations rely on the use of fossil fuels to sustain and accelerate their economic development and to enrich the lives of their people.

The importance of the fossil fuels for the modern society and economies, notwithstanding the mining, transportation, and combustion of these fuels, has also brought significant environmental problems. Chief among these is the CO_2 emissions and the associated global climate change (GCC). An additional concern is that the amount of fossil fuels on this planet is finite, and since they are used at rates much higher than their replacement rates, it is expected that at some point in the future, the supply of fossil fuels will be exhausted. Before their depletion, a plan must be developed for the replacement of fossil fuels as the principal energy source and one of the chief catalysts for the economic development of nations.

4.1 Heating Value of Fuels

Fossil fuels contain large quantities of chemical energy, which is released as heat during combustion. The amount of heat released primarily depends on (a) the type of the fuel, (b) if the combustion is complete or part of the reactants do not burn, and (c) whether or not any water that is formed from the combustion of hydrogen is in the liquid or the vapor form. In most of the industrial combustion processes, the water/moisture produced is in the vapor form, as for example, in the vehicle exhaust gases, the exhaust of a jet engine, and the gas turbine exhaust. In such cases, the heat released during the combustion of a unit mass of fossil fuels is the lower heating value (LHV) of the fuel. If at the end of the combustion process, the water is condensed, a higher amount of heat is recovered, the higher heating value (HHV). Sometimes, the LHV is referred to as the *heat of reaction, heat of combustion*, and *heat content* of the fuel. Table 4.1 gives the chemical equations for the combustion of several types of common fuels and their LHV in the Système International and the British system of units [3].

Air usually supplies the oxygen for the combustion of fuels. The standard composition of air is 21% oxygen and 79% nitrogen by volume. On a mass basis, the composition of air is 23.3% oxygen and 76.7% nitrogen. Thus, 5 kg of air will supply a combustion chamber with $5 \times 0.233 = 1.17$ kg of oxygen.

It is apparent from Table 4.1 that the chemicals listed, most of which are derived from fossil fuels, primarily contain carbon and hydrogen. The last two transform chemically to carbon dioxide and water. During the combustion process, the chemical energy is released as heat. Of significance is that the heat in the engines is released in a controlled way, by controlling the supply of the fuel, and that the combustion process may be designed to achieve the high temperatures that allow engines to operate at higher efficiency.

Table 4.1 gives the LHV when the complete combustion of the fuel is achieved, and this usually happens when oxygen/air is supplied in excess of the needed quantity. In the

TABLE 4.1

Common Fuels, Combustion Equations, and LHV

Fuel	Combustion Equation	LHV, kJ/kg	LHV, Btu/lb
Carbon	$C + O_2 \rightarrow CO_2$	32,770	14,100
Hydrogen	$H_2 + 0.5O_2 \rightarrow H_2O$	119,950	51,610
Methane	$CH_4 + 2O_2 \rightarrow CO_2 + 2H_2O$	50,020	21,520
Ethane	$C_2H_6 + 3.5O_2 \rightarrow 2CO_2 + 3H_2O$	47,480	20,430
Propane	$C_3H_8 + 5O_2 \rightarrow 3CO_2 + 4H_2O$	46,360	19,950
Butane	$C_4H_{10} + 6.5O_2 \rightarrow 4CO_2 + 5H_2O$	45,720	19,670
Pentane	$C_5H_{12} + 8O_2 \rightarrow 5CO_2 + 6H_2O$	45,357	19,542
Hexane	$C_6H_{14} + 9.5O_2 \rightarrow 6CO_2 + 7H_2O$	44,752	19,281
Octane	$C_8H_{18} + 12.5O_2 \rightarrow 8CO_2 + 9H_2O$	44,430	19,110
Acetylene	$C_2H_2 + 2.5O_2 \rightarrow 2CO_2 + H_2O$	48,220	20,740
Ethylene	$C_2H_4 + 3O_2 \rightarrow 2CO_2 + 2H_2O$	47,160	20,290
Benzene	$C_6H_6 + 3.5O_2 \rightarrow 6CO_2 + 3H_2O$	40,580	17,460
Methyl alcohol	$CH_3OH + 1.5O_2 \rightarrow CO_2 + 2H_2O$	19,920	8,570
Ethyl alcohol	$C_2H_5OH + 3O_2 \rightarrow 2CO_2 + 3H_2O$	26,800	11,530
Sulfur	$S + O_2 \rightarrow SO_2$	9,163	3,947

case of incomplete combustion, e.g., when the fuel is burned without the full amount of required oxygen, the combustion products will contain carbon monoxide CO or elemental carbon C. The heat released in an incomplete combustion is significantly lower than the LHV shown in Table 4.1.

In more efficient combustion processes, the water produced in the combustion reaction is condensed and exhausted as liquid water. In this case, an additional quantity of heat, the latent heat of the produced water h_{fg}, is released, and the amount of heat that becomes available is the HHV. The relationship between the LHV and HHV is

$$HHV = LHV + x \times h_{fg},$$

where x denotes the kilograms of water produced per kilogram of the fuel (e.g., $x = 36/16$ for methane). At typical condensation temperatures, the latent heat of water is $h_{fg} \approx$ 2,300 kJ/kg or $\approx$990 Btu/lb.

Example 4.1

A new 60 MW gas turbine with a bottoming cycle has 46% thermal efficiency and uses methane (CH_4) as a fuel. The exhaust gas is cooled by the bottoming cycle so that all the water produced condenses. (a) What is the HHV of methane? (b) Determine how much methane fuel (in kg) the gas turbine consumes per day.

Solution:

a. From the combustion equation for methane, $CH_4 + 2O_2 \rightarrow CO_2 + 2H_2O$, when 16 kg of methane reacts, 36 kg of water is formed. Therefore, $x = 36/16 = 2.25$, and the HHV of methane is $50,020 + 2.25 \times 2,300 = 55,195$ kJ/kg or 55.2 MJ/kg.
b. With 46% thermal efficiency, the gas turbine consumes $60/0.46 = 130.4$ MW of heat. During a day, the turbine consumes $130.4 \times 60 \times 60 \times 24 = 11,266,560$ MJ. This heat is produced from the combustion of methane, and hence, the gas turbine consumes 204,104 kg of methane.

4.2 Types of Fossil Fuels

The fossil fuels of today were living plants and organisms in a previous geological period. After several chemical processes that lasted for millennia, fossil fuels now exist underground in the form of solids, liquids, or gases. Depending on their physical state and chemical composition, we differentiate the following types of naturally occurring fossil fuels.

4.2.1 Coal

The word *coal* is a generic term that applies to all solid, rocklike, and organic materials composed of a complex, heterogeneous group of substances, including moisture. All types of coal are rich in amorphous elemental carbon. Coal is found in different depths and sometimes very close to the surface of the earth. It is classified according to its carbon content as anthracite—the highest quality of coal deposits—bituminous coal, subbituminous

TABLE 4.2

Types of Coal and Their Characteristics

Type	% Moisture	% Carbon	LHV, kJ/kg	LHV, Btu/lb
Anthracite	2	86–98	24,370–31,334	10,500–13,500
Bituminous	2–15	50–86	26,692–32,494	11,500–14,000
Subbituminous	20–30	40–50	19,264–26,692	8,300–11,500
Lignite	30–50	<40	14,622–19,264	6,300–8,300
Peat	50–95	Varies	<11,500	<5,000

coal, lignite, and peat [4,5]. Peat is the lowest-quality coal, with very low carbon content, several other impurities, and very low heating value. Table 4.2 shows the characteristics of the several types of coal, including the range of their heating values [4,5].

Peat is not officially ranked as coal according to the American Society for Testing and Materials, but it is invariably mentioned as a solid fuel because it is abundant; it has a fairly high LHV and is used in several countries for cooking and, sometimes, for the generation of electric power [5,6].

The reason some forms of coal have higher LHV than pure carbon is because they often contain hydrocarbons in gaseous form—primarily methane—as well as small amounts of hydrogen. Bituminous and subbituminous coals constitute the bulk of the coal reserves of most countries including the United States. China's coal minerals mainly consist of lignite, which has a lesser amount of carbon and significantly lower LHV. Because in many countries, it is not considered economical to transport lignite over long distances, this fuel is primarily used for the production of electricity, and the lignite-powered generation units are constructed close to the mining sites.

The chemical compositions of coal deposits in the world differ and depend very much on the type of plants and organisms from which the coal is derived; the time for the transformation; and the prevailing pressures and temperatures. Pressure and temperature highly depend on the depth at which the coal deposits are found. Sulfur is also commonly found in several coal deposits, it reacts with air to form SO_2 in the burner and is typically removed in scrubbers as $CaSO_3$ before the exhaust gases are released (Section 3.2.2 and Figure 3.2). Traces of heavy metals, such as mercury, uranium, thorium, and rhodium, have also been detected in coal deposits. Other pollutants that appear as traces in coal are arsenic, boron, beryllium, and lead. Environmental regulations have been put in place in most countries for coal power plant operators to take measures and to eliminate or limit the discharge of these pollutants to the environment.

Carbon is the principal component of coal that is burned to produce heat in a boiler or a burner. Other combustible materials in the coal—sulfur, hydrogen, and mercury—also react with air and produce oxides. When calculating the air required for the combustion products and the heat of combustion, one must take into consideration all the ingredients in the coal mixture. Since the combustion of 1 kmol of C coal requires 1 kmol of oxygen, and the molecular weights of the two elements are 12 and 32, respectively, 1 kg of carbon requires 2.67 kg of oxygen to react and form 3.67 kg of CO_2. At 23.3% mass fraction in the air, this quantity of oxygen is supplied by 11.5 kg of air. Table 4.3 gives the mass of oxygen, the mass of air, and the volume of air at standard conditions (1 atm pressure and 298 K) for the combustion of 1 kg of carbon, hydrogen, and sulfur, the three most common elements in fossil fuels. The density of air at standard conditions is 1.186 kg/m³.

In order to achieve complete combustion of fossil fuels, air is usually supplied in excess of the stoichiometric quantity, which is denoted in the reaction equation. The *excess air* is

TABLE 4.3

Mass of Oxygen, Mass of Air, and the Volume of Air Needed for the Stoichiometric Combustion of Fuels, per Kilogram of the Fuel

Element	Kilogram of O_2/Kilogram of Fuel	Kilogram of Air/Kilogram of Fuel	Cubic Meter of Air/ Kilogram of Fuel
Carbon	2.67	11.5	9.70
Hydrogen	8	34.33	28.94
Sulfur	1	4.3	3.63

expressed by a percentage. For example, if 25% excess air by mass is used for the combustion of 1 kg of C, the total amount of oxygen supplied is $1.25 \times 2.67 = 3.34$ kg of oxygen per kg of carbon. This quantity of oxygen is contained in 14.32 kg of air or 12.07 m³. Sometimes the amount of excess air is reported as percent *theoretical air*: 25% excess air is 125% theoretical air.

Because the composition of coal varies and depends on the location where the coal is mined, daily analysis of the fuel is performed in coal power plants to determine the composition of the fuel and its heating value. The *ultimate analysis* of coal makes available the mass fractions of the different chemical constituents in the coal, in terms of the elements that are present. Example 4.2 elaborates on some of the calculations for the determination of the LHV and the combustion products of bituminous coal.

Example 4.2

The ultimate analysis (by mass) of a bituminous coal is C, 75%; H, 4%; S, 2%; ash, 12%; and moisture, 7%. Determine (a) the LHV of this coal; (b) if the coal is burned with 40% excess air (140% theoretical air), the total mass of air required per kilogram of coal for the combustion; (c) the amount of coal (in tons) used daily in a 400 MW power plant with 39% efficiency; and (d) the volume of air supplied daily to the combustion chamber at standard conditions.

Solution:

a. The elements that are burned in this type of coal are C, H_2, and S. Using the LHV values of Table 4.1, the LHV of this mixture of elements is $0.75 \times 32,770 + 0.04 \times 119,950 + 0.02 \times 9,163 = 29,559$ kJ/kg. The ash and the moisture do not contribute to the LHV.

b. From the combustion equations of Table 4.1 and from Table 4.3, 1 kg of C burns with 2.67 kg of O_2; 1 kg of H_2 burns with 8 kg of O_2; and 1 kg of S (molecular weight: 32 kg/kmol) burns with 1 kg of O_2. Therefore, for the combustion of 1 kg of the bituminous coal, $0.75 \times 2.67 + 0.04 \times 8 + 0.02 \times 1 = 2.34$ kg of O_2 is used. This quantity of O_2 is contained in $2.34/0.233 = 10.04$ kg of air. Because 40% excess air is used, the air supplied to the combustion chamber is $10.43 \times 1.40 = 14.06$ kg of air per kilogram of coal.

c. The power plant uses $400/0.39 = 1,026$ MW rate of heat. During an entire day, the plant uses $Q = 1,026 \times 60 \times 60 \times 24 = 88.65 \times 10^6$ MJ $= 88.65 \times 10^9$ kJ. This heat is provided by Q/LHV kg of coal: $88.65 \times 10^9/29,559 \approx 3,000,000$ kg of coal (3,000 t). The amount of air supplied to the combustion chamber is $3,000,000 \times 14.06 = 42,180$ t.

d. The density of air at standard conditions (s.c.) is 1.186 kg/m³. Hence, the volume of the air supplied daily to the power plant is $42,180 \times 10^3/1.186 = 35.6 \times 10^6$ m³.

Based on the LHV values of the three most common elements in coal—carbon, hydrogen, and sulfur—one may derive the following approximate expression for the LHV of any coal:

$$LHV = 32,770 \times C + 119,950 \times (H - O/8) + 9,163 \times S \text{ in kJ/kg,}$$
$$LHV = 14,100 \times C + 51,610 \times (H - O/8) + 3,947 \times S \text{ in Btu/lb.}$$

(4.1)

The symbols C, H, O, and S denote the mass fractions of the corresponding elements in the coal. The subtraction of the oxygen fraction in the two parentheses accounts for the presence of water that does not burn to produce heat. If the hydrogen is reported separately as free hydrogen and the water content as moisture, the oxygen content does not need to be subtracted from hydrogen and the water content does not come into the equation.

4.2.2 Petroleum (Crude Oil)

Petroleum, which literally means oil from stone, is a liquid mixture of complex hydrocarbons—aliphatic, alicyclic, aromatic, and olefinic—spanning a wide range of molecular weights. The constituent compounds of petroleum were formed over millions of years primarily from the anaerobic decomposition of aquatic organisms—plankton and algae—trapped at depths of very high pressures and significantly high temperatures. Hydrocarbons are the main constituents of petroleum. Depending on the location of petroleum extraction, petroleum also contains varying amounts of sulfur (from traces to 8% by weight), nitrogen in the form of ammonia, traces of oxygen, vanadium, and nickel. Because petroleum is in a liquid state, it may be conveniently transported in pipelines.

The petroleum naturally exists at high pressures in underground porous rock reservoirs, the *oil fields*. Vertical and horizontal wells drilled into the reservoirs allow a fraction of the petroleum to flow under the local higher pressure. When 15–20% of the petroleum is extracted, the reservoir pressure decreases and the petroleum production diminishes. A secondary recovery method is then employed. This involves the drilling of additional wells and the placement of downhole extraction pumps, which boost the rate of petroleum production. Secondary methods may recover an additional 25–30% of the total petroleum in the oil field. At the end of this stage of production, the flow is significantly impeded by the lower reservoir pressure and the high viscosity and surface tension of petroleum. Further, petroleum production from a field may be achieved by tertiary recovery methods that include (a) the injection of surfactants (detergents or liquid carbon dioxide) that reduce the surface tension of the flowing oil and (b) the decrease in viscosity by increasing the local temperature. The latter is achieved by the injection of high-temperature steam in the reservoir (steam flooding) or by burning some of the petroleum *in situ* with a controlled supply of air (fire flooding). The secondary and tertiary recovery methods add to the cost of production and, consequently, to the price of petroleum and its products.

The chemicals that constitute the petroleum, or crude oil, have significantly different boiling points. In their pure states, these chemicals may be gases, liquids, or solids. The chemicals are separated in refineries by a process that includes complete evaporation and partial condensation at different temperatures, which is known as *petroleum distillation*. This process separates the several constituents of petroleum into groups of useful hydrocarbons, such as synthetic gas, gasoline, diesel, kerosene, lubricants, asphalt, and paraffin. Other chemical processes in a petroleum refinery, known with the general term *cracking*,

TABLE 4.4

Characteristics of Commonly Used Liquid Fuels

Fuel	Density, kg/m³	LHV, kJ/kg	LHV, Btu/lb	LHV, kJ/L
Gasoline	730–780	43,200	18,600	32,600
Diesel oil	830–840	42,600	18,350	35,600
Heavy oil	900–920	40,600	17,500	36,950
Light oil	820–880	42,500	18,300	36,120

Note: The LHVs are representative of the several mixtures of hydrocarbons used. The volumetric LHV (kJ/L) is based on the average density of the fuel.

break up the long molecules of solid hydrocarbons, which have limited practical use and low market value, to shorter hydrocarbons that form the mixtures of fuels our society uses. Table 4.4 shows the properties of the most common fuels used at the beginning of the twenty-first century:

As with the coal, the composition of a liquid fuel is also described in terms of the ultimate analysis that provides the mass fractions of the several constituent elements of the fuel. Example 4.3 illustrates these calculations based on the ultimate analysis of a liquid fuel.

Example 4.3

A heavy oil fuel, with density of 910 kg/m³, has the following composition by mass: C, 86%; H, 12%; and S, 2%. The fuel is burned in an engine with 70% excess air. Determine (a) the LHV of the fuel and (b) the mass and the volume of air used for the combustion of 200 L of this fuel.

Solution:

a. The LHV of this fuel is $0.86 \times 32,770 + 0.12 \times 119,950 + 0.02 \times 9,163 = 42,760$ kJ/kg.
b. The mass of oxygen required for the stoichiometric combustion of the three elements is $0.86 \times 2.67 + 0.12 \times 8 + 0.02 \times 1 = 3.27$ kg of oxygen, which is contained in 14.05 kg of air. The combustion chamber uses 70% excess air (170% theoretical air), and hence, the amount of air used is $14.05 \times 1.7 = 23.89$ kg of air per kilogram of the fuel. Since the density of air at s.c. is 1.186 kg/m³, the volume of air needed for the combustion of 1 kg of this fuel is $23.89/1.186 = 20.14$ m³ of air at standard conditions.

The 200 L of this fuel contain 182 kg of fuel. Therefore, for the combustion of 200 L of this fuel, approximately $20.14 \times 182 = 3,666$ m³ of air must be supplied to the combustion chamber.

4.2.3 Natural Gas

Natural gas, in particular, methane, is also formed during the anaerobic bacterial decomposition of organic matter, including human waste. Natural gas is composed of CO and lighter, gaseous hydrocarbons, primarily methane. The origins of natural gas are similar to that of coal and crude oil, with the difference that the natural gas has been formed at significantly higher geological depths, where the temperatures are higher and the cracking of the heavier hydrocarbons naturally occurred. Oftentimes, natural gas is extracted from the upper part of petroleum fields, having separated from the denser liquid by buoyancy. Most of the natural gas deposits contain traces of ammonia, nitrogen, hydrogen sulfide, and carbon dioxide. In comparison to the other fossil fuels, natural gas burns cleaner and is very easily transported in pipelines at relatively high pressure. Long-haul gas pipelines of compressed natural gas bring this energy resource to central stations that are close to urban centers and then crisscross the cities to supply the households and businesses. Liquefied natural gas is also transported at lower temperatures in large, specially designed tankers from one continent to another.

The conventional natural gas resources are mined from underground deposits, where they exist at higher pressure. In addition to the conventional deposits, there is a very large quantity of *hydrates*—primarily methane hydrates—gases that have been trapped in the arctic permafrost and the deep ocean sediment and have formed weak bonds with water molecules. If utilized as a fuel, the total quantity of existing methane hydrates may provide enough energy that has been estimated to last for 350–3500 years. However, the hydrates are spread over very large areas and are very much diffused as an energy source. The current technology does not have an economical method to extract, transport, and use the methane hydrates. For this reason, they are not characterized as energy reserves.

4.2.4 Oil Shale and Shale Gas

Shale is a fine-grained, sedimentary rock composed of clay minerals and tiny fragments of other minerals, especially quartz and calcite. Oil and gas exists in the micropores of the rock in relatively small quantities, typically from 100 to 10 L/t of the shale. For the recovery of the oil shale, the ore is mined and is processed above ground (*ex situ*), where the small amounts of the trapped hydrocarbons are recovered. An alternative to processing the entire mass of the sedimentary rock is to increase the temperature and pressure inside the underground shale, by pumping and burning natural gas [6]. The increased temperature breaks the longer hydrocarbon molecules and produces lighter and less viscous fluids that naturally flow to production wells and from the wells to the surface. A third, more recent method is to pulverize the shale rock and burn it in a fluidized bed reactor (FBR). Because oil shale needs a great deal of processing, it is significantly more expensive to produce and utilize. This has not deterred several corporations and countries with no other energy resource and with significant shale oil deposits from extracting and using this fuel. For example, Estonia has always been a leader in the use of shale oil: between 1970 and 2007, the power plants in the Narva region of this country generated more than 90% of its electric power using fuel derived from shale.

Similar to shale oil, the shale gas is trapped in a matrix of sedimentary rock. Because a very small amount of shale gas can be extracted with traditional wells, two methods have been recently developed for its extraction: horizontal drilling and hydraulic fracturing (fracking). In horizontal drilling, the direction of the well changes from vertical to almost horizontal and follows the shale mineral deposits. This gives the pipeline access to a larger part of the energy resource. During the hydraulic fracturing process, a mixture of water, chemicals, and fine sand is injected in the well at very high pressure to induce cracks and fissures in the shale. While the high pressure opens and widens the fissures in the shale formation, the chemicals and sand in the fracking fluid keep the fissures open for the gas to flow out to the well. The two methods have enabled the oil and gas industry in the United States to recover a great deal of shale gas (particularly in the Barnett Field of North Texas, where fracking was first applied). Shale gas production in the United States skyrocketed from 1% of the domestic supply in 2000 to 39% in 2012 [7].

The practice of fracking is not limited to shale gas but extends to other reservoirs of oil and gas. The traditional methods of oil and gas production leave a great deal of the energy resource inside hard to reach "pockets" in the underground field where natural gas and oil are "trapped" and do not flow to the wells. Hydraulic fracturing and flooding the fields with water or liquid CO_2 "push" the resources out of the reservoirs, increase the production rates, and allow for the extraction of a higher fraction of these primary energy sources.

4.2.5 Tar Sands

Tar sands are sandstones infused with highly viscous and heavy hydrocarbons. The hydrocarbon content of tar sands ranges from almost zero to 20% by volume. The most extensive deposits of tar sands are known to exist in Canada, where their extraction and processing has been ongoing for decades. Because the "tar sand" is very viscous and almost solid-like and does not easily flow up the well, several methods to produce significant amounts of hydrocarbons from this resource have been developed, including the following:

1. Surface mining, where both sand and oil are excavated and then separated. This *ex situ* separation of oil entails significant energy input and cost.
2. Deep-well production of oil and sand, where the pockets formed from the extracted sand facilitate the local flow of more oil.
3. Steam injection, which increases the pressure and temperature. The high temperature decreases the viscosity of oil and the high pressure forces it to the extraction wells.
4. Extraction by solvents.
5. Partial combustion in the well or heating with electromagnetic waves (microwaves). This practice increases the downhole temperature, reduces the viscosity, and makes the oil–sand slurry easier to flow up the well.

Some of these methods have also been used for the extraction of extra heavy oil deposits, such as the ones in the Orinoco basin, Venezuela [6]. The extraction, purification, and refinement of tar sands and oil shale deposits—the *unconventional petroleum resources*—consume a great deal of energy and are more costly than other types of petroleum reserves. It is profitable for oil corporations to develop and extract unconventional oil, only when the price of petroleum is high enough. If the price of oil dips (as it happened in 2015–2017), bringing the unconventional petroleum to the market can be a financial burden for an oil corporation, because the cost of extraction and processing is higher than the market price of the product.

The extraction of unconventional oil has a significant, adverse environmental impact: The *ex situ* processing of this resource brings into the biosphere toxic materials that are often discarded next to the processing sites. The other methods of extraction also bring a lesser amount of pollutant materials from the underground deposits to the biosphere. Although attempts are made to locally confine these materials in disposal sites, in the long run, the toxic materials may leach and contaminate surface streams, lakes, and underground water reservoirs.

Example 4.4

Four fossil fuel power plants are rated at 100 MW and operate for 88% of the year. The four operate with different fuels and have thermal efficiencies as follows:

1. Anthracite, LHV = 28,200 kJ/kg, η = 38%.
2. Lignite, LHV = 21,300 kJ/kg, η = 36%.
3. Heavy fuel oil, LHV = 41,500 kJ/kg, η = 39%.
4. Natural gas, LHV = 1,065 kJ/scf, η = 39%.

Determine the amount of fuel used per year (in tons of the fuel or million scf).

Solution: For each power plant, the rate of heat input is $100/\eta$ MW. Given that each power plant operates for 88% of the year, their annual heat input is $0.88 \times 100 \times 60 \times 60 \times 24 \times 365/\eta$ MJ = $2.78 \times 10^9/\eta$ MJ = $2.78 \times 10^{12}/\eta$ kJ.
  The quantity of fuel needed annually is $2.78 \times 10^{12}/(\eta \times LHV)$ kg or scf. Therefore,

1. The anthracite plant will use 259.4×10^3 t of anthracite.
2. The lignite plant will use 362.5×10^3 t of lignite.
3. The heavy oil fueled plant will use 171.8×10^3 t of heavy oil.
4. The natural gas plant will use $6,693 \times 10^6$ scf of gas.

4.3 Physicochemical Fuel Conversions

Fossil fuels, which are extracted (mined) from underground fields, are primary energy sources. Several of these natural resources (e.g., crude oil) are unsuitable to be used in their natural, mineral form. Others, if used directly in combustion processes, produce a great deal of pollutants that make them environmentally unacceptable. For these reasons, a high fraction of the fossil fuel minerals are converted to other types of fuels through a complex series of physicochemical processes, to commercial products such as gasoline and diesel fuel, which are more suitable to be used or more benign to the environment. The products of the fossil mineral resources are referred to as *refined* or *secondary* energy sources.

4.3.1 Petroleum Refining

Petroleum (crude oil) is a mixture of a large number of chemicals, primarily hydrocarbons, that cover a wide range of carbon chains and molecular masses. The exact composition of petroleum varies with the location of its extraction and (often) with the timing of the extraction—petroleum fields produce lighter hydrocarbons in the first stages of the extraction process and heavier hydrocarbons in the later stages. The constituent chemical elements exhibit a narrow range of composition by weight as Table 4.5 shows [8].
  The chemicals that constitute the petroleum reserves are extracted as an almost homogeneous mixture of more than 500 chemical compounds, primarily hydrocarbons. In their pure form, these compounds span all three states of matter: under ambient temperature and atmospheric pressure, they would be gases, liquids, or solids. In general, the higher the molecular mass of a chemical compound, the higher its boiling point is, with the heavier compounds being solids. For example, the light methane and ethane in the crude oil mixture are gases; heptane and octane are liquids; and the much heavier waxes and tars are

TABLE 4.5

Elemental Composition of Petroleum

Carbon	83–85%
Hydrogen	10–14%
Nitrogen	0.1–2%
Oxygen	0.5–1.5%
Sulfur	0.05–6%
Metals	<0.1%

solids. When mined, the crude oil mixture has the properties of a very viscous (thick) liquid that may be transported in pipelines with high pressure drop per unit length.

Our engines do not use petroleum in the natural form it is extracted. We use gasoline and diesel for IC engines, kerosene for aviation fuel, heavier liquids for lubricants, and synthetic gas (methane and the other gases in crude oil) for cooking. The crude oil mixture is "refined" to generate these commercial products. The refining takes place by distillation in very high fractionating columns as shown in Figure 4.1. At first, the crude oil is heated in a conventional heat exchanger (the furnace or evaporator) to approximately 400°C (750°F). At this temperature, most of the crude oil constituents are vaporized and exist as less viscous liquids. The liquid–gas mixture is then directed to the 30–40 m high-fractionating column, where a temperature gradient is maintained using heat exchangers with cooling water. The lower part of this column is at approximately 400°C, while the top of the column is closer to the ambient temperature. The liquid part of the heated crude oil separates by gravity, at the bottom of the fractionating column, and is removed by side outlets. When cooled, this fraction is composed of the solids or solid-like, paraffin, wax, asphalt, etc. The remaining part is gaseous and ascends the fractionating column, where several horizontal plates lead to side outlets for the extraction of the various liquid products.

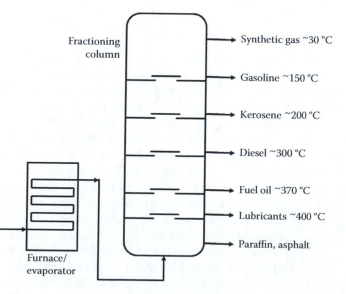

FIGURE 4.1
Petroleum refinement process.

As the temperature drops along the path of the gas, the other crude oil fractions condense on the plates and are removed at the sides.

In the fractionating column, the hydrocarbons that constitute the several lubricants are removed at approximately 400°C; the heavy fuel oil, at about 370°C; diesel, at 300°C; kerosene, at 200°C; and gasoline (petrol), at about 150°C. The remaining gases make up the *synthetic gas* or *syngas*. A fraction of the syngas is used on site as the fuel for the furnace. The rest is either pumped to be sold or further processed to produce hydrogen and additional gasoline components, as, for example, in the octane-producing reaction

$$8CH_4 \xrightarrow{\text{heat}} C_8H_{18} + 7H_2. \tag{4.2}$$

The crude oil refinement produces most of the hydrogen currently used in fuel cells, for research and for industrial processes. One of these processes occurs in the refineries and involves the breaking up of the very heavy hydrocarbons to fuels: oftentimes, there is not sufficient demand for the solid parts of the distillation, paraffin, wax, and asphalt. These heavy hydrocarbons are then broken at very high temperatures with the addition of hydrogen in a process that is called *cracking* (or catalytic cracking when catalysts are used), as for example in the following reaction of eicosipentane ($C_{25}H_{52}$), whose products are two constituents of gasoline:

$$C_{25}H_{52} + 2H_2 \xrightarrow{\text{heat}} 2C_8H_{18} + C_9H_{20}. \tag{4.3}$$

Steam may also be used in the catalytic cracking processes and oxygen is produced as a by-product. The gasoline produced in the refineries is a mixture of several (more than 400) hydrocarbons with 5–12 carbon atoms in their chemical formulae. When sold, it is branded with an *octane rating*. The latter does not represent the actual octane content of the gasoline, but the effects of the combustion of the fuel in the cylinders of an IC engine. Octane ratings differ from country to country and, within the United States, from state to state.

4.3.2 Coal Liquefaction and Gasification: Synfuels

Liquid and gaseous fuels are much easier to transport and more convenient to use in everyday applications. While the demand for liquid fuels is significant, most of the known fossil fuel resources are extracted in the solid form, coal, which is almost exclusively used for the generation of electricity. Electricity, a tertiary energy form, is conveniently used in most domestic and transportation applications, e.g., electric cars, railroad engines, lights, and household ovens. However, the energetic cost of the several conversions of energy from coal to electricity is very significant: by the time coal is converted to electricity and transported to the consumers, more than two-thirds of the initial chemical energy in the coal is lost in the several conversion processes that take place. An alternative to electricity conversion is to convert the coal into liquid or gaseous fuels that are easily transported by pipelines and used in domestic applications or in IC engines. The conversions from coal to the other liquid and gaseous fuels are known as coal liquefaction and coal gasification.

The basic reactions for coal gasification were known since the beginning of the industrial revolution. However, large-scale coal liquefaction and gasification did not start until the first half of the twentieth century. The widespread use of gas fuel at homes for heating and cooking, and the gasoline shortages that occurred in several countries during the Second World War, facilitated the development of chemical processes and equipment for the conversion of the chemical energy in the solid coal to the chemical energy of gaseous and liquid fuels. The starting reaction for all the coal gasification and liquefaction processes is the reaction of carbon with steam at high temperature, which primarily produces hydrogen and carbon monoxide:

$$C + H_2O \rightarrow CO + H_2. \tag{4.4}$$

The high temperature for this reaction is achieved by the partial combustion of coal or reaction by-products. The hydrogen produced in this basic reaction may be used with the further addition of coal to produce gaseous and liquid hydrocarbons, such as CH_4 and C_8H_{18}. The pressure and temperature conditions as well as the choice of catalysts determine the kinetics and yields of the several possible reactions that may occur with C, CO, and H_2 and the final mixture of the hydrocarbon products. Synthetic transportation fuels, polymers, anhydrites, and acrylates are a few examples of useful products from such reactions.

The production of gas and liquid fuel from CO and H_2 or CH_4 is a system of chemical reactions that are known by the generic name: *Fischer–Tropsch* (F-T) *process*. Solid catalyst particles that are commonly used are iron-based or silica–alumina particles, with the latter having several advantages because they fluidize easier. The F-T process is highly exothermic, and this implies that a high rate of heat must be removed from the reactor vessel by cooling water. In several F-T plants, the heat generated is at high temperature and is used for the production of steam, which, in turn, produces electric power in a Rankine cycle.

In the 1950s, the South African Synthetic Oil Limited Corporation developed a high-yield and very efficient recirculating chemical reactor for the F-T process, which became the prototype for the development of other similar chemical reactors. The reactors have a *riser* section, where most of the chemical reactions occur, and a *downer*, which returns all the catalyst particles and unconverted solid fuel particles to the inlet, where they are mixed with the reactants feed and are recirculated in the chemical reactor section, the riser, as shown in Figure 4.2. A heat exchanger, which is incorporated in the design of

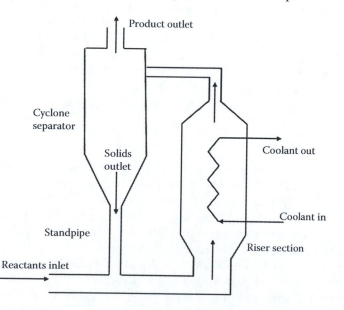

FIGURE 4.2
Schematic diagram of a recirculating reactor for the F-T coal gasification process.

the riser, removes the heat at the middle section of the riser and provides temperature control. This type of chemical reactors operates at intermediate pressures, between 10 and 17 atm. Typical reactor temperatures are in the range of 250–400°C. The volumetric fraction of solids ϕ in the riser section of this type of reactors is rather low, between 3% at the top and 10% at the bottom. By adjusting the temperature, pressure, and catalyst composition in the riser, a wide range of hydrocarbon products, from gases such as methane and ethane to liquids such as octane and benzene, may be produced in large quantities [9,10].

The produced gas is separated from the liquid products, the *synfuel*, by gravitational settling at the outlet of the reactor. Significant control of the reaction rates and of the yield of the various products may be achieved by the choice of solid catalysts, which recirculate in the reactor. Parts of the products may be fed back into the reactor to match the fuel supply with the seasonal variations of the demand for gaseous and liquid fuels. For example, during the summer months, when syngas is not needed for the heating of buildings, a controlled portion of the gaseous products is fed back into the reactor to increase the amount of the liquid hydrocarbons, which are principally used for transportation.

The gasification and liquefaction of coal are two of the most common engineering methods to produce synthetic fuels. Synthetic fuels (synfuels) may be produced at elevated temperatures from oil shale, tar sands, wood by-products, agricultural waste, and other forms of biomass. In most common processes for the production of synfuels, a fraction of the primary fuel is burned in a reactor to maintain the high temperatures. The heat required for the synfuel production process may also be derived from an advanced nuclear reactor or from concentrated solar power using a catalyst, such as cerium oxide [11]. Such systems with outside energy supply conserve the reactants to be used for the production of the synfuel rather than burn them onsite.

Underground coal gasification (UCG) is an environmentally friendly method to produce syngas without extracting the coal from its underground seams. Moist air is supplied to the underground coal, and ignition is initiated. The heat produced by the partial combustion of air increases the underground temperature to produce *in situ* syngas at high pressure, which is removed by gas wells. The syngas produced is free of most of the pollutants associated with coal and may be directly utilized in a power plant. Although this method has been known since the nineteenth century, there are not many UCG plants, primarily because the underground partial combustion of coal and the overall gasification process are not as efficient as the better controlled overground gasification processes.

4.3.3 Fluidized Bed Reactors

Fluidized bed reactors (FBRs) is a generic term that encompasses a variety of engineering systems, including chemical reactors, combustors, boilers, gasifiers, calcifiers, driers, and sulfur removers. The FBRs are efficient chemical converters that are increasingly used in the energy industry for the combustion, gasification, and liquefaction of coal; for the removal of sulfur from the combustion products; and for the better control and abatement of the NOX gases that are formed during combustion.

The common characteristic of the FBRs is the injection of the solid particles (typically pulverized coal, catalysts, and limestone) and the injection of an oxidizing fluid (typically air) that suspends and lifts the particles and carries them within the FBR, where they react. During the fluidization process, the drag force on the solid particles F_D must be higher than the weight of the particles F_G. For particles with an equivalent radius a and drag coefficient C_D, this condition yields the following expression for the minimum air velocity that would lift the solid particles and keep them in suspension [12,13]:

$$F_D > F_G \Rightarrow V_{min} > \sqrt{\frac{8\alpha g}{3C_D} \frac{\rho_p}{\rho_f}}, \tag{4.5}$$

where g is the gravitational acceleration, 9.81 m/s²; ρ_f is the density of the fluid (air); and ρ_p is the density of the particles. Fluid vortices that are formed in the interior of the FBR facilitate the mixing of the air with the particles; the circulating fuel particles come to better contact with the oxygen in the air; and the combustion of the fuel particles occurs faster and with higher completion rates.

In addition to the faster and more efficient combustion, and because all the reaction products are confined in the interior of the FBR, other chemical reactions may be initiated that would remove potential pollutants. For example, limestone particles, which contain $CaCO_3$, are mixed with coal particles in the FBRs to remove *in situ* sulfur dioxide from the combustion gases:

$$CaCO_3 + SO_2 + 1/2O_2 \rightarrow CaSO_4 + CO_2. \tag{4.6}$$

The geometry of FBRs is designed to be optimal for the desired reactions. Particle interactions and fluid movement are designed to induce a circular motion that amplifies the fluid vortices, the turbulence, and the mixing of the reactants. The fluid vortices and turbulence enhance the heat and mass transfer processes in the reactor interior. Figure 4.3 depicts two generic types of FBRs: the flat bed and the spouted bed reactor. The general flow pattern in both reactors is that particles rise to the top of the reactors primarily through the center, move laterally to the sides, and settle close to the walls. In the spouted FBR, the settling particles also move laterally in the spout area toward the center of the

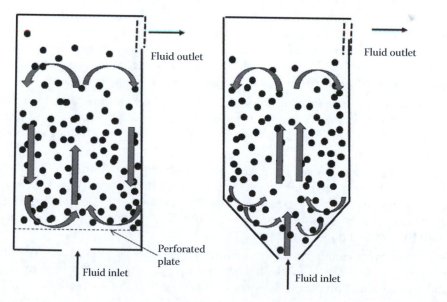

FIGURE 4.3
Principal fluid and particle motion patterns and the mixing of reactants in flat bed and spouted FBRs.

FBR. The broad arrows in the figure indicate the principal flow patterns of the fluid–solids mixture. Cyclone particle separators, not shown in the figure, are attached to the outlets in addition to the filters to capture the solid particles and avoid environmental particulate pollution. The cyclone separators remove the entrained particles from the carrier fluid and feed them back into the FBR, while allowing the gaseous products of the reaction to exit.

The inherent advantages of FBRs, compared to other chemical reactors, are [9,13] as follows:

1. They continuously operate and do not make use of batch processes that require stoppage of the reaction process. The FBR is an open system, with inlets that may be continuously supplied with reactants and outlets that continuously supply the desired products. The continuous and uniform operation of the FBRs enables a more efficient and uniform production process, because the startup and finish conditions are eliminated.

2. There is no delay in the production process for the removal of the products and the supply of the reactants.

3. Significantly higher mixing of both solids and fluid. Because the particles vigorously move inside the FBR, they facilitate the mixing of the fluid. The effect of the higher mixing in the fluid is to develop more uniform concentration and temperature fields. This avoids the creation of "hot spots" and "cold spots" in the reactor, which invariably result in incomplete reactions and product deterioration.

4. Because of the higher turbulence and mixing, the FBRs also have significantly higher heat and mass transfer coefficients. This implies less surface requirements for a given reaction and rate of heat transfer.

5. The vigorous mixing causes the minimization of the temperature and concentration gradients. In particular, the radial and azimuthal gradients may be almost eliminated in a FBR. This invariably enhances the effectiveness of the reactor, which produces a more uniform product and higher quantities of heat from the complete combustion.

6. In comparison to other reactors and burners, the flow of the solids in the FBRs causes more uniform particle mixing and more uniform reactions. The superior mixing of the solid particles produces a uniform product, typically of superior quality, which is more difficult to achieve with batch reactors.

7. Since the volumetric heat capacity of solids (in J/m^3) is much higher than that of the fluid/air ($c_s \rho_s \gg c_f \rho_f$), the solids provide a very effective way to add or remove heat out of the reactor and to maintain constant temperature for both endothermic and exothermic reactions.

8. The control of the reaction rates in a FBR has been significantly improved in the last 40 years. As more research is performed and the engineering experience with FBRs is increasing, it is becoming feasible to design better-performing FBRs, with more accurate controls of the reactions that produce the desired product(s) with the desired quality and yield.

9. The addition of suitable reactants inside the FBR enables the removal of pollutants from the outlets. For example, the acid rain precursor SO_2 may be removed from the coal combustion products by the addition of limestone as in Equation 4.6.

10. Results from continued research on solid–fluid phase interactions and better computer modeling and simulations are used to produce better and more efficient

designs that serve increasingly more applications. This research has helped create an expanding market for the FBRs in the chemical industry and the power production industry.

The inherent disadvantages of FBR systems include [9–11,13] the following:

1. More expensive design for the overall heat transfer system, because the FBRs include several moving parts as well as the fluid injection and removal systems.
2. Stoppage of the equipment if the higher fluidization pressure is lost, e.g., because of a blower or compressor malfunction. This may cause product degradation or unexpected and undesirable reactions. Such malfunctions are avoided by the better design and more frequent maintenance of the equipment.
3. The higher gas velocities, which are inherent in FBRs, result in the entrainment of a larger amount of solid particles, especially fine particles, which may be carried by the fluid outside the reactor and cause the unnecessary waste of reactants. Better reactor design, filters, cyclone separators, and suitable entrainment elimination technologies reduce the solids entrainment problem. For example, one or two cyclone separators with a filter for ultrafine particles almost eliminate the escape and loss of fine particles from the FBR.
4. They represent relatively new technology. All aspects of the operation of the FBRs are not fully understood. Despite the enormous research efforts on FBRs since the 1970s, several aspects of their operation are not well understood and are poorly modeled. A concerted international effort has been underway to better model the operation of FBR systems and remove the design uncertainties [14].
5. Because of the lack of complete understanding of the FBR technology, pilot plants for new processes and products must be built. Since the FBRs have multiple length scales that are important, the scale-up of pilot plants to fully producing facilities is undocumented in the open literature. Oftentimes, the results of full-scale facilities do not reflect what was experienced in the pilot plant trials.
6. Related to the preceding paragraph and to the high economic value of FBRs, a great deal of engineering experience with them is proprietary. Several parts and processes of FBRs are covered by patents. For this reason, the "best engineering practices" are not always well known and, if they are patented, may not be duplicated.
7. Particulate flow inside the reactor causes higher rates of erosion and wear on the reactor vessels. Erosion may be minimized by (a) using erosion-resistant materials; (b) better design of the flow patterns in the FBR; or (c) more frequent and expensive maintenance for the reactor vessel and piping replacement.

The inherent advantages of the FBRs by far outweigh their disadvantages. Through extensive research, the physicochemical phenomena associated with the operation of FBRs are well understood [9,13,14], and the design of FBRs has significantly improved, since the 1980s. In the recent years, the engineering community has seen the widespread use of specialized FBRs in fossil fuel processes, fuel conversions, and cleaner combustion. The use of FBRs rather than other types of burners or chemical reactors is considered environmentally favorable, because of the better control of the reactions and the potential reduction of harmful emissions.

4.4 Fossil Fuel Resources and Reserves: Peak Oil

The subjects of energy resources and reserves were discussed in detail in Section 2.6. It is worth recalling a few facts that are applicable to fossil fuels:

1. The fossil fuel reserves represent the amount of fuel that may be economically extracted. The reserves depend on the total available quantity of the fuel, the state of technology, and the price of the fuel.
2. When the price of a fuel increases, more of that fuel may be extracted economically and the reserves of the fuel increase.
3. At the current and expected rates of consumption, it is certain that all fossil fuel resources will be depleted at some point in the future. What is unknown is the exact time that each fossil fuel will be depleted.
4. Before the fossil fuels are completely depleted, our society must derive ways to substitute them and satisfy its energy needs with alternative energy sources.

The determination or estimate of the timeframe of the eventual shortage and depletion of fossil fuels is of immense importance for the transition of the global community to other energy sources. It is also very important for the financial sector because the high price swings associated with scarce commodities help individuals and corporations derive immense profits from speculation on the scarce commodity. However, the prediction of the timeframe of the exhaustion of fossil fuels with any degree of certainty is a goal that so far has eluded both scientists and economic oracles.

4.4.1 Hubbert Curve

Hubbert [15,16] empirically modeled the process of domestic US petroleum production and its eventual depletion using a logistic curve. Hubbert's empirical model was extended to other natural resources, such as other fossil fuels and minerals. According to this model, during the first stages of the utilization of any resource, its consumption exponentially rises because its use is accelerated by two factors:

1. Relatively low nominal price.
2. Inventions of new technologies and engines that utilize this resource to perform several societal functions and tasks, which are either new or were performed using another resource in the past. In the case of petroleum, the IC engine and the jet engines were developed for the use of petroleum products. Such inventions have significantly increased the demand of petroleum products between 1900 and 2017 and have made petroleum-derived fuels a necessity for the transportation industry.

In the initial stages of the utilization of a resource, there is an increasing, exponential growth of its consumption. During the first stages of its consumption, the resource is defined as a *new* or *emerging resource*.

Given the finite amount of all materials and resources, an exponential growth of consumption of any commodity cannot be sustained indefinitely. This ushers the second stage of the resource utilization, when the growth of the resource slows down because of two other factors:

1. There are no more new applications and new uses of this resource.
2. The price of the resource starts increasing because of profit taking. Producers increase its price because they realize that the resource will be depleted and will become more valuable in the future.

At this stage, the resource is called a *mature resource,* and its consumption is governed by its price, which, in turn, is determined by demand and supply. The total production and consumption of the resource may start decreasing during this stage, depending on the price level. The rate of production and consumption of the resource passes through a maximum. In the case of petroleum, the maximum is often referred to as *Hubbert's peak* or *peak oil* [17].

The third stage in the theoretical life cycle of the energy resource occurs when it is realized that there are few reserves of the resource left. At this stage, the resource is classified as a *rare* or *depleting resource.* Because the resource is scarce, the real and nominal prices of the resource rise continuously and significantly. The widespread use of the resource becomes uneconomical, and new technologies are developed to substitute this resource with other, more abundant, longer-lasting, more secure, and more affordable resources. This is sometimes called the *substitution effect.* The immediate consequence of the substitution effect, which is driven by the increase in the price of the commodity, is that the consumption of the resource slows down and finally diminishes when the resource is about to be depleted. The three stages of the development and depletion of an energy resource may be modeled by the following equation [18,19]:

$$P(t) = \frac{\omega Q_\infty \exp\left[\omega(\tau - t)\right]}{\left[1 + \exp\left(\omega(\tau - t)\right)\right]^2} = \frac{\omega Q_\infty}{4\left[\cosh\left[\dfrac{\omega}{2}(\tau - t)\right]\right]^2}, \qquad (4.7)$$

where $P(t)$ is the production level during any year t; Q_∞ is the total available quantity of the resource, often called ultimate recoverable reserves, or *ultimate reserves*; ω is the inverse of the characteristic time of the exponential increase and decay of the resource production; and τ is the time/year at which the production of the resource reaches its maximum and starts decreasing. The annual rate of production $P(t)$ in Equation 4.7 is the derivative of a commonly used function in economics, the *logistic equation.*

Figure 4.4 shows the production function and the three stages of the life cycle of petroleum production in the United States, which was the subject of Hubbert's original model [15]. The three curves in the figure represent (a) the best fit to the original Hubbert's curve, which predicts a peak in 1971; (b) the modification of this curve if the US petroleum resources were twice as much as in the original model; and (c) a similar production curve with twice the resources, but with the production peak occurring in 1995 instead of 1971. It is apparent in this figure that by modifying the parameters of Hubbert's curve (total available resources and timescale), one may derive entirely different scenarios for the production and consumption of petroleum and any other resource, where the logistic equation applies. Hubbert's peak, which has been often used for future price predictions, shifts by several decades, according to the parameters used for the generation of the curve.

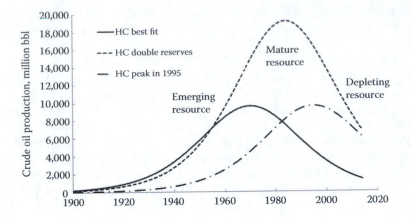

FIGURE 4.4

Annual petroleum production in the United States: best fit of the original Hubert's curve (*solid line*), optimized Hubert's curve with double the originally assumed petroleum resources (*dashed line*), and Hubbert's curve peaking in 1995 (*solid–dot line*). HC, Hubert's curve.

Hubbert's model is frequently referenced in the popular press; it has been widely used for the predictions of future fossil fuel supply and prices and deserves careful examination. It must be noted that *Hubbert's curve* and *Hubbert's peak* are not part of a physical law, which has universal validity, but only a hypothesis that was promulgated in the 1970s and is based on the limited empirical production data of that time. While the early part in the consumption of energy resources may be well approximated by an exponential curve, all the empirical data since the 1980s prove that the production and consumption of fossil fuel resources does not follow this curve. Figure 4.5 depicts the historical and the more recent data for the US petroleum production together with an optimized Hubbert curve and the real price of petroleum in 2015 US dollars. It is observed in this figure that while the initial domestic petroleum production curve followed an exponential growth, the growth stopped in the early 1970s and the domestic production declined for a number of years,

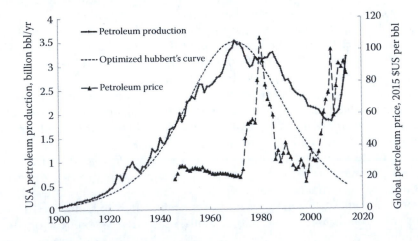

FIGURE 4.5

Domestic petroleum production in the United States (*solid line*) and global petroleum prices, in constant 2015 US dollars. An optimized Hubbert's curve (*dashed line*) is also shown.

only to increase again after 2008, following the dramatic rise of the petroleum prices in the first decade of the twenty-first century. The decline of the US domestic petroleum production in the period of 1971–2008 is due to the importation of cheaper petroleum from abroad, principally from the Middle East countries and Venezuela and not to any signs of depletion [20].

It is observed in Figure 4.5 that the global petroleum price is a much better parameter to explain the fluctuations of the domestic US petroleum production after 1970. The data prove that the US domestic petroleum production follows the global prices with a time lag of 3–4 years—the time it takes for new production facilities to be planned, designed and constructed and to start producing [20]. The global petroleum prices, which are an important variable in petroleum production, are very much influenced by global political and military events, especially events in the Middle East, as it is also apparent in Figure 2.10 [21,22].

A close examination of the global petroleum and natural gas data in the period of 1960–2015 proves that because of the development of international markets and the extensive trade of the two fossil fuels, a model for the depletion of the fossil fuels would apply to the global, rather than the national or regional production of petroleum and natural gas. International trade and the capability of nations to substitute their domestic production with imports have made the global petroleum production and petroleum reserves more significant parameters to consider as the principal variables in an energy resource model [20].

4.4.2 Life Cycle of Fossil Fuels: New Models for the Depletion of a Resource

All mineral resources, including fossil fuels, exist in the crust of the earth in finite amounts. The number of molecules and the total mass of coal, petroleum, and natural gas are finite. In the last three centuries, the rate of consumption of the fossil fuel resources by far outpaced their rate of formation. It is apparent that if we continue mining and using fossil fuels at their current levels, these resources will be exhausted in the future. The questions "when will the fossil fuels be depleted?" or "when will the petroleum be depleted?" have very high political, social, and economic significance and may only be answered by detailed mathematical models that include all the factors, which influence the global demand of the several types of fossil fuels. Most important of these factors are as follows:

1. *The total global amount of the resource*: This is not only the currently proven reserves, which fluctuate with the price of the fuel, but also the entire amount of the energy resource, proven, inferred, and speculated, often called the *ultimate reserves* Q_∞. When the demand for a scarce resource is not satisfied by the supply, its price increases to higher levels that allow the exploration and mining at places that were not considered and exploited at the lower prices. Recent examples of this effect are the offshore petroleum drilling and deep-sea exploration and petroleum production in the North Sea, the Gulf of Mexico, and off the coasts of Angola and Nigeria. Offshore production was not economically justified during the period of lower petroleum prices in the period of 1950–1972, and most offshore petroleum fields were not considered as viable reserves. In the second decade of the twenty-first century, the ultimate reserves include the unconventional fossil fuel resources—principally tars sands and shale and oil and gas. The annual production of these resources is determined primarily by price and secondarily by regional and national energy security considerations.

2. *Technological advances*: New or improved technology allows for the exploration and extraction of fossil fuels in remote parts of the world, offshore locations, the more systematic secondary and tertiary recoveries of the fuel resources, and the extraction of the resource from tight rock formations (shale) that was impossible to do before. For fossil fuels, the following are among the significant technological advances that occurred after 1960:

   a. Strip mining that allows easy access to coal deposits and more thorough extraction

   b. Offshore drilling and production for gas and petroleum using floating platforms and underwater pipelines

   c. Directional drilling for petroleum and gas wells that enable the better penetration of the production zone with horizontal and inclined wells

   d. Invention of bottom hole pumps and motors, which are placed at the bottom of wells and increase the rate of production

   e. Hydraulic fracturing (fracking) that enables the extraction of petroleum and natural gas from tight shale formations

   f. Secondary and tertiary petroleum recovery methods, including (a) injecting a penetrating and displacing fluid, e.g., carbon dioxide, in the production zone to displace and recover the petroleum and (b) heating high-viscosity petroleum reserves by burning part of the resource with injected air or by microwaves

   g. Tar sand extraction and oil recovery from tar sands

3. *Environmental regulations*: Concern for the environment and related international, national, and regional regulations and treaties have affected the production of fossil fuels since the late part of the twentieth century. Acid rain mitigation regulation changed the exhaust systems of coal burners and boilers. The effect of the Kyoto protocol was to significantly reduce the amount of coal consumption and production in the countries of the European Union. A potential binding international agreement to mitigate GCC effects of fossil fuels will significantly reduce the production and consumption of coal and, very likely, of petroleum and natural gas.

4. *Political and military events*: Wars, revolutions, social upheavals, and regional political instabilities, especially instabilities in petroleum-producing countries. The 1973 Arab–Israeli war ushered the oil crisis of the 1970s. The 1979 revolution in Iran and the taking of the hostages in the American Embassy significantly affected both the price and the domestic (United States) production of petroleum. Although very short in duration, the Gulf War of 1991 had an impact on the petroleum price and the petroleum production. The 2003 war in Iraq and subsequent political instability in that country had long-lasting impacts on petroleum production and prices. And the "Arab Spring" of 2011 that disrupted oil and gas production in several Middle Eastern countries sustained the very high oil prices and boosted conservation as well as the domestic production in the United States and several other countries. On the contrary, the relative stable political environment in the oil-producing countries of the Middle East after 2014 ushered the lower petroleum prices of 2015–2017. The lower prices were followed by the reduction of domestic investment in oil rigs, which resulted in reductions of the domestic petroleum production in the United States and Canada [22,23]. The influence of

such political events on the global price of petroleum is apparent from the data of Figure 2.10, which shows the clear relationship between political and military events in the oil-exporting countries and the global price of petroleum.

5. *Substitution effects*: At the beginning of the twentieth century, coal and wood were the dominant fuels used for cooking and the heating of buildings. When the cleaner natural gas became available and was piped into urban dwellings, it substituted coal and most of the wood for domestic use. Gasoline and diesel substituted biofuels in IC engines. Some of the natural gas functions in buildings were later substituted by electric power. Kerosene, acetylene, and natural gas substituted whale oil and wax for the lighting of buildings at the end of the nineteenth century. All three fuels were substituted by the more convenient electric energy during the first part of the twentieth century. In the 1970s, the cheaper coal substituted and almost eliminated the petroleum derivatives for the production of electricity in most of the petroleum-importing countries. A society that consumes several types of fuels will substitute the use of one fuel for another if the latter is cleaner, more convenient to use, or cheaper.

6. *The state of global trade*: An implicit assumption in the balance of demand and supply for primary energy forms is that the international trade of energy resources that balances the deficit of one state with the surplus of another [24] continues to be unimpeded. Any event that interrupts or impedes global trade also causes energy shortages—permanent or temporary—that must be met by increasing the domestic production (and prices) or by finding other routes of international trade. The temporary closure of the Suez Canal in 1973 caused energy shortages in Western Europe and enormous energy price increases that distressed European economies. Severely restricted international trade caused petroleum shortages in both world wars, disruptions of all the economic activities, and the search for alternative liquid and gaseous fuels derived from coal and wood gasification and liquefaction.

Based on these six factors, and the available historical and recent data for petroleum and natural gas global production, a new model was developed for the global production of fossil fuels [20]. The new model is based on the following conclusions that emanate from the empirical data for the global production of petroleum and natural gas:

1. The initial part of the production curve follows an exponential growth. At this stage, the resource is characterized as an emerging resource.

2. The second part of the curve follows a linear growth. At this stage, the resource is characterized as a mature resource. Most of the fossil fuel resource is produced and consumed during this period of linear production growth.

3. There are no empirical data on the trends and the shape of the production curve when the energy resource becomes a depleting resource and production continuously decreases on a global scale. In the history of civilization, there has not been an energy resource that was thoroughly depleted on a global scale by human consumption. Because of the currently high rates of consumption, fossil fuels may become the first such energy resource that will be depleted on a global scale. Given the lack of data, the new model makes the assumption that the area under the depleting resource (the total amount of the energy consumed during this period) is equal to the area under the emerging resource curve.

These observations yield the following equations for the production curve of petroleum and natural gas:

1. For the time period of the emerging resource,

$$P(t) = \frac{P(T_1)}{\cosh^2\left[\dfrac{\omega_1}{2}(T_1 - t)\right]}, \quad t \le T_1, \tag{4.8}$$

where ω_1 is the inverse of the timescale of the emerging resource and is obtained from the empirical production data during the emerging resource stage.

2. For the time period of a mature resource, where the linear growth occurs,

$$P(t) = P(T_1) + \frac{dP}{dt} t, \quad T_1 \le t \le T_2. \tag{4.9}$$

3. For the time period of a depleting resource,

$$P(t) = \frac{P(T_2)}{\cosh^2\left[\dfrac{\omega_2}{2}(T_2 - t)\right]}, \quad T_2 \le t, \tag{4.10}$$

where ω_2 is the inverse of the timescale of the depleting resource. Figure 4.6 depicts the three parts of this model. The final equation that determines the time of peak oil T_2 is [20]:

$$\frac{1}{2}\left[\frac{dP}{dt}\right][T_2 - T_1]^2 + P_{T_1}[T_2 - T_1] + \frac{4P_{T_1}}{\omega_1} - Q_\infty = 0, \tag{4.11}$$

where P_{T_1} is the annual production rate during the year T_1.

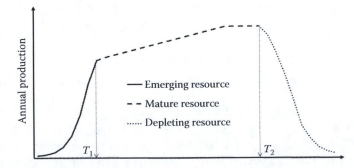

FIGURE 4.6
New model for the life cycle of a fossil fuel resource that emanates from empirical data of petroleum and natural gas production.

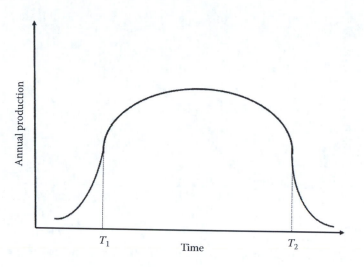

FIGURE 4.7
Schematic of the variation of the new model of a fossil fuel resource with gradual consumption decrease.

A more elaborate variation of this model takes into account the gradual decrease in petroleum and natural gas production because of rising prices and the substitution effect. Very high prices are expected when the fossil fuel resources are close to becoming depleting resources. The production curve of this model is schematically shown in Figure 4.7. The two areas that represent the quantity of the resource produced during the emerging and depleting resource periods are still the same, but the linearly, increasing part of the production curve is followed by a symmetrically decreasing part, when prices rise and substitution effects start taking place gradually. Under this scenario, the governing equation for the production of the fossil fuel resources and the determination of the time of peak oil T_2 becomes

$$\frac{1}{6}\left[\frac{dP}{dt}\right][T_2 - T_1]^2 + P_{T_1}[T_2 - T_1] + \left[\frac{4P_{T_1}}{\omega_1} - Q_\infty\right] = 0. \qquad (4.12)$$

Equations 4.11 and 4.12 are quadratic algebraic equations of the form $Ax^2 + Bx + C = 0$. An inspection of the coefficients proves that $A > 0$, $B > 0$, and $C < 0$. The discriminants of both equations ($B^2 - 4AC$) are positive, and each equation always has two real roots, one positive and the other negative. The positive root is the only realistic root that determines the time period $T_2 - T_1$, which must be positive.

From the empirical data of the period of 1960–2015 for the global production of petroleum and natural gas, the following values are obtained for the coefficients of the two equations [20]:

For petroleum: $T_1 = 1{,}973$; $dP/dt = 0.334$ Gbbl/year^2; $P_{T1} = 20.6$ Gbbl/year; $\omega_1 = 0.0748$ year^{-1}; and $Q_\infty = 3{,}000$ Gbbl.

For natural gas: $T_1 = 1973$; $dP/dt = 5.332$ Tscf/year^2; $P_{T1} = 112.9$ Tscf/year; $\omega_1 = 0.180$ year^{-1}; $Q_\infty = 14{,}900$ Tscf.

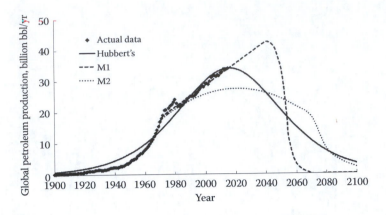

FIGURE 4.8
Historical data of global petroleum production, optimized Hubberts curve and the two new models.

Figure 4.8 depicts the three production models for the global petroleum production. Hubbert's curve is optimized with Q_∞ = 3,000 Gbbl and ω based on the available data. The curve for model M1 represents the linearly increasing model when the resource is a mature resource [20]. The curve for model M2 represents the modified parabolic model. The historical and recent data on global petroleum production are also depicted in the figure. It is observed that the linear model M1 represents the up-to-date data and the recent trends more faithfully than the other models.

It must be noted that there is high uncertainty with all these variables, and especially with the ultimate recoverable reserves Q_∞, which is the dominant variable in all the fossil fuel production models. This happens because there are several unexplored regions on the earth and new technological developments that may be applied to fossil fuel extraction worldwide. As fossil fuel resources become depleting resources and their prices rise, the higher exploration activity and the application of new technologies will bring up Q_∞ to much higher levels than current estimates.* The estimates of the models place the peak oil year between 2032 and 2070 and the *peak natural gas* between 2033 and 2076, with possible extensions until the beginning of the twenty-second century, if higher energy conservation measures are adopted and the fossil fuel demand is moderated. Before this stage is reached, it would be prudent for the human community to seek and develop alternative energy sources for the future.

4.5 Environmental Effects

The most significant and adverse environmental effect of fossil fuels is the emissions of CO_2, which is the most common greenhouse gas (GHG). The SO_2 and NOX emissions that were significant environmental threats in the past are now almost under control, as described in Chapter 3 and as shown in Figure 3.3. The solid materials (ash) produced

---

* The quote "To know ultimate reserves, we must have ultimate knowledge. Nobody knows this, and nobody should pretend to know" is often attributed to Morris Adelman. We can only estimate the ultimate reserves and draw conclusions based on our (uncertain) estimates.

during the coal combustion process are commonly stored or buried near the production site or they are mixed with cement to produce concrete.

Example 4.5

The ultimate analysis of a type of lignite is C, 40%; H, 3%; S, 2%; N, 3%; moisture, 25%; and solids/ash, 28%. The lignite is used in a 150 MW power plant with 36% thermal efficiency; 20% excess air is used to achieve complete combustion. Determine (a) the daily amount of lignite used, (b) the daily amount of effluents in the combustion products, (c) the daily amount of ash to be disposed.

Solution:

a. The LHV of this lignite is $0.40 \times 32{,}770 + 0.03 \times 119{,}950 + 0.02 \times 9{,}163 = 16{,}890$ kJ/kg.

   a. The daily amount of heat input to this power plant is $(150/0.36) \times 60 \times 60 \times 24 = 36 \times 10^6$ MJ $= 36 \times 10^9$ kJ. The daily combustion of $36 \times 10^9/16{,}890 = 2.131 \times 10^6$ kg of lignite (2,131 t) provides the heat input. Based on the composition of this lignite, the masses of C, H, and S burned daily are 852.4×10^3 kg, 63.9×10^3 kg, and 42.6×10^3 kg, respectively.

b. C, H, and S undergo combustion with air. The mass ratios of oxygen to each of these elements are for C, 2.67; for H, 8; and for S, 1. Therefore, for the complete combustion of these elements, the mass of oxygen that is daily used is $2.67 \times 852.4 \times 10^3 + 8 \times 63.9 \times 10^3 + 1 \times 42.6 \times 10^3$ kg $= 2{,}830 \times 10^3$ kg of oxygen, which is contained in $12{,}146 \times 10^3$ kg of air. Because 20% excess air is used, the daily amount of air supplied to the plant is $12{,}146 \times 1.20 = 14{,}575 \times 10^3$ kg, or 12,289 m³ under standard conditions.

   From the preceding paragraph, and the mass balance of the chemicals produced the burner of this power plant, the daily amount of effluents are the following:

- b1: $3{,}125 \times 10^3$ kg of CO_2
- b2: $1{,}047 \times 10^3$ kg of H_2O, 536 from the moisture content and 511 from the hydrogen combustion
- b3: 85.2×10^3 kg SO_2 (discharged unless it is removed at the scrubber)
- b4: 569×10^3 kg of the excess oxygen (not used in combustion)
- b5: $14{,}643 \times 10^3 \times 0.767 = 11{,}179 \times 10^3$ kg of nitrogen, which is part of the air and is not used in the combustion process.

c. The daily amount of ash that is disposed is $2{,}131 \times 10^3 \times 0.28$ kg $= 597 \times 10^3$ kg (597 t).

Several other environmental effects of fossil fuels are related to their mining and transportation.

4.5.1 Coal Mining and Strip Mining

Coal mining has been going on for centuries in several countries. In the traditional mining process, tunnels are drilled in the crust of the earth and the coal is extracted. While in the past, human power did most of the extraction work, more recently, excavation machinery and conveyor belts operated by the miners are used in modern coal mines. Some of the environmental and ecological effects associated with deep earth coal mining are as follows:

1. *Water pollution*: When deep earth mines are abandoned, they fill with water from local underground leakages. Several minerals, including toxic compounds, dissolve into the water. Leakage or a sudden breach from the mine transports the

polluted water to underground aquifers and surface rivers or lakes, which supply drinking water to populations in the surrounding region. The August 2015 spill of 3 million gal of contaminated water from an abandoned gold mine to the Animas River (a tributary of the Colorado River) is a prime example of the risks associated with mining. The toxic sludge released from the mine polluted a river system that supplies drinking water to communities with millions of inhabitants and made this water undrinkable for months.

2. *Land subsidence*: The removal of coal from the interior of the earth creates voids and leaves the entire area susceptible to subsidence.

3. *Site preparation*: This includes the construction of facilities at the mining site, the construction of roads for coal hauling, the removal of vegetation or partial deforestation, and wildlife disturbance by the increased traffic and workers in the area.

4. *Excavation activities*: Primarily, the removal and displacement of soil, which invariably causes the release and suspension of fine particulates (dust) and aerosols in the air.

5. *Mining accidents and hazards*: Deep earth mining has always been a dangerous profession. Accidents and channel cave-ins have claimed the lives of tens of thousands of miners, since the beginning of the industrial revolution. An additional risk of coal mining is later-life cancers due to mesothelioma and black lung diseases. It is estimated [4] that coal mining claims one human life for every 1 billion kWh of electricity produced.

The strip mining of coal was developed in the 1950s and is now the predominant mining method for coal. Strip mining is used for the extraction of seams of coal closer to the surface of the earth. The rock and dirt (sometimes called the "overburden") above the coal seam are removed to expose the coal, which is then extracted by powerful engines that operate on the surface. Strip mining has reduced mining accidents in the deep tunnels, and because it is a very efficient method of mineral extraction, it has significantly reduced the cost and the market price of coal. However, strip mining poses several additional environmental threats among which are the following:

1. *Release of particulates*: Explosives are frequently used for the removal of the overburden. The explosions remove from the ground and disperse in the local environment fine dust particles. The increased concentration of particulates is always detrimental to public health. The environmental and public health damage can be severe if particles of toxic chemicals are released and suspended in the air.

2. *Land reclamation*: At the end of their useful life, several strip mines in the past were abandoned by their operators to become giant mudholes. Legislation has been adopted in most countries (in the United States, the Surface Mining and Reclamation Act of 1977) for the reclamation and restoration of all lands affected by surface mining. Restoration is usually accomplished by replacing the topsoil in strip mines and planting vegetation. However, in several cases, reclamation has not restored the disturbances to the ecosystem and the environment because of lack of sufficient water supply.

3. *Water runoff pollution*: Strip mining exposes underground sulfur to air, which is a constituent of many coals. The sulfur is oxidized in the atmosphere and, with the

abundant water, forms hyposulfuric acid (H_2SO_3) and, to a lesser extent, sulfuric acid H_2SO_4. The two acids are the principal components of acid rain. In the case of strip mines, the acids dissolve in the rainwater and the runoff transports them to surface rivers and lakes, which become significantly more acidic, to the detriment of aquatic life.

4.5.2 Oil Transport and Spills

Improvements in oil drilling technology have allowed the offshore exploration and extraction of petroleum. In recent years, coastal and deep-sea wells extract petroleum from deep reservoirs in the North Sea, the Gulf of Mexico, the coasts of Nigeria and Angola, and several other locations of the globe. Any disruption of the pipeline that conveys the petroleum from the deep sea to the land terminals causes a spill with environmental and ecological damage.

Petroleum and refined petroleum products are transported daily in the sea by a fleet of oil tankers and on land by a group of pipelines that crisscross entire continents as well as by trains and trucks. Any accident, large or small, in the petroleum transportation system will cause an oil spill that will have adverse effects on the local environment and the ecosystems. Three oil spills that attracted global notoriety are as follows:

1. The *Exxon Valdez* oil spill occurred in Prince William Sound, Alaska, on March 24, 1989, when the oil tanker *Exxon Valdez*, which was carrying approximately 1,300,000 bbl of petroleum, struck a reef and spilled an estimated 260,000–900,000 bbl of its cargo to the remote Prince Island Sound. Even though there were no human casualties, the spilled oil covered 28,000 km² of sea; it washed out on 2,100 km of coastline and damaged a pristine ecological area that was habitat to salmon, seals, sea otters, and several species of seabirds.

2. The *Deepwater Horizon* oil platform was owned and operated by Transocean and leased by BP. A surge of natural gas on April 22, 2010, blasted through the seal of a deep sea well, traveled up the pipeline and ignited, causing an explosion that killed 11 workers and injured another 17. The entire oil rig sank and the ruptured pipeline allowed large quantities of petroleum and natural gas to escape uncontrolled for almost 4 months. The volume of liquid petroleum escaping the rupture pipeline reached a maximum of 60,000 barrels per day and created a very large oil spill in the northern part of the Gulf of Mexico. This had very significant, adverse effects on the environment and the coastal ecosystems of Louisiana, Mississippi, Alabama, and Florida, as well as on the economies of the four states.

3. The *Lac Mégantic* accident in the early hours of July 6, 2013, was caused by a runaway train with 74 tank cars full of petroleum. The train, which was automatically operated, was a long-haul petroleum carrier from Montana to the East Provinces of Canada. High speed caused the train to derail after a 7 mi downhill run. The fractured tanks released 1.5 million gal of petroleum that immediately caught fire and incinerated a large part of the Lac Mégantic town, killing 47 of its inhabitants and injuring hundreds. Apart from the human casualties, the inferno caused the death of large populations of animals and birds as well as significant chemical contamination in the surrounding area.

4.5.3 Hydraulic Fracturing (Fracking)

Hydraulic fracturing (fracking) is used to increase the productivity of oil and gas reservoirs. Oil and gas wells in shale rock formations have a vertical section and end in an almost horizontal section that penetrates the low permeability shale formation. Water is first injected into the well under very high pressure, penetrates the production zone, and creates local fissures and cracks in the shale rocks. After the creation of the cracks, the *proppant* (a mixture of water with sand, solid organics, and polymers) is injected. When the high water pressure subsides, the solids in the proppant keep the cracks open and maintain the permeability of the shale rock at higher levels. Fracking allows oil and gas to flow easier and faster into the production wells; it significantly increases the productivity of the wells; and it allows for the extraction of a higher fraction of the existing oil and gas to be extracted from the shale rock.

Although fracking has been experimentally tried by the oil and gas industry since the 1940s as a method for enhanced oil recovery, it has not been extensively used until the early part of the twenty-first century and has become a very controversial issue in the United States and Canada, because of the following:

1. The additive materials used with the water are proprietary and their composition has not been released. Some of these materials are thought to be toxic and carcinogenic.
2. The cracks and fissures in the shale rock alter the stability of the rock and induce low-level seismicity.
3. In a few cases, the induced cracks and fissures have propagated in the local aquifers. This has created a pathway for the volatile gases, most notably methane, to enter the water wells with the potential to contaminate the local water supply.
4. The injection of radioactive tracers, which have been used to determine the extent of the penetration of the proppant in the production zone.
5. Very high quantities of water are needed. The fracking process of a single well may use up to 6 million gal of water (more than 20 million L).

The fracking controversy has pitted a large fraction of the oil and gas industry in the United States and Canada against environmental groups and several local citizens' organizations. As a result of the controversy, the State of New York has banned the use of fracking, while most of the other states and the Canadian provinces allow this practice. Several municipalities in North Texas banned fracking within their city limits, but the Texas legislature passed a law, in 2015, that prohibits such local bans on fracking within the city limits. A report issued by the Texas Academy of Medicine, Engineering and Science falls short from endorsing fracking and recommends more research on the practice [25].

Central to the fracking controversy is the issue of "property rights," what constitutes such rights, and how far such rights can be extended. On one side, the owners of the (underground) oil and gas production fields claim that all the oil and gas that exists underground is their property, and hence, they have the right to use all available methods for the extraction of their mineral resources. On the other side, owners of surface properties (e.g., city dwellers, farmers, and ranchers) in the same area observe that as a result of the fracking activity, water from local aquifers becomes contaminated, and the increased low-level seismic activity disrupts their lifestyle and causes ground movements that induce cracks in their dwellings. Such effects infringe on their own property rights. Fracking is a technological development that certainly not only allows higher oil and gas production, but has also become a social and legal issue of property rights and environmental stewardship. Further research on the effects of fracking and several lawsuits in the United States and Canada are expected to provide technical and legal resolutions to this controversial environmental issue.

4.6 Future of Fossil Fuel Consumption

Since the beginning of the industrial revolution, fossil fuels have been a source of motive power, heat, and lighting. Fossil fuels have been the driving force behind the economic prosperity of several nations, including all Organisation for Economic Co-operation and Development (OECD) countries. However, there are significant impacts of fossil fuels on the environment, most notably GCC, which may lead to regulations for the curtailment of combustion of fossil fuels. Environmental concerns notwithstanding, we still (in 2017) do not have other widely available energy sources that would provide motive power, heat, and lighting at a similar cost and with the same reliability. Fossil fuels are currently very important energy sources that are needed in the foreseeable future to sustain the economic prosperity of nations. Considerations of the available technology and a glance on the global energy demand and supply prove that we do not have other energy sources that can readily substitute for the use of fossil fuels with the same convenience, economy, and reliability. Moreover, a viable plan for the gradual reduction and elimination of fossil fuels on a global scale, which is necessary for the mitigation of GCC effects, does not exist. The 2015 Paris Agreement falls short from being a definitive global plan for the reduction of GHGs.

The developed (OECD and other more affluent countries) are trapped with the use of fossil fuels, because they need to sustain their economies. Developing countries with fossil fuel reserves—the majority of United Nations (UN) countries—have already made plans for the use of these resources to accelerate their own economic development. The frequently pronounced grandiose "plans" for the use of cleaner energy resources in the developing countries are often unrealistic and have failed in the past 40 years. The transition to higher efficiency and cleaner energy sources, e.g., carbon sequestration or solar energy, has proven to be very much capital intensive and unaffordable by many. The absence of sustainable energy infrastructure and energy storage has proven to be an impediment for several nations: the "leap frog" progress paradigm (e.g., starting with cell phones without the growth of a landline system of telecommunications) does not work for the production and transmission of the vast quantities of needed energy.

A characteristic of the global reliance and necessity for fossil fuel combustion in both developed and developing countries is that the International Energy Agency—a branch of the UN—estimates that the consumption of fossil fuels will increase by more than 15% between 2012 and 2035 even under the optimistic "New Policies Scenario" that includes a great deal of conservation and new policies for the promotion of renewable energy [24]. Independently derived recent projections by the US Department of Energy up to the year 2040 give the same indications for the demand of petroleum and natural gas in the near and intermediate future [26].

If our society is to wean itself from the use of fossil fuels, it is necessary to carefully plan for the transition to more sustainable energy forms, in both the developed and developing countries. The global plan to be adopted must be viable and must take into consideration the economic progress of all nations, the availability of other energy resources, the reliability of their supply, and the necessary infrastructure. In order to have a significant mitigating effect on GCC, the transition plan must be acceptable to all the nations and must be enforceable on a global scale. In 2017, the international community is several years—perhaps several decades—behind adopting such an effective global plan. In the absence of a global plan, the transition period for fossil fuels will be long and will probably extend well after the mid-twenty-first century or until the fossil fuels are depleted.

It is inevitable that the transition from fossil fuels to other more sustainable forms of energy will occur at some point in the future. There are two driving forces that will ensure such a transition in the future:

1. The irreversible combustion processes that continuously diminish the finite quantity of fossil fuels
2. Increased global concerns over climate change and economic disruptions

However, no one can predict when this point in the future will be. In the meantime, scientists and engineers must strive to develop the technology that leads to a more sustainable future and to more efficiently and sensibly use the remaining fossil fuel resources.

4.7 CO_2 Avoidance

With the realization that the increased CO_2 concentration in the atmosphere significantly contributes to GCC, environmental organizations, educational institutions, and regional and national governments have encouraged the community to adopt measures for the reduction of fossil fuel consumption. This is primarily achieved by three types of actions:

1. Energy conservation measures
2. Improved efficiency measures
3. Increased use of renewable energy sources

The amount of CO_2 not produced by these actions is referred to as CO_2 *avoidance*. Based on the mass balances of the reactions that produce CO_2 in Table 4.1, one may calculate the CO_2 avoidance by saving from consumption given quantities of fuels. Table 4.6 contains the CO_2 avoidance for several quantities of common fuels. The numbers in the table are based on the most common composition of the fuel.

Example 4.6

After several energy conservation measures in the community, it is no longer necessary to operate the 400 MW power plant mentioned in Example 4.2. Determine the annual CO_2 avoidance from the closing of this power plant.

Solution: From the solution of Example 4.2, the power plant uses 3,008,226 kg of coal daily with 75% carbon content. Therefore, $0.75 \times 3,008,226 = 2,256,168$ kg of carbon burn to produce 8,272,616 kg of CO_2 daily and 3.020×10^9 kg of CO_2 annually. Hence, the annual CO_2 avoidance is 3.020×10^9 kg CO_2.

TABLE 4.6

CO_2 Avoidance for Several Common Fuels

	CO_2 Avoidance, kg	CO_2 Avoidance, lb
1 kg of carbon	3.67	8.07
1 gal of gasoline or diesel	8.71	19.16
1 L of gasoline or diesel	2.32	5.10
1 m³ of methane at standard conditions	2.40	5.28
1 m³ of CO at standard conditions[a]	4.20	9.24
100 ft³ of propane at standard conditions	15.28	33.62
1 t of anthracite with 90% carbon	3,300	7260
1 L (0.789 kg) of ethyl alcohol	0.75	1.65

[a] CO and CH_4 are the two principal constituents of natural gas.

PROBLEMS

1. Determine the HHV of propane.

2. A new gas turbine with a small bottoming cycle uses hexane as its fuel. The cooling of the exhaust gases liquefies 72% of the water that is formed during combustion. Determine the heating value of the fuel for this process in kilojoules per kilogram of the fuel.

3. A 600 MW power plant with 38% thermal efficiency uses bituminous coal with heating value 26,500 kJ/kg. How much coal does this power plant use per day and per year? If the coal contains 8% ash by weight, what is the amount of ash collected per year?

4. The ultimate analysis (by mass) of a bituminous coal is C, 78%; H, 1%; S, 2%; ash, 11%; and 8%, moisture. Determine (a) the LHV of this coal; (b) if the coal is burned with 30% excess air (130% theoretical air), the total air required per kilogram of coal for the combustion; (c) the amount of coal (in tons) used daily in a 400 MW power plant with 39% efficiency and the amount of air supplied to the combustion chamber.

5. The ultimate analysis (by mass) of a type of lignite is C, 35%; H, 1%; S, 1%; ash, 18%; and 45%, moisture. Determine (a) the LHV of this coal; (b) if the coal is burned with 40% excess air (140% theoretical air), the total air required per kilogram of coal for the combustion; and (c) the amount of this type of coal (in tons) used daily in a 400 MW power plant with 39% efficiency and the amount of air supplied to the boiler.

6. It is proposed to operate a 200 MW power plant with 35% thermal efficiency with peat, which has heating value 10,500 kJ/kg and has 22% ash content. Determine (a) how many tons of peat are to be used daily by this power plant and (b) what is the daily and yearly amount of ash produced.

7. A heavy oil fuel with density of 900 kg/m³ has the following composition by mass: C, 82%; H, 15%; and S, 3%. The fuel is burned in an engine with 50% excess air. Determine (a) the LHV of the fuel and (b) the mass and the volume of air used for the combustion of 200 L of this fuel.

8. Gasoline with density of 700 kg/m³ has the following composition by mass: C 81%; H 17%; S 2%. The fuel is burned in an IC engine with 50% excess air. Determine (a) the LHV of the fuel and (b) the mass and volume of air used for the combustion of 40 L of this gasoline.

9. The composition by volume of natural gas from a well is 35% CH_4, 12% C_2H_6, 3% H_2, 15% CO, and 35% N_2. Determine the LHV and the HHV of this gas in kilojoules per kilogram.

10. Determine the fluidization velocity for particles of equivalent radii 3 mm, 5 mm, and 10 mm in air at standard conditions (density: 1.186 kg/m^3). The density of the particles is 800 kg/m^3, and you may assume that the drag coefficient for all particles is constant, 0.24.

11. A small FBR operates with coal particles with density of 860 kg/m^3 and diameter of 1 cm. The interior of the FBR conditions are at 1.1 atm and 700 K. The drag coefficient of these particles is 0.19. Determine the minimum fluidization velocity for this reactor (hint: you will need to calculate the air density).

12. Should hydraulic fracture (fracking) of natural gas fields be banned? Write a 300-word essay supporting your opinion.

References

1. Galbraith, J.K., *The Affluent Society*, (repr. 1975) Pelican Books, London, 1958.
2. Galbraith, J.K., *The New Industrial State*, (repr. 1975) Pelican Books, London, 1967.
3. Moran, M.J., Shapiro, H.N., *Fundamentals of Engineering Thermodynamics*, 6th edition, Wiley, New York, 2008.
4. Gates, D.M., *Energy and Ecology*, Sinauer, Sunderland, MA, 1985.
5. El-Wakil, M.M., *Power Plant Technology*, McGraw-Hill, New York, 1984.
6. Dunlap, R.A., *Sustainable Energy*, Cengage Learning, Stamford, CT, 2015.
7. US Energy Information Administration, *North America Leads the World in Production of Shale Gas*, US Energy Information Administration, Washington, DC, October 2013.
8. Speight, J.G., *The Chemistry and Technology of Petroleum*, 5th edition, CRC Press, Boca Raton, FL, 1999.
9. Michaelides, E.E., *Heat and Mass Transfer in Particulate Suspensions*, Springer, Berlin, 2013.
10. Shingles, T., McDonald, A.F., Commercial experience with synthol CFB reactors, in *Circulating Fluidized Bed Technology II*, eds, Basu, P., Large, J.F., Pergamon Press, Toronto, 1988.
11. Krenzke, P.T., Davidson, J.H., Thermodynamic analysis of syngas production via the solar thermochemical cerium oxide redox cycle with methane-driven reduction, *Energy Fuels*, **28**, 4088–4095, 2014.
12. Michaelides, E.E., *Particles, Bubbles and Drops—Their Motion, Heat and Mass Transfer*, World Scientific Publishers, Hackensack, NJ, 2006.
13. Grace, J.R., Leckner, B., Zhu, J., Cheng, Y., *Fluidized Bed Reactors*, Chapter 17 in *Multiphase Flow Handbook*, 2nd edition, eds, Michaelides, E.E., Crowe, C.T., Schwarzkopf, J.D., CRC Press, Boca Raton, FL, 2017.
14. Syamlal, M., Musser, J., Dietiker, J.F., The two-fluid model in the open-source code MFIX, Chapter 2 in *Multiphase Flow Handbook*, 2nd edition, eds, Michaelides. E.E, Crowe, C.T., Schwarzkopf, J.D., CRC Press, Boca Raton, FL, 2017.
15. Hubbert, M.K., *Nuclear Energy and the Fossil Fuels*, Proceedings of the Spring Meeting of the Southern District, American Petroleum Institute, San Antonio, TX, 1956.
16. Hubbert, M.K., Energy Sources of the Earth, *Sci. Am.*, **224**, 31–40, 1971.
17. Deffeyes, K.S., *Hubbert's Peak*, reissue, Princeton University Press, Princeton, NJ, 2009.
18. Laherrère, J.H., The Hubbert curve: Its strengths and weaknesses, *Oil Gas J.*, February 2000.

19. Cavallo, A.J., Hubbert's petroleum production model: An evaluation and implications for world oil production forecasts, *Nat. Resour. Res.*, **13**, 211–221, 2004.
20. Michaelides, E.E., A new model for the lifetime of fossil fuel resources, *Nat. Resour. Res.*, **26**, 161–175, 2017.
21. Michaelides, E.E., *Alternative Energy Sources*, Springer, Berlin, 2012.
22. US Energy Information Administration, *Petroleum and Other Liquids—History*, US Energy Information Administration, Washington, DC, 2015.
23. US Energy Information Administration, *International Energy Statistics*, US Energy Information Administration, Washington, DC, 2016.
24. International Energy Agency, *Key World Statistics*, IEA-Chirat, Paris, 2015.
25. TAMEST (The Academy of Medicine, Engineering and Science of Texas), *Environmental and Community Impacts of Shale Development in Texas*, TAMEST, Austin, TX, 2017.
26. US Energy Information Administration, *Petroleum and Other Liquids—Annual Projections to 2040*, US Energy Information Administration, Washington, DC, 2015.

5

Nuclear Energy

For a society that strives to wean itself from the use of fossil fuels, nuclear energy provides a very simple and feasible alternative. Nuclear power plants have the capability to produce a high fraction of our electric power at relatively low cost and without any carbon dioxide emissions. A case in point: Had the United States constructed 56 additional nuclear reactors in the 1990s, in addition to the currently operating 103 commercial reactors, the country would have been in compliance with the Kyoto protocol. Several Organisation for Economic Co-operation and Development (OECD) countries, in particular France and Japan, produce more than 70% of their electricity from nuclear power plants. Despite the setbacks in the late twentieth century with the nuclear accidents at Three Mile Island (TMI) and Chernobyl and the more recent one at Fukushima, in the second decade of the twenty-first century, nuclear energy is on the rise. Several power plants are in the construction stage in developing economies, most of them in Asia. More nuclear power plants throughout the globe are in the initial design, financing, or construction stage.

This chapter provides an overview of the use of nuclear energy, starting with the fundamental physics of fission and fusion processes. The nuclear reactions and the radioactive decay of elements are explained in a simple, mechanistic manner. The chain reaction and the neutron cycle in a thermal reactor are also explained. Based on the fundamentals of fission, the essential components of nuclear reactors as well as the operation of the basic types of most commonly used nuclear reactors are described in simple terms. The function and operation of the breeder reactors is also included because if the world is to rely on nuclear energy in the long-term, the more abundant uranium-238 and other fertile nuclear materials must be utilized in breeder reactors, which may become the next generation of reactors. Several useful parameters, which help understand the importance of nuclear reactors and help derive quantitative conclusions on the use of nuclear energy, are computed and presented. Finally, and because past accidents in nuclear power plants have impeded the growth of nuclear energy for decades and have spread fear among the public, three accidents that received the highest notoriety, at the TMI in the United States, at Chernobyl in the former Soviet Union, and at the Fukushima Dai-ichi power plant in Japan, are described. The causes and scenarios of the three accidents are enumerated, and early actions by the plant operators that could have prevented or mitigated the accidents are suggested.

5.1 Elements of Nuclear Physics

Every atom is composed of a *nucleus* (plural: nuclei), and nuclei are composed of *protons* and *neutrons*. Protons and neutrons are often referred to as *nucleons*. The *electrons* of atoms orbit the nuclei. The electrons are very light particles and have negative charges, $e = -1.602 \times 10^{-19}$ Coulomb (C). The protons are heavier particles with positive electric charges of +1.602 × 10⁻¹⁹ C, and the neutrons are heavier particles without any electric charge. The numbers of electrons and protons are the same in all atoms, and this makes every atom electrically neutral. Figure 5.1 schematically shows the structure of the atom of carbon-12 ($_6C^{12}$). The nucleus of this atom is composed of six protons and six neutrons. Six electrons are also present and revolve around this nucleus at two distinct orbits as shown in the figure.

The radii of the electron orbits are very large: if the nucleus were of the size of a tennis ball (approximately 7 cm), the first electron orbit would have been at a distance of 600 m and the second at a distance of 2,000 m. The size of nuclei and the distances of the electron orbits from the nuclei follow similar proportions. When one puts together these dimensions and visualizes the atoms, one reaches the conclusion that matter, as we experience it, is chiefly composed of vacuum. The nuclei, where most of the mass in the universe is contained, occupy a very small fraction of the space.

The unit of mass, by which nuclei and subatomic particles are measured, is the *atomic mass unit* u, which is equal to 1.6604 × 10⁻²⁴ g (1.6604 × 10⁻²⁷ kg). This elemental quantity of mass is defined as equal to 1/12 of the mass of the carbon-12 atom. The masses of the three elementary particles and their respective electric charges are as follows [1]:

• Mass of an electron:	• 0.000549 u	• charge: −1.602 × 10⁻¹⁹ C
• Mass of a proton:	• 1.007277 u	• charge: +1.602 × 10⁻¹⁹ C
• Mass of a neutron:	• 1.008665 u	• charge: 0

All atoms are characterized by their *atomic number* Z and their *mass number* A. The atomic number is equal to the number of protons in its nucleus. Thus, oxygen has eight protons and its atomic number is Z = 8. For carbon, Z = 6; for barium, Z = 56; and for uranium, Z = 92. Since the numbers of protons and electrons are equal, the atomic number of the atom is also equal to the number of the electrons orbiting the atom. The mass number is equal to the sum of the numbers of protons and neutrons in the nucleus. The mass number is an

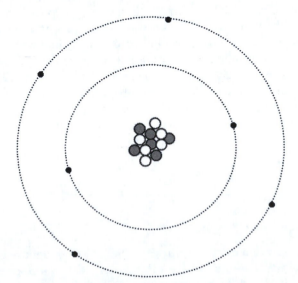

FIGURE 5.1
Schematic diagram of the carbon-12 atom. Protons (*gray*) and neutrons (*white*) are clustered inside the nucleus.

integer, even though the actual mass of an atom may not be an integer number of atomic mass units u.

There are atoms that have the same number of protons (and, hence, the same number of electrons) but differ in their numbers of neutrons. Such atoms are called *isotopes* of an element. Isotopes have the same atomic number Z but different atomic mass A. For example, uranium-235 and uranium-238 are two isotopes of the same element. Their nuclei have 92 protons, but they have different numbers of neutrons. Because they have the same number of electrons, isotopes have the same chemical properties and cannot be separated chemically.

An atom is usually denoted by its chemical symbol (O for oxygen, U for uranium, Pb for lead, etc.) accompanied by the atomic number as a left subscript and the mass number as a right superscript. Thus, the general notation for an atom, with chemical symbol X, is $_Z X^A$. For example, $_{92}U^{235}$ denotes the uranium isotope, which has an atomic number of 92 and mass number of 235, sometimes also referred to as uranium-235. Since the difference $A - Z$ is equal to the number of neutrons, this isotope has 92 protons and 143 neutrons. Similarly, the symbol $_{92}U^{238}$ denotes the uranium isotope with 146 neutrons and mass number of 238 (uranium-238), while $_{92}U^{233}$ denotes the uranium isotope with 141 neutrons and mass number of 233 (uranium-233). Other common isotopes are as follows:

- Of hydrogen: $_1H^1$ (common hydrogen); $_1H^2$ or D (deuterium); and $_1H^3$ or T (tritium), which have 1 proton and 0, 1, and 2 neutrons, respectively.
- Of oxygen: $_8O^{16}$, $_8O^{17}$, and $_8O^{18}$, all of which exist in the atmosphere at different concentrations.
- Of plutonium: $_{94}Pu^{238}$, $_{94}Pu^{239}$, $_{94}Pu^{240}$, and $_{94}Pu^{241}$, all of which have been artificially formed by nuclear reactions.

In our macroscopic world, the mass of the elements is measured in grams (g), kilograms (kg), or gram-atoms. The latter is equal to the atomic mass A expressed in grams. Thus, a gram-atom of $_8O^{16}$ is equal to 16 g, while a kilogram-atom of $_{94}Pu^{239}$ is equal to 239 kg. The number of atoms in a gram-atom is a constant for all the elements and is equal to 6.023×10^{23} atoms per g-atom (or 6.023×10^{26} atoms per kg-atom). This very large number is known as the *Avogadro number*, N_A.

Central to the theory of energy obtained from nuclear reactions is the mass-energy equivalence relation derived by Einstein:

$$E = mc^2, \tag{5.1}$$

where c is the speed of light in the vacuum, approximately equal to 3×10^8 m/s. It follows that the equivalent energy of 1 u is 1.494×10^{-10} J. A commonly used energy unit in nuclear reactions is the electron volt (eV), which is equal to 1.602×10^{-19} J. The equivalent energy of 1 u is 9.31×10^8 eV or 931 MeV. This implies that a mass of 1 u will produce 931 MeV if it were entirely converted to energy. If 1 kg were converted to energy, it would produce 5.62×10^{29} MeV or 9×10^{16} J. This is a very high amount of energy, equivalent to the energy obtained from 3,100,000 t of bituminous coal. Such calculations exemplify the advantage of nuclear fuels in comparison to fossil fuels: a very small amount of nuclear fuel may produce vast amounts of heat.

Table 5.1 shows the values of the atomic masses of a few common isotopes [1].

A glance at Table 5.1 and the masses of protons, neutrons, and electrons prove that the actual mass of an atom is slightly smaller than the sum of the masses of its constituent

TABLE 5.1

Atomic Masses of Common Isotopes, in u

$_1H^1$	1.007825	$_6C^{12}$	12.00000
$_1H^2$	2.01410	$_6C^{13}$	13.00335
$_1H^3$	3.01605	$_7N^{14}$	14.00307
$_2He^3$	3.01603	$_8O^{16}$	15.99491
$_2He^4$	4.00260	$_8O^{17}$	16.99914
$_2He^5$	5.0123	$_8O^{18}$	17.99916
$_3Li^6$	6.01513	$_{92}U^{234}$	234.0409
$_3Li^7$	7.01601	$_{92}U^{235}$	235.0439
$_5B^9$	9.01333	$_{92}U^{238}$	238.0508

protons, neutrons, and electrons. The mass of boron-9 ($_5B^9$) is 9.01333 u, while the sum of the masses of its constituent five protons, five electrons, and four neutrons is 9.07380 u. The small difference between the two numbers 0.06047 u is called the *mass defect* of an atom. One way to understand the significance of the mass defect is to relate it to the equivalent energy, which in this case is 56.3 MeV: If the boron atom were to be constructed by its constituents, there would be a mass defect of 0.06047 u. The equivalent energy 56.3 MeV (or 9.008×10^{-12} J) would be released if this atom were formed by its subatomic particles. The lower energy of the atom in comparison to its constitutive elements signifies that this is a stable atom.

While this quantity of energy for the formation of atoms may appear to be very small, it must be noted that if a gram-atom (9 g) of $_5B^9$ were to be formed from its constituent elements, the amount of energy released is $9.008 \times 10^{-12} \times 6.023 \times 10^{23}$ or 5.43×10^{12} J, and this is a very high amount of energy. This is the "magic" of the very large Avogadro number: the miniscule amount of energy released by an atom translates to a tremendous amount of energy released by a small amount of mass of an element. A high mass defect and a high amount of energy released per nucleon characterize the stable isotopes. In general, the more stable an isotope is, the higher is the mass defect per nucleon.

Example 5.1

What is the mass defect of lithium 7? If 1 kg of this element were to be formed by its constituent subatomic particles, what would be the energy released?

Solution: The atom of $_3Li^7$ has mass 7.01601 u. It is composed of three protons, four neutrons, and three electrons. The total mass of the constitutive elemental particles is $3 \times 1.007277 + 4 \times 1.008665 + 3 \times 0.000549 = 7.05814$ u. Therefore, the mass defect of the atom is $7.05814 - 7.01601 = 0.04213$ u.

One kg of $_3Li^7$ contains approximately $6.023 \times 10^{26}/7 = 0.86 \times 10^{26}$ atoms, for a total mass defect $0.86 \times 10^{26} \times 0.04213 = 3.62 \times 10^{24}$ u.

Since 1 u produces 931 MeV of energy, this mass defect corresponds to 3.37×10^{27} MeV or 5.40×10^{14} J. This is a very high amount of energy and explains why lithium-7 is a very stable isotope.

5.1.1 Nuclear Fission

The concept of nuclear fission was developed in the 1930s after the detection of the neutron. While the protons and electrons were known elementary particles since the late nineteenth century, the neutron was discovered in 1932. Neutrons are electrically neutral particles that are not repelled by the negative charge of the electrons or the positive charge of the nuclei. Thus, neutrons may penetrate the outer orbits of electrons, reach the nuclei, and directly interact with them. It is reasonable to postulate that some neutron–nuclei interactions will cause the splitting (fission) of less stable nuclei. The artificial fission process was first detected in an experiment by Hahn and Strassmann in the late 1930s in Germany: they observed that when uranium-235 ($_{92}U^{235}$) was bombarded with a beam of neutrons, barium-139 was produced. They concluded that the $_{92}U^{235}$ atoms split to produced barium-139 and several other isotopes.

Fission occurs when neutrons are captured by heavy and relatively unstable nuclei, such as $_{92}U^{235}$, $_{92}U^{238}$, and $_{92}Th^{232}$. During the fission process, a neutron is absorbed by a larger nucleus to form a composite and highly unstable nucleus. The unstable nucleus splits into two large fragments and releases a few—two or three—free neutrons. A typical fission process with uranium-235 as the fuel is schematically shown in Figure 5.2. The four stages of this process show how the free neutron and the $_{92}U^{235}$ nucleus produce the unstable nucleus of $_{92}U^{236}$. The latter oscillates almost like a large droplet. It then splits into the isotopes $_{57}La^{147}$ and $_{35}Br^{87}$ (lanthanum and bromine). In the process, two free neutrons are also produced. This fission reaction may be written as follows:

$$_{92}U^{235} + {}_{0}n^{1} \rightarrow {}_{92}U^{236} \rightarrow {}_{57}La^{147} + {}_{35}Br^{87} + 2\,{}_{0}n^{1}. \tag{5.2}$$

Oftentimes, the intermediate step of the production of uranium-236 is omitted in the reaction. Lanthanum and bromine are not the only elements produced by fission. A plethora of isotopes are produced by the interaction of neutrons with the $_{92}U^{235}$ isotope. A second typical fission reaction of the $_{92}U^{235}$ nucleus produces xenon-140 and strontium-94:

$$_{92}U^{235} + {}_{0}n^{1} \rightarrow {}_{92}U^{236} \rightarrow {}_{54}Xe^{140} + {}_{38}Sr^{94} + 2\,{}_{0}n^{1}. \tag{5.3}$$

Plutonium-239 ($_{94}Pu^{239}$), which is artificially produced by the capture of a neutron by the $_{92}U^{238}$ nucleus, may also undergo fission in several ways. One such fission reaction of $_{94}Pu^{239}$ produces barium-137 and strontium-100 in addition to three free neutrons:

$$_{94}Pu^{239} + {}_{0}n^{1} \rightarrow {}_{56}Ba^{137} + {}_{38}Sr^{100} + 3\,{}_{0}n^{1}. \tag{5.4}$$

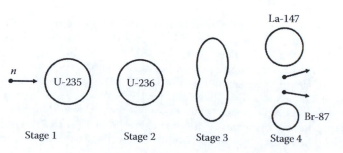

FIGURE 5.2
Stages in the fission process of $_{92}U^{235}$ by neutrons.

The entire fission process takes place very fast, at timescales on the order of 10^{-14} s. The free neutrons produced in a reaction may interact with other nuclei and cause additional fissions and additional free neutrons. The latter cause a new generation of fissions, and in this way, the fission process is repeated for several generations of neutrons.

Conventional nuclear reactors use uranium as their fuel. Natural uranium is mined in several parts of the earth and contains both isotopes $_{92}U^{235}$ and $_{92}U^{238}$. Small amounts of plutonium-239 ($_{94}Pu^{239}$) are always artificially produced in most conventional reactors. When they are split by neutrons, they contribute to the thermal energy produced and add to the neutron flux. The so-called *fast breeder reactors* (FBRs) produce larger amounts of $_{94}Pu^{239}$, which later become the main fuel of the reactor, as will be explained in more detail in Section 5.3.4.

The amount of energy released in the fission of $_{92}U^{235}$ and $_{94}Pu^{239}$ solely depends on the products of the reaction and may be calculated from the mass defects of the particular reaction. The average energy of the fission of $_{92}U^{235}$ has been calculated [1] to be approximately 200 MeV. Most of this energy is in the form of the kinetic energy of the fission fragments and is almost immediately dissipated into heat by atomic and molecular collisions. Additional energy (again in the form of heat) is continuously produced from the radioactive decay of the reaction products, the so-called "daughter isotopes." This thermal energy must be continuously removed from the nuclear reactor by a circulating coolant, typically water or gas, which is called the "primary coolant." During the normal operation of a nuclear power plant, the removal of heat from the reactor by the primary coolant provides the heat to a thermodynamic cycle, a Rankine or a Brayton cycle. The continuously produced heat is converted by the machinery of the cycle to electric energy. If for any reason the cycle stops and the primary coolant does not remove the produced heat, an emergency coolant must be injected in the reactor to ensure the removal of the heat produced. A failure of the cooling systems may result in the overheating of the nuclear reactor and, perhaps, a meltdown, which has catastrophic consequences.

Example 5.2

Use the periodic table of elements to complete the following nuclear reactions by identifying the isotopes X, Y, and Z.

1. $_{90}Th^{232} + _{0}n^{1} \rightarrow _{53}I^{129} + X + 3_{0}n^{1}$

2. $_{92}U^{235} + _{0}n^{1} \rightarrow _{55}Cs^{138} + Y + 2_{0}n^{1} + _{-1}e^{0}$

3. $_{3}Li^{7} + _{0}n^{1} \rightarrow 2Z + _{-1}e^{0}$

Solution: For the first reaction: The atomic number of element X is equal to $90 + 0 - 53 - 3 \times 0 = 37$. Hence, the element is rubidium (Rb). The atomic mass of this isotope is $232 + 1 - 129 - 3 \times 1 = 101$. Therefore, $X = _{37}Rb^{101}$.

For the second reaction: The atomic number of element Y is equal to $92 + 0 - 55 - 2 \times 0 - 1 \times (-1) = 38$. Hence, the element is strontium (Sr). The atomic mass of this isotope is $235 + 1 - 138 - 2 \times 1 - 1 \times 0 = 96$. Therefore, $Y = _{38}Sr^{96}$.

For the third reaction: Because two atoms of element Z are produced from this reaction, the atomic number of element Z is equal to $\frac{1}{2} \times [3 - 1 \times (-1)] = 2$. Hence, the element is helium (He). The atomic mass of this isotope is $\frac{1}{2} \times (7 + 1 - 0) = 4$. Therefore: $Z = _{2}He^{4}$.

5.1.2 Nuclear Fusion

Fusion is the source of energy of the stars including our own sun. Nuclear fusion occurs when the nuclei of lighter isotopes, primarily hydrogen isotopes, combine to form the heavier helium nuclei, as for example in the following simple reaction:

$$4_1H^1 \rightarrow {}_2He^4. \tag{5.5}$$

Table 5.2 [1] shows the masses, in atomic mass units (u) of the most important isotopes that are involved in fusion reactions.

It must be noted that the fusion reaction of Equation 5.5 does not instantaneously occur in the stars or the laboratory, because the simultaneous collision of four hydrogen nuclei is extremely improbable to occur. Rather, the reaction occurs in two stages, with deuterium being the intermediate isotope [2]:

$$\begin{aligned} 2_1H^1 &\rightarrow {}_1H^2 +_{+1} e^0 \quad \text{or simply} \quad 2H \rightarrow D, \\ 2_1H^2 &\rightarrow {}_2He^4 \quad \text{or simply} \quad 2D \rightarrow He^4. \end{aligned} \tag{5.6}$$

Other paths to nuclear fusion are via tritium (${}_1H^3$ or simply T) and helium-3 (${}_2He^3$) [2]:

$$\begin{aligned} {}_1H^3 + {}_1H^2 &\rightarrow {}_2He^4 + {}_0n^1 \quad \text{or simply} \quad T+D \rightarrow He^4 + n, \\ {}_2He^3 + {}_1H^2 &\rightarrow {}_2He^4 + {}_1H^1 \quad \text{or simply} \quad He^3 + D \rightarrow He^4 + p. \end{aligned} \tag{5.7}$$

It is apparent from the values of Table 5.2 that the mass defect of the helium formation is 0.0287 u. Since the energy equivalent of 1 u is 931 MeV, the production of a single nucleus of helium-4 from four light hydrogen atoms releases 26.72 MeV of thermal energy. When 1 kmol of helium-4 is produced (4.0026 kg), the energy released is 257.5×10^{13} J, a very high amount of energy. The energy produced by the formation of 41,000 kmol of helium-4 (using approximately 164 t of hydrogen) would be sufficient to satisfy the entire energy demand of a country such as the United States for an entire year. Several tons per second of helium-4 produced provide all the energy our sun broadcasts to the entire solar system. Fusion reactions have explained the formation and evolution of the universe since the big bang and the life cycles of stars.

Regarding human-made fusion, the very first large-scale demonstration of artificial fusion was the development and detonation of the hydrogen bomb (the H-bomb). Fusion

TABLE 5.2

Atomic Masses of Commonly Occurring Isotopes in Fusion Reactions

Isotope	Mass, u	Isotope	Mass, u
${}_1H^1$	1.007825	${}_2He^4$	4.00260
${}_1H^2$ (D)	2.01410	${}_2He^5$	5.0123
${}_1H^3$ (T)	3.01605	${}_3Li^6$	6.01513
${}_2He^3$	3.01603	${}_3Li^7$	7.01601

in the H-bomb occurs by the combination of two atoms of deuterium ($_1H^2$ or simply D) to form a single atom of $_2He^4$. The fusion reaction in the H-bomb, which takes place at extremely high pressures and temperatures, is crudely accomplished by the detonation of a conventional nuclear weapon (the U-bomb). The blast and the shock waves produced by the detonation of the U-bomb create the conditions for the fusion to occur. It is apparent that if these fusion nuclear reactions occurred in a controlled manner, the tremendous energy released could be used for the peaceful production of electricity.

Controlled nuclear fusion at a scale that would produce significant power has not been achieved yet. The principal reason for this is that for fusion to occur, the hydrogen must be in the state of plasma, at temperatures of a few hundred million kelvins (10^8 K) and pressures of the order of 10^{10} atmospheres for a short time duration, which is known as the *ignition condition* [2]. An ignition condition may also be achieved at lower pressures and temperatures, but the nuclei must be maintained at these conditions for a longer period.

The vast reserves of hydrogen in the oceans of the earth ensure that if controlled fusion is achieved in practice, fusion-derived energy could supply the energy needs of the humans for billions of years. For this reason, many projects on fusion research and technology are conducted in several countries. While this research has high promise, we are still many decades, perhaps centuries, before the development of controlled and safe fusion reactors that would produce significant electric power. Several technological breakthroughs are still needed for the harnessing of fusion energy for peaceful applications. While a great deal of investment is required for these technological breakthroughs to occur, if the technology develops for fusion energy to be harnessed at a commercial scale, the energy needs of the earth's population will be solved!

Example 5.3

What is the mass defect and the energy produced when one atom of $_1H^1$ and one atom of $_1H^3$ fuse to form an atom of $_2He^4$? What is the energy produced when 1 kmol of each of the two isotopes undergo fusion?

Solution: This nuclear reaction may be simply written as $_1H^1 + _1H^3 \rightarrow _2He^4$. From Table 5.2, the combined mass of the hydrogen and tritium atoms in the left-hand side is 4.023875 u and the mass of helium on the right-hand side is 4.00260 u. Hence, the mass defect of this reaction is $\Delta m = 0.021275$ u. Since 1 u produces 931 MeV of energy, the fusion of the two atoms will produce 19.807 MeV of energy.

1 kmol has 6.023×10^{26} molecules. Since the hydrogen molecule has two atoms, each kilomole of the two isotopes has $2 \times 6.023 \times 10^{26} = 12.046 \times 10^{26}$ atoms.

The mass of the two isotopes is approximately 2 kg of hydrogen and 6 kg of tritium. The energy produced when the 2 kmol (1 kmol of each of the two isotopes fuse) is $19.807 \times 12.046 \times 10^{26} = 2.386 \times 10^{28}$ MeV or 3.818×10^{15} J. This is a very large amount of energy to be produced by about 8 kg of fuel. Approximately 25,000 such reactions would produce the equivalent of 90 Q which is the total primary energy annually consumed in the United States.

5.1.3 Radioactivity

Several chemical elements have a number of isotopes, some stable and others unstable.* The unstable isotopes undergo spontaneous transformations to different elements or to

---

* There are approximately 110 known elements and more than 1500 isotopes (nuclides). Of these, only 279 are stable.

isotopes of the same element. This spontaneous process is called *radioactivity*. The original nucleus is often referred to as the *parent* nucleus, and the product of the transformation, as the *daughter* nucleus. The unit of radioactivity is one transformation per second or 1 Bq. A second radioactivity unit is oftentimes used, the curie (1 Ci), which is equal to 3.7×10^{10} Bq or 3.7×10^{10} transformations per second. The unit 1 Ci represents the radioactivity of one gram of radium-226.*

In general, stable isotopes (especially those with atomic numbers less than 40) have approximately the same number of protons and neutrons. When there is a significant imbalance in the numbers of protons and neutrons in the nucleus, the atom is unstable, undergoes a nuclear transformation, and forms a different element. The types of transformations are categorized as follows:

1. *Alpha decay*: During this transformation, an *alpha particle* (i.e., the nucleus of $_2\text{He}^4$) is emitted from an isotope. Alpha decay typically occurs with the heavier nuclei of atomic numbers higher than 82, as in the conversion of plutonium-239 to uranium-235:

$$_{94}\text{Pu}^{239} \rightarrow {}_{92}\text{U}^{235} + {}_2\text{He}^4. \tag{5.8}$$

In an alpha decay, the daughter nucleus has an atomic mass 4 units less than that of the parent nucleus and an atomic number 2 units less than that of the parent nucleus.

2. *Beta decay*: During this transformation, a neutron inside the nucleus becomes a proton, and an electron (beta particle) is simultaneously emitted. This type of radioactive decay occurs in nuclei where the number of neutrons is significantly higher than the number of protons. A very small subatomic particle, which is called a neutrino and denoted by ν, accompanies the emission of the electron as in the following example of the conversion of indium-115 to tin-115:

$$_{49}\text{In}^{115} \rightarrow {}_{50}\text{Sn}^{115} + {}_{-1}e^0 + \nu. \tag{5.9}$$

The parent and daughter nuclei of the beta decay have the same atomic mass number. The atomic number of the daughter nucleus is higher by 1.

3. *Positron emission decay*: The positron is a light particle of the same mass as the electron, but has positive charge and is denoted by $_{+1}e^0$. Positron decay occurs when the nucleus has too many protons in comparison to the number of neutrons as in the following example of the conversion of iron-53 to manganese-53:

$$_{26}\text{Fe}^{53} \rightarrow {}_{25}\text{Mn}^{53} + {}_{+1}e^0 + \nu. \tag{5.10}$$

The parent and daughter nuclei have the same mass number. The atomic number of the daughter nucleus is one unit less than that of the parent nucleus. The emitted positron is a short-lived antimatter particle among the plethora of electrons that surround the nuclei.

---

* Radium-226 and the phenomenon of radioactivity were discovered by Marie and Pierre Curie, who were students of Professor Henri Becquerel at the Grande Ecole de Physique et Chimie Industrielles, in Paris. The three shared the Nobel Prize in Physics in 1903.

Positrons are fast annihilated after combining with electrons, producing two *gamma particles*, which are very strong electromagnetic radiation (strong X-rays):

$$_{+1}e^0 + _{-1}e^0 \rightarrow 2\gamma. \tag{5.11}$$

Oftentimes, the daughter nucleus is unstable and transforms to another daughter nucleus. Thus, a parent nucleus may cause a series of transformations called a *radioactive decay series*. Such decay series are typical of the transuranium radioactive elements, with $Z > 92$ that produce a long series of daughter, mostly radioactive, elements. The last element in this radioactive decay series is usually lead-206 or another stable isotope. The series of nuclear reactions in Equation 5.12 shows the radioactive decay of uranium-238 as a series of alpha and beta decay processes, which involve the several intermediate isotopes of thorium (Th), protactinium (Pa), radium (Ra), radon (Rn), polonium (Po), and bismuth (Bi):

$$
\begin{aligned}
_{92}U^{238} &\rightarrow _{90}Th^{234} + _2He^4, &\text{alpha decay,} \\
_{90}Th^{234} &\rightarrow _{91}Pa^{234} + _{-1}e^0, &\text{beta decay,} \\
_{91}Pa^{234} &\rightarrow _{92}U^{234} + _{-1}e^0, &\text{beta decay,} \\
_{92}U^{234} &\rightarrow _{90}Th^{230} + _2He^4, &\text{alpha decay,} \\
_{90}Th^{230} &\rightarrow _{88}Ra^{226} + _2He^4, &\text{alpha decay,} \\
_{88}Ra^{230} &\rightarrow _{86}Rn^{222} + _2He^4, &\text{alpha decay,} \\
_{86}Rn^{222} &\rightarrow _{84}Po^{218} + _2He^4, &\text{alpha decay,} \\
_{84}Po^{218} &\rightarrow _{82}Pb^{214} + _2He^4, &\text{alpha decay,} \\
_{82}Pb^{214} &\rightarrow _{83}Bi^{214} + _{-1}e^0, &\text{beta decay,} \\
_{83}Bi^{214} &\rightarrow _{84}Po^{214} + _{-1}e^0, &\text{beta decay,} \\
_{84}Po^{214} &\rightarrow _{82}Pb^{210} + _2He^4, &\text{alpha decay,} \\
_{82}Pb^{210} &\rightarrow _{83}Bi^{210} + _{-1}e^0, &\text{beta decay,} \\
_{83}Bi^{210} &\rightarrow _{84}Po^{210} + _{-1}e^0, &\text{beta decay,} \\
_{84}Po^{210} &\rightarrow _{82}Pb^{206} + _2He^4, &\text{alpha decay.}
\end{aligned}
\tag{5.12}
$$

Some of these transformations are very fast while others may take millennia to occur. A sample of the mass of a radioactive isotope consists of a very large number of nuclei, which continuously decay. If a sample initially has a number N_0 nuclei, the number of nuclei that remains after a time t is given by the following expression:

$$N = N_0 e^{-\lambda t}. \tag{5.13}$$

The constant λ in the last equation may be given in terms of the *half-life* $T_{1/2}$ of an isotope, by the following equation:

$$\lambda = \frac{\ln 2}{T_{1/2}} = \frac{0.693}{T_{1/2}}. \tag{5.14}$$

TABLE 5.3

Half-Lives and Radioactivity Emitted by Some Common Isotopes

Isotope	Half-Life	Radioactivity
Uranium-235	7.1×10^8 years	α, γ
Uranium-238	4.51×10^9 years	α, γ
Plutonium-239	2.44×10^4 years	α, γ
Thorium-232	1.41×10^{10} years	α, γ
Krypton-87	76 minutes	β
Strontium-90	28.1 years	β
Barium-139	82.9 minutes	β, γ

The half-life of an isotope represents the time for the number of nuclei to be reduced to half their original number. Table 5.3 shows the half-lives of several common isotopes [3]. It is apparent that the half-lives span several orders of magnitude and range from millions of years to a few seconds. A glance at the table and the last two equations proves that isotopes with small half-lives are rapidly depleted, while isotopes with very long half-lives are very slowly depleted.

Radioactivity is a natural phenomenon that occurs regardless of the human-made nuclear reactors and atomic weapons. Several radioactive elements in the environment contribute to the *natural* or *baseline radioactivity* of the earth. Cosmic rays also carry a few radioactive isotopes to the earth's atmosphere. Some of the naturally occurring radioactive isotopes are absorbed by humans and animals and become part of their metabolism. For example, potassium-40 exists in the human body and contributes approximately 0.1 µCi of radioactivity (3700 transformations per second).

While low-level exposure of humans may be harmless, the exposure to high levels of radioactivity is very harmful, because the emitted particles mutate and destroy cells in the human body. The radioactivity that followed the explosions of the two atomic bombs in Hiroshima and Nagasaki as well as the Chernobyl nuclear power plant accident dispersed high doses of radioactive materials in the environment with catastrophic consequences for the local human and animal populations [4].

Several practical units of radioactive exposure have been adopted. The most meaningful and most commonly used unit of radioactive exposure is the sievert (Sv), which takes into account not only the amount of radiation received by the human body, but also the type and severity of the radiation. In this system of measurements, X-rays, γ-rays, and β particles (electrons) are weighted by a factor of 1; neutrons, by a factor of 10; and α particles, which cause the most harm to living tissue, are weighted by a factor of 20. The radioactive exposure (dose) of a person on the surface of the earth by natural radioactivity is approximately 2.2×10^{-3} Sv per year. Short-term (e.g., over a week) doses of 1 Sv cause radiation sickness to humans and doses of 3–5 Sv cause acute radiation sickness that may lead to death. Short-term doses over 5 Sv have proven to be lethal.

Example 5.4

After the accident at Chernobyl, a great deal of strontium-90 was deposited on the soil of several parts of Russia, Belarus, and Ukraine. Part of this isotope was absorbed by the soil, and another part was washed by rain runoff. Of the original amount that was absorbed, what are the percentages of strontium-90 that remained after 1 year, 5 years, and 20 years from the accident? When is the absorbed isotope expected to reach 10% of its initial concentration?

Solution: From Table 5.3, the half-life of strontium-90 is 28.1 years. Hence, for this isotope, $\lambda = 0.0247$ year^{-1}. Using Equation 5.13, we obtain the following for the fraction N/N_0 of the remaining isotope:

- After 1 year: $N/N_0 = 0.976$ or 97.6% of strontium-90 remained.
- After 5 years: $N/N_0 = 0.884$ or 88.4% of strontium-90 remained.
- After 20 years: $N/N_0 = 0.611$ or 61.1% of strontium-90 remained.

The strontium-90 deposits will reach 10% of their initial concentration when $N/N_0 = 0.1$. Equation 5.13 gives $t = 93.35$ years for this to happen. Since the Chernobyl accident occurred in 1986, this event is expected to occur in 2079.

Example 5.5

Uranium-238 is one of the longest-living constituents of nuclear waste. If 1 t of this material is stored now, how much of it remains 1000 years from now? How long does it take for a quantity of uranium-238 to decrease to 20% of its mass by radioactivity?

Solution: The half-life of uranium-238 is 4.51×10^9 years. and therefore, for this isotope, $\lambda = 1.537 \times 10^{-10}$ year^{-1}. From Equation 5.13, after 1000 years, 99.99998% (or almost all) will remain. For the amount of uranium-238 to decrease to 20% of its original mass, $0.2 = \exp(-1.537 \times 10^{-10} \times t)$, which implies that this time t will be after 10.5 billion years.

5.1.4 Chain Reaction

All the current nuclear reactors operate with the fission of uranium isotopes by neutrons. The entire operation of fission nuclear reactors depends on the production of neutrons and the maintenance of a high flux of neutrons, a process known as *chain reaction*. A schematic diagram of a chain reaction is shown in Figure 5.3. Two neutrons that are initially present

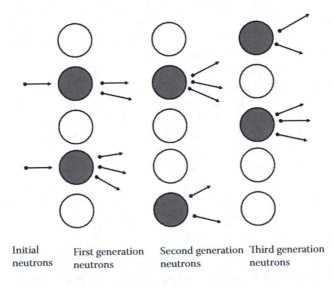

| Initial | First generation | Second generation | Third generation |
| neutrons | neutrons | neutrons | neutrons |

FIGURE 5.3
Chain reaction: The nuclei in gray undergo fission and produce two or three neutrons. The neutrons generated either cause new fissions or are absorbed by other nuclei or leak outside the reactor.

in the reactor interact with the nuclei of the first generation and cause two fissions, from which five neutrons are produced. Of the latter, three neutrons are absorbed or leak outside, but two interact with the nuclei of the second generation and cause two additional fissions, producing five new neutrons. Two of these neutrons interact with the nuclei of the third generation and so on. The necessary flux of two neutrons is maintained in the nuclear reactor by the neutrons produced during the fissions that took place during a previous stage. The neutron fluxes in actual reactors are on the order of 10^{20} neutrons/(m²·s) and cause equivalent numbers of nuclei fissions. If not enough neutrons are produced, or if too many neutrons are absorbed by the several types of nuclei in the reactor, or if neutrons leak outside the reactor, the chain reaction slows down and stops and the nuclear power plant stops producing electric power.

A key parameter of the chain reaction is the *reproduction constant k* defined as the number of neutrons in a generation divided by the number of neutrons in the preceding generation. When $k = 1$, the chain reaction is maintained. When $k < 1$, the neutron flux and, hence, the fission reaction rate are decreasing. If the number of fission reactions continues to decrease, the deficiency of neutrons causes less fission reactions, and the production of heat decreases. When $k > 1$, the neutron flux increases, the fission reaction rate increases and the reactor produces more heat power. If the condition $k > 1$ is sustained for a few seconds, the large amount of heat produced inside the reactor may cause steam explosion with catastrophic consequences, as it happened during the Chernobyl accident in 1986.

While it is rather easy to visualize the chain reaction, its practical realization with fuels other than pure $_{92}U^{235}$ poses several difficulties. The main problem is that neutrons produced from a fission reaction are fast neutrons with energies approximately 2 MeV, while fissions are primarily caused by slower, thermal neutrons with energies 0.025 eV or lower. The retardation of neutrons in a reactor is caused by a large number of collisions of the neutrons with the nuclei of lighter atoms, such as hydrogen ($_1H^1$), deuterium ($_1H^2$), or carbon ($_6C^{12}$). The first two types of atoms are common in water and heavy water and the last in graphite. The materials that slow down the neutrons—and their conversion from fast neutrons to thermal neutrons—are called *moderators*.

The conversion of fast neutrons to thermal neutrons is accomplished by a large number, typically 20–50, of collisions of the neutrons with the atoms of the moderator. As a consequence, before a neutron can cause the fission of a uranium atom, it must interact and collide with 20–50 other atoms. Some of these interactions may result in the absorption or escape (leakage) of a fraction of the neutrons that are produced. One of the principal challenges of nuclear reactor design is to slow down the neutrons produced from fissions and yield as many thermal (slow) neutrons as possible, while avoiding the leakage and capture of neutrons by other materials during the retardation process [1,3].

Contrary to popular thinking, conventional nuclear reactors that operate with natural and slightly enriched uranium cannot explode as atomic bombs. The conventional reactors simply do not produce the large number of thermal neutrons that are necessary to cause a nuclear explosion. Instead, the design of the reactors is optimized to "thermalize" (retard) the neutrons, while avoiding their capture, to ensure that a high fraction of the produced thermal neutrons survives the collisions and interactions with other materials in the reactor, and that the thermal neutrons cause enough fissions of the fuel atoms to generate enough new neutrons that will sustain the chain reaction.

5.2 Essential Components of Nuclear Reactors

In the schematic diagram of the typical power cycle of Figure 1.5, the nuclear reactor creates the high-temperature heat reservoir that supplies heat to the cycle. If a steam cycle is used in the power plant, the reactor and its associated heat exchangers become the steam boiler. If a gas cycle is used, the reactor replaces the burner (or combustion chamber). In all cases, the reactor supplies the needed heat to the power plant via its coolant. It is important to note that power is continuously produced in the reactor, and it must be continuously removed by a fluid, the reactor coolant, which is typically water, heavy water, or a gas. Several other components are added to the nuclear power plants to facilitate the transfer of heat to the working fluid of the cycle and enhance the safety of the reactor.

In 2017, there were approximately 440 nuclear reactors in operation globally, of several types and designs. Depending on the energies of the neutrons used for the chain reactions, the reactors are categorized as *thermal reactors* and *fast reactors*. The vast majority of the reactors in operation are thermal reactors, using the slow thermal neutrons that are retarded by the moderator. The reactors are also categorized according to the type of coolant used as *water-cooled* and *gas-cooled* reactors. Most of the water-cooled reactors use common water as coolant, while a small number (most notably the Canada deuterium uranium [CANDU] reactors) use heavy water. The gas-cooled reactors operate with a compressed gas as their coolant, most commonly air, helium, or argon.

The main components of the nuclear reactor that produce or withstand the highest amount of radioactivity are enclosed inside the *reactor vessel*, a cylindrical vessel with radius of 5–8 m and height of 8–12 m. The reactor vessel is made of stainless steel with a wall thickness of 6″ (15 cm) and is designed to contain all the radioactive materials that are produced by the nuclear reactions. In addition, this vessel has sufficient strength to withstand the high pressure of the cooling water and moderate reactor malfunctions.* The following are necessary components for sustaining the chain reaction. These components are common in all types of the currently operational reactors.

1. The reactor fuel, which undergoes fission and supplies the heat: Natural uranium, which is mined in several parts of the earth, primarily contains the isotopes $_{92}U^{235}$ and $_{92}U^{238}$ in proportions 1:139 atoms or 0.715%:99.285% of the total uranium atoms [3]. The naturally mined uranium is processed and slightly enriched in centrifuges to produce the fuel for most types of nuclear reactors. This fuel contains 1.5–3.5% of $_{92}U^{235}$.† A few thermal reactors in operation use $_{92}U^{233}$ and $_{90}Th^{233}$ as alternative fuels, while $_{94}Pu^{239}$ is the most common fuel of breeder reactors.

---

* During the Three Mile Island accident, the reactor vessel withstood a partial meltdown of the reactor core at local temperatures above 2500 K.
† At these low enrichment levels, the reactor fuel cannot produce an atomic weapon. Concentrations of more than 90% are needed for weapon-grade uranium.

2. The moderator, which slows down the produced fast neutrons and converts them to slower thermal neutrons: Water and heavy water are the most commonly used materials for moderators, and they double as coolants. Graphite and beryllium are the most common moderators in gas-cooled reactors.

3. The reactor coolant, which circulates in the reactors and removes the heat that is continuously produced: Water and heavy water are the most common coolants for pressurized and boiling water reactors. Helium, carbon dioxide, and a mixture of gases are the most common coolants in gas-cooled reactors. A liquid metal, such as sodium or potassium (or a mixture of two), is the coolant of the breeder reactors, which are significantly more compact. The circulation of coolant in the reactor is accomplished with a series of pumps or compressors. The failure of the coolant circulating systems* may cause severe overheating of the reactor with possible catastrophic consequences.

4. The control systems that monitor the state of the reactor; modify the neutron flux and the rate of fission in the reactor; and, in general, ensure the smooth operation of the reactor within the design guidelines.

5. The safety devices and systems, which include emergency pumps or compressors, diesel generators for the uninterrupted operation of pumps and compressors, safety valves, and tanks of cooling water. These systems are located outside the reactor vessel and inside a large containment building.

6. The radiation shield: This constitutes the outer layer of the reactor vessel and is composed of a thick layer of lead, a material that reflects or absorbs most of the radioactivity produced in the reactor. The radiation shield and the thick reactor vessel ensure that only a very small fraction of the radioactivity escapes the nuclear reactor.

In addition to the preceding systems that are necessary for the operation of a nuclear reactor, for the vast majority of the nuclear reactors built in OECD countries, the reactor vessel and all the auxiliary systems are placed within the *reactor containment building*, a large cylindrical building with a hemispherical dome. The containment building houses the reactor vessel; all the equipment needed for the circulation of the coolant, including pumps, emergency pumps, motors, emergency motors, and emergency generators; the pressurizer; several tanks with additional cooling water for emergencies; flooding tanks; and the heat exchangers—also called steam generators—where steam is produced and is subsequently fed to the turbines of the power plant. A typical layout of a reactor containment building for a pressurized water reactor (PWR) with its major components and systems is shown in Figure 5.4. For added safety, the outside walls of the containment building are made of reinforced concrete, which may withstand internal pressures (e.g., from steam escaping the reactor vessel) up to 14 atm. In the United States, the reactor containment buildings are made of reinforced concrete with a thickness approximately 90 cm (3 ft) and are designed to withstand the direct impact of a fully loaded, large commercial jetliner.

---

* This is frequently referred to as loss of coolant accident (LOCA).

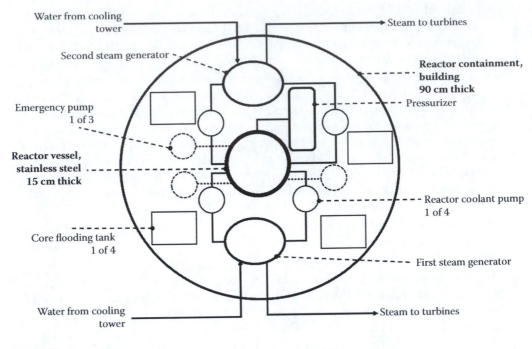

FIGURE 5.4
Layout of the major components in the reactor building in a PWR nuclear power plant.

5.3 Reactor and Power Plant Classifications

The reactor core is the heat source of the nuclear power plant. Heat generated from nuclear fissions and secondary reactions is continuously removed by the coolant, water, or a gas. The reactor core supplies the heat to a Rankine steam cycle or a Brayton gas cycle via the coolant. The coolant is either directly fed to the turbines of the power plant or (for added safety) it is directed to secondary heat exchangers—the steam generators—that produce steam in a secondary circuit. The generated steam is subsequently fed to the power plant turbines; it is condensed in the condenser and returns to the secondary heat exchanger.

As the national nuclear programs evolved in different countries, several reactor types and power plant types have been developed throughout the world. The nuclear power plants are categorized according to the reactor types, the coolants, and other significant characteristics. The most common categories and acronyms of nuclear reactors are as follows:

1. LWR: light water reactor (PWR and BWR are subcategories of LWR)
2. PWR: pressurized water reactor
3. BWR: boiling water reactor
4. GCR: gas-cooled reactor
5. AGCR: advanced gas-cooled reactor
6. HTGR: high-temperature gas reactor
7. HTAGR: high-temperature advanced gas reactor

TABLE 5.4

Typical Dimensions and Characteristics of the Most Common Types of Thermal Reactors

Type	Reactor	Thermal Power, MW	Core Diameter, m	Core Height, m	Core Volume, m³	Power Density, MW/m³	Fuel Rating, kW/t
LWR	PWR	3,800	3.6	3.8	40	95	39
	BWR	3,800	5	3.8	75	51	25
AGCR	Hincley	1,500	9.1	8.3	540	2.8	11
HWR	CANDU	3,425	7.7	5.9	280	12	26
RBMK	Chernobyl	3,140	11.8	7	765	4.1	15.4
FBR	Phenix	563	1.4	0.85	1.4	406	149
FBR	Prototype	612	1.5	0.9	1.6	380	153

8. PBR: pebble bed reactor

9. CANDU: Canadian heavy water reactor

10. HWR: heavy water reactor

11. RBMK (acronym in Russian): channel, graphite-moderated boiling-water reactor

12. VVER (acronym in Russian): the Union of Soviet Socialist Republics' (USSR) version of the PWR

13. FBR: fast breeder reactor

14. LMFBR: liquid metal fast breeder reactor

Thermal reactors, primarily water-cooled reactors (PWR, BWR, CANDU) and gas-cooled reactors (AGCR, HTGR), are the main categories of the reactors currently in operation and are often called conventional reactors. The few FBRs that were constructed are still in the testing stage and account for a very small percentage of the nuclear capacity around the world.

The power density of a reactor is the amount of power, in megawatts, produced per unit volume of the reactor. The fuel rating is the amount of power produced per unit mass of the fuel inside the reactor. Typical dimensions, power densities, and fuel ratings that are characteristics for the most common reactor types are presented in Table 5.4 [3,5]. It is apparent that among the thermal reactors, PWRs have the highest power density and provide the most thermal power per unit mass of fuel. FBRs are the most compact reactors, with power density and fuel ratings exceeding by far those of the thermal reactors. It is this very high power density of the FBRs that makes them more cumbersome and more risky to operate.

The main characteristics and the advantages and disadvantages of the nuclear power plants that utilize these reactors are summarized in the following subsections.

5.3.1 Pressurized Water Reactors and Boiling Water Reactors

Both of these types are light water reactors (LWRs), and they are the most commonly used nuclear reactors. The main difference between the two is that the PWRs produce liquid water at very high pressures (15.5–25 MPa) and moderately high temperatures (280–320°C), while in the BWRs, the water evaporates, and superheated steam is produced inside the reactor, sometimes reaching 450°C. The liquid water in the PWR is fed to steam generators, as shown in Figure 5.4, where its heat is transferred to another stream of water at lower pressure, which evaporates and is fed to the turbines.

PWRs make up for 75% of all the reactors in the United States and the majority of the nuclear reactors in France and Japan. The fuel in the PWRs is usually slightly enriched uranium containing 1.5–3.5% of the isotope $_{92}U^{235}$. Liquid water, which functions both as the moderator and as the coolant, enters the reactor vessel through a number (two to four) of nozzles and exits at a much higher temperature, which is low enough to be below the saturation temperature. Therefore, the water does not boil in the PWR reactor, but exits as liquid at high temperature. The high pressure in the reactor vessel is maintained by the *pressurizer*, a large vessel filled with a mixture of steam and water, which is pressurized by the action of electric heaters. A system of *coolant pumps* (at least four and as many as eight) circulates the cooling water through the reactor in what is called *the primary loop* or the *nuclear steam supply system*.

Because the PWR reactor coolant is in the liquid state and practically incompressible, small changes in its volume may cause significant changes in the pressure of the reactor. If the liquid water expands, e.g., by excess heat power production and the subsequent temperature rise, the systems in the reactor and the surrounding reactor vessel would be subjected to enormous internal pressure changes that may undermine the structural integrity of the reactor vessel and cause fracture. Also, if the pressure significantly decreases, boiling will occur in the bulk of the reactor, and this would cause pump cavitation and reactor malfunction.

The avoidance of large pressure fluctuations and the effective control of the reactor pressure are achieved by the operation of the *pressurizer*, an essential element of the PWR power plants. The function of the pressurizer is to accommodate the large volume changes in the reactor vessel that cause high-pressure fluctuations. The pressurizer is a large, closed vessel, which contains approximately equal volumes of water and steam. Since water is in equilibrium with steam, the temperature in the pressurizer vessel is equal to the saturation temperature that corresponds to the nuclear reactor pressure. Several electric heaters at the bottom of the pressurizer vessel supply heat to increase the amount of steam when needed. Spray nozzles at the top of the pressurizer supply colder water from the primary cooling loop, to condense a fraction of the steam. The lower part of the pressurizer is connected to the hot part (hot leg) of the primary cooling loop and through this to the reactor.

The BWR system, where water boils inside the reactor, is considerably simpler. Water boiling inside the nuclear reactor simplifies the design of the power plant, eliminates the use of the pressurizer and the additional steam generators, and lowers the cost of the entire nuclear power plant. Steam is generated inside the BWR and is directly fed to the turbine. Effectively, a BWR power plant substitutes the boiler of a conventional Rankine cycle with the nuclear reactor. Because the steam generated in the reactor carries a percentage of moisture (droplets) with it, BWR plants use a *steam separation* section, to separate liquid water from the steam, which is subsequently fed to the turbines of the power plant.

Light water (H_2O or natural water) is abundant, very inexpensive, and well suited to be both a reactor moderator as well as coolant. The only disadvantage of natural water is that it absorbs a relatively high fraction of the produced neutrons. This makes it necessary for both PWRs and BWRs to use slightly enriched uranium, which is produced by the gasification of natural uranium (as UF_6) and a series of centrifugal processes. The fuel enrichment process adds to the cost of the nuclear fuel and, adds to the amount of nuclear waste produced. On the contrary, heavy water (D_2O, which is composed of deuterium atoms) absorbs a very small fraction of the produced neutrons. Reactors that are moderated and cooled with D_2O operate with natural uranium fuel, which is cheaper than enriched uranium.

However, D_2O is expensive to produce, because the deuterium isotope must be separated from natural water, where it exists at a molecular ratio of 1:6,000.

The *CANDU* is a heavy water reactor (HWR) that was developed by the Canadian nuclear program and utilizes heavy water (D_2O) as coolant and moderator. The CANDU is essentially a PWR. The use of D_2O as coolant and moderator allows the reactor to operate with natural uranium instead of the more expensive enriched uranium. The Canadian nuclear program with the CANDU reactors essentially substitutes the more expensive enriched fuel with the more capital intensive D_2O as coolant and moderator in the reactors. The CANDU reactors have higher capital cost—associated with the production of D_2O—but lower fuel cost because they use natural uranium.

One of the shortcomings of the PWR and CANDU (and to a lesser extent of the BWR) power plant cycles is that the temperature of the steam supplied to the turbines is rather low, close to 300°C, and there is typically very low superheating. Comparable temperatures in fossil fuel plants, which use superheated steam, are close to 550°C. Simple second law of thermodynamics considerations prove that because of the lower temperatures, the thermal efficiency of the nuclear power plants is significantly lower than that of fossil fuel plants. Cycle efficiencies of the nuclear power plants are in the range 30–34%, while those of modern fossil fuel steam cycles are in the range 40–45%.

While the lower cycle efficiencies may be a thermodynamic drawback for the LWR and HWR nuclear power plants, it is not a practical disadvantage, because the electricity production industry strives for lower costs per kilowatt hour produced, rather than higher thermal efficiency. The price of nuclear fuels (in dollars per British thermal unit or megajoules produced in the reactor) has been historically significantly lower than that of fossil fuels. Nuclear power plants have higher capital (fixed) costs and lower fuel (variable) costs. The lower cost of the nuclear fuel makes nuclear plants very much economically competitive, once the initial higher capital costs have been amortized. Several nuclear power plants in the United States, which have been in operation for more than 30 years, and their capital cost has been amortized, produced significantly cheaper electric energy (close to $0.01/kWh) than fossil fuel plants in 2017.

5.3.2 Gas-Cooled Reactors

Gas-cooled reactors (GCRs) have been primarily developed and built in Britain, where the *Magnox* and the *advanced gas reactor* (AGR) types were originally constructed. The *high-temperature gas reactor* (HTGR) design has been developed in France and Britain. All reactor types are cooled by a gas (CO_2, Ar, and He are the most common) and use graphite as their moderator. A typical arrangement for a GCR is to have blocks of graphite with long bores drilled. The fuel is natural uranium for the Magnox type and slightly enriched uranium for the AGR and the HTGR types. The fuel, clad in canisters, and the control rods, typically made of boron carbide, are inserted into the graphite bores. A high fraction of the channels in each graphite block are reserved for the circulation of the gas coolant. The coolant, which is at high temperature, is directed to a steam generator, where it produces steam that is fed to the turbines.

The main advantage of the GCRs is that the temperature of the coolant is not limited by the saturation pressure. The temperatures achieved in GCRs are significantly higher than those of the water-cooled reactors. Typical coolant temperatures leaving the GCRs are 650°C, and this allows superheated steam to be produced at approximately 400–500°C in the AGRs and at even higher temperatures in the HTGRs. Thus, the temperature of the generated steam is significantly higher than that of the PWR, the BWR, and HWR plants.

Because of this, the thermodynamic efficiency of the GCR power plants is higher, typically in the range 33–38%. A second advantage of the GCR power plants is that they may operate with natural or slightly enriched uranium, while their graphite moderator is not as expensive as the D_2O of the CANDU reactors. Because of the lower heat conductivity of the coolant gases, higher areas and, hence, higher reactor volumes are necessary for the transfer of the generated heat power. The dimensions and the volume of the GCRs are significantly higher than those of comparable water-cooled reactors, and hence, their power density is significantly lower. A disadvantage of the GCRs is that the circulation of the compressible coolant gas requires higher electric power for the gas blowers and compressors (the parasitic power) than the water pumps of the water-cooled reactors.

5.3.3 Other Thermal Reactor Types

The *pebble bed reactor* (PBR) is essentially a GCR. The fuel in the PBR is a mixture of uranium and thorium oxides (UO_2 and ThO_2), and the moderator is graphite. Thorium is gradually converted to the $_{92}U^{233}$ isotope, which undergoes fission. Thus, the PBR breeds some of its own fuel. The fuel and moderator are fused in 6 cm spheres (slightly smaller than tennis balls), which are randomly placed in the reactor vessel. Helium gas at 40 bar enters from the top of the reactor and leaves at the bottom at approximately 750°C. The helium is directed to a steam generator, where it produces superheated steam at 550°C. An innovative feature of the PBR is that the spent fuel elements are easily removed from the bottom of the reactor by gravity, to be reprocessed. Reprocessed balls are inserted at the top of the reactor.

One of the main advantages of this reactor is that small-size reactors, on the order of 100 MW, may be produced and operate economically. The PBR is continuously recharged and does not need to contain a large amount of nuclear fuel. Since the reactor is small in size and a very small amount of fuel is inside the reactor at any time, a loss-of-coolant accident (LOCA) in the PBR will not cause a reactor meltdown. Because of this, PBRs are better controlled and do not pose the danger of a large-magnitude environmental disaster.

The RBMK (channel, graphite-moderated boiling-water reactor) was designed and operated in the former Soviet Union as an inexpensive, modular reactor. This type of reactors became notorious following the Chernobyl accident. The reactor is composed of a large number (1661 at Chernobyl) of long stainless steel channels with five fuel rods inside. Each channel is inserted in a graphite matrix, which acts as the moderator. The main advantage of the RBMK reactor is its low cost and the modular style that facilitates the expansion of the reactor. However, the lack of a reactor vessel and other inherent safety features makes this type of reactors one of the less safe to operate. For this reason, no new RBMK-type reactors were designed or built after the Chernobyl accident.

The VVER reactor is essentially a PWR reactor that was designed and operated in the Soviet Union. Several countries that were part of the former Soviet Union and their trading partners (e.g., Lithuania, Russia, Kazakhstan, Poland, and Bulgaria) still operate RBMK and IGG reactors.

5.3.4 Breeder Reactors

Natural uranium consists of 0.715% $_{92}U^{235}$ and 99.285% $_{92}U^{238}$. The thermal nuclear reactors of the current generation primarily utilize the isotope $_{92}U^{235}$ and a very small fraction of $_{92}U^{238}$. A very large portion of the uranium fuel is unused in thermal reactors. The unused fuel becomes nuclear waste that needs to be stored for millennia. With the finite amount of nuclear fuel reserves on earth, the utilization of natural uranium by conventional thermal reactors is a significant waste of a scarce resource.

In addition to uranium-238, another naturally occurring fertile material, thorium-232, $_{90}Th^{232}$, may be converted to the fissile isotope $_{92}U^{233}$ in breeder reactors. The two fertile isotopes $_{92}U^{238}$ and $_{90}Th^{232}$ are converted to the fissile isotopes $_{94}Pu^{239}$ and $_{92}U^{233}$ according to the following nuclear reactions:

$$_{92}U^{238} + _{0}n^{1} \rightarrow _{92}U^{239} \rightarrow _{93}Np^{239} + _{-1}e^{0} \quad (T_{1/2} = 23\,\text{min}),$$

$$_{93}Np^{239} \rightarrow _{94}Pu^{239} + _{-1}e^{0} \quad (T_{1/2} = 55.2\,\text{hours}),$$

$$_{90}Th^{232} + _{0}n^{1} \rightarrow _{92}Th^{233} \rightarrow _{91}Pa^{233} + _{-1}e^{0} \quad (T_{1/2} = 22\,\text{min}),$$

$$_{91}Pa^{233} \rightarrow _{92}U^{233} + _{-1}e^{0} \quad (T_{1/2} = 27.4\,\text{days}).$$

(5.15)

The importance of breeding fuel becomes apparent when one considers the finite amount of natural uranium and thorium reserves. It has been estimated that North America has uranium minerals sufficient to supply all the energy needs of the continent for approximately 100 years, if only $_{92}U^{235}$ is utilized [5]. The same minerals would be sufficient to supply the North American continent with energy for 5500 years if the more abundant $_{92}U^{238}$ isotope were to be converted to $_{94}Pu^{239}$ and used as fissile material in breeder reactors [5,6]. Similar considerations apply to other parts of the world, in particular countries such as Russia, India, France, the United Kingdom, and the People's Republic of China [7]. The time period of nuclear fuel depletion becomes even longer if the thorium reserves are also included in the total nuclear resources. One may easily conclude that the widespread use of breeder reactors may solve the energy needs of the entire world for millennia.

The conversion and fission processes that take place in breeder reactors consume *two* neutrons instead of the one neutron consumed in thermal reactors: First, a neutron must be used to convert the fertile material to a fissile material as Equation 5.15 shows. After the conversion, a second neutron is necessary to cause the fission of the fissile material. In a breeder reactor, the average number of neutrons produced by the fission reactions must be higher than 2 to allow for the typical neutron losses, such as capture by the moderator and coolant nuclei in the reactor, as well as for other parasitic losses. This requirement places a constraint on the design of breeder reactors: they must be compact to conserve the produced neutrons. A fortuitous characteristic of breeder reactors is that the conversion reactions and the fission of $_{94}Pu^{239}$ are accomplished with fast rather than thermal neutrons (and this is the reason for being called "fast"). Therefore, a moderator is not necessary to be used in FBRs.

The core of an FBR is significantly smaller than the core of a thermal reactor. The consequence of this is a very high rate of volumetric heat production in the reactor core, on the order of 120 MW/m^3. Water and gases are not effective coolants at such rates of heat production, because of their lower thermal conductivity. Liquid metals with very high thermal conductivity, such as molten sodium (Na) or potassium (K) or a mixture of the two, are used for the removal of heat from FBRs. The melting point of these light metals at the operating pressures is slightly below 100°C. The breeder reactors are designed to always maintain temperatures sufficiently high to keep the metal coolants in the liquid state and avoid solidification and the clogging of the pipelines.

The irradiation of sodium with neutrons produces the highly radioactive isotope $_{11}Na^{24}$, which decays to $_{12}Mg^{24}$ by the emission of an electron. Because of the high radioactivity

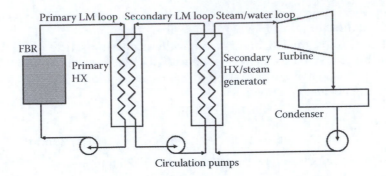

FIGURE 5.5
Schematic diagram of a liquid metal FBR nuclear power plant.

in the core of the FBR, two cooling loops are necessary for steam generation in FBR power plants as schematically shown in Figure 5.5. The primary loop extracts heat from the reactor core and, via a closed circuit primary heat exchanger, transfers it to a secondary loop, where a high-conductivity metal is also used. The secondary loop fluid emits significantly lesser amounts of radioactivity because it does not enter the nuclear reactor and does not come in contact with the high neutron flux and the other radioactive elements of the reactor. The liquid metal in the secondary loop transfers heat to produce steam in a second closed circuit steam generator. The steam produced in the second heat exchanger is fed to the turbines, and after it condenses, it is recirculated to the steam generator.

The main difference of the FBR power plants from thermal reactor nuclear power plants (PWR, BWR, or AGR) is the secondary loop of the liquid metal, whose function is to keep most of the radioactivity confined to the reactor and to the primary loop. The rest of the FBR power plant is similar to the conventional thermal power plants and is composed of turbine(s), condenser(s), feed-water heater(s), pumps, safety equipment, etc. For simplicity, not all the generic equipment (e.g., feed-water heaters and pumps) are included in Figure 5.5, which only depicts the equipment that are necessary for the operation of the basic Rankine cycle.

5.4 Useful Parameters for Nuclear Energy

The fundamental equation of mass–energy equivalence, $E = mc^2$, is at the center of all calculations for nuclear power. The conversion of 1 u (1.6604×10^{-27} kg) of mass produces 931 MeV or 1.49×10^{-10} J of energy. The fission of nuclei, such as $_{92}U^{235}$ and $_{94}Pu^{239}$, generates a mass defect for the reaction products, which on the average is equivalent to 200 MeV and is released as thermal energy in the reactor. Based on these numbers, we may calculate several useful parameters for the operation of the nuclear reactors and power plants.

A typical large-scale nuclear reactor produces approximately 1000 MW of electric power. Because the temperatures in nuclear reactor cycles are kept lower than the corresponding temperatures of fossil fuel plants, the overall thermal efficiencies of nuclear power plants are lower, in the range of 30–35%. For a typical reactor, the rate of heat production is 3000 MW, and this is converted to electric power in the turbine generator system. In order to distinguish the thermal from the electric power output of the reactor, the units of these two numbers are sometimes denoted as MW-e and MW-t.* Hence, the power produced by the power plant would be denoted as 1000 MW-e; and the rate of heat produced by the reactor, as 3,000 MW-t. The latter is often referred to as the *rating* of the reactor.

Given that the average fission reaction produces 200 MeV of heat, at this thermal power, the number of reactions per second inside the reactor of the 1000 MW-e (3,000 MW-t) is $3,000 \times 10^6/(200 \times 1.6 \times 10^{-13}) = 9.375 \times 10^{19}$ or approximately 10^{20} fissions/s.

While this appears to be a very large number of nuclear reactions, when compared to the Avogadro number (6.023×10^{23} atoms/g-atom), it represents a very small mass of the uranium-235 fuel: 10^{20} fissions/s correspond to $10^{20}/6.023 \times 10^{23} = 1.56 \times 10^{-4}$ g-atom/s or 0.037 g/s of $_{92}U^{235}$. The complete consumption of 1 g-atom (235 g of uranium-235) in this 3000 MW-t reactor will take $6.023 \times 10^{23}/9.375 \times 10^{19} = 6,424$ s or 1.78 hours. During this period the reactor consumes the 235 g of uranium-235, it produces $1.88 \times 3,000 = 5,340$ MWh of heat, and approximately 1,780 MWh of electrical energy.

A single gram of pure $_{92}U^{235}$ produces 22.7 MWh (81.8×10^9 J) of heat. For comparison, in a coal power plant, this amount of heat is produced by the combustion of approximately 2.5 t of carbon. The vast difference in the masses of fuel for the two types of thermal power plants is astounding: 0.001 kg uranium-235 corresponds to approximately 2500 kg of anthracite and a mass ratio of 1:2,500,000.

During a full day, the 3000 MW-t reactor consumes $235 \times 24/1.78 = 3169$ g of uranium-235; and during a year, $3.169 \times 365 = 1,157$ kg of uranium-235 (a little more than a ton of uranium-235).

Because only 0.715% of the natural uranium comprises the fissile isotope $_{92}U^{235}$, 1 g of uranium-235 is contained in approximately 140 g of natural uranium (mass ratio 1:139). During one day, the reactor would consume the equivalent of approximately $3.169 \times 140 = 443.7$ kg of natural uranium (less than half a ton per day). The yearly consumption of natural uranium fuel by this reactor is $443.7 \times 365 = 161,900$ kg (161.9 t) per year. The consumption of an equivalent thermal power plant that uses coal is approximately 2.9 million t of carbon.

Under normal operating conditions, a typical nuclear reactor is fueled (charged) every 18–24 months and continuously operates day and night. Let us assume that the reactor is initially charged with 1000 t, or with 52.4 m³, of natural uranium. With continuous operation, this reactor would consume only 16.3% of its fuel in an entire year. The actual amount of mass deficit Δm, which is converted to energy, is by far smaller than the amount of fuel consumed. During a whole year, the mass deficit for this reactor would be, from $E = mc^2$,

$$3,000 \times 10^6 \times 365 \times 24 \times 60 \times 60 / (9 \times 10^{16}) = 1.05 \text{ kg.}$$

This mass deficit is negligible compared to the original weight of 1000 t of fuel the reactor has been charged.

A quantity of interest for nuclear power plants is the *burnup*, or burnup rate, of the nuclear fuel. The burnup is defined as the thermal energy released per metric ton of fuel. During one day, when the typical reactor produces $3,000 \times 24 = 72,000$ MWh-t of thermal energy and consumes 0.4437 t of nuclear fuel, the burnup rate of the reactor is $72,000/0.4437 = 162,272$ MWh/t of natural uranium. Enriched uranium has a higher percentage of the fissile $_{92}U^{235}$ nuclei, and its burnup rate is higher.

---

* In the absence of the designator (-e or -t), the engineering convention is that MW denotes electric power.

TABLE 5.5

Useful Parameters for the Thermal Power (Heat) Produced by a Nuclear Reactor

Fissions per second: 9.375×10^{19}

Consumption of uranium-235: 0.037 g/s; 3.169 kg/day; 1,157 kg/year

Consumption of natural uranium:[a] 5.14 g/s; 447.6 kg/day; 161.9 kg/year

Mass deficit (mass converted to energy) for an entire year: 1.05 kg

Burnup: 162,272 MWh/t of natural uranium

Ratio of masses in comparison to carbon: 0.001 kg of uranium-235 $\leftrightarrow$ 2,500 kg of carbon

Note: All quantities pertain to the operation of a 3000 MW-thermal nuclear reactor.
[a] Assumes all the heat is produced from the uranium-235 fission. When other fissionable materials in the reactor produce heat, the consumption is reduced, as in example 5.6.

Table 5.5 provides a summary of the important parameters for a reactor that produces 3000 MW-t.

It must be noted that all calculations in this section are based on the assumption that only $_{92}U^{235}$ undergoes fission in the reactor. In all thermal reactors, a small fraction of the $_{92}U^{238}$ undergoes fission by fast neutrons and, in addition, a corresponding amount of fissile $_{94}Pu^{239}$ are formed from $_{92}U^{238}$ conversion and produces additional energy. Unlike the breeder reactors, the amount of plutonium produced in the conventional reactors is not enough to replenish the consumed fissile $_{92}U^{235}$. As a consequence, the amounts of the original fuel used per day and per year are lower than the preceding numbers. Their precise values very much depend on the design of the particular reactors and on the amounts of $_{94}Pu^{239}$ that are produced and undergo fission.

Example 5.6

The two nuclear reactors of the Comanche Peak Power Plant in North Texas produce 2,467 MW of electric power at 31.7% overall thermal efficiency; 35% of the total heat produced comes from the fission of $_{92}U^{238}$ and $_{94}Pu^{239}$. Calculate (a) the heat power produced, (b) the amount of natural uranium the power plant consumes in a year, and (c) the burnup of the reactors.

Solution:

a. The heat power of the two reactors is 2,467/0.317 = 7,782 MW.
b. Of this rate of heat, the heat produced by the fission of uranium-235 is 7,782/1.35 = 5,764 MW. Based on the useful parameters of this section, the two reactors will consume 5,764/3,000 × 161.9 = 311 t of natural uranium yearly.
c. The burnup of the pair of reactors is 7,782 × 24 × 365/311 = 219,197 MWh/t.

Example 5.7

The Superphénix breeder reactor was designed to produce 1,200 MW of power at 40% efficiency. Determine (a) the annual amount of natural uranium required for the operation of such a reactor and (b) the burnup of the reactor.

Solution:

a. Breeder reactors convert and utilize the $_{92}U^{238}$ nuclei in the natural uranium. This reactor produces 1,200/0.4 = 3,000 MW of heat. With 200 MeV per fission, $3,000 \times 10^6/(200 \times 1.6 \times 10^{-13}) = 9.375 \times 10^{19}$ fissions/s occur in the reactor at the expense of $9.375 \times 10^{19}/6.023 \times 10^{23} = 1.56 \times 10^{-4}$ g-atoms of uranium per second. Since the natural uranium is almost exclusively composed of uranium-238, the mass of natural uranium needed per year is $1.56 \times 10^{-4} \times 238 \times 60 \times 60 \times 24 \times 365 = 1{,}171{,}000$ g or 1.711 t of uranium.

b. The burnup of the Superphénix breeder reactor is $1{,}200 \times 24 \times 365/1.171 = 8.98 \times 10^6$ MWh/t.

One may compare the last two examples and note that the Superphénix breeder reactor has approximately half the capacity of the pair of the thermal reactors. However, the fuel requirement of the breeder reactor is 300 times less than that of the pair of the thermal reactors. The low fuel consumption of the Superphénix breeder reactor is also reflected in the ratio of the burnup quantities, which is approximately 40.

Example 5.8

It is proposed that five coal power plants with a total capacity 2,000 MW-e and average thermal efficiency of 38% be substituted by two PWR nuclear power plants with 32% efficiency. Determine (a) the annual nuclear fuel consumption of the nuclear reactors if 40% of the heat is produced from isotopes other than $_{92}U^{235}$ and (b) the annual amount of CO_2 emission avoidance from this substitution.

Solution:

a. The rate of heat by the two reactors is $2{,}000/0.32 = 6{,}250$ MW-t. Since the average fission reaction produces 200 MeV of heat, at this thermal power, the number of fissions per second is $6{,}250 \times 10^6/(200 \times 1.6 \times 10^{-13}) = 1.953 \times 10^{20}$, and the number of fissions per year is 6.159×10^{27}.

Since 40% of the heat is produced from other isotopes, the quantity of heat produced from the fission of $_{92}U^{235}$ is $6{,}250/1.4 = 4{,}464$ MW-t, and the number of $_{92}U^{235}$ fissions per year is 4.399×10^{27}. This number of fissions corresponds to the consumption of 7.30 kmol of $_{92}U^{235}$ per year or 1,717 kg, which is contained in approximately 243,000 kg of natural uranium.

b. The rate of heat produced in the five coal power plants is $2{,}000/0.38 = 5{,}263$ MW-t. Assuming that all the heat is produced from the combustion of carbon (LHV = 32,770 kJ/kg from Table 4.1), the coal power plants use 160.6 kg of coal per second, or 5.065×10^9 kg of coal per year. Since 12 kg of coal produces 44 kg of CO_2 upon combustion, this amount of unused coal corresponds to 18.6×10^9 kg (18.6 million t) of CO_2 avoidance.

5.5 Notorious Nuclear Power Plant Accidents

Nuclear energy has always been considered with trepidation and fear by the general public. The first perception of nuclear energy by the public was one of utter destruction following the bombing of Hiroshima and Nagasaki at the end of World War II. This perception was exacerbated by the fear of nuclear war in the Cold War period of 1945–1989. Because of this, the general public has always been apprehensive and mistrustful of nuclear fission, even when it is applied for peaceful purposes. The mistrust of the public toward nuclear power plants was significantly inflated in the aftermath of the few accidents that attracted a great deal of publicity and notoriety, worldwide. The accidents were caused by inadequate cooling of the nuclear fuel and resulted in the spread of radionuclides to the environment.

Regardless of the type of the nuclear reactor, the adequate and continuous cooling of the reactor core is the most essential safety feature of any nuclear power plant. Unlike fossil fuel power plants, where the fuel supply may be interrupted and the power produced instantaneously stops, nuclear reactors are closed systems that contain the entirety of their fuel inside the reactor vessel. Even after the reactor is shut down, there is still a fraction of residual neutron flux that causes fissions and produces heat. In addition, all radioactive materials in the reactor decay and continue to produce a significant amount of heat power several days after the shutdown. Figure 5.6 shows the power produced in the nuclear reactor of the typical 1000 MW-e nuclear power plant. The time is in seconds after the nominal shutdown. It is observed that the reactor produces 170 MW 10 s after the shutdown and continues producing 2.8 MW 120 days (10^7 s) after the shutdown. If the thermal power produced is not promptly removed from the reactor by the continuous circulation of coolant, the reactor temperature increases fast, to the point that melting of the materials inside the reactor occurs. The continuous production of heat may also cause partial or total melting of the reactor vessel and the concrete foundation of the reactor building. Such a reactor *meltdown* will bring the radioactive material of the reactor into the environment with uncontrolled and catastrophic consequences for the human and animal populations in the region.

In all nuclear power plants, the LOCA represents an unacceptable risk because it may have catastrophic consequences for the population of a vast geographic region in the vicinity of the reactor. Nuclear power plants are designed with a primary cooling circuit and one or two separate and independent emergency cooling circuits, designed to continuously operate, even when the plant does not produce electric power. The reliability and effectiveness of the primary and emergency cooling systems of the reactor are of paramount importance to the safe operation of any nuclear power plant.

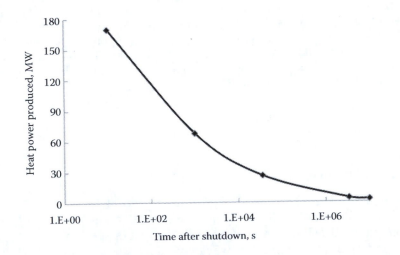

FIGURE 5.6
Power produced in a nuclear reactor after it is shut down.

5.5.1 Accident at Three Mile Island

The TMI power plant, which had three nuclear reactors, is located in a small island on the Susquehanna River, approximately 10 mi from Harrisburg, the capital of the state of Pennsylvania, United States. The accident occurred in reactor 2 of the power plant (TMI-2), a PWR reactor, which supplied approximately 2,700 MW-t thermal power and produced approximately 900 MW-e of electric power.

A few days before the accident, an emergency readiness test was performed. During the test, the feed-water valves were closed and remained inadvertently closed after the test, thus making the emergency feed-water system inoperative. Following a series of errors by the operators of the nuclear reactor in the early morning hours of March 30, 1979, the upper third of the reactor core lost its cooling water supply. The reactor was shut down but, because of the continuous generation of heat, the water boiled; the upper part of the reactor was exposed to the low conductivity steam and overheated to approximately 2100°C [3]. At this temperature, the fuel cladding (zirconium alloy) partly melted and chemically reacted with the steam to produce hydrogen gas according to the chemical reaction:

$$Zr + H_2O \rightarrow ZrO + H_2. \tag{5.16}$$

It was calculated that 1000 kg of hydrogen gas was produced and formed a "hydrogen bubble" at the upper part of the reactor vessel. This raised fears for a strong hydrogen explosion that would have destroyed the reactor and would have widely spread radioactivity in the region. During the week that followed the accident, the hydrogen gas was slowly bled off by a pumping system that circulated cold water in the reactor. The water dissolved the hydrogen and was fed to a regenerator where the dissolved hydrogen combined with oxygen in a controlled reaction. With this process, the "hydrogen bubble" shrank and the explosion was avoided.

It was estimated that the partial melting of the zirconium alloy cladding exposed at least 70% of the reactor fuel and that 30–40% of the fuel melted because of the high temperatures. The 6″ (15 cm) thick reactor vessel contained all the molten material in their entirety and prevented further meltdown. In the case of TMI, all the damage was contained in the now-defunct reactor, and the only environmental consequence was the release of small amounts of radioactive steam from the containment building.

5.5.2 Accident at Chernobyl

The Chernobyl power plant is located in Ukraine, approximately 80 km to the north of the capital Kiev and very close to the border of what is now the country of Belarus. At the time of the accident, both Ukraine and Belarus were parts of the USSR. The power plant had four units, each producing approximately 1000 MW of electric power and the accident occurred in unit IV. The reactor of that unit was an RBMK reactor with graphite as the moderator and with 1661 channels composed of 9″ stainless steel tubes with the fuel assemblies inside.

The accident in the Chernobyl nuclear power plant occurred on April 26, 1986, following a test to supply emergency power to the reactor and a series of operator errors. As a result of the errors, the thermal power of the reactor increased from 200 to 530 MW-t in 3 s and to an estimated 300,000 MW-t in another 5 s (approximately 100 times more than the rated power of the reactor). At this rate of thermal power, the water in the channels violently evaporated; the temperature and pressure in the reactor channels rapidly rose causing a steam explosion; and a large part of the reactor was destroyed. As in the TMI, the high temperature also caused the steam to chemically react with the zirconium alloy. The zirconium–steam reaction produced large quantities of hydrogen in the reactor building. An explosive mixture of hydrogen and oxygen was quickly formed and caused a hydrogen explosion a few seconds later that completed the destruction of the reactor building, exposed the interior of the reactor core to the atmosphere, and released a high fraction of the radionuclides to the environment. The radioactive materials were driven upward by the high-temperature plume and were carried away by the weather patterns. A high fraction of these radionuclides was deposited in the area around the power plant. The remainder were scattered by the local wind currents and were finally deposited in several regions throughout the European continent [3,8,9].

The effects of the Chernobyl disaster touched many: There were 31 immediate fatalities at the power plant and 24 other persons permanently disabled. An area of 25 km around the plant was evacuated; 131,000 citizens of the former USSR were displaced officially (and many others unofficially). The released radiation, which has long-term effects, is expected to cause between 100,000 and 1,000,000 additional cancers in Europe. And 600,000 citizens of the former USSR were placed in a special register to be monitored throughout their lives for radiation-related illnesses [4,8,9].

The environmental impact of the Chernobyl accident by far exceeded that of the accident at TMI-2. The main reason for this is the safety equipment of the PWR that include the 15 cm thickness reactor vessel and the 90 cm thickness reactor containment building. The two contained most of the released radioactivity at TMI-2. The effect of these structures in containing the released radioactive materials becomes apparent in Table 5.6, which shows the fraction of radioactivity released by the two accidents to the environment and to the containment building at TMI-2. While both reactors were severely and permanently damaged, the environmental and human impact of the TMI-2 accident was minimal because most of the released radioactive elements were trapped in the reactor containment building, were contained there, and were subsequently removed.

By examining with today's hindsight, the accidents at Chernobyl and TMI, one may conclude that both accidents were preventable: Both accidents happened either during a test or a short time after a test. In both cases (and especially in the Chernobyl accident), there were a number of warnings signs for operator actions that should not have happened. There were several actions of the operators, which were in direct violation of operating procedures, design standards, engineering judgment, and even common sense (especially in the Chernobyl case). Should the operators have not overridden all the emergency systems and

TABLE 5.6

Radioactivity Released at TMI-2 and Chernobyl

	TMI-2 to Containment Bld., %	TMI-2 to Environment, %	Chernobyl to Environment, %
Noble gases	48	1	100
Iodine	25	0.00003	20
Cesium	53	0	12
Uranium	0.5	0	8
Rare-earth metals	0	0	3

allowed the emergency systems to automatically shut down the two reactors, the accidents would not have occurred. The overconfidence of the operators also played an important role. Several years of successful reactor operations, from the 1950s to 1979 and 1986, had lulled those in operating rooms into a false sense of security that the reactors are indestructible. Overconfidence with such complex, highly nonlinear systems and indifference for operating procedures can only be destructive. A thorough study of the devastation that followed the Chernobyl accident and the scenario of what could have followed a meltdown at TMI-2 makes it apparent that in a world of ever-increasing technical complexity, with tremendous economic pressures for higher efficiency, higher automation, and higher profitability, our society cannot afford to relax its vigilance on the operation of the complex engineering systems that nuclear reactors are.

5.5.3 Accident at Fukushima Dai-ichi

The Fukushima Dai-ichi power plant (Dai-ichi means "one" in Japanese) was composed of six units, each with its own nuclear reactor. The units were operated by the Tokyo Electric Power Company, and all reactors were of the BWR type. On March 11, 2011, an earthquake, measuring 9.0 in the Richter scale, struck the coastal area of Japan. At the time of the earthquake, the first three units (I, II, and III) were in operation, and the last three units (IV, V, and VI) were shut down and undergoing seasonal maintenance. Immediately after the earthquake, the three operating units were shut down—according to established operating procedure—and the emergency generators were switched on to remove the decay heat from the reactors, which did not show any damage because of the earthquake.

The earthquake was followed by a 14 m tsunami, which exceeded the safety design (6.5 m) of the nuclear power plants. The tsunami caused a widespread flooding in the entire area of Fukushima and disabled the emergency generators. A second emergency system in the power plants was powered by batteries, which run out of power after a few hours. Without adequate cooling, the water in the reactors boiled and generated steam. This caused higher pressure in the reactors and triggered the relief valves to open and vent the radioactive steam. At the same time, the liquid water level in the three reactors significantly dropped, the temperature rose locally to high levels, and this caused the partial meltdown of all three reactors that were initially in operation. It also caused the production of hydrogen. With the absence of hydrogen removal efforts, low-intensity hydrogen explosions were noticed in units I and III, and very likely in unit II. The partial meltdown and the explosions damaged the three units and made them inoperable for the future.

Partial cooling water flow in the reactors was restored several hours after the accident, and this stopped a wider meltdown of the nuclear reactors. Unit I remained without cooling water supply for 27 hours, and unit II, for 7 hours. A further complication of the accident arose because the reactor buildings also contained the spent fuel (waste) of the reactors in local water pools, which also requires cooling. The absence of adequate cooling in the spent fuel pools caused a significant increase of the temperature and a destructive fire in the fourth reactor unit of the power plant.

A great deal of the released radiation was contained in the four containment buildings. However, the venting of the steam and the fires that followed the accident significantly increased the radiation levels near the reactor and in the entire Fukushima region. The atmospheric releases from the accident are estimated to be approximately one-tenth of the releases from the Chernobyl accident [10]. Most of the radioactive releases were dispersed over the North Pacific Ocean.

The government of Japan reacted fast and declared first a 10 km and subsequently a 20 km evacuation zone. Isolated pockets of high radioactivity outside the 20 km radius were also included in the evacuation.

Unlike the accidents at TMI and Chernobyl, the Fukushima accident was not caused by operator error. At Fukushima, the engineering systems performed according to their design specifications. The natural disaster—the tsunami that followed the earthquake—was too powerful for the nuclear power plant to endure. However, overconfidence also played a role in this accident too. According to the International Atomic Energy Agency (IAEA) [10], "a major factor that contributed to the accident was the widespread assumption in Japan that its nuclear power plants were so safe that an accident of this magnitude was simply unthinkable. This assumption was accepted by nuclear power plant operators and was not challenged by regulators or by the Government. As a result, Japan was not sufficiently prepared for a severe nuclear accident."

5.6 Environmental Effects: The Nuclear Fuel Cycle

Uranium and thorium ores exist in several parts of the earth. Australia, Russia, Kazakhstan, Tajikistan, the United States, Canada, South Africa, Namibia, Mali, and Brazil have significant deposits of the two ores, while other countries, including the Western European countries, China, and India, have deposits that can supply them with nuclear fuel for several decades. In the United States, uranium ores are primarily located in the states that border the Rocky Mountains (Utah, Texas, Wyoming, Colorado, and New Mexico).

5.6.1 Mining, Refining, and Enrichment

Uranium metal, which is the most common nuclear fuel, naturally exists with other chemicals in several minerals. Rich uranium ores contain up to 2% uranium metal and medium-grade ores 0.5–1% of the metal. At these compositions, the mined uranium needs to be refined to be used as a fuel. Most of the uranium mines are surface mines, where the ore is extracted with giant earth excavation systems. The uranium ore is then crushed and ground to a coarse powder. Sulfuric acid (H_2SO_4) is used to form sulfate salts with the metals in the ore and to separate the metals from the rest of the chemicals of the ore. The uranium is then separated from the other metal sulfides and by conversion to uranium oxide U_3O_8, which precipitates in the acid solution. This solid oxide is often commonly called *yellowcake* and is shipped to the processing or enrichment facilities.

The solid U_3O_8 has the natural uranium concentration of 0.715% $_{92}U^{235}$ and 99.285% $_{92}U^{238}$. The CANDU reactors are designed to use natural uranium and accept uranium with this composition as a fuel, but the other reactor types require enriched uranium as their fuel. The latter contains a higher fraction of $_{92}U^{235}$ atoms, in the range of 1–5%, depending on the type of the reactor. Because $_{92}U^{235}$ and $_{92}U^{238}$ isotopes have the same chemical properties, the uranium enrichment process cannot be achieved by chemical reactions. Instead, the enrichment processes use the very small difference of the mass of the two isotopes in diffusion and centrifuging: Lighter atoms and molecules in the gaseous form diffuse faster through membranes. Also, the lighter atoms and molecules experience a lower centrifugal force and concentrate toward the inner boundaries of centrifugal separators.

In the centrifuging processes, the uranium metal is first converted to UF_6 (uranium hexafluoride), which is gaseous at very moderate temperatures, above 56°C. This gas is fed to the centrifuge system, which spins at very high angular velocity. The heavier $^{238}UF_6$ molecules are forced toward the outer surface of the centrifuge, while the slightly lighter $^{235}UF_6$ molecules concentrate toward the inner surface. Gas close to the inner surface, which has a higher concentration of $^{235}UF_6$, is extracted and fed to a second centrifuge, where the process is repeated. The gas at the inner surface of the second centrifuge has an even higher concentration of the $^{235}UF_6$ molecules. The centrifuging process is repeated until the desired concentration (enrichment) of $^{235}U_{92}$ is reached.

In the membrane diffusion enrichment process, the uranium is also converted to the gaseous UF_6 compound. The gas is then forced (by pressure) through the pores of a series of membranes. Because the $^{235}UF_6$ molecules are lighter, they diffuse faster than the $^{238}UF_6$ molecules and reach the end of the pores faster. Therefore, at the end of a membrane (or stage on the diffusion process), the concentration of the lighter $^{235}UF_6$ molecules is higher. Passage through a second stage of membranes increases the concentration of the $^{235}UF_6$, and after a number of diffusion stages, the desired enrichment is achieved.

After the enrichment process, the gaseous UF_6 is converted to the solid uranium dioxide UO_2. The latter is a solid, ceramic material with very high melting point, approximately 2300°C. Thus, UO_2 may withstand very high temperatures without melting. The UO_2 is formed into small pellets (approximately 1 cm in diameter and 1.5 cm high) which are then loaded in the fuel rods and placed in the reactor. The UO_2 fuel pellets will remain in the reactor for at least two refueling cycles, 36–48 months, after which they are considered "spent fuel" and are reprocessed.

5.6.2 Reprocessing of Spent Fuel; Temporary and Permanent Storages

When the fraction of $^{235}U_{92}$ atoms in the fuel is low, the fission reactions slow in the reactor, the thermal power produced gradually decreases, and this signals that the fuel must be replenished. At refueling, the spent fuel rods are replaced with new ones. At the same time, the partially spent fuel rods are moved toward the center of the reactor, where the rates of fission are higher, and the remaining $^{235}U_{92}$ atoms are likely to undergo fission. Almost fully spent fuel rods are removed from the reactor and form the "waste" products. At first, this waste, which is highly radioactive, is stored close to the reactor and within the containment building in water pools. All the heat and most of the radioactivity released by the nuclear waste are absorbed by the water and remain within the confines of the containment building.

After the level of radioactivity of the spent fuel has considerably declined, a process that takes 10–60 years, the remaining materials are either reprocessed or transported for permanent storage to a designated facility. During reprocessing, the spent fuel is dissolved again in sulfuric acid, where the uranium and any plutonium that has been created are chemically separated. The fissile plutonium is mixed with natural uranium to form new fuel for reactors. The recovered uranium, which is very rich in the fertile but nonfissile $^{238}U_{92}$, is gasified and further enriched and then processed to form UO_2 pellets, which again become reactor fuel.

The permanent storage facility must also store all the waste from the enrichment, processing, and reprocessing of the minerals (sulfuric salts, excess uranium-238, nonfunctioning membranes, etc.) as well as the spent fuel. The waste is typically stored in stainless steel tanks and artificially cooled. When the level of radioactivity in the waste is very low, the waste is processed to a fine solid powder (the calcination process) and then vitrified (glassified) and poured in steel containers, and is ready to be stored in deep geological formations for long-term storage. Such long-term storage facilities have been constructed in western Europe. The Yucca Mountain facility in the United States was designated by the Congress in 2002 as a long-term nuclear waste storage facility. However, in 2009, the US Department of Energy announced that the location "is not considered" as a central nuclear waste facility any more. In 2017, the United States is lacking a permanent site for the storage of nuclear waste, most of which is located close to the nuclear power plants. The lack of a permanent nuclear storage site in the United States and the uncertainty and risks associated with the ownership, stewardship, and storage of the nuclear waste produced in the country are significant deterrents to private investments in new nuclear power plants.

5.6.3 Environmental and Health Effects of the Fuel Cycle

The most important environmental impact of the nuclear fuel is the radioactivity, which is associated with both uranium and thorium. The natural deposits of the two elements also contain radioactive decay products (the daughter elements) such as actinium, radium, and radon, which are very radioactive and pose health hazards to the miners. Deep-earth mines trap some of these elements and pose significantly higher health risks to the miners, while open earth (surface) mines allow radon and the radioactive dust to diffuse into the atmosphere and are considered to be safer. Regardless of the type of mining, the inhalation of any radioactive elements poses a significant health hazard, and for this reason, suitable masks must be worn at all stages of the mining and ore processing operations.

The transportation, processing, and reprocessing of the nuclear fuel bring the fuel in contact with workers and pose significant threats to human health and the environment. For this reason, safety considerations during these processes are of paramount importance. The majority of these processes, which are highly regulated and always well monitored, are automated and performed by robots.

The temporary storage of the nuclear waste within the perimeter of nuclear power plants poses moderate environmental concerns. This nuclear waste has low levels of radioactivity and may become dangerous only if it becomes airborne and is inhaled by humans and animals. In the early part of the twenty-first century, when terrorist attacks are frequent in OECD countries, the bombing of temporary nuclear storage facilities and the atmospheric dispersion of radioactive nuclear waste has become a significant safety concern. The long-term or permanent disposal of the nuclear waste represents a lesser environmental threat because of the following:

1. The permanently stored nuclear waste is processed and contains lower levels of radioactivity.
2. The material is vitrified or otherwise compacted in stable solid formations and is not prone to become airborne or to leach in water streams.
3. Permanent storage facilities are centralized, well designed, well guarded, and well protected from terrorist attacks.
4. In contrast to temporary storage facilities, which are inside the nuclear power plants and close to urban populations, permanent storage facilities are located far from high-population areas. Any accident or any other mishap in permanent storage facilities poses a threat to only a small numbers of humans and animals.

5.7 Economics of Nuclear Energy

In 1955, the then president of the United States Dwight Eisenhower inaugurated the first commercial nuclear power plant in the United States. He declared that nuclear energy would be the way of the future and that nuclear power would produce "... electricity too cheap to measure." President Eisenhower's and similar predictions did not materialize. Actually, the cost of nuclear reactors dramatically rose in the United States and, with it, the price of electricity they produce. The accident at TMI, in 1979, the adverse public opinion this generated, and the new safety guidelines that followed the investigation of the accident had a momentous and destructive impact on the US nuclear industry: Reactors that were completed before 1979 had construction costs in the range $1,500–2,600 per installed kilowatt, and their completion times varied between 4 and 10 years. Reactors that were in construction before 1979 and were completed after that year had construction costs between $2,000 and $11,000, and their completion times rose to between 6 and 15 years [11]. The two effects fundamentally changed the economics of nuclear reactors in the United States, and several corporations that owned nuclear reactors in the construction stages declared bankruptcy. As a result of the new economic reality, no new nuclear reactors were planned in the United States (and several other OECD countries) between 1979 and 2010.

The accident at the Chernobyl power plant, in 1986, had a similar but not as dramatic effect on the construction of new reactors in Europe. For French reactors that were completed before 1986, the construction duration was between 4.5 and 7 years. For reactors in France that were under construction in 1986 or their construction started after this year, the construction duration increased between 6 and 13 years. However, the cost of the French nuclear reactors did not significantly increase as a result of the Chernobyl accident [11]. Figure 5.7 shows the historical average cost of nuclear reactors in the United States, France, Germany, Canada, India, Japan, and South Korea (in 2010 US dollars adjusted for purchasing parity). The figure shows the dramatic effect of the TMI accident on the cost of the US reactors: reactors that began construction between 1971 and 1978 and were in a midconstruction

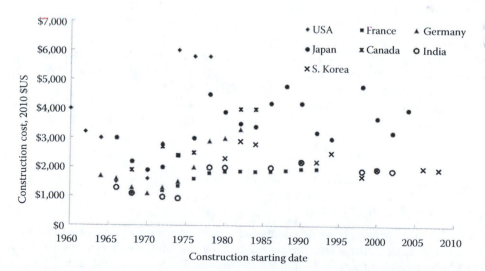

FIGURE 5.7
Historical average cost per megawatt of nuclear reactors in United States, France, Germany, Canada, India, Japan, and South Korea (in 2010 US dollars per megawatt adjusted for purchasing parity).

stage during the TMI accident show a significant increase in cost. In contrast, the reactors in France, India, and South Korea indicate lower and almost constant cost.

Clustering of similar reactors at the same location helps with the overall cost. Countries that have adopted nuclear reactor design standardization and countries that consistently build reactors in pairs or larger groups at the same site—France, South Korea, Canada—also have experienced significantly lower construction cost for new nuclear power plants [11,12].

While the capital cost for the construction of new nuclear reactors is very high, their variable costs are the lowest in the power production industry. Once the reactor is paid and the capital investment is amortized, the existing power plants operate at very low costs and are effectively "cash engines" for their owners [13]. The operations, maintenance, and fuel costs (including used fuel management) of existing units are among the lowest in the power production industry, slightly higher than the costs of the hydropower plants. Data published by the Nuclear Energy Institute in the United States show that, in 2012, nuclear power plants generated energy at an average cost of $24/MWh. The corresponding cost of coal was $32.7/MWh, and of natural gas, $34/MWh [13]. The variable costs of producing electricity from these three primary energy sources indicate very similar trends in the entire period of 1995–2012 [13]. Because of the low variable costs, regardless of whether or not the capital costs are amortized or depreciated, nuclear power plants are used for base-load power generation.

A recent experience that balances the high capital investment cost of conventional nuclear power plants is the longer life span of these plants. Originally, the nuclear power plants in the United States were constructed with 40-year operating licenses. This implies that nuclear units built in the 1970s and 1980s would have had to be decommissioned in the period of 2010–2030. Detailed engineering assessments of these power plants have shown that their life expectancy can be extended, and the US Nuclear Regulatory Commission has granted 20-year extensions to the operating licenses of more than 75 nuclear reactors. With good maintenance and life extensions, the operating licenses of most Japanese reactors have been extended to 70 years. Russia and other countries that developed nuclear power early have also extended the operating licenses of their nuclear reactors by 15–20 years. The extension of the life cycle of nuclear reactors makes the capital amortization period longer and offsets part of the higher cost of the nuclear power plants.

The production of electricity from nuclear energy has very high initial cost for the construction of the power plants and very low fuel, operational, and maintenance costs. To these, one must add the decommissioning costs for nuclear reactors, which are 10–15% of the construction costs, and the cost of storing the nuclear waste, for which estimates widely vary. Nuclear reactor design standardization and constructing nuclear reactors in groups at the same site introduce economies of scale that reduce the overall costs of nuclear energy.

5.8 Future of Nuclear Energy

The construction of nuclear power plants accelerated during the 1960s and 1970s in OECD countries, the USSR, and Eastern Europe: from a few megawatt hours in 1960, nuclear power plants produced approximately 1000 billion kWh in 1980 and more than 2000 billion kWh in 1990 worldwide. Countries with few other energy resources, such as France, Japan, and parts of the USSR, developed ambitious nuclear programs and produced a high fraction of their electric power from nuclear energy. By 1986, 18% of the electric energy produced in the world came from nuclear power plants. However, the notoriety following the accidents at TMI and Chernobyl put a stop in the further development of nuclear energy and the construction of new nuclear power plants. In the late 1980s, following the accident at Chernobyl, citizens of several OECD countries—Germany, Austria, Holland, and Belgium are among them—voted for moratoria on the further construction of nuclear reactors. In the United States, licensing and construction of nuclear power plants effectively stopped between 1985 and 2012. If in the 1950s, the citizens welcomed nuclear power plants as a source of needed electric power, in the 1990s, they became skeptical and their collective attitude was summarized as "not in my back yard."

Figure 5.8 shows the total electricity production in the world since 1970 and the fraction of the electricity produced by nuclear power globally [14]. It is apparent that the electric energy produced has remained almost constant since 1998 and fluctuates between 2,400 and 2,600 billion MWh. There was a notable increasing trend in the production of electric energy between 2004 and 2011, following the significant increase in energy prices in the early 2000s, but the trend stopped in 2011, and the energy produced noticeably dropped in the aftermath of the Fukushima Dai-ichi accident and the decision to temporarily shut down most of the nuclear reactors in Japan and permanently 8 of the 17 reactors of Germany. It is also remarkable that the fraction of global electric power produced has peaked in the late 1980s, at approximately 19%, and that this fraction has slowed down since 1989. This decline is primarily due to the increased electricity production in the world and the recent (since 2011) reduction of operational nuclear power plants.

Since the early 1980s, the construction of nuclear power plants has all but stopped in most OECD countries (except in France and Japan), and any increase in electricity production from nuclear energy is almost entirely due to new plants in developing countries. In 2015, the People's Republic of China had 23 nuclear reactors under construction, India had 6, Korea had 4, and Russia another 9 [15]. In contrast, the United States had 5 reactors under construction; France has only 1; and the United Kingdom, Germany, Sweden, and Holland had no reactors under construction. The trends show that at least in the near future, the fraction of electric power produced from nuclear power plants will significantly decrease in OECD countries and will most likely increase in the developing countries, primarily in Asia.

The main reason for the retreat of nuclear power in OECD countries is the increasing number of safety codes and standards that have been attached to the construction and

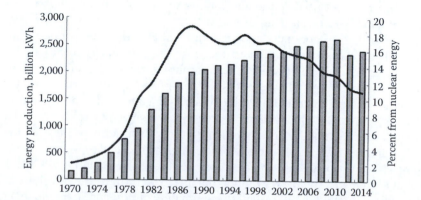

FIGURE 5.8
Total global electric energy production from nuclear power plants (*bars*) and the fraction of total electricity produced (*continuous curve*). (Data from IAEA, *Trends in Electricity Supplied*, IAEA, Vienna, 2015.)

operation of the nuclear reactors since the early 1970s. In 1978 (before the accident at the TMI-2 reactor), these statutory requirements were increasing at the rate of 1.3 new regulations per day. As a result, the construction cost of new reactors skyrocketed. The construction of a 1000 MW nuclear power plant in the United States in 1970 cost approximately $100 million; at the time, there were close to 100 codes and standards applicable to the nuclear power plant, and the entire construction took about 4 years. By 1992, the cost of the same nuclear power plant increased to more than $3.5 billion, the applicable codes and standards were more than 2,000; and the construction lasted close to 11 years [16]. The cost of building a 1000 MW nuclear power plant in 2015 in the United States is close to $8 billion.

In addition to the rising capital costs of new nuclear reactors, the world is faced with two other issues that are related to science and technology:

1. How safe are the current nuclear power plants.
2. The fate of the accumulated and produced nuclear waste materials, which are radioactive and must be reliably stored for several hundreds of thousands of years. The expected decommissioning of the older nuclear power plants will significantly increase the amount of nuclear waste, because several equipment—including reactor vessels, heat exchangers, tubing, pumps, and turbines—have become contaminated with radioactive materials and need to be stored as radioactive waste.

The scientific community has been divided on the first issue: while many scientists advocate that the current PWR, AGR, and BWR nuclear power plants are safe and that the accumulated experience of their operation warrants the safe operation of these plants in the future, there are others who point out similar statements that were made only a few months before the TMI-2, Chernobyl, and Fukushima Dai-ichi accidents. A few scientists advocate that the risk associated with a severe accident is too high for a society to bear and that nuclear power plants should be shut down. A nuclear reactor meltdown would contaminate the aquifers of a very large geographic region and has the potential to make thousands of square miles of land uninhabitable. From the statistics of the currently operating and planned nuclear power plants, it appears that the societies in most of OECD countries are not willing to accept such an enormous risk. The societies in the developing nations are more willing to accept the nuclear energy risk since they plan to construct and operate more nuclear power plants in the near future.

The energy shortages experienced globally since 2005 and the accompanying enormous spike in fossil fuel prices have mitigated the unwillingness of several OECD nations to accept the safety risks of nuclear energy. This is manifested in the number of nuclear reactors that have been proposed (the numbers for "proposed" plants are separate from those "under construction" or "planned"): 20 in the United States, 6 in the United Kingdom, 6 in Canada, and 10 in Italy [15]. However, these new power plants are in the early planning stages and far from construction, licensing, and power production. The citizenry in OECD countries is apprehensive and suspicious of the nuclear industries, which are in a precarious position. The actual construction and completion of these plants in OECD countries, as well as that of many of the power plants that are to be built in the developing countries, may be summarily cancelled, not only if there is a new serious accident, such as the TMI-2 or the Chernobyl accident, but also if there is any incident that has *the appearance, perception by the public, or portrayal as a serious accident by the information media*. The 2011 accident in the Fukushima Dai-ichi power plant is such a serious accident. In the wake of this accident, several OECD governments have cancelled or put on hold the expansion of their nuclear programs, while Germany announced that it will abandon nuclear energy by 2022. In the wake of this accident, the International Energy Agency (IEA) halved its estimate of additional nuclear-generating capacity to be built by 2020 and reduced by 30% its estimates for additional capacity in the years 2035 and 2040 [17].

Regarding the second issue of nuclear waste disposal, the scientific community has provided a viable solution: nuclear waste may be safely processed and stored in underground caverns and subterranean tunnels. The processing of this waste may require densification and vitrification (i.e., glassification). Several European countries, including France, Belgium, and Germany, have central waste processing and storage facilities, such as those in Dessel, Belgium, and the Ahaus salt mine in Germany, but no permanent disposal facilities in operation. Only the United Kingdom has a permanent low level-storage facility, in Sellafield, the site of the 1957 AGR Windscale power plant accident. The radioactive waste is first vitrified and then stored in this facility for the long term.

In the United States, where there are 103 nuclear reactors in operation, most of the nuclear waste materials are stored in temporary facilities (water pools) close to the reactor. The nuclear waste storage in the United States has become more of a political problem than a scientific or engineering problem. It has been proposed, and was approved by an act of Congress, in 2002, that a permanent federal repository facility be constructed in the Yucca Mountain, Nevada. A great deal of research and several feasibilities studies had shown in the past that the repository at Yucca Mountain would be safe to operate and would not pose a major environmental threat to the region. However, local groups opposed the construction of such a repository on the grounds that the preliminary studies are not complete, while other groups have raised objections to the transportation of thousands of tons of radioactive waste from their current locations to Yucca Mountain. After several years of studies, the spending of at least $4 billion, a great deal of judicial and political activities and several legislative acts, the US Department of Energy announced in March 2009 that the Yucca Mountain ". . . is not considered as an option for storing nuclear waste." Given the lack of viable alternative solutions for the storage of nuclear waste, very few consider this statement to be the final decision on this project.* This was a political decision that did not consider the scientific reports and the national need to permanently store the continuously accumulating nuclear waste.

Despite the safety and nuclear waste issues, nuclear energy has made a lot of new advocates in the early twenty-first century because it is an alternative to fossil fuels and is one of the viable solutions for the mitigation of global climate change. Nuclear power plants do not emit carbon dioxide, do not contribute to methane releases, and do not produce NOX gases and smog. Their carbon footprint is almost zero, and the chemical pollution from them is significantly lower than that of coal and natural gas power plants. If the safety and storage problems are solved in a satisfactory way, it is very likely that the world will experience a revival of the nuclear industry. A binding international agreement for the reduction of CO_2 emissions may become a factor for renewed optimism and growth of nuclear energy in the future: if such an agreement is binding to the participating nations and if it is put into effect by the participant states, then the production of electricity from nuclear energy will become one of the most promising and reliable ways to fulfill the national CO_2 reduction quota.

It must be noted that sufficient nuclear fuel resources exist globally to ensure the uninterrupted and reasonably priced supply of nuclear fuel to the reactors. Figure 5.9 shows

---

* A low-radioactivity storage site in Andrews County, West Texas, is being developed in 2015 by a private corporation, but no large-scale storage licenses have been issued for this facility.

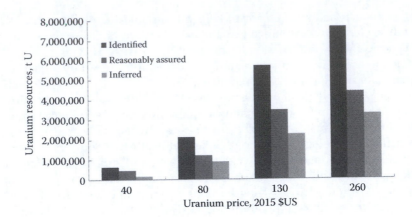

FIGURE 5.9
Identified global recoverable uranium resources based on the price of uranium. (Data from Nuclear Energy Agency/IAEA, *Uranium 2016: Resources, Production and Demand*, A Joint Report by the Nuclear Energy Agency and the International Atomic Energy Agency (the Red Book), Organisation for Economic Co-operation and Development, Paris, 2016.)

the identified global recoverable uranium resources as reported by the IAEA as a function of the fuel price [18]. At the current rate of uranium consumption, the identified uranium reserves are expected to last for more than 250 years, even with the globally planned expansion of nuclear power plants. The utilization of thorium and the identification of additional reserves will extend this timeframe to five to eight centuries. Unlike fossil fuel resources that may be depleted in the twenty-first century, all indications are that nuclear fuel reserves will last much longer.

5.8.1 To Breed or Not to Breed?

It is apparent from Section 5.3.4 that breeding resolves the two most important issues of nuclear industry:

1. It significantly extends the time period for nuclear energy by extending the amount of nuclear fuel. The breeding of known nuclear resources may provide the entire energy of the world's population for more than 5000 years.
2. Most of the current and future radioactive waste, which primarily consists of the $_{92}U^{238}$ isotope, will be consumed in breeder reactors. This isotope is converted to the fissile $_{94}Pu^{239}$ in the breeder reactors and is subsequently consumed as fuel.

FBRs would not only use all the material in the natural uranium and thorium ore, but would also convert some of today's nuclear waste to tomorrow's nuclear fuel. While this conversion is possible and its feasibility has been demonstrated in several experimental and commercial breeder reactors, FBRs that have been constructed have had many engineering and safety problems, among which are the following:

1. FBRs produce $_{94}Pu^{239}$, which is the primary material used in nuclear weapons. One of the risks associated with the operation of FBRs is the formation of a critical mass of $_{94}Pu^{239}$ within the reactor that may lead to a small-scale nuclear explosion. While the typical thermal reactor, such as the PWR, BWR, and AGR, does not contain

nuclear fuel in concentrations that may become "critical" for a nuclear explosion to occur, the isotope $_{94}Pu^{239}$ is formed almost indiscriminately in a FBR at a rate that is difficult to control. Because of this, it is possible to form a sufficient amount of the $_{94}Pu^{239}$ fuel within the reactor to attain critical mass. While this explosion may not have the full power of an atomic bomb, it would be sufficient to destroy the reactor and disperse an enormous amount of radionuclides in the environment.

2. Related to the preceding paragraph, the production of $_{94}Pu^{239}$ contributes to global political and military instabilities and poses a threat by itself. A FBR produces sufficient mass of $_{94}Pu^{239}$ annually for the construction of several nuclear weapons. The danger of nuclear weapon proliferation alone would dictate the severe restriction of the use of FBRs and the export of FBR technology to several nations, as well as the strict accounting and regulation of the FBR fuels and products. This is very difficult to accomplish on a global scale under the auspices and the infrastructure of the existing international institutions, such as the UN, the IEA, and the IAEA.

3. The very high-power density of FBRs and the necessity to use Na or K as coolants can be the cause of accidents. Minor leakages of the coolant fluid occur in large-scale heat power removal systems, such as the ones used in nuclear reactors. The leakage of water or gases, which are commonly used as coolants in thermal reactors, has occurred on several occasions in conventional power plants, but was always contained and did not pose a major threat. However, Na and K violently react with air and water, and any leakage of them causes local fires that can become catastrophic in the nuclear reactors. The reaction

$$2Na + 2H_2O \rightarrow 2NaOH + H_2 \tag{5.17}$$

and its equivalent for potassium, are highly exothermic, may cause local fires, and release enough heat to melt steel pipes. The melting of the piping releases more of the light metal coolant; feeds the fire; raises the local temperature to a higher level; and, thus, starts a catastrophic effect that may cause a severe accident.

4. Even though FBRs have been constructed since the 1950s for experimental purposes, our experience with them is not sufficient to warrant their safe operation. Of the more than 300 FBRs that have been built for commercial or experimental purposes, only a handful are in operation in 2017, and only for experimental purposes. Commercial reactors (most notably the *Phenix* FBR in France and the *Jojo* FBR in Japan) have ceased operations because of safety concerns [7].

5. There have been several unexpected accidents and notable "incidents" with other FBRs: The experimental EBR1, EBR2, and Fermi-1 reactors in the United States have had several problems recorded and were shut down. In 1973, France's 250-MW prototype FBR at Marcoule went critical, followed, in 1974, by the British prototype FBR at Dounreay. There is surveillance satellite evidence that a severe accident occurred in the Shevchenko plant in the former Soviet Union, although the Soviet authorities at the time gave no details or explanation for it. The operation of the 375 MW Clinch River Breeder Reactor in the United States was marred with several mishaps until, in 1980, President Carter issued an executive order that the plant be shut down and the Breeder Reactor program in the United States be discontinued. Similar incidents contributed to the shutting down of FBRs in other countries. In 1995, flow vibrations in the secondary loop of the *Monju* FBR power plant in Japan caused a sodium leakage,

which started a fire that melted part of the sodium piping. This rendered the reactor inoperable for more than a decade and required a decision by Japan's Supreme Court to clear the way to its restarting [7]. The reactor has not restarted because it was uncertain whether or not adequate measures have been taken to avoid a similar accident, and in 2017, it is actually in the process of being decommissioned. Such incidents in FBR power plants do not inspire confidence for the widespread use of the FBR reactors, at least in the first part of the twenty-first century. Clearly, more tests, more definitive research and several technological advancements are needed to render the FBRs safe for the extensive production of electric energy.

FBRs may hold the solution to the energy challenges the humans face at the beginning of the twenty-first century. Not only are they able to use the available nuclear resources in their entirety and to satisfy the energy needs of humanity for several millennia, but they may also eliminate the current nuclear waste. However, these reactors have a variety of severe operational problems and may contribute to nuclear weapons proliferation. At present, the risk of their operation is significantly higher than that of the conventional thermal reactors, and their commercial development has stopped. Given the advantages of the use of FBRs, the community of nations may have to seriously debate the nuclear solution to the energy problem, to make the technological advances that are needed for the safer operation of FBRs, to weigh the advantages and disadvantages of FBRs, and to answer the hard question for the future: *to breed or not to breed?*

5.9 Myths and Reality about Nuclear Energy

Myth 1: We all know what happened in Hiroshima. Nuclear reactors can explode at any time.

Reality: Nuclear weapons contain almost pure uranium-235 and are designed to form a critical mass of this material that leads to the nuclear explosion. In contrast, the fuel of the conventional thermal power plants is a slightly enriched (up to 5%) uranium-235. Because this fuel is primarily composed of the nonfissile uranium-238, a critical mass of uranium-235 is impossible to be formed in the reactor, even after a meltdown. As a consequence, it is impossible for the thermal reactors to cause nuclear explosions as the atomic weapons that were dropped in the two Japanese cities. This notwithstanding, a LOCA, which may result in reactor meltdown or a chemical explosion that would contaminate a very large area, is possible with conventional reactors.

Myth 2: Nuclear energy will produce electricity too cheap to be measured.

Reality: Section 5.7 explains in detail the economics of nuclear power at present. While the cost of the nuclear fuel is low, the capital cost associated with the construction of modern nuclear reactors is very high. When the capital cost is amortized into the price of electricity produced by nuclear power plants, the cost becomes comparable to the cost of electricity currently produced from fossil fuels.

Myth 3: Breeder reactors are very safe! It is only public opposition to nuclear energy that delays their development.

Reality: The global experience with FBR prototypes proves that these reactors cannot be considered as safe. Several of the experimental reactors that were constructed in the

United States, France, Japan, and the USSR have developed significant operational problems and were shut down for safety reasons. The two commercial reactors that have operated for a few years—the *Phenix* FBR in France and the *Jojo* FBR in Japan—have been shut down for safety reasons too. The public opposition to the further development of FBRs and the construction of additional FBR power plants is well founded.

Myth 4: Nuclear reactors produce electric power without any carbon dioxide emissions. We must substitute all fossil fuel power plants with nuclear.

Reality: While it is correct that the nuclear power plants do not produce any CO_2 and that nuclear energy is a possible solution to the national and global reductions of CO_2 emissions, the substitution of *all* fossil fuel power plants with nuclear units is impractical because the power produced by most nuclear power plants of current designs is constant and cannot fluctuate to meet the power demand during short time periods. Nuclear power plants cannot be turned on and off at will as gas turbines do. For this reason, nuclear power plants are currently used as *base-load* plants in the electric grids to supply the constant amount of electric power that is always demanded (Chapter 7). It is possible for nuclear power plants to substitute the other base-load plants (primarily large coal units), and this would have a significant effect on CO_2 emission reduction, nationally, regionally, and globally.

Myth 5: Nuclear waste is not an issue in the development of nuclear power. After all, we can send all the nuclear waste to the moon or to the outer space.

Reality: It costs more than $25,000 to lift 1 kg of mass into space, and nuclear reactors produce several tons of waste per year. When this disposal expenditure is added to the price of electricity from nuclear reactors, nuclear energy becomes very expensive. Additional considerations of the adverse effects of "space debris" to spacecraft, satellites, and the global communications network make this a prohibitive option.

Myth 6: There is not enough uranium-235 to power our nuclear reactors. If we do not immediately switch to breeder reactors, we will shortly run out of nuclear fuel.

Reality: As it is apparent in Figure 5.9, at the current (2017) rate of uranium consumption, which is approximately 5,600,000 t per year, the existing uranium reserves in the world are large enough to guarantee the continuous supply of nuclear fuel for more than 250 years. In addition to the proven and inferred reserves, other resources and potential resources exist that may extend this time frame to 600–700 years [18]. Therefore, there is not an immediate urgency to switch to breeder reactors, even if our nuclear fuel consumption doubles or triples. It will be good though to develop safe breeder reactor technologies in the future because the breeder reactors can guarantee our future electric power supply for millennia, rather than centuries.

Myth 7: Nuclear reactors are the first step for countries to develop nuclear weapons. We must stop the proliferation of weapons by shutting down their nuclear reactors.

Reality: Most of the countries that currently possess nuclear weapons have developed these weapons several years before they developed nuclear energy for peaceful purposes. A few of these countries do not even have nuclear power plants for the production of electricity. In addition, several of the countries that currently use nuclear energy for the production of electricity do not possess nuclear weapons and do not plan to develop nuclear weapons. Therefore, one cannot argue that nuclear energy is the first step for countries to develop nuclear weapons in the future. The nuclear fuel cannot be used in nuclear weapons, because the latter need very high-grade uranium-235 or plutonium-239, which cannot be obtained from a peaceful nuclear energy program. While the development of nuclear reactors may help with knowledge and instrumentation, a different technology altogether is needed for the production of nuclear weapons.

PROBLEMS

1. Use the periodic table of elements to identify the elements with the following atomic numbers: 8, 13, 21, 34, 55, and 89.

2. Complete the following reactions (identify the elements X, Y, Z, and W):

$$_{92}U^{238} + _{0}n^{1} = X + _{-1}e^{0} = Y + 2 _{-1}e^{0},$$

$$_{92}U^{235} + _{0}n^{1} = _{38}Sr^{90} + Z + 3 _{0}n^{1},$$

$$_{3}Li^{7} + _{0}n^{1} = 2_{2}He^{4} + W.$$

3. What is the energy equivalent of 1 lb (0.453 kg) of mass? What is the mass equivalent of the electric energy produced by a 1000 MW nuclear power plant for an entire year?

4. If four nuclei of $_{1}H^{1}$ were to fuse and produce a single nucleus of $_{2}He^{4}$, what is the mass defect and the corresponding energy released? What is the energy in kilojoules that would be released from the fusion of 1 kg of hydrogen?

5. It is suggested that nuclear waste, which includes $_{92}U^{238}$, be stored under controlled conditions in mountainous caverns. If 1 t (1000 kg) of $_{92}U^{238}$ is stored in the year 2020, how much of the material would remain in the year 3000 and in the year 5000? How long would it take for the amount of uranium stored to reach 10% of its initial amount?

6. The Chernobyl accident occurred on April 26, 1986, and most of the strontium-90 in the reactor was released to the environment. What is the percentage of this isotope that remained on January 1, 2012, in the environment?

7. Identify 12 isotopes that are produced in a thermal nuclear reactor.

8. A nuclear reactor provides 3600 MW of thermal power. The reactor operates with 2% enriched uranium. Assuming that only $_{92}U^{235}$ reacts in the reactor, determine (a) the daily consumption of $_{92}U^{235}$ in the reactor and (b) the burnup rate of the reactor.

9. Nuclear energy is often touted as a solution to the problem of CO_2 emissions and global warming. A 1000 MW coal power plant with 41% efficiency is to be replaced by a 1000 MW nuclear plant with 36% efficiency. Assuming that the coal plant burns only carbon (C) determine the following:

   a.   The amount of carbon (in metric tons) used by the coal plant during a year

   b.   The amount of CO_2 (in tons) produced during a year

   c.   The amount of natural uranium to be used by the nuclear plant in a year

10. In a short essay (250–300 words), explain why the moderator is an essential part of the nuclear reactor.

11. What are the advantages and disadvantages of the PWR compared to the BWR?

12. What are the advantages and disadvantages of the GCR compared to the LWR?

13. Is it possible for an accident similar to the Chernobyl accident to occur in a typical PWR or a BWR? Explain carefully your reasons.

14. Identify the issues associated with nuclear waste disposal and why there has been no permanent solution for the storage of nuclear waste in the United States.

15. In your opinion, what is the best way to dispose of the nuclear waste and why.

16. "Fission reactors are only a short-term solution to the energy problem, and it is not worth debating and pursuing this option. We should move full-steam ahead with the breeder reactor option." Comment on this statement.

References

1. Bennet, D.J., *The Elements of Nuclear Power*, Longmans, London, 1972.
2. McCracken, G., Stott, P., *Fusion—The Energy of the Universe*, Elsevier, Amsterdam, 2005.
3. Michaelides, E.E., *Alternative Energy Sources*, Springer, Berlin, 2012.
4. Warner, F., Harrison, R.H., *Radioecology after Chernobyl*, Wiley, New York, 1993.
5. El-Wakil, M.M., *Power Plant Technology*, McGraw-Hill, New York, 1984.
6. Hewitt, G.F., Collier, J.G., *Introduction to Nuclear Energy*, Taylor & Francis, New York, 2000.
7. Cochran, T.B., Feiveson, H.A., Patterson, W., Pshakin, G., Ramana, M.V., Schneider, M., Suzuki, T., von Hippel, F., *Fast Breeder Reactor Programs: History and Status*, Program on Science and Global Security, Princeton University, Princeton, NJ, 2010.
8. Medvedev, Z., *The Legacy of Chernobyl*, Norton, New York, 1992.
9. Gould, P., *Fire in the Rain: The Democratic Consequences of Chernobyl*, Johns Hopkins University Press, Baltimore, MD, 1990.
10. IAEA (International Atomic Energy Agency), *The Fukushima Daiichi Accident—Report by the Director General*, IAEA, Vienna, 2015.
11. Lovering, J.R., Yip, A., Nordhaus, T., Historical construction costs of global nuclear power reactors, *Energy Policy*, **91**, 371–382, 2016.
12. World Nuclear Association, *The Economics of Nuclear Power* (updated January 2017), http://world-nuclear.org/uploadingfiles/info/pdf/economicsnp.pdf (website).
13. IAEA, *Trends in Electricity Supplied*, IAEA, Vienna, 2015.
14. European Nuclear Society, *Nuclear Power Plants, World-wide*, European Nuclear Society, Brussels, 2015.
15. International Energy Agency, *World Energy Outlook—2010*, International Energy Agency, Paris, 2010.
16. Smil, V., *Energy—Myths and Realities*, The AEI Press, Washington, DC, 2010.
17. International Energy Agency, *World Energy Outlook—2014*, International Energy Agency, Paris, 2014.
18. Nuclear Energy Agency/IAEA, *Uranium 2016: Resources, Production and Demand*, A Joint Report by the Nuclear Energy Agency and the International Atomic Energy Agency (the Red Book), Organisation for Economic Co-operation and Development, Paris, 2016.

6

Renewable Energy

As their name implies, renewable energy sources are essentially inexhaustible and may provide energy for several millennia. Hydraulic (hydroelectric or hydro), solar, wind, biomass, and geothermal energy are the most widely used renewable energy sources. Less frequently used forms are tidal, wave, sea current, and ocean–thermal energy conversion (OTEC). Renewable energy sources pose a lesser threat to the environment and ecosystems than the other energy sources; they do not contribute to global warming and, for this reason, have been at the forefront of development since 2000. In particular, solar and wind energy systems for the production of electricity as well as solar energy for space heating and water have grown very rapidly in the period of 2010–2015. Table 6.1 gives the existing global capacity at the end of 2015 and the annual globally averaged growth rate of the capacity of renewable energy installations [1].

It is apparent from Table 6.1 that the utilization of solar and wind energies is galloping at very fast rates, while the rest of the renewables show very healthy annual growth rates. The faster rates of growth are due to the previously low contributions of renewables in the global energy production and to current favorable national policies. Solar, wind, and geothermal sources combined produced 6.3% of the global electricity consumption in 2014, while hydroelectric power plants produced approximately 16.4% of the total [2]. When the contribution of all the renewables to the total primary energy sources (TPESs) of the world is scrutinized, the combined contribution of all renewable energy was less than 8% in 2015, and most of this contribution came from the combustion of wood/biomass [2]. A promising portent for the future growth and utilization of renewables is that because of greenhouse gas (GHG) emissions and global climate change concerns, there are currently several national and international policies that will help maintain the growth rates of renewables in the TPES, especially in the production of electricity. Projections of the Energy Information Administration estimate that the contribution of renewables in the global TPES supply will be between 25% and 33% by the year 2040 [2]. This implies that the utilization of renewable energy sources (and especially that of wind and solar energies) will significantly increase in the near future, and the renewables industry will experience high rates of growth in the next 20 years.

6.1 Hydroelectric Energy

The potential and kinetic energies of water has been used as a source of power with watermills for thousands of years. At present, much bigger power plants convert the energy of several rivers to electricity. A river with relatively high mass flow rate is restricted by a dam, which is constructed at a location where a significant drop in the river elevation occurs, often in a region of waterfalls. The dam creates an artificial water reservoir

TABLE 6.1

Global Capacity and Annual Rates of Growth of Renewable Energy Sources (2015)

	Global Capacity	Annual Growth Rate, %
Solar electric—PVs	227 GW	42
Solar electric—thermal	4.8 GW	35
Solar thermal—heating	435 GW	12
Wind—electric	433 GW	17
Geothermal electric	13.2 GW	3.7
Geothermal heating	22 GW	12.3
Hydroelectric	1,064 GW	2.9
Biomass—ethanol	98×10^9 L/year	3.0
Biomass—biodiesel	30×10^9 L/year	6.5

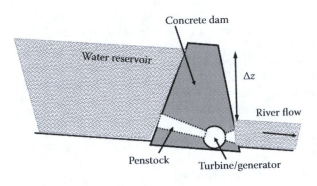

FIGURE 6.1
Schematic diagram of a hydroelectric power plant.

upstream, which may store billions or trillions of cubic meters of water.* The capacity of artificial water reservoirs is measured in acre feet of water, with 1 acre ft being equal to 326,000 gal or 1,239 m³ of water. In addition to electric power generation, dams and reservoirs regulate the seasonal water flow in the river and have become parts of large-scale irrigation projects.

A schematic diagram of a typical hydroelectric power plant is shown in Figure 6.1. Water from the artificial water reservoir enters a large pipe, the *penstock*, and is directed to one or more hydraulic turbines, which are connected by shafts to electric generators and produce power. The discharged water from the turbines exhausts downstream to the same river at a lower elevation. Frictional and other losses in the penstock are typically very small. The power produced by the flowing stream with mass flow rate $\dot{m}$, when its elevation changes by the height difference Δz, is

$$\dot{W} = \eta \dot{m} g\, \Delta z, \tag{6.1}$$

where η is the efficiency of the turbine-generator system, typically in the range of 75–85%. A large hydroelectric power plant is constructed with several penstocks, typically 10–30,

---

* One cubic meter of water contains approximately 1 t, or 1000 kg, of mass.

TABLE 6.2

Hydroelectric Plants Currently in Operation

Name of Plant	River/Country	Power, MW
Hoover Dam	Colorado/US	1,500
Niagara Falls	Niagara/US	1,950
Grand Coulee	Columbia/US	6,500
La Grande	St Lawrence/Canada	10,000
Itaipu	Parana/Paraguay–Brazil	12,600
Three Gorges	Yangzee/China	22,500
Krasnoyarsk	Yenisei/Russia	6,400
Guri	Caroni/Venezuela	10,300

which lead to a number of independent turbine-generator systems that operate in parallel to produce the total power of the plant. With this arrangement of independent penstocks and turbines that may be open or shut at will, the number of operating penstocks and turbines that produce power is continuously controlled by the operators of the plant. Hence, the power of the hydroelectric plant may be controlled to match the demand for electricity.

Table 6.2 shows several of the larger hydroelectric power plants in the world that are currently in operation as well as their power generation capacity, in megawatts [3]. Most of these are older installations, with some, e.g., Krasnoyarsk and Hoover dams, operating since the 1930s. The Three Gorges plant in China is the only recent plant, with its construction completed in 2013.

One may observe in this table that the electricity-generation capacity of the hydroelectric power plants is significantly higher than that of nuclear, coal, and gas-fired power plants.* The very high power of the larger hydroelectric plants is provided by the vast amount of water carried by large rivers. The flow rates in the large rivers with hydroelectric plants are in the range of 10^4–10^5 m^3/s, which are equivalent to mass flow rates of 10^7–10^8 kg/s. Equation 6.1 shows that with 75% turbine-generator efficiencies, this flow rate of water will produce between 2,200 and 22,000 MW of electricity, a very significant amount of electric power. Smaller rivers with lower flow rates would produce lesser electric power.

Example 6.1

A small river passes through the outskirts of a town carrying on average 45 m^3/s of water. The river has a small waterfall downstream, and the town is considering building a damn and a hydroelectric power plant. The damn will increase the height of the fall to 9.5 m. It is estimated that the turbine-generator system to be installed will have 80% efficiency. Determine how much power the town can get from this hydroelectric plant.

Solution: The 45 m^3/s volumetric flow rate of water is equivalent to a mass flow rate of 45,000 kg/s. From Equation 6.1, the average power produced by this plant will be 0.8 × 45,000 × 9.81 × 9.5 = 3.36 × 10^6 W or 3.36 MW.

Note that this is sufficient electric power to supply about 3,500 homes. The addition of this relatively small electric installation can make a small town self-sufficient with electric power.

---

* A nuclear reactor typically produces 1000 MW, a coal power plant 400 MW and a gas turbine 5-100 MW.

TABLE 6.3

Leading Countries in Hydroelectric Power Production

Country	Installed Capacity, GW	Energy Produced, TWh	Percentage of Total Electric Energy
Brazil	86	391	69
Canada	76	392	60
PRC	194	920	17
India	40	142	12
Japan	49	85	8
Norway	31	129	98
Russia	50	183	17
US	102	290	7
Venezuela	15	84	68

6.1.1 Global Hydroelectric Energy Production

Approximately 4,000 TWh of electric energy were produced globally in 2013 from hydroelectric power plants with total generating capacity of 1,034 GW [2]. Table 6.3 shows the largest producing countries of hydroelectric energy. An interesting observation in this table is that a few smaller countries, such as Norway, have the capacity to produce almost their entire electricity demand from their hydroelectric power plants. Others, such as Canada, Brazil, and Venezuela, produce more than half of their electricity demand from this renewable energy source. This is not the case with the larger electricity–producing countries—United States, Japan, Russia, and the People's Republic of China (PRC)—that produce only a smaller fraction of their electricity from hydroelectric energy but, nevertheless, produce very high quantities of electric energy from hydroelectric units annually.

One of the advantages of hydroelectric power plants for the national power grids is that they are available to produce electric power at any time. This is accomplished by regulating the water flow through the penstocks to the turbines. For this reason, hydroelectric power plants are used as intermediate and peak electric demand units. A glance at Table 6.3 proves that the hydroelectric power plants in the United States have the capacity to produce $102 \times 24 \times 365 = 893.5$ TWh of electric energy annually. Instead, they produced 290 TWh in 2013. This implies that the total capacity of the US hydroelectric power plants is utilized for 32.4% of the time. The corresponding fraction for Norway is 47.5% and that of the entire world is 42.8%. Clearly, if it becomes necessary, a great deal more electric energy may be produced globally with the existing hydroelectric installations.

Of all the categories of electricity generating units, the hydroelectric power plants are the most lasting and durable. Hydroelectric units that were constructed in the 1920s and 1930s (e.g., the Niagara project and the Hoover Dam in Nevada/Arizona in the United States) are still in operation (2017). The main mechanical parts, the turbines, are frequently maintained and may be replaced if necessary. Dams, penstocks, and water gates do not have any moving parts and last much longer. For this reason, hydroelectric installations operate and generate electricity for very long times.

One of the limiting factors with the life of a hydroelectric power plant is the silting of the reservoir upstream the dam: there is a stagnation region for the water in front of the dam, water comes to a standstill, turbulence is significantly reduced, and smaller silt particles fall to the bottom (sedimentation). Excessive siltation in hydroelectric reservoirs may be avoided by diverting part of the water from the bottom, special weirs, special water projects upstream that reduce soil erosion and particulate entrainment, and periodic dredging of the reservoir [3]. A well-maintained water reservoir will last for more than 100 years. Their longevity makes the hydroelectric units one of the least expensive systems for the production of electricity.

6.1.2 Environmental Impacts and Safety Concerns

Hydroelectric energy is a very clean energy source with zero emissions. However, the construction of the hydroelectric power plants, especially those of very large scale, may cause environmental and ecological problems, among which are the following:

1. *The construction of the dam and the reservoir behind it.* Dams often exceed 50 m in height and 20 m width and require a very high volume of materials, primarily concrete, to construct them. The construction of the Hoover Dam in United States, which has power rating of 2,080 MW and produces more than 4 TWh of electricity annually, required as much concrete as a 3,200 km four-lane highway. The reservoir of the Hoover Dam has 35.2×10^{12} m^3 capacity.

2. *Dams cause the flooding of large areas upstream, which become the water reservoirs.* This may cause the displacement of communities, the abandonment of towns, and the relocation of inhabitants. The recent construction of the Three Gorges Dam in PRC was for a long time a very controversial project, because it resulted in the displacement of an estimated 1,240,000 persons, the abandonment of several towns, extensive ecological change, and the flooding of cultural and archaeological monuments.

3. *The dams prevent fish migration and restrict the waterways.* This ecological impact may be mitigated by the installation of artificial "fish ladders," where the fish are able to swim upstream in a bypass waterway. Because the turbines harm the salmon spawn, in several reservoirs along the Columbia River, young salmon is transported downstream by barges during the spawning season.

4. *The dam–reservoir system alters the downstream water environment.* Because a great deal of sedimentation happens in the reservoir and oftentimes water passes through sieves, the water that comes from the turbine outlets contains a smaller amount of suspended sediment. This prevents the river bottom nourishment downstream with fresh sediment and may lead to the scouring of the riverbeds and partial loss of riverbanks.

5. *Riverbed erosion downstream.* The frequent opening and shutting of the penstocks— for the control of the flow and power output—causes fluctuations of the river flow and contributes to the erosion downstream.

6. *Downstream oxygen deprivation.* Because of the high swirl in the turbines, the dissolved air in water nucleates, forms bubbles, and leaves the water. Hence, the oxygen content of the water decreases downstream. This has a detrimental effect on the fish population and the ecological systems downstream.

It is apparent that the environmental and ecological impacts of the hydroelectric dams and power plants are relatively few and rather benign in comparison to the impacts of the fossil fuel and nuclear power plants.

Safety concerns to communities are posed by dam failures, which have caused some of the principal human-made disasters: The collapse of a cascade of dams in Southern China including the Banqiao Dam, during a flood in 1975, was reported to have caused the deaths of 171,000 people and to have left millions homeless [4]. The failure of the Vajont Dam in Italy, which was built in a geologically unstable region, caused the deaths of 1,917 persons in 1963.

Smaller dams and smaller hydroelectric power plants, which are touted as ideal for the future, pose a lesser threat to the populations downstream. However, even the small plants are not immune from failures and disasters and must be well maintained. The Kelly Barnes, a small hydroelectric dam in Georgia, United States, failed in 1,977 killing 39 persons. Good design and careful construction in a geologically stable location, followed by frequent and meticulous inspections and maintenance, will avoid dam failures and disasters in the future.

Finally, large dams are tempting industrial targets for terrorism and wartime sabotage. In the early twenty-first century, appropriate security measures to thwart terrorist activities and sabotage are becoming part of the routine operations in all the large dams of the world.

6.1.3 Planned Hydroelectric Installations and Future Expansion

Large hydroelectric power plants are simple in operation, inexpensive to construct and, most importantly, do not use fuel for the production of electricity. Our experience with units that were built early in the twentieth century proves that large hydroelectric power plants may operate for more than 100 years and produce electricity at significantly lower cost than thermal power plants.

A great deal of the prime large-capacity hydroelectric resources in Organisation for Economic Co-operation and Development (OECD) nations has already been utilized with hydroelectric plants that have been in operation for more than 50 years. Because of this, almost all the future expansion of hydroelectric power is expected to occur in developing countries, especially in PRC and the continents of South America and Africa. Controversies that surround the construction and safety of the dams and their environmental impacts downstream have delayed the construction of several of these projects, as for example, the *Grand Ethiopian Renaissance Dam* on the Blue Nile, which is very close to the Sudan–Ethiopia border. This project, which in its final stage will provide 6,000 MW of electric power, has drawn the ire and threats of sabotage by the Egyptian government, because of its effect on the flow of the River Nile [5]. Despite this kind of opposition for a few projects, and because of the increasing demand for electricity worldwide, it is expected that most of the planned and the under-construction large hydroelectric power plants will be built in the next two decades.

A glance at Equation 6.1 proves that smaller dams in smaller rivers with volumetric flow rate of 10^3 m^3/s and elevation drop of 4 m will produce approximately 30 MW with typical turbine-generator pairs that have 75% efficiencies. Smaller units at rivers with volumetric rates of 100–500 m^3/s would produce electric power in the range of 3–15 MW, enough to meet the electricity demand of a small town. OECD nations that have already utilized most of their large-scale hydroelectric resources currently put their efforts in the construction of such *small hydro* (less than 30 MW) and *microhydro* (less than 1 MW) plants. These plants will serve a small community or will be dedicated to provide electric power to a nearby industrial plant. There is a multitude of locations in Europe, North America, Japan, and Korea where such resources exist and where small hydroelectric and microhydroelectric power plants may be constructed to produce more electric energy from this clean and renewable source.

The small-scale hydroelectric resources are abundant around the globe. Their development for the production of electricity may quadruple the hydroelectric energy produced in OECD countries. An additional advantage of such *small hydro or microhydro* units is that because the units are very small, it is not necessary to construct large dams, and hence, their safety and environmental and ecological impacts are minimal. Oftentimes, the hydroelectric projects may be combined with flood control measures for the region, agricultural irrigation, and recreational projects that enhance their utility and their benefits to the local communities.

6.2 Solar Energy

The earth receives 1.73×10^{14} kW of power in the form of solar radiation from the sun. This amount by far surpasses all the power demand of the entire human population. The continuously received power from the sun, called *incident solar radiation*, or in short, *insolation*, sums up to total energy of 5.46×10^{21} kJ/year, and this is 11,000 times the total primary energy that was consumed by all the nations in 2015. Solar energy is abundant, free of charge, and available to all the nations and all the inhabitants of the planet. At present, only a very small fraction of insolation is used by the earth's population in the form of (passive) solar heating in buildings and for the production of electricity with photovoltaics (PVs) and thermal electric systems.

Daily and seasonal variations of insolation are caused by the rotation of the earth. The rotation of the earth around its axis causes day and night and the daily variation of insolation. The rotation of the earth around the sun, which has a period of 1 year, occurs in a slightly elliptical orbit. The earth is closest to the sun in January (perihelion) and furthest from the sun in July (aphelion). The daily rotation of the earth occurs around its polar axis—the axis from the north to the south poles—which is at a constant angle of 23.45° with respect to the plane of the elliptical orbit around the sun. As a result of the combined motion, the earth turns one of its hemispheres toward the sun, the northern hemisphere in June and the southern hemisphere in December. The hemisphere that faces the sun receives a higher amount of insolation than the other and has summer. The results of the variable insolation are the four seasons and the seasonal dependence of insolation, the seasonal temperature variations, the local wind patterns, and the local weather.

Because the motion of the earth and its position relative to the sun are predictable with high accuracy, solar energy is considered a *periodically variable* energy source. The power emitted by the sun is homogeneous in all directions. A fundamental measure that characterizes the insolation on the earth is the *solar constant*. This is the power received by a unit area, perpendicular to the solar rays and located outside the earth's atmosphere. The solar constant S_{const} is approximately equal to 1.353 kW/m^2 or 4.871 MJ/(hour·m^2) or 438 Btu/(hour·ft^2) or 1.940 Ly/minute. Between the boundary of the atmosphere and a specific location on the surface of the earth, several factors moderate the power received by a plate (e.g., a solar collector or a PV cell) on the surface of the earth:

1. The position of the earth: Maximum insolation occurs at the solar noon, and zero insolation, during the night.
2. The atmospheric gases and vapors that absorb some of the radiation energy.

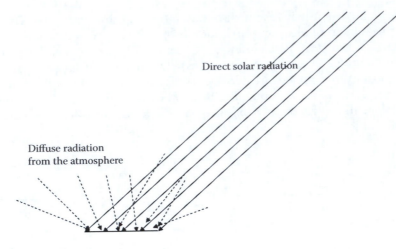

FIGURE 6.2
Direct and diffuse solar radiation on a horizontal surface.

3. The overhead clouds that reflect some of the radiation.

4. The location of the collector: On a clear day at noon, the power that reaches the surface of the earth varies from approximately 1 kW/m² on the equator, to less than 0.2 kW/m² in northern locations, such as Canada and the United Kingdom.

The clouds, several of the gases in the atmosphere, and the aerosol particles cause the diffusion of a fraction of the insolation. The diffusion is equivalent to a secondary weaker emission from within the atmosphere, the *diffuse radiation*, which contributes to the total power received by a collector on the surface of the earth. Figure 6.2 depicts a schematic diagram of the direct and diffuse radiation on the flat, horizontal surface of the collector. The *direct radiation* directly comes from the sun's disk, while the diffuse radiation comes from all directions. The sum of the two is the *total insolation* on a horizontal surface. It must be noted that the diffuse radiation is a significant fraction of the total insolation, even on a clear day: In Austin, Texas, it accounts for 12% of the total radiation during a clear-sky summer day. The diffuse radiation may account for more than 90% of the total insolation on a cloudy day in London, and it accounts for 50–55% of the yearly average radiation in most of the northern European countries.

6.2.1 Variability of Solar Radiation

The solar energy at all points on the surface of the earth is periodically variable. On a clear day, a small quantity of insolation appears at dawn, and the insolation steadily increases until it reaches a maximum at solar noon, when the sun is at its zenith—the highest point on the horizon. The insolation decreases to zero at dusk and remains at zero through the night hours. The daylight hours, hours when insolation is finite, are more during the summer and less during the winter. Because of this, there is significantly more insolation to be harvested during the summer, than in the winter. Figure 6.3 depicts the total amount of radiation (direct and diffuse) that falls on a horizontal surface in Austin, Texas, during the two solstices and the vernal (spring) equinox [6,7].* The daily and seasonal

---

* The solar time is determined by the position of the sun in the sky and is different than the local time. The solar noon in Austin, Texas, is at 1:33 PM on the summer solstice.

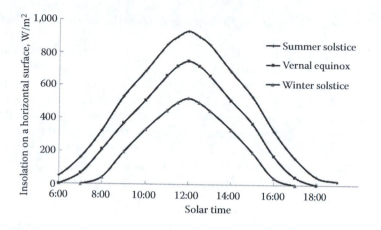

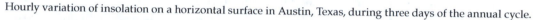

FIGURE 6.3

Hourly variation of insolation on a horizontal surface in Austin, Texas, during three days of the annual cycle.

variations of solar insolation is apparent in this figure as well as an experience of everyday life: Insolation is a periodically variable energy source, with a daily and an annual cycle.

The incident solar energy on a surface during an entire day is given by the integral of the total insolation:

$$E = \int_{t=0}^{24} (S_{sd} + S_D)A\,dt \approx \sum_{1}^{24} (S_{sd} + S_D)A, \tag{6.2}$$

where S_{sd} is the insolation directly received from the sun's disk and S_D is the diffuse insolation. Both are functions of time and location. When S_{sd} and S_D are measured in watts per square meter and the time in hours, the total energy is measured in watt hours per day (Wh/day). Referring to Figure 6.3, the daily amount of incident solar energy on a 1 m² horizontal surface in Austin, Texas, is approximately 8,000 Wh/m² on the summer solstice, 5,800 Wh/m² on the vernal equinox, and 3,500 Wh/m² on the winter solstice. The daily averaged (averaged over 24 hours) power received by a horizontal surface in Austin is 332 W/m², 240 W/m², and 148 W/m², respectively, during the three days depicted in Figure 6.3.

It is apparent that because of the daily variation of the incident radiation, the average power received by a surface is significantly lower than the maximum power, which occurs at the solar noon. The daily and annual variabilities of the incident radiation pose several challenges in the design and reliability of solar energy systems, which must be designed to accommodate this highly variable input. The nominal power of the solar energy systems is typically the power they produce during the hour of the highest insolation of the year, but these systems produce energy when the insolation is significantly less than the maximum.

Solar energy collectors and PV panels are composed of flat or curved plates. For a stationary flat surface, to receive the maximum insolation during an entire year, the surface must be tilted toward the sun at an angle with respect to a horizontal plane approximately equal to the latitude of the site/region where the surface is located. In the northern hemisphere, the flat surface faces south, and in the southern hemisphere, it faces north. If the tilt angle of the surface is equal to the latitude, the surface is perpendicular to the sun's rays at the solar noon of the two equinoxes. To maximize the insolation on the flat surface in the summer, when the sun is higher on the horizon, the angle of tilt should be less than the latitude. To maximize the insolation in the winter, the angle of tilt should be greater than the latitude.

The spatial and temporal variability of insolation is demonstrated in Table 6.4, which shows the amount of energy collected on several flat surfaces located at three northern latitudes of 30°, 40°, and 50° during four days of a year, March 21, June 21, September 21, and December 21. The surfaces are placed in the following orientations: horizontal; vertical; faced south and tilted at an angle equal to the latitude of the location; and normal to the sun's rays, following the relative position of the sun in the sky. The data are for "clear-sky" days, when there is no cloudiness [6,7].

The data in the table apply to latitudes of the southern hemisphere, with the December and June dates reversed. One may draw the following conclusions from the data of this table:

1. The amounts of energy collected during the March and the September dates are almost equal because the dates are very close to the annual equinoxes.

2. There is significant variability of the incident energy on the horizontal and vertical surfaces during the year. The variability of the tilted surface facing the south is less.

TABLE 6.4

Energy Collected Daily at Differently Oriented Surfaces in Three Northern Latitudes on Simulated Clear-Sky Days

Date	Horizontal	Vertical	Tilt = Latitude	Normal	Units
30° Latitude					
March 21	24.3	13.8	33.3	36.3	MJ/(m²·day)
June 21	29.8	4.0	35.8	38.6	MJ/(m²·day)
September 21	23.1	12.9	31.0	33.5	MJ/(m²·day)
December 21	13.9	21.1	24.7	29.7	MJ/(m²·day)
Annual average	22.8	12.9	31.2	34.5	MJ/(m²·day)
Average power	263.7	149.4	361.4	399.7	W/m²
40° Latitude					
March 21	21.1	16.5	33.1	34.9	MJ/(m²·day)
June 21	28.5	8.4	32.3	33.5	MJ/(m²·day)
September 21	20.4	15.9	32.1	33.9	MJ/(m²·day)
December 21	8.6	17.7	19.3	22.1	MJ/(m²·day)
Annual average	19.7	14.6	29.2	31.1	MJ/(m²·day)
Average power	227.7	169.0	338.4	359.8	W/m²
50° Latitude					
March 21	15.6	16.6	27.0	27.6	MJ/(m²·day)
June 21	28.3	11.7	35.0	36.1	MJ/(m²·day)
September 21	16.3	18.9	32.2	33.5	MJ/(m²·day)
December 21	3.2	10.2	10.2	11.4	MJ/(m²·day)
Annual average	15.9	14.3	26.1	27.1	MJ/(m²·day)
Average power	183.6	165.9	302.8	314.0	W/m²

Source: Wilcox, S., *National Solar Radiation Database 1991–2010 Update: User's Manual*, NREL/TP-5500-54824, NREL, Golden, CO, 2012; NREL, *NREL—National Solar Radiation Data Base: 1991–2010 Update*, NREL, Golden, CO, http://rredc.nrel.gov/solar/old_data/nsrdb/1991-2010/ (website), last visited August 16, 2016.

3. The ratio of the energy collected by a horizontal surface to that of a vertical surface decreases with the latitude and from summer to winter. At 50° latitude in the winter, a window (vertical surface) will collect more than three times the solar energy than a flat horizontal rooftop. In the summer, the ratio is almost reversed. This observation is used in the more efficient design of solar heating systems for buildings.

4. A stationary, tilted flat surface facing the south collects more than 90% of the maximum energy that could be collected by a surface, which follows the position of the sun. This implies that for small projects, there is little economic justification in investing on control systems that allow the solar panels to follow the sun.

5. The total daily amount of energy received is significantly less than the energy received on June 21, which is very close to the maximum energy.

6. The ratio of the energy collected during the winter to the energy collected in the summer decreases with the latitude of the location. The energy produced in the latitude of 50° during the winter days is very low, primarily because winter days at the northern latitudes are very short.

The data in Table 6.4 pertain to clear-sky days. The radiation in clear-sky days is periodically variable and predictable with accuracy. Weather phenomena—cloudiness, rain, strong winds, dust storms, etc.—and the microclimate of a region affect the insolation at a given location. The weather phenomena temporarily affect the insolation and may be averaged during the course of a year. The effects of such weather phenomena on the average annual insolation are shown in the map in Figure 6.4, which depicts the contours of the annually averaged local insolation on a flat horizontal surface [3]. It is apparent that several local factors, the microclimate, significantly affect the average power received and that this

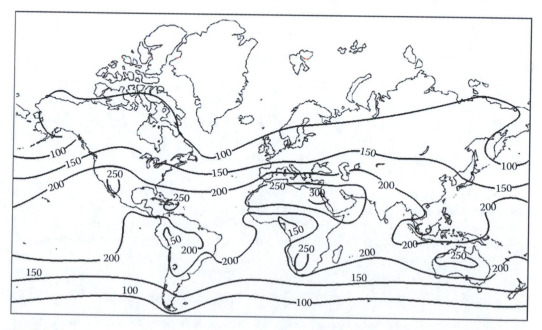

FIGURE 6.4
Contours for the annual average solar radiation on a horizontal surface in watts per square meter.

power is less than the clear-sky insolation. The following conclusions and observations may be drawn from this figure:

1. Despite the fact that the solar radiation at the edge of the atmosphere is approximately $S = 1,353$ W/m^2, and the maximum incident radiation at an equatorial location is close to 1,100 W/m^2, when the insolation is averaged over the entire day and night and over the whole year (8,760 hours), the best locations on earth receive an average of only 250–300 W/m^2.

2. The deficit of water vapor and general lack of cloudiness in arid regions (e.g., Arizona, Sahara, the Arabian Peninsula, Kalahari, and the interior of Australia) accounts for the higher average insolation in these regions. Other regions at the same geographic latitudes have significantly lower average insolation.

3. While the best locations for solar energy utilization are near the equatorial zone, there are several locations in the temperate zones where the annually averaged power is close to that in the equatorial zone. Most notable among these locations are North Mexico and the southwest of the United States, the Mediterranean countries, South Africa, Australia, Argentina, and Chile.

4. The geographical area from Pakistan to Japan and Indonesia, where 60% of the earth's inhabitants currently reside and is experiencing high economic and high-energy demand growth, is within the contours of 150–200 W/m^2. This signifies a high potential for solar energy utilization. The development of solar energy and the realization of the solar energy potential in this region will ease the global TPES demand as well as the adverse environmental effects of fossil fuel combustion.

Several insolation parameters are used to determine the potential of a location to convert solar power to useful energy. Among these parameters are as follows:

1. The maximum power on a horizontal or a tilted surface (W/m^2).
2. The total energy received by a stationary horizontal or tilted surface (J/m^2), during a day, a month, or a year.
3. The total energy received by a surface that always points to the sun disk (J/m^2) during a day, a month, or a year.
4. The annually averaged incident power on a horizontal surface (W/m^2). The average also includes the nighttime when the insolation vanishes.
5. The annual energy received by a horizontal surface (W/m^2).

A solar installation that converts the incident solar power to thermal or electrical energy operates throughout the year, and therefore, its merits are better to be evaluated based on the total energy it produces for an entire year. Data, such as those in Figure 6.4 as well as regional and national databases [6,7], are useful in assessing the total annual energy produced by a solar energy system.

Example 6.2

A PV electricity generation system is planned to be built in Senegal. If the area of the PVs is 160 m^2 and their conversion efficiency is 18%, estimate how many kilowatt hours this system will produce annually.

Solution: It appears from the map that the average insolation in Senegal, at the tip of West Africa, is approximately 250 W/m². Therefore, the PV system to be installed will receive on average 250 × 160 = 40,000 W of solar power, and with the conversion efficiency of 18%, it will produce 7,200 W of electricity (7.2 kW). Since this is an average power produced, in a year (8,760 hours), the system will produce 63,072 kWh annually.

6.2.2 Thermal Collectors

Everyday experience teaches us that solar radiation through transparent surfaces will keep the interior of systems at temperatures higher than that of their surroundings: the interior of a car is much hotter than the surrounding air on a sunny day; the greenhouses maintain higher temperatures than their environment throughout the year, and the sun will raise the temperature of a covered glass of water by 10–20°C even on a cold winter day. The higher temperature indicates that part of the solar energy is absorbed by the system. Therefore, we may use solar energy to heat up systems instead of using other primary sources of energy for heating. The process of heating up systems with insolation is sometimes called *passive heating* or *passive solar energy*, and the systems used are relatively simple.

Solar heating in buildings is caused by the *greenhouse effect*: Glass and transparent polymeric materials allow most of the solar radiation to pass through them, but trap the high-wavelength, infrared radiation. Since infrared radiation does not escape the transparent material, the temperature in the enclosure rises. The "trapped energy" may be used to keep a building warmer during the winter days, to provide hot water to the building, or to provide process heat—usually at moderate temperatures—to a factory. Such enclosures are called *solar collectors* or simply *collectors*. The solar collectors are placed on the roofs of buildings, facing the prevailing direction of the sun—south in the northern hemisphere and north in the southern hemisphere—and are most commonly used for water and space heating. Well-insulated collectors reach temperatures 40–60°C higher than the ambient.

Figure 6.5 depicts the schematic diagram of a solar collector. The collector is tilted at an angle, which is usually the angle for maximum annual insolation and is approximately equal to the latitude of the location. The inside material of the collector is coated with radiation-absorbing paint, usually of black color. Solar energy enters the top of the collector through a double-glass window or a similar material and heats up the enclosed fluid, typically a water-based solution. Depending on the design of the collector, the temperature of the insolation-absorbing fluid may reach 30–60°C above the ambient. Circulating

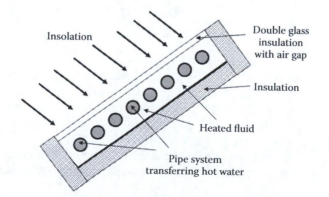

FIGURE 6.5
Schematic diagram of a solar collector.

water in a system of internal pipes picks up the energy from the collector and transfers it inside the building. The circulating water may be stored in a tank to supply the hot water for the building or it may dissipate this energy to the building via space heaters to maintain a comfortable interior temperature.*

Solar collectors trap both direct and diffuse solar radiation. They operate even during the cloudy autumn and winter days, when there is greater need to supply heat to buildings. Because the higher-temperature water may be stored for later use in insulated tanks, solar heating for buildings is not limited only to the sun-up hours and is extended into the night hours. Well-designed solar collector systems may provide space heat and hot water for buildings in the temperate zones for most of the year and, thus, substitute boilers and gas burners that use petroleum products or natural gas. This substitution saves fossil fuels and reduces the CO_2 emissions.

The amount of solar energy that enters the solar collector is less than the local insolation S on a surface normal to the sun, because of reflections on the glass surface and the angle of the collector. The loss of solar power that enters the collector is accounted by a radiation-loss factor β_r. If the temperature of the collector is T_{col}, the ambient temperature is T_{amb}, and the rate of energy/heat transferred by the circulating fluid to the building is $\dot{Q}$, the following energy balance equation defines the steady-state operation of the solar collector:

$$\beta_r SA = UA(T_{col} - T_{amb}) + \dot{Q}, \tag{6.3}$$

where U is the overall heat transfer coefficient of the collector and A is the area of the collector facing the sun [3]. The efficiency of the solar collectors is defined as the rate of energy/heat removed by the water circuit to the local insolation:

$$\eta_{col} = \frac{\dot{Q}}{SA} = \beta_r - \frac{U(T_{col} - T_{amb})}{S} \tag{6.4}$$

If the solar energy that enters the collector is not removed, then $\dot{Q} = 0$; the temperature in the enclosure rises, and all the absorbed energy is transferred to the surroundings by convection. This condition is referred to as "loss of coolant," the efficiency of the collector vanishes, and the maximum possible temperature is reached in the interior of the collector. From Equation 6.3, the maximum temperature a collector may reach is when $\dot{Q} = 0$:

$$T_{max} = T_{amb} + \frac{\beta_r S}{U} \tag{6.5}$$

Typical overall heat transfer coefficients for well-designed collectors are $U = 4.5$ W/m^2 K; radiation-loss factors, $\beta_r = 0.80$; and ambient temperatures, $T_{amb} = 20°C$. Table 6.5 gives several values for the efficiency under different rates of insolation and interior collector temperatures. The maximum temperature that may be achieved in the collector under the given insolation is also shown in the last column of the table.

One may observe in Table 6.5 that the efficiency of solar collectors decreases as the collection temperature increases. This happens because more heat leaves the collector by

---

* The commonly used terminology must be noted in this figure: insolation is the energy from the sun, while insulation refers to materials that prevent the loss of heat.

TABLE 6.5

Collector Efficiency and Maximum Attainable Temperature

S, W/m²	T_{col}, °C	η_{col}, %	T_{max}, °C
500	45	57.5	109
800	45	65.9	162
1000	45	68.8	198
500	80	26.0	109
800	80	46.3	162
1000	80	53.0	198

Note: $U = 4.5$ W/(m²·K), $\beta_r = 0.80$, and $T_{amb} = 20$°C.

convection at higher T_{col}. When the collector temperature is equal to T_{max} the collector does not supply any heat to the building, and the collection efficiency vanishes. Solar collectors may very well be used for low-temperature applications, such as the heating and hot-water needs of buildings, food processing, and pasteurization. However, common solar collectors are not capable of supplying the heat at the high temperatures needed in many industrial applications, as for example, the production of steel, petroleum refinement, and glass production. For the latter processes, one may use solar concentrators (lenses) to increase the value of S and achieve higher collection temperatures. This practice is associated with high heat losses; it is currently very expensive and is not commercially used.

Example 6.3

A solar collector is designed to provide 3 L/s of hot water at 50°C to a large office building during the day hours only. The average daylight insolation in this location is 640 W/m², and the water is supplied at 18°C. The average ambient temperature is 15°C, and the overall heat transfer coefficient for the collector is $U = 4.5$ W/(m² K). Determine how much area is needed for a solar collector with a reflectance loss factor $\beta_r = 85\%$.

Solution: The rate of heat needed to heat 3 L/s (approximately 3 kg/s) of water from 18°C to 50°C is 3 × 4.184 × (50 − 18) = 402 kW = 402,000 W (the specific heat capacity of water is 4.184 kJ/(kg K)). From Equation 6.3, we derive the following expression for the area of the collector:

$$\beta_r S A = U A (T_{col} - T_{amb}) + \dot{Q} \Rightarrow A = \frac{\dot{Q}}{\beta_r S - U(T_{col} - T_{amb})}.$$

With the given values, the area needed for the collector is 1005 m², approximately an area 31.7 × 31.7 m². The collector area may be reduced if the collector is better insulated and if the coefficient β_r is higher. For example, if $U = 1.5$ W/(m² K) and $\beta_r = 93\%$, then the area of the collector needed is 735 m² or 27.1 × 27.1 m².

It must be noted that the use of solar collectors for the heating of buildings and hot water supply is not restricted to regions of high insolation. Some of the central and northern European countries have developed excellent programs for solar heating. One of them, the region of Upper Austria, which is located at high altitude in the Alps, has developed a very good domestic market for solar energy products that work well in the region. Solar energy accounts for 46% of the household heating needs in this region [8].

The use of *solar stoves* has become a popular solar energy application in the early twenty-first century: the solar stoves use solar concentrators (lenses) and better insolation than the solar collectors. As a result, their interior temperatures are high enough for the preparation of food.

An often neglected form of solar energy use is in the drying of clothes: A gas or electric clothes dryer consumes 5–6% of the total household energy in Europe and the United States. Consequently, drying the clothes under the sun would save on the average 5–6% of the total household energy. One of the impediments, most notably in urban United States, is local policies and regulations against clothes drying in the open for "aesthetic reasons." National policies that prohibit such energy-wasting local regulations would result in primary energy savings by using a higher fraction of the free solar energy.

Recent experience has shown that most, if not all, of the heating needs in buildings may be provided by thermal solar systems. Well-designed solar collectors supply heat and hot water to buildings for most of the year. Cooking and clothes drying may also be accomplished with solar energy. The wider application of passive solar heating in commercial and residential buildings worldwide will significantly reduce the use of fossil fuels; it will save these primary energy sources for other industrial uses and will reduce all forms of pollution associated with the combustion of fossil fuels.

6.2.3 Thermal Solar Power Plants

It is apparent from the fundamentals of thermodynamics that any available heat source may be used for the production of electricity by employing a suitable thermodynamic cycle. The sun is such a source of heat, and several small (10 kW to 20 MW) power plants operate around the globe using solar energy. There are two classes of solar power plants: the *thermal* that use the sun as a source of heat and convert this heat to electric energy and the PV that use the direct conversion of insolation to electricity. The *Rankine cycle* depicted in Figure 1.8 is the cycle most often used for solar thermal power plants. A few engines that work on a version of the *Stirling cycle* have also been used for the conversion of solar radiation to electricity [3,9,10]. Figure 6.6 is a schematic diagram of a solar thermal power

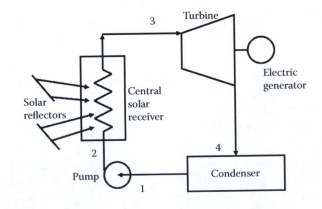

FIGURE 6.6
Components of a solar thermal power plant using a Rankine cycle.

plant that utilizes the Rankine cycle. The four essential components of this unit are very similar to those of a steam power plant that utilizes fossil fuels:

1. The pump that circulates the working fluid, typically water
2. The system of *solar reflectors* and a *central receiver* that receive the solar energy and convert the working fluid to vapor (steam); the solar reflectors cover a wide area around the receiver
3. The steam turbine, directly connected to an electric generator
4. The condenser

The components of a thermal solar power plant are identical to those of the fossil fuel power plants with the exception that the boiler is substituted by the reflector–receiver system.

Figure 6.7 is an aerial photograph of *Solar Two*, a 10 MW pilot thermal power plant near Barstow, California. One may see the multitude of solar collectors (heliostats) on the ground that reflects the solar radiation to a central receiver, which stands on a tower at a height of approximately 150 m. With an area of 40 m^2 for each heliostat, the 1926 heliostats of this thermal power plant have a total surface area of 82,750 m^2. Each heliostat is mounted on a control system with two degrees of freedom and is designed and programmed to directly reflect the sun's radiation to the central receiver. Thus, all solar radiation that falls on the heliostats is reflected to the central receiver. The turbine-condenser-generator housing building is on the ground next to the tower. In order to mitigate the effects of short-term insolation fluctuations (e.g., because of passing clouds), Solar Two used a pool of molten salts (60% NaNO$_3$ and 40% KNO$_3$) as the temporary heat storage medium that receives the solar radiation. The pool of molten salts also allowed the power plant to produce electric power up to 3 hours after sunset. Solar Two was a pilot project financed by the US

FIGURE 6.7
Aerial photograph of Solar-Two, a 10 MW solar thermal plant near Barstow, California. (Courtesy of Sandia National Laboratories, Livermore, CA.)

Department of Energy and was decommissioned in 2009. Its successor *Solar Tres* (later renamed *Gemasolar thermosolar plant*) was commissioned in 2011 in Andalucía, Spain. Solar Tres is bigger, rated to produce 19.9 MW of peak power.

The thermal efficiency $\eta = W_{net}/Q_{in}$ of the practical Rankine and Stirling cycles is approximately 40%. In addition, there are significant insolation losses associated with receiving and transmitting the solar energy to the central receiver (primarily reflection, shadowing and atmospheric absorption). The overall thermal efficiency (ratio of the power produced to the total insolation on the area of the receivers) of thermal solar power plants is in the range of 15–28% [3,10]. This is not necessarily a disadvantage of solar power plants because solar energy is free of cost to the plant operators, and the efficiency of a solar thermal installation does not represent a benefit–cost ratio, as explained in Section 1.4.

Solar ponds have also been proposed for harnessing solar energy. A solar pond is essentially a large pool of water, approximately 1 m deep, with a transparent cover that does not allow water evaporation and creates a "greenhouse effect" within the water mass [3,11]. The water temperature in a well-designed pond may reach 90°C during the summer months. The pond may be considered as the heat source for a thermodynamic cycle that produces electricity. Experimental solar ponds have produced small amounts of electric power [11,12]. However, because of the low temperature of the heat source, solar ponds convert significantly less solar energy to electricity than PVs and solar thermal power plants [12]. Since the capital cost of the solar ponds is comparable to the other two solar energy conversion plants, the cost of electricity from solar ponds is significantly higher. As a result, there are very few solar ponds in operation—mostly experimental pilot plants—and there is currently no significant effort to advance this technology.

6.2.4 Solar Cells and Photovoltaics

PV cells or *solar cells* directly convert the energy of the sun to electricity. Solar cells have been successfully used since the 1950s to power satellites, the International Space Station, buoys, and platforms at sea, as well as remote locations on earth, difficult to be reached by the electric grid. Since solar energy is abundant and the sun will always shine during daytime in the foreseeable future, the potential of PV devices to produce electric energy is immense and will not be exhausted. Solar cell systems, combined with energy storage for nighttime, may provide the entire needs of electric power in households. If solar cells could be economically deployed in communities (with suitable energy storage for the night hours), they may potentially supply the entire energy needs of humanity, including thermal and electric energies, for millennia. However, and despite the recent advances in the manufacturing of solar cells, in the second decade of the twenty-first century, electricity from PV devices is more expensive to produce than electricity from fossil fuel and nuclear power plants. As shown in Table 6.1, the annual growth rate of PV systems is very high, 42%, but the produced electricity globally is still less than 1% of the total [1,2].

Four factors in the early part of the twenty-first century may help shrink the price gap between PVs and fossil fuels:

1. The depletion of fossil fuels, which causes their prices to increase
2. New knowledge in materials science and technological advances, which are applied to the manufacturing of PVs and cause the solar cell prices to decrease

3. New materials for battery systems and an increased use of hydrogen-based systems for the storage of electric energy

4. Several financial incentives by most governments for the development and purchase of renewable energy equipment

With these developments, there may be a time in the future when electricity produced by PV devices will be first competitive and then cheaper than electricity produced by other means. At that time, humanity will enter an era when most of its energy needs are provided by the sun.

Figure 6.8a is a schematic diagram of a PV cell: A very thin layer (on the order of 1 μm) of a p-type semiconductor is joined to a thicker stratum (on the order of 1000 μm) of a n-type semiconductor. Joining the two types of semiconductors creates a *p–n junction*. When solar radiation falls on the face of the p-type semiconductor and penetrates to the p–n junction, a voltage difference V is generated across the junction and across the external circuit, similar to the voltage produced by a battery. This voltage drives a current I in an external circuit of resistance R, and thus, the PV cell produces electric power:

$$\dot{W} = VI = \frac{V^2}{R}. \tag{6.6}$$

The photons of the sunlight cause the continuous excitation of electrons in the two types of semiconductors. As long as there is solar radiation on the PV cell, there is a continuous excitation of electrons around the p–n junction, which maintains the voltage difference and the supply of power to the external circuit. The voltage V produced by the PV cell is

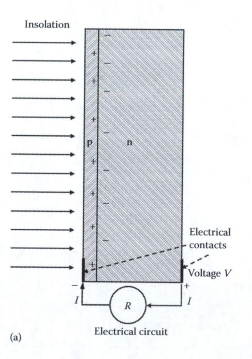

(a)

FIGURE 6.8

(a) p–n junction for the production of electric power. *(Continued)*

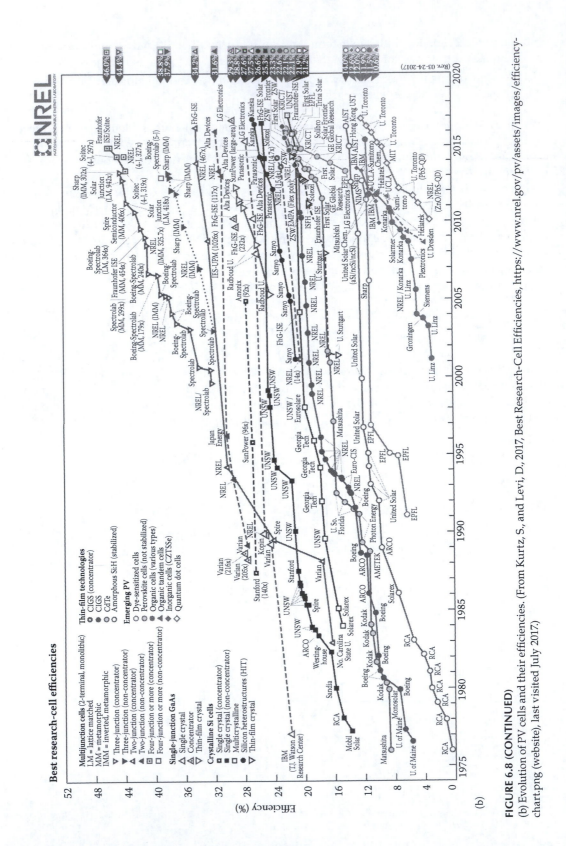

FIGURE 6.8 (CONTINUED)

(b) Evolution of PV cells and their efficiencies. (From Kurtz, S., and Levi, D., 2017, Best Research-Cell Efficiencies, https://www.nrel.gov/pv/assets/images/efficiency-chart.png (website), last visited July 2017.)

monotonically increasing with the amount of insolation and reaches a maximum at the solar noon, when the cell produces maximum power. Of course, when there is no insolation during the night hours, the solar cell does not produce any power.

While the thermal solar systems convert solar energy to heat and subsequently to electric energy, PV cells directly convert the solar energy to electricity, without the intermediate step of heat production. For this reason, they are called *direct energy conversion (DEC)* devices. A significant advantage of the DEC is that it is not subjected to the thermal (Carnot) efficiency limitations that are associated with thermal energy conversion (Equation 1.6). Because of this, a DEC device may, in principle, convert all the energy it receives to electricity. However, the materials technology in the existing and the projected generations of solar cells show significantly lower conversion efficiencies. The theoretical efficiency of the single-junction solar cells (defined as the ratio of the electric power produced to the total insolation) is approximately 47% [3,13] and practical single-junction solar cell efficiencies are in the range of 8–39% [2,14,15]. Multijunction PV cells are not subjected to this limit and their recent (2017) efficiencies have exceeded 45%. A comprehensive pictorial compilation of the time evolution of PV cell efficiencies since 1975 was completed by the National Renewable Energy Laboratory (NREL) and is shown in Figure 6.8b [15]. The compilation includes the type of technology used in the PV cell, the laboratory of manufacture of the PV cells, and information on the number of junctions. The efficiencies in the chart have been determined at 25°C. Typically, PV cell efficiencies drop by approximately 1% for every 4–6°C temperature above 25°C [16].

It is observed in Figure 6.8b that the efficiency of all types of solar cells has increased continuously and significantly since the 1970s, when concerted research on PVs started. It must be noted, however, that this figure pertains to research-grade cells that are illuminated and operated under ideal conditions. Less expensive, commercial-grade PV cells of the same categories have lower conversion efficiencies. As noted in Section 1.4, the concept of the efficiency as revenue-to-cost ratio does not apply to renewable energy sources and solar cells in particular, because the incident solar energy is free of cost. A better figure of merit for solar energy is the cost of electricity produced, in dollars per kilowatt hour, which is largely determined by the capital cost of the PV installations. The apparent lower efficiency of PV cells is not an impediment to the growth of this type of renewable energy, provided that the cost of the PV cells is low enough for the overall cost of electric energy to be less than the cost from other sources. For the more widespread use of energy from PVs, the solar energy technology will have to produce tangible results primarily in two areas:

1. Lower generated electric energy costs, which will have to be reduced to less than $0.1/kWh in the United States.

2. Affordable and reliable storage systems for solar energy to become available during the nighttime as well as during periods of lower insolation.

The power ratings (or plate capacity) of both PV and solar thermal power plants are based on the maximum power they may deliver. Because of the daily and annual variabilities of the available insolation, the average power produced by a solar power plant is always significantly less than the rated power. For example, the *Blue Wing* Solar Project, in San Antonio, Texas, which operates with PV cells, has a power rating of 14.4 MW. If continuously operating at this rating, the power plant would generate annually $14,400 \times 365 \times 24 = 126,144,000$ kWh. However, the plant produced only 23,688,000 kWh in 2013 and 23,141,000 kWh in 2014, and this suggests that it operated at a capacity that is significantly lower than its rated power for most of the year. The capacity factor (Section 7.1.1) of the plant is approximately 18%. Thermal solar power plants use thermal storage (e.g., molten salts) to prolong their operation and have higher capacity factors: the Gemasolar thermosolar plant in Andalucía, with a power rating of 19.9 MW, has the capability to store a fraction of solar radiation and produces approximately 110,000,000 kWh/year, with a capacity factor of 63%.

As with solar collectors, light concentrators (lenses) may be used to amplify the insolation on the PV cells. The concentrators may be common Fresnel lenses or parabolic dishes. Typical light concentrator systems make use of two lenses to better focus and distribute the light on the entire surface area of the PV cells. The PV cells used with concentrator systems (concentrator photovoltaic [CPV]) are usually polycrystalline cells, which are more expensive, but have higher efficiency than monocrystalline cells. The higher insolation helps further increase the CPV cell efficiency. Because the insolation is significantly higher and the area of the cell is small, the CPV cells must be fitted with heat sinks—long fins at the back or the circulation of a coolant in channels at the back of the cells—to dissipate the heat produced and to avoid overheating. CPV systems mostly utilize the direct solar radiation because the diffused part of the insolation is not concentrated by the lenses. For this reason, CPV systems perform best in sunny regions, where the direct solar radiation is high and the diffuse radiation is low. The Sahara region, the southwestern part of the United States and northwestern part of Mexico, the Middle East, Australia, and Central Asia are regions best suited for the CPV systems.

6.2.5 Solar Power Data and Solar Energy Calculations

The ratings of 14.4 and 19.9 MW for the Blue Wing and Gemasolar thermosolar power plants in San Antonio and Andalucía, which were mentioned in the last section, are the maximum power they may produce. Similarly, a 5 kW thermal solar collector will collect 5 kW of heat, when the sun is at the zenith. Because of the daily and annual variabilities of insolation, solar energy systems produce a lesser amount of energy annually than their ratings indicate. The annually averaged contours of insolation depicted in Figure 6.4 take into account the diurnal and seasonal variabilities of insolation and better represent the annual energy a solar system may produce. However, because of the lack of detail on such global maps, information from them may be used only for approximate calculations of the annual energy.

For the economic/financial decision making processes, it is necessary to conduct more accurate computations for the power and the total energy a given solar system will produce. Because the price of electricity in deregulated energy markets varies according to the demand, the development of a commercial solar system is based on detailed calculations of the hourly amount of power the solar system is expected to produce over the period of 1 year. To calculate these needed parameters, we must have details of the insolation on the location of the proposed energy system. Such information is available through detailed spreadsheets in several countries. In the United States, the *NREL—National Solar Radiation Data Base* contains hourly isolation data for 239 sites that cover the entire country [6,7]. The data, which are based on local measurements and were generated by two analytical models, are available in spreadsheets. Among other specific information, the sets of data include the following:

1. Date and hour
2. Position of sun (azimuth and zenith angle) relative to the location
3. Direct and diffuse insolation
4. Insolation on horizontal and vertical surfaces

TABLE 6.6

Excerpts from NREL—National Solar Radiation Data Base for Atlanta, Georgia

Date	Time	Horizontal, W/m²	Direct, W/m²	Diffuse, W/m²
4/17/2009	6:00	0	0	0
4/17/2009	7:00	16	0	16
4/17/2009	8:00	214	416	94
4/17/2009	9:00	423	571	144
4/17/2009	10:00	622	638	200
4/17/2009	11:00	824	833	159
4/17/2009	12:00	934	896	140
4/17/2009	13:00	953	849	172
4/17/2009	14:00	940	873	154
4/17/2009	15:00	858	875	136
4/17/2009	16:00	708	829	126
4/17/2009	17:00	516	756	110
4/17/2009	18:00	299	603	93
4/17/2009	19:00	85	265	50
4/17/2009	20:00	1	17	1
4/17/2009	21:00	0	0	0

Source: NREL, *NREL—National Solar Radiation Data Base: 1991–2010 Update*, NREL, Golden, CO, http://rredc.nrel.gov/solar/old_data/nsrdb/1991-2010/ (website), last visited August 16, 2016.

Excerpts of the NREL data set for the city of Atlanta, Georgia, United States, are shown in Table 6.6. The columns show the date (April 17, 2009), the hour, the total amount of insolation on a horizontal surface, the direct insolation from the sun's disk, and the diffuse radiation. The diurnal variation of the data is apparent, and it is typical of a spring day in the south part of the United States.

Data, such as those of Table 6.6, are necessary for any detailed calculations of the total energy produced by solar energy systems and may lead to investments in solar energy systems. The following two examples illustrate how these calculations are performed.

Example 6.4

A solar panel of PVs with dimensions of 2.5 × 3 m² is placed horizontally on the roof of a skyscraper in Atlanta. The efficiency of the solar cells is $\eta_{pv} = 18\%$. Determine the hourly average power produced on April 17, 2009, the average power over the 24 hours, and the total energy produced on that day.

Solution: The data of Table 6.6 (from the database of NREL [7]) pertain to this situation. The power produced at any time by this panel is

$$\dot{W} = \eta_{pv} S A,$$

with the area $A = 7.5$ m² and $\eta_{pv} = 0.18$. The energy is equal to the product of the power and the time (1 hour). Using the values of Table 6.6, we develop the following table for the average hourly power generated by this panel and the energy produced during the respective hour (the insolation is zero at nighttime).

Time	Power, W	Energy, kWh
6:00	0	0
7:00	21.6	0.0216
8:00	288.9	0.2889
9:00	571.05	0.57105
10:00	839.7	0.8397
11:00	1,112.4	1.1124
12:00	1,260.9	1.2609
13:00	1,286.55	1.28655
14:00	1,269	1.269
15:00	1,158.3	1.1583
16:00	955.8	0.9558
17:00	696.6	0.6966
18:00	403.65	0.40365
19:00	114.75	0.11475
20:00	1.35	0.00135
21:00	0	0

The average power of the solar panel is obtained by summing the second column of the last table and dividing by the 24 hours of the day. The average power in this case is 416 W. The total energy produced during the day is the sum of the third column: 9.98 kWh.

Similar calculations may be performed using this database for other days or months or for the entire year (8,760 hours). The National Solar Radiation Data Base [7] and similar databases provide sufficient information to perform detailed calculations on surfaces at any angle to the horizontal and facing in any direction of the compass.

Example 6.5

A solar collector with area 10.2 m² and efficiency 65% is placed horizontally on the surface of the skyscraper of Example 6.4 in Atlanta. The radiation collected is used to heat water from 18°C to 45°C. Determine the total mass and the volume of water that is heated on the day described by Table 6.6.

Solution: The total amount of energy collected is

$$Q = \sum_{6:00}^{21:00} \eta S_i At = \eta At \sum_{6:00}^{21:00} S_i = 0.65 \times 10.2 \times 60 \times 60 \times \sum_{6:00}^{21:00} S_i \quad \text{in joules,}$$

where S_i is the average hourly insolation from Table 6.6. The total energy collected is 238 × 10^6 J. Since the specific heat of water is 4,184 J/kg K, the mass of water that may be heated from 18°C to 45°C (difference of 27°C, which is equivalent to 27 K) is 238 × 10^6/4,187/27 = 2,107 kg. The volume of this mass of water is approximately 2.110 m³ or 2,110 L (557 gal).

6.2.6 Environmental Impacts of Solar Energy

Solar energy is one of the most benign forms of primary energy. The foremost environmental advantage of solar energy is that its utilization does not involve chemical or nuclear reactions, and hence, it does not contribute to any harmful chemical emissions (e.g., carbon dioxide and NO_x gases) or radionuclides to the environment. The most significant environmental impact of solar energy is associated with the production of the materials for the PV cells. Silicon, germanium, phosphorus, and the rare-earth metals used in the PV cells are produced, refined, and purified using a few polluting chemicals, such as sulfuric acid and cyanide. All the pollution associated with the production of materials for PVs is localized and may be easily contained in the production facility.

A level of thermal pollution is associated with all solar energy operations. Solar thermal power plants reject a significant amount of heat through their condenser. The rate of waste heat rejected by a solar installation, which produces electric power $\dot{W}$, with an overall thermal efficiency η is

$$\dot{Q}_{rej} = \dot{W}\frac{1-\eta}{\eta}. \tag{6.7}$$

A 10 MW thermal power plant operating with 40% thermal efficiency would reject 15 MW of heat to the environment. Since all the heat received by a solar power plant emanates from solar radiation, this heat would have been received as solar radiation anyway. In addition, and because solar thermal power plants occupy a much larger surface area than similar fossil fuel installations, the waste heat is dissipated over significantly larger areas to minimize environmental impact. PV arrays also have rather low efficiencies and reject in the environment all the incident radiation that is not converted to electricity. A 100 kW solar panel array with an efficiency of 20% would reject a total of 400 kW of heat in its surroundings. The waste heat from PV cells cannot also be considered as thermal pollution, because the heat originates in the incident solar radiation and, in the absence of the PV array, would have affected the locality anyway. Because part of the incident solar radiation is converted to power, the PV arrays actually have a cooling effect in the area they operate.

Land use is another environmental effect of solar energy utilization. Solar energy is diffuse. The production of electricity necessitates a very large land area, significantly more than the area of fossil or nuclear power plants. The Gemasolar themosolar plant in Andalucía, Spain, occupies 210 ha (520 acres); it is rated at 19.9 MW and produces 110×10^6 kWh of electricity annually. For comparison, a 2,000 MW nuclear power plant occupies a lesser area and produces close to 17×10^9 kWh annually. For this reason, inexpensive, underutilized, and usually deserted areas have been selected for the location of solar thermal power plants as well as for PV installations.

A beneficial environmental effect of all forms of solar energy used in urban settings is that a fraction of the incident solar radiation is utilized as electric power inside the buildings, and therefore, it is not emitted and absorbed in the immediate vicinity of the buildings. The solar systems may cause a modest cooling effect in urban environments, which slightly lessens the need for building air-conditioning in the summer.*

An often cited problem for PV cells is the high total energy required to produce them: the purification of silicon and the mining, refinement, and shipment of the other chemicals used in the manufacturing of PV cells consumes a great deal of energy. An energy figure of merit that applies to PV cells is the energy payback time (EPBT), defined as the time it

---

* Because all the electric power is finally dissipated as heat in the buildings, the modest cooling effect in the area emanates from not having electricity transmitted from other regions to be consumed and dissipated in the urban setting.

takes for PV cells to produce all the energy used in their life cycle. The EPBT has dramatically decreased since the 1970s, when the first PV cells were manufactured and their EPBT was close to 40 years. Depending on the method of manufacturing, the EPBT for the commonly used silicon-based PV cells is in the range of 2–5 years. Other types of PV cells (e.g., cadmium–tellurium and polymeric cells) have EPBT of less than 1 year. With all the technological progress since the early period of PVs, PV cells in their lifetimes produce several times more energy than the energy used in their manufacturing processes.

6.3 Wind Energy

Wind energy is a by-product of solar energy because wind is generated by the uneven heating of different parts of the globe. Hot air rises in regions of higher insolation and this creates a small pressure differential, which induces colder air from the surrounding regions to rush in and fill the vacuum. Approximately 2% of the total solar radiation received by the earth, 3.46×10^{12} kW, is converted by the differential heating to wind power. The total solar radiation converted into the mechanical energy of the wind is more than 1.1×10^{23} J/year. This energy is 218 times more than the total primary energy the entire human population used in 2015. Wind power is an important renewable energy source, and if suitably harnessed, it is capable of supplying a significant fraction of the global energy demand.

Depending on their origin and effects, the air currents are classified as *planetary* and *local*. Planetary currents affect very large regions of the earth. They encompass large masses of air and are primarily caused by the higher amount of insolation on the land mass near the equator. Figure 6.9 is a schematic diagram of the prevailing planetary wind patterns: The hotter air near the equator rises and moves toward the poles. This motion of the air is affected by the rotational motion of the earth, which generates the *Coriolis force* on the rising air masses. The air that rushes from the temperate zones to fill the relative vacuum in the tropics develops a velocity component from the east to the west. The westerly motion of the planetary currents causes the northeasterly *trade winds* in the northern hemisphere and the southeasterly trade winds in the southern hemisphere.* Early navigators, such as Columbus and Magellan, mastered the effect of the planetary wind patterns and accordingly planned their western journeys. As the rising warm air moves from the tropics toward the poles, it cools in the upper atmosphere and descends at approximately 30° latitude (the *horse latitudes* in the nautical terminology) in both the northern and the southern hemispheres. The downward motion at these latitudes and the Coriolis force develops an air pattern in the temperate zones from the west to the east. This causes the *prevailing westerlies*, in the northern and the southern hemispheres between 30° and 60° latitude. The pattern of the planetary winds is completed with the *polar easterlies* in both the south and the north polar regions.

The effect of local winds is confined to a small area, and they are usually generated by the uneven heating of neighboring masses of land or of land and sea. Very common among local winds is the *sea breeze*, which is caused by the higher temperatures of the land in comparison to the neighboring water surface. Breezes are ubiquitous in the coastal areas, near the lakes, and near large rivers.

---

* Winds are named from the direction they come from: A wind blowing from west to east is a westerly. A wind blowing from north to south is a northerly wind.

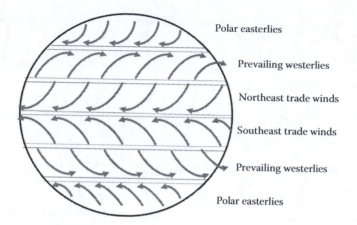

FIGURE 6.9
Planetary winds on the surface of the earth.

Another, lesser-known mechanism of air current development is caused by the differential heating on the sides of hills and mountains in comparison to the lower ground. During the early part of the day, the mountainsides receive more insolation; they become hotter and air rises creating an *anabatic flow*. At dusk, the sides of the mountains and hills cool faster than the horizontal ground, and this creates a downward, *katabatic flow*. Mountains and hills also create their own wind patterns by deflecting the planetary currents.

6.3.1 Fundamentals of Wind Power

Wind power has been extensively used for several millennia. The ancient Chinese, Indians, Egyptians, and Hittites have used large sails to propel their ships in coastal waters. Thousands of Greek ships sailed to conquer Troy around 1100 BCE in an expedition that was recorded in the epic poems of Homer: *Iliad* and *Odyssey*.* Until the midnineteenth century and the advent of the steam engine, wind propulsion by sail and labor (rowing) were the only forms of seagoing power. The land use of wind power with windmills was recorded in the tenth century by writers in the regions that are now part of Iran and Afghanistan. The windmills were transplanted to Europe, where they were used for the grinding of wheat as well as small manufacturing and processing tasks.

Wind power is mechanical power, based on the kinetic energy of the moving mass of air, which is converted to electricity by wind turbines. The power of the air stream that becomes available for conversion to a wind turbine is

$$\dot{W}_{av} = \frac{1}{2}V^2\dot{m} = \frac{1}{2}A\rho V^3 = \frac{\pi}{8}D^2\rho V^3, \qquad (6.8)$$

where $\dot{m}$ is the mass flow rate of the air that passes through the turbine; V is the wind velocity; A is the area swept by the blades of the turbine; D is the turbine diameter; and ρ is the density of air, approximately 1.2 kg/m^3. When Système International units are used (m for D; m/s for V; and kg/m^3 for ρ), the power is in watts (W). The wind velocity is often

---

* Wind energy is oftentimes referred to as *aeolic energy*, named after *Aeolos*, the ancient Greek god of the winds.

quoted in kilometers per hour (1 km/hour = 0.278 m/s); in miles per hour (mph) (1 mph = 0.447 m/s); and in knots (1 kn = 1,852* m/hour = 0.514 m/s).

An actual turbine does not produce all the power that is available to it. The power produced by a turbine may be calculated from Equation 6.8 by multiplying the available power by the efficiency of the turbine η_T:

$$\dot{W} = \eta_T \dot{W}_{av} = \frac{1}{2}\eta_T A\rho V^3 = \frac{\pi}{8}\eta_T D^2\rho V^3. \tag{6.9}$$

The turbine efficiency is also referred to as the power coefficient and is denoted by the symbol C_p. The efficiency of a wind turbine depends on the type of the turbine and the wind velocity. Typical values for this variable are in the range of $0.2 < \eta_T < 0.3$.

The wind power is proportional to the cubic power of the wind velocity and to the square of the diameter of the wind turbine. When the wind velocity doubles, the power increases by a factor of 8. In choosing the location of wind turbines, regions with high and steady wind velocities are preferable. Trebling the length of the blades would increase the power produced by a factor of 9. For this reason, a great deal of research and development effort, since the 1970s, has been devoted to the construction of wind turbines with longer blades that cover larger areas. The development of light and strong composite materials since the 1980s has considerably assisted in the construction of the long blades currently used in wind turbines and the global growth of wind power.

One of the complications in harnessing the wind power is that the velocity of the wind is not uniform. The wind velocity almost vanishes at the ground and monotonically increases with the distance from the ground. Fluid dynamics has provided an approximate expression for the ratio of the wind velocities at levels z_1 and z_2 from the ground, the so-called 1/7th power law:

$$\frac{V(z_2)}{V(z_1)} = \left(\frac{z_2}{z_1}\right)^{1/7}. \tag{6.10}$$

For uniformity in wind velocity measurements, the instruments (anemometers) at meteorological stations are placed at height 30 ft (9.1 m) from the ground. The reported wind measurements are at this level, which is the reference level for all meteorological data. From these measurements, one may calculate the air velocity at different heights at the location where a wind turbine is installed. If the meteorological measurement for the wind velocity $V(z_1 = 9.1\text{ m})$ is measured to be 15 km/hour, then the wind velocity at the hub of a wind turbine at $z_2 = 120$ m above the ground is $15 \times (120/9.1)^{1/7} = 21.7$ km/hour.

Example 6.6

At three different times, the wind velocity measured by a meteorological station is 8, 12, and 16 m/s. A wind turbine is to be constructed at the vicinity of this station. The tower of the turbine will be 130 m high, the blades of the turbine will be 45 m long and the efficiency of the turbine is 28%. Determine (a) the corresponding velocity at the turbine hub; (b) the power produced by the turbine; and (c) the total energy produced, in kilowatt hours, if the wind velocity were to remain steady for 24 hours.

---

* The distance of 1852 m is a fundamental length scale on earth. It is equal to 1 minute of arc latitude along all meridians on the surface of the earth.

Solution:

a. The velocities 8, 12, and 16 m/s are measured at a height of 9.1 m. The hub of the turbine is at 130 m. Using Equation 6.10, we obtain the following expression for the first velocity, 8 m/s:

$$\frac{V(130 \text{ m})}{V(9.1 \text{ m})} = \left(\frac{130}{9.1}\right)^{1/7} = 1.462 \Rightarrow V(130 \text{ m}) = 8 \times 1.462 = 11.70 \text{ m/s}.$$

Similarly for 12 and 16 m/s, the wind velocities at 130 m are 17.55 and 23.39 m/s.

b. The air velocities at the hub are used for the calculation of the power produced by this turbine. Since the wind blade is 45 m, the turbine diameter is 90 m. With air density of 1.2 kg/m³ and efficiency of 0.28, Equation 6.9 yields for the power produced by the turbine at the first velocity of 11.70 m/s:

$$\dot{W} = \frac{\pi}{8} \eta_T D^2 \rho V^3 = \pi/8 \times 0.28 \times 90^2 \times 1.2 \times 11.70^3 = 1,710,888 \text{ W} \approx 1,711 \text{ kW}.$$

Similarly, the power produced at the other two velocities is 5,774 and 13,670 kW. It is observed that the power increases by a factor of 8, when the ground velocity doubles.

c. The energy in kilowatt hours is obtained by multiplying the power produced in kilowatts by the number of hours. During 24 hours at this velocity, the wind turbine would produce 41,064, 138,576, and 328,080 kWh, respectively, at the three velocities stated in this problem.

The wind intermittently flows, and its velocity is unpredictable over long periods. Current meteorology is capable of predicting with relative accuracy the wind velocity at a given location in the next 1–12 hours, but the accuracy and reliability of wind velocity predictions diminishes as the timeframe becomes longer. Our everyday experience tells us that there are periods within a year when the wind is very strong and periods of calmness. While it is certain that within an entire year (or even within a month), there will be several periods when the wind velocity on a Mediterranean island exceeds 20 km/hour, it is not certain at all when exactly these periods will be or whether or not on a certain date (e.g., July 17) the wind velocity will exceed 20 km/hour. While it is almost certain that on the same date next year, there will be "some wind" in the Texas Panhandle, it is not certain at all what will be the velocity of the wind on that date. Therefore, it is impossible to predict how much is the power a wind turbine will produce at a given location on a given day and time. This has an important consequence for the electric energy produced by the wind turbines: wind energy is not *dispatchable*, which implies that the producer of wind energy is unable to guarantee energy delivery to the electric network on a specified day and hour. The calculations on the wind velocity and the energy harnessed from the wind are probabilistic and are based on the observation that while the wind velocities at a given point and time are unpredictable, the yearly averages of the wind velocity at a given location on the surface of the earth are almost invariable.

When decisions are made for the construction and operation of wind turbines at a given location, the magnitude of the wind velocity as well as the hours within a year that ranges of velocities occur are important. The *wind distribution curves*, also known as *exceedance curves*, *endurance curves*, and *duration curves*, are used for the calculation of the annual amount of wind energy that may be produced by the wind turbines. Such curves show

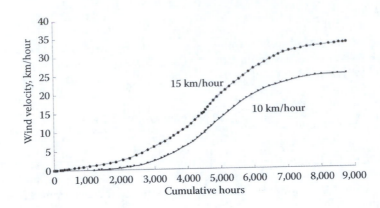

FIGURE 6.10
Wind distribution curves at two locations with annual average velocities 10 and 15 km/hour.

how many hours the local wind velocity exceeds a certain value annually. The wind velocity is the abscissa of these diagrams and the hours of the year—0–8,760—is the ordinate. Figure 6.10 depicts the wind distribution curves at two locations with average wind speeds of 10 and 15 km/hour. The exceedance curves in this figure signify that the wind velocity exceeds 3 m/s for 6,520 hours of the year at the location with of 15 m/s average velocity and that the wind velocity is below 20.5 m/s for 6,150 hours per year in the location with 10 m/s average velocity. Such statistical information provides valuable input on the time-averaged wind velocity and power to be produced and assists in the decision-making process of wind energy projects.

Another widely used method to report the annually averaged velocity at a location is with histograms that portray the time periods or the relative frequency when velocities in the different ranges/bins (e.g., 0–2 m/s, 2–5 m/s, and 10–15 m/s) are measured. Figure 6.11 is such a histogram and portrays the relative frequency for several wind velocity ranges. The sum of the relative frequencies is 1 (or 100% if expressed in percentages). Figure 6.11 indicates that wind velocity in the range of 14–16 m/s occurs at 11% of the time (964 hours in a year) at that location. One may use information from exceedance curves and histograms to estimate the yearly averaged power that is available and expected to be annually produced at a given location. It is of significance that the histograms and the exceedance

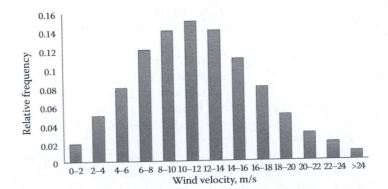

FIGURE 6.11
Histogram of wind velocities. The annually averaged wind velocity is approximately 11.5 m/s.

curves provide only annually averaged information and, for that matter, may be used to calculate annually averaged parameters only.

More detailed, hourly data that show the wind velocity, the direction of the wind, and several other parameters that are useful for the power production calculations are also available from several meteorological stations. Table 6.7 encompasses small parts of such a set of data for a meteorological station in Canyon, Texas, where several wind farms have been built since 2005. The velocity reported is the average wind velocity from several measurements during the hour. The standard deviation pertains to all the velocity measurements during that hour, and the direction of the wind is in degrees following international convention.

Such detailed data enable us to calculate the expected hourly energy production and provide more accurate calculations of the total daily, monthly, and annual energy expected to be produced by wind turbines in a wind energy farm.

Data for the predominant wind direction at a location assist engineers in the placement of the entire group of wind turbines within a wind farm. Wind location data are often displayed in a *wind rose* chart, such as the one depicted in Figure 6.12, which was generated from data for the month of July 2011 at the meteorological station of Canyon, Texas. The wind rose is a polar coordinate diagram. The direction of the wind is depicted on the angular axis, and the hours (during a month or a year) the wind flows from the indicated direction, on the radial axis. In the diagram of Figure 6.12, the wind blew from the

TABLE 6.7

Hourly Averaged Wind Data from Canyon, Texas

Date	Hours	Temperature, °C	Velocity, m/s	Standard Deviation	Direction
7/3/2011	0:00:00	24.8	8.1	0.6	158
7/3/2011	1:00:00	23.6	8.4	0.6	154
7/3/2011	2:00:00	24.3	10.6	0.6	176
7/3/2011	3:00:00	23.6	9.9	0.7	188
7/3/2011	4:00:00	25.2	8.8	1.2	211
7/3/2011	5:00:00	23.7	4.5	1.1	177
7/3/2011	6:00:00	24.1	4.4	0.4	176
7/3/2011	7:00:00	27.2	5.8	0.6	238
7/3/2011	8:00:00	29.9	7.7	0.8	259
7/3/2011	9:00:00	32.2	7.3	0.8	260
7/3/2011	10:00:00	34.6	6.5	1.2	236
7/3/2011	11:00:00	36.3	5.4	1.3	211
7/3/2011	12:00:00	37.5	6	1.4	194
7/3/2011	13:00:00	38	6.5	1.7	191
7/3/2011	14:00:00	38	6.1	1.5	192
7/3/2011	15:00:00	38.9	6	1.7	178
7/3/2011	16:00:00	38.4	7.2	1.2	178
7/3/2011	17:00:00	38.4	6.3	1.3	198
7/3/2011	18:00:00	37.6	5.8	1.1	202
7/3/2011	19:00:00	36	6.8	0.7	192
7/3/2011	20:00:00	31.4	11.2	1.8	159
7/3/2011	21:00:00	22.5	14.1	1.2	28
7/3/2011	22:00:00	22	4.4	0.9	79
7/3/2011	23:00:00	23.2	5.6	0.9	105

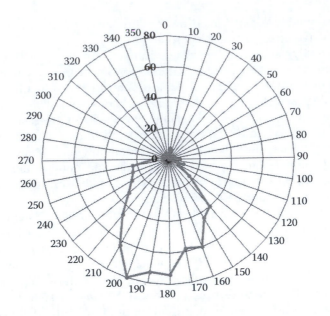

FIGURE 6.12
"Wind rose" chart for Canyon, Texas, for July 2011. The predominant wind velocity in the area is in the range of 180–200° (south–southwest).

200° direction for 80 hours during the month of July 2011 and from the direction 160° for 60 hours during that month. It is apparent from this figure that the predominant wind direction at this location is south–southwest, in the compass range of 180° to 200°. Engineers will use this information to place the turbines in a wind farm in rows that face this predominant wind direction.

6.3.2 Wind Turbines

The wind turbines are rotating engines with very long blades. The larger and most commonly used wind turbines operate with two or three long blades that rotate on an almost vertical plane. The axes of rotation of these turbines are horizontal, and they are commonly called *horizontal axis turbines*. There are also a number of turbine designs with vertical axes, which are typically smaller and produce a lesser amount of power. Figure 6.13a shows a group of horizontal axis wind turbines at an offshore location in Belgium, and Figure 6.13b shows some of the components of a turbine as they are lifted into position in the nacelle. The main parts of a horizontal axis wind turbine are as follows:

1. The *tower* may reach 150 m in height to allow for the clearance of the rotating blades and to take advantage of the stronger wind velocities at the higher elevations. The towers are made of reinforced concrete and provide the structural support for the entire system. Towers are thicker at the base, for stability, and they are typically hollow to allow for an internal shaft and ladders that are used to access for maintenance and repairs in the nacelle of the turbine, which contains the gearbox and the electric generator.

2. The *rotor blades* are one of the most important parts of the electricity generating system, where the wind energy is imparted to the engine. The blades of a wind turbine

(a)

(b)

FIGURE 6.13
(a) Group of wind turbines at an offshore location in Belgium. (Courtesy of Hans Hillewaert.) (b) Components of wind a turbine as they are lifted into position. (Courtesy of Paul Anderson.)

are long, typically 30–75 m, and are made of light composite materials. The wind power turbine blades are very similar in shape with blades of airplane propellers, because their actions are similar: wind turbine blades extract power from a stream of air and propeller blades convert the power of the airplane engine to create the stream of air that propels the airplane. The design of the blades for wind turbines conforms to the principles of turbomachinery with pitch angles that vary along the blade. The blades are bolted to the turbine hub, which extends to a horizontal steel shaft that transmits the power. A mechanism to adjust the pitch of the blades is positioned between the hub and the blades. The pitch mechanism, supported by a system of sensors and actuators, adjusts the blade angle of attack according to the wind speed for maximum power and controls the rotational speed of the blades.

3. The *yaw bearings and yaw break* are necessary because the wind turbine rotates to face the direction of the wind. The entire power producing system is pivoted on bearings that allow the entire nacelle of the turbine to rotate around a vertical axis so that the blades face the wind direction. The aerodynamic drag force on a rudder (in smaller engines) or a small *yaw motor* provides the force for this rotation. In order to avoid overshooting and fast rotation that may damage the engine, the yaw break system dampens the rotational motion.

4. The *gearbox:* In order to minimize the centrifugal forces, the rotational speed of the blades is designed to be fairly low, typically in the range of 15–25 revolutions per minute (rpm). A gearbox with an electronic *controller* steps up the rotational speed of the prime mover to reach 1800 rpm, which produces electric power at 60 Hz (1500 rpm and 50 Hz in Europe) in the generator. Because a small fraction,

less than 5%, of the power is dissipated in the gearbox by friction, larger wind turbines use cooling systems for their gearboxes.

5. The *electric generator* uses either permanent magnets or electromagnets (exciters) and converts the mechanical/rotational energy of the prime mover to electricity. The generators of the more modern and larger engines may produce a few mega-watts of power and include power electronics, such as *variable speed constant fre-quency* devices. Power spikes in the system are usually absorbed by the inertia of the blade–rotor–gearbox system.

The *solidity* of wind turbines describes the fraction of the swept area that is covered by the blades. It is equal to the ratio of the total area of the blades and the area of the circle swept by them. The blades of turbines with high solidity—e.g., the free-tail that is used for pumping water—highly interact with the incoming wind at low angular velocities. High solidity and lower angular velocities for these turbines generate higher torque in the central shaft that is needed to compensate for the higher starting torque the water pump requires. The blades of turbines with low solidity that are used for the production of elec-tric power—e.g., the ones in Figure 6.13a and b—rotate faster to optimally interact with the incoming wind. These turbines have two or three blades and their optimum angular velocities are significantly higher. Because of the higher angular velocities of the blades and the steel shaft, the gear mechanisms of the two- and three-blade turbines are smaller and simpler, because a lower gear ratio is needed to match the angular speed of the rotor to that of the electric generator.

6.3.3 Wind Power Generation

If one were to convert the entire available power of the wind, as expressed by Equations 6.8 and 6.9, in a wind turbine to electricity, the wind would have to impart all its kinetic energy to the turbine and leave with vanishing kinetic energy. In this case, the wind would come to a stop downstream the wind turbine, something that does not happen in nature. For the continuous operation of a wind turbine, the air velocity downstream the turbine must be finite. As a consequence, the power extracted by the turbine would be less than that given by Equation 6.9. One may calculate the actual amount of power a turbine will generate from a wind with incoming velocity V_i following an analysis of the airflow near the wind turbine, based on the fundamental principles of fluid dynamics.

Let us consider a control volume that surrounds the wind turbine. The cross-section of the control volume at the turbine is equal to the area described by the rotating blades A_b ($=\pi D^2/4$). A schematic diagram of the pressure and velocity distributions within this control volume is given in Figure 6.14. The inlet and outlet of the open system are far enough from the turbine so that the influence of the rotating blades on the static pressure is vanishingly small. Therefore, the static pressure at the inlet and the outlet of the control volume is equal to the prevailing atmospheric pressure P_a, while the static pressure varies within the control volume as depicted in Figure 6.14. Observations have shown that the static pressure variations in the entire control volume are small enough for the air den-sity to be considered constant. For this simple analysis, it is also assumed that the control volume is at a level significantly higher than the ground and that the velocity is almost uniform and directed toward the turbine. The velocity variation along the control volume is as depicted in Figure 6.14, with the inlet velocity of the air stream denoted by V_i and the outlet velocity by V_e. The latter is always less than V_i, because the turbine extracts a frac-tion of the kinetic energy of the wind.

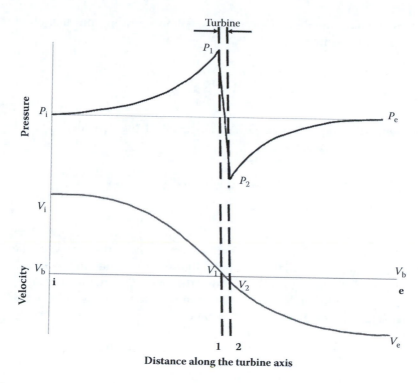

FIGURE 6.14
Pressure and wind velocity variations in the vicinity of the wind turbine.

The cross-sectional area of the control volume for this problem is not constant. This area varies along the control volume in order to enclose the entire mass of air that passes through the turbine, to accommodate the local variation of the air velocity, and to ensure that the flow streamlines remain within the control volume. The mass flow rate,

$$\dot{m} = \rho A V, \tag{6.11}$$

is constant throughout any cross section of the control volume, and the mass continuity equation is satisfied in all cross sections of the control volume.

In the diagram of Figure 6.14, the wind turbine occupies segment 1-2. In that segment and because of the extraction of mechanical power from the wind, the static pressure drops from P_1 to P_2 and the air stream velocity from V_1 to V_2. The static pressure and the velocity continuously vary within the control volume. However, the first derivative of the static pressure is not continuous, because of the extraction of mechanical power at the turbine. The first derivative of the velocity is continuous, as it is shown in Figure 6.14. Because the control volume is defined by the air streamlines, no air mass escapes or is added to it, and the mass conservation in Equation 6.11 is satisfied in the direction of the streamlines. Throughout the control volume, there is not any source of significant friction, and hence, the flow is considered to be frictionless. Under these conditions, the momentum conservation equation in the control volume is the same as the Bernoulli equation for incompressible fluids and may be written as follows for the front section of the control volume in Figure 6.14, section i to 1:

$$\frac{P_i}{\rho} + \frac{1}{2}V_i^2 = \frac{P_1}{\rho} + \frac{1}{2}V_1^2. \tag{6.12}$$

Similarly, Bernoulli's equation applies at the back of the control volume from 2 to e, where it may be written as

$$\frac{P_2}{\rho} + \frac{1}{2}V_2^2 = \frac{P_e}{\rho} + \frac{1}{2}V_e^2. \tag{6.13}$$

Bernoulli's equation in this form is not valid in region 1-2, because work/power is extracted by the blades of the turbine. This is a very thin part of the cross section of the control volume, because the thickness of the blades is very small in comparison to the length of the control volume. One may subtract the last two equations, apply the condition $P_i = P_e = P_a$, and derive the following expression for the difference of the static pressure $P_1 - P_2$, immediately before and after the turbine.

$$P_1 - P_2 = \frac{1}{2}\rho\left(V_i^2 - V_1^2 + V_2^2 - V_e^2\right). \tag{6.14}$$

From an inspection of the velocity variation in Figure 6.14 and from the fact that both the velocity as well as its first derivative are continuous functions, one may conclude that points 1 and 2 are very close, and hence, the velocities immediately before and after the wind turbine are approximately equal. If we define the velocity at the center of the wind turbine blade as V_b, we have the condition $V_b \approx V_2 \approx V_1$. Therefore, the last equation yields the following expression for the static pressure drop before and after the turbine blades:

$$P_1 - P_2 = \frac{1}{2}\rho\left(V_i^2 - V_e^2\right). \tag{6.15}$$

The axial force (thrust) developed on the wind turbine blades is equal to the product of the pressure difference and the area swept by the blades of the turbine A_b:

$$F_x = A_b(P_1 - P_2) = \frac{1}{2}A_b\rho\left(V_i^2 - V_e^2\right). \tag{6.16}$$

Since there are no other forces acting on the control volume, the axial force is the net force developed on the entire control volume. From the principle of momentum conservation, this force is equal to the momentum change of the entire mass flow rate through the control volume, given by Equation 6.11. Since the turbine blade cross section is $A_b = \pi D^2/4$, and the air velocity at the turbine is V_b, the momentum conservation equation is reduced to a second expression for the axial force (thrust):

$$F_x = (\dot{m}V)_i - (\dot{m}V)_e = \dot{m}(V_i - V_e) = \rho A_b V_b (V_i - V_e). \tag{6.17}$$

A comparison of the last two equations yields an expression for the unknown velocity at the turbine blades V_b:

$$V_b = \frac{1}{2}(V_i + V_e). \tag{6.18}$$

Equation 6.18 proves that the air velocity at the turbine blades is equal to the arithmetic average of the velocities at the inlet and the outlet of the control volume. Since the power developed by the turbine is equal to the product of the axial force F_x and the local wind velocity V_b, an expression for the power developed by the turbine with diameter D is

$$\dot{W} = \frac{\pi D^2}{16} \rho (V_i + V_e)^2 (V_i - V_e) = \frac{\pi D^2}{16} \rho (V_i + V_e)\left(V_i^2 - V_e^2\right). \tag{6.19}$$

For the optimization of the operation of the wind turbine and the production of maximum power, one realizes that the inlet air velocity is the incoming wind velocity and cannot be altered. The outlet velocity V_e may be modified with the design of the turbine blades and their geometric characteristics. The wind turbine and its operation may be designed in a way that the exit velocity V_e is optimized to yield maximum power, according to Equation 6.19. A simple way to perform this optimization is to set the first derivative of the power with respect to the adjustable parameter V_e equal to zero:

$$\frac{\partial \dot{W}}{\partial V_e} = 0 \Rightarrow 3V_e^2 + 2V_e V_i - V_i^2 = 0 \Rightarrow V_e = \frac{1}{3}V_i. \tag{6.20}$$

The last equation implies that for the optimum operation of a given turbine and the production of maximum power, the downstream wind velocity should be one-third of the incoming velocity. Under this condition, the maximum power produced at steady and continuous operation is

$$\dot{W}_{max} = \frac{8}{27} \rho A V_i^3 = \frac{2\pi}{27} \rho D^2 V_i^3. \tag{6.21}$$

The efficiency* of the wind turbines is defined as the ratio of the actual power they produce to the available power in the wind stream:

$$\eta = \frac{\dot{W}}{\dot{W}_{av}}. \tag{6.22}$$

From Equations 6.8, 6.21, and 6.22, it follows that the maximum efficiency a wind turbine may have is $\eta_{max} = 16/27 = 0.593$. Therefore, even an ideal wind turbine will convert less than 60% of the available power of the wind to electric power. The maximum wind turbine

---

* In some publications, the turbine efficiency is called *power coefficient* and is denoted by C_p.

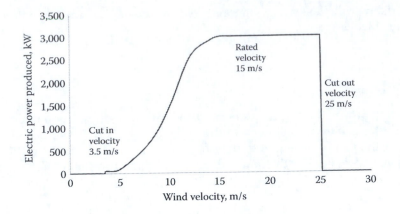

FIGURE 6.15
Power characteristics of the Vestas V90-3.0 MW wind turbine.

efficiency (59.3%) is sometimes referred to as *the Betz limit* or *Betz's law*. Actual wind turbines have lesser efficiencies, which depend on the magnitude of the wind velocity.

Figure 6.15 depicts the typical relationship between the power produced by an actual wind turbine and the prevailing wind velocity. The figure is obtained from data in the specifications of the V90-3.0 MW turbine manufactured by Vestas Inc. [17]. As observed in this figure, the turbine does not produce any power when the wind velocity is below 3.5 m/s. This is the *cut-in velocity* of this turbine. From 3.5 m/s to 15.0 m/s, the power produced by the turbine increases closely following the cubic dependence $\dot{W} \sim V^3$, according to Equation 6.9. The velocity 15.0 m/s is the *rated velocity* for this turbine, and the power produced in the range of 15 m/s < V < 25 m/s is almost constant, called the *rated power* of the turbine, 3.0 MW in the case of the *Vestas* turbine. The *cut-out velocity*—25m in this case—is the upper limit of the wind velocity range, where the engine may safely operate and generate electric power. When the wind velocity exceeds the cut-out velocity, the pitch of the blades is adjusted for minimum drag, the blade rotation slows down, and the power produced by the turbine vanishes. This procedure is called *feathering* the turbine blades, and it is followed when very high winds (e.g., during a storm) prevail in order to avoid damage to the blades and the entire system of the wind turbine.

It is apparent that when the wind velocity increases beyond the rated velocity, which is a design parameter, the turbine does not take advantage of the increased kinetic energy of air to generate more power. A large fraction of the available power from the wind is not harnessed when the turbine operates at its rated power, in the velocity range defined above the rated velocity and below the cut-out velocity. This has a significant impact on the time-averaged power produced by a wind turbine.

Table 6.8 lists several key operating parameters for four commercially available turbines. The data are from Vestas [17], GE [18], and Siemens [19]. It is observed in this table that the cut-in, the cut-out, and the rated velocities are within narrow ranges in the four types of commercial turbines. The rated velocity of the turbines is in the range of 10–15 m/s, and the turbines do not produce any power when the wind velocity is more than about 25 m/s (55.9 mph).

TABLE 6.8

Key Parameters of Four Commercially Available Wind Turbines

Turbine	GE 1.5sle	Vestas V90-3.0 MW	Vestas V110-2.0 MW	Siemens SWT-2.3
Rated power, kW	1,500	3,000	2,000	2,300
Rated velocity, m/s (mph)	14 (31.3)	15 (33.6)	10 (22.4)	12 (28.3)
Cut-in velocity, m/s (mph)	3.5 (7.8)	3.5 (7.8)	3.0 (6.7)	3.0 (8.7)
Cut-out velocity, m/s (mph)	25 (55.9)	25 (55.9)	20 (44.7)	25 (55.9)
Rotor diameter, m	77	90	110	108
Tower height, m	65–80	65–80	80–95	80
Efficiency at rated velocity	19.5%	23.3	20.3	24.3

Source: Vestas, Aarhus, Denmark, https://www.vestas.com/en/products/turbines/v90-3_0_mw# (website), last visited July 2017; GE, Boston, MA, https://geosci.uchicago.edu/~moyer/GEOS24705/Readings/GEA14954C15-MW-Broch.pdf (website), last visited July 2017; Siemens, Berlin, https://www.siemens.com/wind (website), last visited July 2017.

6.3.4 Average Power and Annual Energy Production

Wind turbines do not always produce at their rated power, because the incident wind velocity significantly varies. For any decision-making process on the development of wind power, the total energy produced in a given period is of importance, because it is this energy that will be sold by the owner of the wind turbine(s). Such calculations may be performed when the turbine characteristics as well as the local wind conditions are known. The following example illustrates the calculation of the annual energy produced by a wind turbine.

Example 6.7

Calculate the energy, in megawatt hours, produced by a wind turbine annually, when the turbine is placed in a location where the wind velocity is 0.0 m/s $<V<$ 3.5 m/s for 7% of the total hours in the year; 3.5 m/s $< V <$ 8 m/s for 23% of the total hours; 8 m/s $< V <$ 12 m/s for 30% of the total hours; 12 m/s $< V <$ 16 m/s for 16% of the total hours; 16 m/s $< V <$ 20 m/s for 12% of the total hours; 20 m/s $< V <$ 25 m/s for 10% of the total hours; and 25 m/s $< V$ for 2% of the total hours. The diameter of the turbine is $D = 90$ m, the cut-in velocity is 3.5 m/s; the rated velocity is 12 m/s; and the cut-out velocity is 25 m/s. The turbine efficiency is constant, $\eta_T = 24\%$, and the air density is 1.2 kg/m³.

Solution: At first it is noted that because the cut-in velocity of the turbine is 3.5 m/s and the cut-out velocity is 25 m/s, this wind turbine produces zero power when the wind velocity is in the ranges of 0.0 m/s $< V <$ 3.5 m/s and 25 m/s $< V$. This occurs at 8760 × (0.07 + 0.02) = 788 hours annually.*

Secondly, during the wind velocity range of 12 m/s $< V <$ 25 m/s, the turbine produces constant power, which may be calculated from Equation 6.9:

$$\dot{W} = \frac{\pi}{8}\eta_T D^2 \rho V^3 = \frac{3.14}{8} \times 0.24 \times 90^2 \times 1.2 \times 12^3 = 1.58 \times 10^6 \text{ W} = 1.58 \text{ MW}.$$

---

* A year of 365 days has 8,760 hours.

The wind turbine operates at this power for $8{,}760 \times (0.16 + 0.12 + 0.10) = 3{,}329$ hours annually, and during this period, it produces $1.58 \times 3{,}329 = 5{,}260$ MWh.

Thirdly, during the wind velocity range of 3.5 m/s $< V <$ 12 m/s, the turbine produces variable power given by Equation 6.9: at 3.5 m/s, it produces 0.04 MW; at 8 m/s, it produces 0.47 MW; and at 12 m/s, it produces 1.58 MW.

The approximate average power the turbine produces when the wind velocity is in the range of 3.5 m/s $< V <$ 8 m/s is $(0.04 + 0.47)/2 = 0.26$ MW. Since this velocity range occurs for 23% of the hours in a year, the turbine produces $0.23 \times 8{,}760 \times 0.26 = 524$ MWh in this range of velocities. It must be noted that it is more accurate to use the average of the power in the calculations than the power at the average velocity.

When the wind velocity is in the range of 8 m/s $< V <$ 12 m/s, the approximate power the turbine produces in this range is $(0.47 + 1.58)/2 = 1.03$ MW. Since this velocity range occurs for 30% of the hours in a year, the turbine produces $0.3 \times 8{,}760 \times 1.03 = 2{,}707$ MWh.

The following table summarizes the average power and the total energy produced by this turbine during a year:

Wind Velocity, m/s	Average Power, MW	Total Hours	Energy Produced, MWh
$0 < V < 3.5$	0	613	0
$3.5 < V < 8$	0.26	2015	523
$8 < V < 12$	1.03	2628	2,707
$12 < V < 16$	1.58	1402	2,215
$16 < V < 20$	1.58	1051	1,661
$20 < V < 25$	1.58	876	1,384
$25 < V$	0	175	0

From this last table and the calculations, the total energy produced by this turbine annually is 8,490 MWh. This is the energy that can be used for the calculation of the annual revenue from this turbine.

It is apparent that the computations in the wind velocity range of 3.5 m/s $< V <$ 12 m/s, when the turbine output is proportional to the cube of V, are approximate. More accurate calculations may be accomplished if the wind velocity and duration were known with more detail. The higher accuracy of the calculations is achieved when one uses data sets such as the one in Table 6.7, where the average wind velocity, its direction, and its standard deviation are given at hourly intervals for the entire year.

6.3.5 Wind Farms

Groups of wind turbines are installed at windy sites and are called *wind farms*. An offshore wind farm is shown in Figure 6.13a. The first criterion for the choice of the sites of wind farms is high average wind velocity. This may imply that wind farms are located in remote and, often, very difficult to access sites, which are far from the urban centers and the consumers of electricity. The second criterion for the choice of the wind farm location is proximity to high-voltage electricity transmission lines that will transmit the power produced to the consumers. The practice of wind power generation corporations is to lease large areas and place a number of wind turbines—from tens to hundreds—in such remote sites. The large number of wind turbines generates economies of scale by reducing maintenance and operational costs and by reducing the cost for the connection of the group of turbines to the high-voltage electricity grid.

Naturally, electricity generation corporations wish to minimize the leased areas. A moment's reflection, however, proves that the area of a wind farm—the *footprint* of the wind farm—cannot be minimized at will, because the turbines will be very close to one another. As Equation 6.20 shows, the wind velocity aft of a wind turbine is merely one-third of the prevailing wind velocity. Therefore, if a second turbine is placed directly behind the first, it would generate only 1/27 of the power of the upwind turbine, an unacceptable result. The wind velocity recovers to (almost) the upwind value far downstream of the first turbine, at distances 5–10 times the rotor diameter. Judging from the characteristics of the wind turbines in Table 6.8, the distance of the downstream turbine would be several hundreds of meters. This implies that the wind turbine farms have a very large footprint. If we place a large number N of wind turbines in a square arrangement with the distance between the turbines $L = \beta D$, the total ground area occupied by the group of turbines is $NL^2 = N\beta^2 D^2$. One may define the *area power density* of the group of turbines as the power produced per unit surface area (per square meter or per hectare):

$$\frac{\dot{W}}{A} = \frac{\frac{1}{8}\pi\eta_T \rho ND^2 V_i^3}{N\beta^2 D^2} = \frac{\pi\eta_T \rho V_i^3}{8\beta^2}. \tag{6.23}$$

At $V_i = 15$ m/s and $\beta = 7.5$, the area power density of the wind farm is approximately 5 W/m². This number is by far lower than the area power density of thermal (nuclear and fossil fuel) power plants and significantly lower than that of solar power plants, which is close to 50 W/m². It appears at first that the area needed for wind power generation is very large. However, the high area is not a significant impediment for the development of wind farms and wind energy in general: the actual footprint of a wind turbine on the surface of the earth is the footprint of their towers, approximately 5×5 m². The rest of the land in the wind farm may be returned to its original use, typically agricultural or animal grazing. Since several wind farms are located offshore, even this smaller footprint does not present a disadvantage for the development of wind power. The only realistic drawback with the large area needed for wind farms is the lengthy connections that must be made between each turbine and the electricity grid.

6.3.6 Environmental Impacts of Wind Energy

As with solar energy, wind energy is one of the most benign and most environmentally friendly forms of electric energy production. Electricity from the wind does not produce harmful emissions or thermal pollution. The materials that make the towers and the components of the engines are commonly used structural and engineering materials. Their production involves very limited environmental impacts. While the construction and operation of wind turbines does not inflict any environmental threat, there are a few minor environmental and ecological issues associated with wind power:

1. *Noise pollution*: A rotating engine always produces noise, and wind turbines are no exception, especially the parts that operate at higher revolutions per minute. Noise pollution is a limiting factor for the expansion and more widespread use of smaller wind turbines in urban and suburban environments. Larger wind turbines are located in remote, rural areas with low population densities, and this mitigates the effect of noise to the local communities. However, noise may have

a significant effect on the wildlife of the area and force animals to migrate, thus disturbing the balance of the ecosystem.

2. *Bird injuries and mortality*: Flying birds are often killed by the rotating blades. The motion of the blade and the local pressure reduction immediately upstream of the rotating blades detract the flight of birds and often kill them. Concerns for bird populations, which migrate from Canada to the shores of the Gulf of Mexico, have put severe restrictions on wind power development projects near the Gulf, where the sea breeze is always strong.

3. *Aesthetic pollution*: The picturesque landscape of remote pristine areas is often disturbed by the placement of wind farms with their high towers. The aesthetic pollution has raised opposition to the development of wind power in many suburban areas, especially those with tourist attractions.

4. *Radio and TV signal interference*: Many wind turbines are located near the top or the sides of hills and mountains, and their operation interferes with the transmission of electromagnetic waves including signals from radio, television, and cellular telephones. Better location design and stronger signals will counteract this effect.

While not insignificant, the environmental problems associated with wind power are much less harmful to the environment and the ecosystems than the effects of most other energy sources, especially those of the fossil fuel plants. The further expansion of wind power and the substitution of conventional power sources by wind are on the whole benevolent to the environment.

6.4 Geothermal Energy

Geothermal energy is the primordial energy of the earth. It is produced as heat in the core of the earth by the disintegration of radioactive nuclei that were formed and trapped there in the early stages of the formation of the planet. The heat is convected by the magma to the crust of the earth, and this process creates the geothermal gradient—the local temperature in the crust of the earth increases with depth. In several locations, where the magma pockets reach closer to the surface, the geothermal gradient is high enough for the temperature to be close to 200°C at moderate depths, in the range of 1.5–2 km from the surface. While these premium high-temperature gradient locations are close to the tectonic boundaries, other lower temperature resources are abundant in all geographic locations.

Underground aquifers near the high-temperature spots are heated to higher temperatures. The water in the aquifers typically remains in the liquid state because the static pressure is very high at such depths. For the utilization of geothermal energy, geothermal wells are drilled in the hot aquifers to extract the high-temperature water. When the hot water ascends to the surface, inside the wells, its static pressure decreases. If the static pressure drops below the saturation pressure, the water "flashes" to produce a mixture of steam and water, which may be used for the production of electricity in flashing power plants. If the water does not flash, the well produces liquid water at higher temperature, which is typically used for the heating of buildings or the production of electricity in binary power plants. Most geothermal power plants are smaller and simpler in construction than fossil or nuclear power plants, i.e., they have a lesser number of components.

6.4.1 Fundamentals of Geothermal Energy

Approximately 44 TW (44×10^{12} W) of heat power is transferred from the core to the surface of the earth. Of this amount, 30 TW is continuously generated by the radioactive decay of elements in the core of the earth and the rest from the cooling of the earth's interior, where the maximum temperature has been estimated to approach 6,000 K. The continuous cooling of the earth's interior by the difference of the two numbers, 14 TW, is an indication that the interior temperature of the earth decreases and that geothermal power will decrease in the future and will eventually be exhausted. However, this cooling process is expected to take place in a geological timescale, on the order of billions of years. Because of this very large timescale, geothermal energy is considered as a renewable source of energy.

Given that the average electric power consumed by the entire population of the earth is approximately 2.4 TW and the corresponding average power for all the energy needs is 18 TW, one may conclude that geothermal power alone is capable of satisfying all the energy requirements of the human society. However, in comparison to fossil fuel combustion, geothermal energy is a low-temperature resource and is converted to electricity with significantly lower thermal efficiency. In addition, a great deal of the geothermal power is dissipated in the oceans, which make up more than 70% of the surface of the earth. For this reason, geothermal energy may produce electricity and help satisfy the energy needs of small countries with high geothermal potential—such as Iceland and Costa Rica—but may not be reasonably relied upon to provide a very high fraction of larger countries.

Geothermal energy has been used since the ancient times as thermal energy. Among others, ancient Greeks and Romans used the high-temperature water from hot springs and fumaroles for baths and for the heating of public places. In the eighteenth century, therapeutic properties were attributed to hot springs, and a whole thermal resort industry was developed around them, which peaked in the late nineteenth century, when several resort towns with hot springs were developed in Europe as tourist attractions. Geothermal water from hot springs filled individual or communal baths in the resort towns; provided high moisture air in steam baths; and, in the winter months, provided heating for the buildings of the resort towns.

The harnessing of geothermal energy for the production of electricity commenced in 1904 in Larderello, a small town in central Italy, where the local count constructed a shallow well and used a small steam turbine to produce enough electric power for his household. In the 1920s, the development of the Geysers field started in United States, to the north of San Francisco, California, and continued in the 1950s and 1960s with the installation of more than 2,200 MW of electric power. Several other electric generation projects from geothermal energy around the world, for example, in Wairakei, New Zealand; Tsukuba, Japan; and Reykjavik, Iceland, were completed after 1960. In 2015, there were a total of more than 13,200 MW of installed geothermal electric capacity in 24 countries [20]. Another 74,000 MW of geothermal energy were used in 80 countries as heat for buildings, fresh water production, snow melting, horticulture, aquaculture, and agriculture.

Figure 6.16 shows the regions on the surface of the earth with high geothermal temperature gradients. These regions are predominantly located at the boundaries of tectonic plates where magma intrudes into the crust of the earth and creates hot spots. Among these regions are the western Pacific Rim, which encompasses the United States; Mexico; all central American countries, Ecuador, Peru, and Chile; the eastern Pacific Rim with the countries of Japan, Indonesia, New Zealand, and the Philippines; in Europe, the Mediterranean countries, Madeira, and Iceland; and the eastern African countries of Ethiopia, Eritrea, Kenya, and Tanzania. These countries currently have a high fraction of

FIGURE 6.16
High geothermal activity regions are predominantly at the boundaries of tectonic plates.

the existing geothermal projects for the production of electricity and geothermal heat utilization. Several of these countries and their installed geothermal capacity are listed in Table 6.9. For some of the smaller countries, e.g., El Salvador, Guatemala, Nicaragua, and Iceland, geothermal energy provides a high percentage of the total electric power demand.

Figure 6.17 is a schematic diagram of an entire geothermal field. The high-temperature magma intrusion causes the transfer of a great deal of energy (heat) to the underground

TABLE 6.9

Leading Countries and Installed Geothermal Capacity in MW, in 2014

Country	Installed Capacity, MW
China	25
Costa Rica	210
El Salvador	205
Guatemala	45
Iceland	660
Indonesia	1,380
Italy	940
Japan	540
Kenya	600
Mexico	1,005
New Zealand	970
Nicaragua	88.0
Philippines	1,915
Russia	95
United States	3,525

Source: Geothermal Energy Association, *2015 Annual US & Global Geothermal Power Production Report*, Geothermal Energy Association, Washington, DC, 2015.

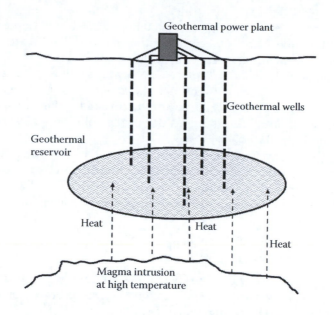

FIGURE 6.17
Schematic diagram of a geothermal reservoir–wells–plant system for the production of electric power.

aquifer, which attains a higher temperature and becomes the geothermal reservoir. Several geothermal wells drilled into the reservoir draw the hot water and, via a system of over-ground pipelines, supply the geothermal power plant with liquid hot water, steam, or a mixture of steam and water. The power plant produces the electric power and transmits it to the electricity demand sites. The wellheads and the system of pipelines, which are typically 1–5 km long, cover a large surface area, the geothermal field. Because the static pressure in the reservoir is high, the hot fluid naturally ascends in the wells and is conveyed to the power plant at moderate pressures.

Surface water that seeps through the ground and permeable geological strata replenishes the water supplied by the geothermal wells. In addition, liquid water that has been used in the power plant is injected back to the aquifer via *reinjection wells*. With these filling mechanisms, the water in the geothermal reservoirs is continuously replenished. It has been observed that in a period of a few decades, the pressure and volumetric flow rate of water produced by a specific well decline. In this case, a *downhole pump* is inserted at the bottom of the geothermal well to augment the water pressure at the well-bottom and supply a higher mass flow rate of hot water to the geothermal installation on the surface. After a few decades, wells with very low flow rates are shut, and new wells are drilled in the same aquifer to supply the geothermal power plant. While individual wells may be depleted, the geothermal aquifer is continuously replenished and supplies the power plant with geothermal fluid for very long times. The Larderello field in Italy has continuously been in operation for more than 100 years, the Geysers field in California since the 1920s, and the Wairakei field in New Zealand since the 1960s.

The geothermal fluid, which comes out of the wells, is primarily a mixture of vapor (steam with a mixture of volatile gases) and a liquid water solution. Depending on the location of the aquifer, the geothermal fluid contains several minerals, sometimes in significant quantities. For this reason, the geothermal fluid is often referred to as *brine*. NaCl, KCl, and $CaCl_2$ are among the most common minerals in geothermal fluids [21]. The first two are soluble in water, but $CaCl_2$ is essentially insoluble at lower temperatures. If the temperature drops in the pipelines that carry the brine, $CaCl_2$ will deposit on the walls to cause clogging of the pipe. Care is taken in geothermal installations to keep brines with higher $CaCl_2$ content at the higher temperatures that do not cause precipitation and clogging. Geothermal fluids also contain a number of water-soluble gases, among which are CO_2, CH_4, the malodorous H_2S, and oftentimes traces of the radioactive radon (Ra), which is produced from the nuclear disintegrations in the core of the earth. A very high fraction of the gases in the geothermal fluids comes out of the solution when the static pressure decreases and steam is produced.

When modeling the flow and characteristics of geothermal wells, the properties of pure water are typically used as an approximation. More accurate modeling requires the exact composition of the geothermal fluid as well as knowledge of the influence of all the constituents/impurities on the properties of the water substance [21].

6.4.2 Geothermal Resources

The quality of geothermal resources—the specific energy, and exergy that may be extracted to produce work—depends on the type of fluid the geothermal reservoirs produce. Geothermal resources are classified in the following categories [3,22]:

1. *Dry steam*: The brine in these aquifers is at a significantly high temperature, typically in the range of 200–280°C. As the brine flows through the permeable rocks of the aquifer zone, its static pressure drops and steam is produced inside the aquifer by local flashing. The steam, with a small fraction of droplets, is carried in the geothermal well, where its pressure drops further, most or all the water droplets evaporate, and dry steam is conveyed by the pipeline system to the power plant. Wells from a dry steam resource supply the power plant with saturated or superheated steam, which is directly fed to a turbine for the production of power. The fields at Larderello and the Geysers are primarily dry steam geothermal fields. The specific energy, enthalpy, and the exergy of the dry steam resources are significantly higher than the energy of the other types of geothermal resources and for this reason, dry steam resources are considered to be of very high quality.

2. *Two-phase fluid*: When the reservoir temperature is lower, the brine does not flash in the reservoir, and liquid hot water at high pressure enters the bottom of the wells. As the water rises in the wells, the static pressure decreases and the water flashes (evaporates) to produce steam. Because the static pressure continuously decreases in the well, the fraction of steam in the geothermal fluid continuously increases until it reaches the wellhead. These wells produce a mixture of steam and liquid water, the *two-phase fluid*, which is transported to the power plant. The latter may produce additional steam by flashing. The steam is separated from the liquid water and directed to one or two turbines that produce electric power. The dryness fraction (quality) of the geothermal fluid supplied to the power plant depends to a great extent on the pressure and temperature of the geothermal reservoir as well as on the depth of the well. Several geothermal fields that have been developed since the 1950s—e.g., those in the Imperial Valley of California, Mexicali in Mexico, and Wairakei in New Zealand—produce a two-phase fluid. Fields with temperatures more than 180°C are considered as high-quality geothermal resources.

3. *Geopressured*: The temperatures in the geopressured reservoirs are lower, in the range of 100–150°C. These geothermal reservoirs are typically located at greater depths where the aquifer pressure is significantly higher than dry steam resources. Because of the higher downhole pressure, flashing does not occur in the geothermal wells and water vapor is not produced. The wells produce hot, liquid water and convey it to the power plant, which typically uses the energy of the hot water in a binary cycle. The specific energy of the brine produced by the geopressured resources is significantly lower than that of steam and of the two-phase fluid resources. Because of this, geopressured resources are considered to be of lower quality.

4. *Hot, dry rock*: As the name implies, these resources are hot rocks, located deep inside the earth's crust, which do not have a local aquifer to produce geothermal brine. Despite the fact that there is no local fluid to be carried to the surface, one may develop an engineering system to harness the thermal energy of the rock by drilling a number of wells that terminate at the hot rocks, injecting a fluid to the wells to transfer the heat, bringing this fluid to the surface, and using the higher thermal energy of this fluid for the production of electric power. The working fluid in such a system is not necessarily water. It may be another fluid with convenient thermodynamic properties, such as butane, pentane, or a refrigerant [23]. Two pilot power plants using this type of system have been constructed in Soultz-sous-Forêts, Alsace, France, and Landau, Pfalz, Germany [24].

While the dry steam geothermal resources are the most desirable for the production of electric power, these resources are scarce. The vast majority of dry resources on earth have already been utilized. Similarly, a high fraction of the available high-temperature, two-phase fluid resources have also been utilized. The few remaining two-phase fluid resources are planned to be developed in the near future. Geopressured resources and hot, dry rock resources are abundant on the earth's crust, and only a small fraction of these resources has been developed. Since geopressured resources represent the overwhelming majority of the geothermal resources on the planet, and they have the capacity to produce economically high quantities of electric and thermal power in several regions, they represent a growth area for geothermal energy utilization [25]. Engineering systems that utilize the geopressured resources for the production of heat or electric power are expected to multiply and will become more common in the future.

6.4.3 Electric Power Production

Because the geothermal fluid is extracted at moderate temperatures, the operating temperatures in geothermal power plants rarely exceed 220°C. For this reason, geothermal units are constructed differently than fossil fuel, and nuclear power plants, with an emphasis on extracting a high fraction of the exergy of the geothermal fluid. From Equation 1.23, the maximum power one may obtain from the geothermal resource is [3,22,26]

$$\dot{W}_{max} = \dot{m}\left[(h - T_0 s) - (h_0 - T_0 s_0)\right] = \dot{m}(e - e_0), \tag{6.24}$$

where e is the exergy of the geothermal fluid. It follows that the type of the available geothermal resource that supplies the brine dictates to a large extent the type of power plant that is constructed. A short description of the most common types of geothermal power plants is given in the following subsections [3,22,26].

It should be noted that unlike solar and wind power, geothermal energy is not periodically variable or intermittent. The geothermal fluid is continuously supplied to the units that convert its energy to electric power, and hence, these units may operate 24/7 without interruptions. Electric power produced from geothermal power plants is dispatchable, and this is a significant advantage of geothermal power plants.

6.4.3.1 Dry Steam Units

Dry steam units are the simplest geothermal power plants and utilize the high-temperature resources that produce dry steam. The original unit at Larderello, Italy, several of the units at the Geysers geothermal field in California, United States, and the units in Kamojang, Indonesia, are of this type of geothermal power plants. A schematic diagram of a dry steam geothermal power plant is depicted in Figure 6.18. The network of geothermal wells produces dry (or almost dry) steam, which is fed to the *separator* or *steam drier* unit. Small water droplets and particles that are carried by the steam are removed in the separator by gravity. The steam is then directed to a turbine generator system, which produces the electric power. The spent steam from the turbine is fed to a condenser. In most cases, the condensate, which is almost pure water, is pressurized by a pump and reinjected back to the geothermal reservoir. If local environmental regulations allow it, the condensate may be used as a fresh water source, or as cooling water for the power plant. The power produced by a dry steam power plant is

$$\dot{W} = \dot{m}(h_1 - h_2), \tag{6.25}$$

where 1 and 2 are the states of the steam at the entrance and exit of the turbine and $\dot{m}$ is the mass flow rate of steam supplied to the turbine.

Dry steam geothermal power plants are the simplest of all steam units for the production of electricity. They require a modest amount of capital to construct and are very inexpensive to maintain. The electric power produced by the dry steam geothermal units

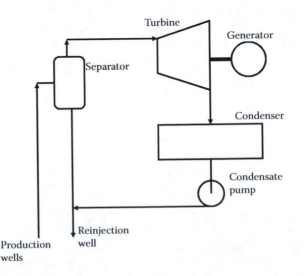

FIGURE 6.18
Schematic diagram of a dry steam geothermal unit.

is significantly less expensive than the power produced by most of the other alternative energy sources and oftentimes less expensive than that generated by fossil fuel and nuclear units.

6.4.3.2 Single- and Dual-Flashing Units

Flashing power plants are currently the most common geothermal installations and include units in Cero Prieto, Mexico; Imperial Valley, California; Wairakei, New Zealand; and Miravalles in Costa Rica. A schematic diagram of a single-flashing geothermal power plant is shown in Figure 6.19. Geothermal fluid from the wells enters the power plant at state 1. At this state, water may be liquid at high pressure and temperature or a water–steam two-phase mixture. The geothermal fluid enters a flashing chamber, where its pressure is significantly reduced and more steam is produced. The production of steam from the pressure reduction is called *flashing*. Typically, 10–25% of the mass of the geothermal fluid is converted to steam in the flashing chamber. The total amount of steam produced is separated from droplets and small particles in the *steam drier* and then fed to the turbine, at state 2″, where it expands to the pressure of the condenser, at state 3. An electric generator, connected to the turbine, produces electric power. Flashing units have two liquid effluents, the first from the condensate, at state 4 in the figure, and the second from the residue liquid water in the flashing chamber, which is at state 2′. States 2′ and 2″ are at the same temperature, typically in the range of 105–130°C. In most flashing geothermal units, both the condensate and the liquid effluent of the flashing chamber are reinjected into the geothermal aquifer.

The flashing process is isenthalpic. Hence, the dryness fraction of the steam produced from the geothermal fluid entering the flashing chamber is

$$h_1 = h_2 = (1 - x_2)h_{2'} + x_2 h_{2''} \Rightarrow x_2 = \frac{h_1 - h_{2'}}{h_{2''} - h_{2'}}. \tag{6.26}$$

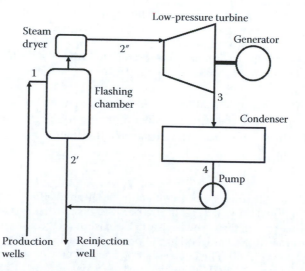

FIGURE 6.19
Schematic diagram of a single-flashing geothermal unit.

The subscripts indicate the states of the fluid as they are depicted in Figure 6.19. The optimum flashing temperature T_2 is the average of T_1 and the condenser temperature T_3, which is approximately 40°C $[T_2 = (T_1 + T_3)/2)]$. If the total mass flow rate of the geothermal fluid from the well is denoted by $\dot{m}$, the amount of steam fed to the turbine is equal to $\dot{m}x_2$ and the total power produced is calculated from the expression

$$\dot{W}_{act} = \dot{m}x_2(h_{2''} - h_3).$$ (6.27)

Because only 10–25% of the geothermal fluid is converted to steam and most of the fluid is reinjected at moderate temperatures several *dual-flashing geothermal power plants* have been constructed to better utilize the geothermal resources. In the dual-flashing units, the effluent of the flashing chamber is directed to a second, lower-pressure flashing chamber, where the pressure is further reduced and an additional amount of steam is produced. This lower pressure steam is fed to a second turbine generator system to produce additional power [3,22,26]. Dual-flashing units utilize 15–35% of the mass of the geothermal fluid and have higher thermal efficiencies than single-flashing units.

Example 6.8

The input to a 20 MW single-flash geothermal power plant is saturated liquid water at 200°C. The dryness fraction (steam percentage) at the exit of the flashing chamber is 15.6% and the enthalpy difference at the entrance and exit of the steam turbine is 360 kJ/kg of steam. Determine (a) the mass flow rate of steam to the turbine, (b) the total mass flow rate of geothermal water input to the flashing chamber, and (c) the mass flow rate of the liquid water rejected from the bottom of the flashing chamber.

Solution:

a. The power produced by the turbine is the product of the steam mass flow rate and the work per unit mass: $\dot{W} = \dot{m}_s w_T$.
   Hence, $\dot{m}_s = 20{,}000/360 = 55.6$ kg/s.
b. Because the 55.6 kg/s fed to the turbine represents 15.6% of the liquid water input to the flashing chamber, the total saturated water input to the power plant is 55.56/0.156 = 356.1 kg/s.
c. The mass flow rate of the liquid water rerejected from the flashing chamber is the difference of the two: 356.1–55.6 = 300.5 kg/s.

It is observed in this example that single-flash geothermal units use a great deal more water than the steam input to the turbine. Most of the geothermal fluid (84.6% in this case) is reinjected back to the aquifer from the exhaust of the flashing chamber.

Example 6.9

A 40 MW, dual-flash geothermal power plant has a thermal efficiency of 15% and uses 620 kg/s of geothermal brine at a cost of $0.5 per 1000 kg of brine. (a) Determine the heat input to this power plant, the waste heat, and the cost of the geothermal water per day. (b) It is *proposed* that 10 of such geothermal power plants are constructed to phase out a 400 MW coal power plant. The latter has 35% thermal efficiency and uses anthracite with heating value of 29,300 kJ/kg, which costs $89 per 1000 kg. Compare the "fuel" costs of the two alternatives.

Solution:

a. The heat input to the geothermal power plant is $\dot{Q}_{in} = 40/0.15 = 266.7$ MW. The waste heat $\dot{Q}_{out} = 266.7 - 40 = 226.7$ MW.

   Since the plant uses 620 kg of brine (the fuel of the geothermal unit) per second, the cost is $0.62 \times 0.5 = \$0.31$/second. The daily cost of the geothermal water (fuel) is $\$0.31 \times 60 \times 60 \times 24 = \$26{,}784$. Ten such power plants will have a daily cost $\$267{,}840$.

b. The 400 MW coal power plant operates with heat input $\dot{Q}_{in} = 400/0.35 = 1{,}142{,}857$ kW. With the given heating value for anthracite, the power plant uses $1{,}142{,}857/29{,}300 = 39.0$ kg of anthracite/s or 3,370,000 kg/day. The daily cost of this quantity of anthracite to the operator is $\$299{,}930$.

   The fuel cost of the proposed 10 geothermal units is lower than that of the coal power plant.

It is observed in this example that fuel cost considerations favor the unit with the lower thermal efficiency. This happens because the cost of the geothermal fuel is very cheap. This example is another indication that thermal efficiencies of power plants should be used as figures of merit for comparison among similar types of power plants and not to compare units that use different types of energy sources.

6.4.3.3 *Binary Units*

Most binary units have been constructed in the United States and include those in East Mesa, California, and Raft River, Idaho. The construction of binary units is recommended when the geothermal fluid entering the power plant from the wells is liquid and at lower temperatures (120–160°C). At this range of temperatures, the quantity of steam produced by the flashing process is very low, and consequently, the electric power produced is very low too. For such geothermal resources, it is better to use a heat exchanger for the transfer of thermal energy from the geothermal fluid to a *secondary fluid*, often called the *working fluid*. The secondary fluid is an organic fluid and is chosen to have a lower boiling point. It evaporates at lower temperatures in the heat exchanger, and the vapor produced is fed to a turbine generator system, where it expands to produce electric power. Effectively, the heat exchanger becomes the boiler of a simple Rankine cycle that uses the secondary fluid for the production of power. A schematic diagram of a binary geothermal power plant is shown in Figure 6.20. The geothermal fluid enters the heat exchanger from the production wells and exits at the reinjection well(s). Heat is transferred from the geothermal (primary) to the secondary fluid in this heat exchanger, which becomes the "boiler" for the secondary fluid. The latter exits the heat exchanger as vapor and is directed to a turbine generator system, where power is produced. The condenser liquefies the secondary fluid vapor and the pump pressurizes the condensate and feeds it back to the heat exchanger. Effectively, the secondary fluid, which is organic, undergoes a thermodynamic Rankine cycle, which is often called an *organic Rankine cycle* (ORC).

While the primary geothermal fluid is always a water solution (brine), the secondary fluid may be chosen from among many substances that possess desirable properties for the ORC. Suitable choices are fluids with high saturation pressures and low corresponding saturation temperatures. Among the fluids that have been used and considered to be used in ORCs are butane and isobutene, pentane, hexane, ammonia, and several refrigerants [3,22,26,27].

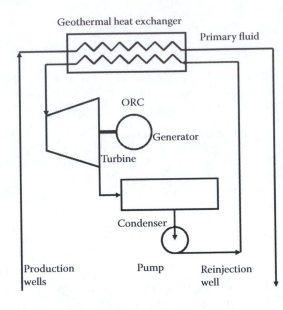

FIGURE 6.20
Schematic diagram of a binary geothermal power unit.

As it may be observed in Figure 6.20 of the binary unit, the geothermal fluid is reinjected into the earth in a closed-loop system; it does not come in contact with the atmosphere and does not release any harmful materials to the environment. The organic fluid also undergoes a closed cycle and does not come in contact with the environment. Because of this, binary units do not have any impact on the environment, except for thermal pollution. In general, binary geothermal units utilize a higher percentage of the energy in the geothermal fluid than the other units. Among the disadvantages of the binary geothermal power plants are (a) the high cost of the heat exchanger (b) the higher cost of the secondary fluid turbine, and (c) the lower temperatures in the primary fluid side may cause significant scaling in the heat exchanger pipes that could interrupt the continuous operation of the unit. Scaling is easily avoided by the periodic scrubbing and cleaning of the primary-fluid side of the heat exchanger.

6.4.3.4 Hybrid Geothermal–Fossil Power Units

The thermal efficiency of geothermal power plants significantly increases when the temperature of the steam produced is higher. This may be accomplished by the burning of fossil fuels to superheat the steam produced by the flashing units or the vapor of the secondary fluid produced by the binary units. The superheated steam or vapor produces significantly more power. While this type of geothermal units has higher efficiency and may produce power economically, there are very few hybrid units worldwide in 2017. The main reasons for this are the higher capital cost for the unit; the cost of transportation of fossil fuels to the remote locations, where geothermal units have been built; and regulations that may reduce or eliminate the tax advantages of renewable source power plants if fossil fuels are used in tandem with a renewable energy source.

6.4.4 District Heating

Since the antiquity, providing hot water for baths and spas has been one of the earliest and better-known uses of geothermal energy. Hot water from hot springs has supplied high-temperature water for public baths and private houses for several centuries in Iceland. When geothermal resources are available near the population centers, this practice continues today with the heating of entire districts of houses and businesses from a single or multiple geothermal wells. This practice is called *district heating* and is now utilized in several locations including Iceland, Japan, France, New Zealand, and Oregon, in the United States.

The engineering system for district heating is conceptually very simple and is shown in Figure 6.21. The geothermal brine from the production well passes through a heat exchanger, where it transfers energy to a cleaner secondary fluid, often fresh water. The latter is distributed via well-insulated pipes to the buildings in the district and keeps the interior temperature of the buildings at the desired temperatures. Valves at the entrance of the main heat exchanger control the amount of geothermal water used and balance the demand for heating with the supply of geothermal brine. Since the geothermal fluid does not exit the heat exchanger, any dissolved gases and solids in the brine are reinjected and do not pose environmental problems. It is possible that because of the lower temperatures at the exit of the heat exchanger, some of the dissolved solids may come out of solution and deposit at the walls of the pipes. For this reason, the periodic cleaning of the main heat exchanger—e.g., during the summer season when heating is not needed—is highly recommended.

The use of geothermal water for district heating is an excellent example of how exergy and primary energy sources are saved by the use of renewable energy. Let us consider two alternatives for the heating of a group of buildings, which require a heating load of 10 MW

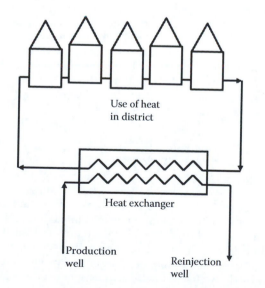

Use of heat
in district

Heat exchanger

Production
well

Reinjection
well

FIGURE 6.21
Schematic diagram of a district heating system. The geothermal water from the production well(s) is used to heat up houses in the entire district.

(thermal). The first provides heat by the combustion of a hydrocarbon fuel, such as methane, and the second involves the use of a low-temperature geothermal resource that provides geothermal brine at 90°C and is cooled in the heat exchanger of Figure 6.21 to 45°C.

If the heating load were met solely by the combustion of methane in conventional burners with 70% combustion efficiency and given that the heat of combustion of methane $-\Delta H$ is 50,020 kJ/kg, we would need 0.29 kg/s of methane to maintain the buildings at the desired temperature. This adds up to 1,028 kg/hour and 24,676 kg (37,832 m³ or 1,336,025 scf) of methane per day. Alternatively, if the geothermal brine is used for the heating of the buildings, from the expression for the heat rate

$$\dot{Q} = \dot{m}_w c_p (T_{in} - T_{out}),\tag{6.28}$$

with c_p = 4.184 kJ/(kg·K) for water and the 45 K temperature difference, it follows that a geothermal well with 53.1 kg/s supply of brine would accomplish this task. Such a well would save the burning of 24,676 kg of methane per day, a significant saving of a primary energy source. The CO_2 avoidance is 67,859 kg/day. Apparently, the substitution of fossil fuels with geothermal water for district heating saves precious natural resources (methane in this case) and prevents environmental pollution.

It must be noted that the hot water for district heating may be stored for several hours in insulated tanks without losing a great deal of its enthalpy. Because geothermal wells may be damaged with frequent adjustments to the mass flow rate produced, a geothermal district heating system typically stores a fraction of the water produced during the days, when temperatures are higher, and uses the stored water during the colder nights.

Example 6.10

A geothermal well supplies a group of 1,500 houses with 85°C water during the colder months of the year. The average heating demand for each house is 5 kW during the 8 hours of sunlight and 7 kW during the 16 hours when it is dark. The geothermal water is cooled to 45°C in the central heat exchanger that supplies clean hot water to the houses. (a) Determine the average mass flow rate the geothermal well must produce. (b) If 12% of the heat stored during the daylight hours is lost, how much mass flow rate is required and what is the amount of water that must be stored during the daylight hours?

Solution:

a. Without any heat losses from storage, the 1,500 houses demand 1,500 × (8 × 60 × 60 × 5 + 16 × 60 × 60 × 7) kJ = 821 × 10⁶ kJ/day (24 hours). The average heat rate demanded is $\dot{Q}_{av}$ = 1,500 × (8 × 5 + 16 × 7)/24 kW = 9,500 kW.

   This rate of heat is supplied with an average water mass flow rate: $\dot{m}_w$ = 9,500/[4.184 × (85 − 45)] = 56.8 kg/s. This average mass flow rate of geothermal water will supply the entire 821 × 10⁶ kJ that are demanded during the 24-hour period.

b. With the heat storage during the daylight hours and the 12% heat loss, the geothermal rate of water demanded is higher. If we denote the total heat stored during daylight as Q_{st} and the average rate of heat produced by the geothermal water as $\dot{Q}_{av}$, we derive the following expressions for the total energy demand during the daylight hours and during the nighttime:

$$Q_{day} = 1,500 \times 8 \times 60 \times 60 \times 5 \text{ kJ} = \dot{Q}_{av} \times 8 \times 60 \times 60 - Q_{st},$$
$$Q_{night} = 1,500 \times 1 \times 60 \times 60 \times 7 \text{ kJ} = \dot{Q}_{av} \times 16 \times 60 \times 60 + 0.88 Q_{st}.$$

This is a system of two linear equations with two unknowns, Q_{st} and $\dot{Q}_{av}$. Solving the system yields $Q_{st} = 60.1 \times 10^6$ kJ and $\dot{Q}_{av} = 9{,}587$ kW, slightly higher from the value in part a. Hence, the new geothermal water mass flow rate is $9{,}587/[4.184 \times (85 - 45)] = 57.3$ kg/s.

6.4.5 Environmental Impacts of Geothermal Energy

Geothermal energy is a relatively clean and environmentally friendly source of energy. The geothermal power plants that currently operate throughout the world cause minimal environmental impact. The main source of atmospheric pollution from geothermal units is the discharge of the noncondensable gases, primarily CO_2. This is the CO_2 that may be dissolved in the geothermal brine and is several orders of magnitude less than the mass of CO_2 produced by an equivalent fossil fuel plant. Geothermal power plants produce between 1,000 and 5,000 times less CO_2/kWh than fossil fuel plants [28]. Hydrogen sulfide H_2S is another gas that is sometimes released from the condenser of a geothermal power plant. At very small concentrations, less than 1 part per million, H_2S is malodorous (it smells like rotten eggs) and leaves a distinct odor near the geothermal units. At higher concentrations, H_2S desensitizes the olephatic (smelling) nerves and is not detected by its odor. Sulfur abatement methods may be used to eliminate the release of H_2S [28]. The Geysers field, north of San Francisco, California, uses such methods to eliminate a very high fraction of H_2S from its emissions and comply with the air standards of the state of California, which are among the strictest in the world. Other gases that are released by geothermal power plants, such as NH_3, are in traces and do not pose environmental problems.

Soil subsidence may become a problem in the vicinity of geothermal power plants. As steam or water is removed from the aquifers, the open cracks shrink and the permeability of the rocks and aquifers decreases. The surrounding rocks and soil are displaced to fill the voids, and the soil surface may subside, but not always in a uniform manner. Uneven soil subsidence on the surface may cause problems to nearby buildings. Water reinjection and rainwater seepage mitigates to a large degree the soil subsidence. Since most of the geothermal resources and most power plants are located far from population centers, soil subsidence does not pose a problem to large populations. A beneficial effect of the subterraneous fluid volume reduction is that mechanical stresses, caused by geological plate movements, are released. This stress release alleviates earthquakes or significantly mitigates their strength and their destructive effects on the structures and the population.

As with all the other thermal power plants, thermal pollution caused by the waste heat released by geothermal electric units, has an impact to the environment. Because, geothermal power plants operate at lower temperatures, their thermal efficiency (W/Q_{in}) is significantly lower than that of fossil fuel and nuclear power plants. As a result, the thermal pollution per unit energy produced (kJ of heat rejected per kWh produced) is higher—in the range of 1.7–2.5 times more than that of the fossil fuel plants.

Finally, noise, which is always a problem with all thermal power plants, is also an environmental concern for the geothermal units. Ejectors of noncondensable gases, especially steam ejectors, significantly add to the noise pollution of the locality. Since the vast majority of geothermal power plants are located at the sites of geothermal resources—in isolated and far from population center sites—noise pollution does not affect the human populations and does not pose a significant environmental concern.

6.5 Biomass Energy

The general term *biomass* encompasses all organic plant matter as well as organic waste derived from plants, humans, animals, and the aquatic life. The term is very broad and includes, among others, the following energy sources:

1. Fire wood
2. Methanol derived from wood
3. Alcohol made from grains or tree products
4. Methane derived from the anaerobic decomposition of waste
5. Aquatic plants, such as seaweed and kelp
6. Algae
7. Animal and municipal waste
8. Liquid fuels from the reprocessing of spent engine oil or kitchen oil

Wood was the predominant fuel at the beginning of the industrial revolution, when households used it for domestic heating and cooking. As the energy consumption rose and the population of the earth dramatically increased, wood combustion led to the partial deforestation of the land. In most nations, the use of wood for domestic energy needs was replaced first by coal, and then by oil, gas, or electricity. The few areas that did not secure other energy resources to satisfy the growing energy needs of their populations (several nations in the central Asian plateau are in this group) have suffered severe biomass depletion and deforestation, with adverse environmental, ecological, and microclimatic implications.

6.5.1 Heating Value of Biomass

Biomass is a natural method of solar energy storage in the form of chemical energy. Plants use the substance *chlorophyll* in their leaves as catalyst and large quantities of solar energy to convert atmospheric carbon dioxide and water into the complex molecules of glucose and fructose. The chemical reaction for the formation of glucose during photosynthesis is

$$6CO_2 + 6H_2O \rightarrow C_6H_{12}O_6 + 6O_2.$$

This reaction has several intermediate stages with chlorophyll as the catalyst. The reaction is highly endothermic with the chemical energy (ΔG^0) stored in the glucose molecules being 480,000 kJ/kmol (2,667 kJ/kg). From glucose and fructose, the plants form the more complex organic compounds, such as sucrose, starch, and other organic molecules with high chemical energy content. Plants form complex carbohydrate molecules by capturing atmospheric CO_2 from the atmosphere and releasing O_2 according to the following general chemical reaction:

$$xCO_2 + xH_2O \rightarrow C_xH_{2x}O_x + xO_2.$$

Two important conclusions from this reaction that help with the quantitative calculations are as follows:

1. For every 1 kg mass of CO_2 removed from the atmosphere, the plant combines with $18/44 = 0.409$ kg of water, forms carbohydrates with mass $(30/44) = 0.682$ kg, and rejects 0.727 kg of oxygen to the atmosphere.

2. Since grown trees are largely composed of carbohydrates, the amount of carbon "trapped" in them is $12/30 = 0.4$, or 40% of their mass. Thus, a large eucalyptus tree that weights 2 t contains approximately 800 kg of carbon. If all this carbon is absorbed from the atmosphere, the CO_2 removed is approximately 2930 kg.

Because the carbon in the plants is removed from the atmosphere, the combustion of plants and plant residues does not add to the atmospheric carbon dioxide, as the fossil fuels do. The combustion of biomass is neutral to the environment and does not add to the anthropogenic carbon footprint. Plants may be considered as engines that receive solar energy, convert it, and store it as chemical energy. Their conversion efficiency is in the range of 0.1–1.5%. While such values are significantly lower than that of PV solar cells and human-made power plants, the capital cost associated with this conversion and storage is much lower.

Biomass is considered a renewable energy source because the plants and biomass crops grow at shorter timescales in comparison to the human lives. As a renewable energy source, the time to regenerate the various forms of biomass becomes important. Table 6.10 includes the time, in years, for several forms of biomass to reproduce or regenerate [29]. One may conclude from the timescales of this table that while agricultural crops, municipal solid waste (MSW), and fast growing timber may be considered as renewable energy sources, natural forest trees and especially those grown in the northern latitudes do not regenerate fast enough and that their classification as renewable energy sources is questionable.

Fossil fuels, such as coal and petroleum, are also products of plants and animals, but they may not be considered as renewables, because the time to produce or to regenerate the fossil fuels is of the order of millions of years. This is by far longer than the human lifetime or planning timeframe, and fossil fuels are considered depletable and not renewable energy sources. Table 6.10 also implies that it is not prudent to excessively use timber from forests for the production of energy, especially in the northern zones. The excessive use of

TABLE 6.10

Time for the Generation of Biomass Forms in Years

Biomass Form	Time to Form or Regenerate, Years
Switch grass, corn and sugar cane	0.5
Fast-growing timber, sycamore	2–3
Forest/timber, southern temperate zone	25
Forest/timber, northern temperate zone	91
1 t MSW in US per person	1.2
1 t MSW in the EU per person	4.3

Source: Tester, J. W. et al., *Sustainable Energy*, MIT Press, Cambridge, MA, 2005.

TABLE 6.11

LHV or Energy Density of Biomass

Biomass	LHV, kJ/kg	LHV, Btu/lb
Rice hull, 0% moisture	13,905	5,970
Rice hull, 30% moisture	9,175	3,940
Bagasse, 0% moisture	17,266	7,414
Bagasse, 30% moisture	11,038	4,740
Dry wood (average)	18,500	7,934
Wood chips	13,600	5,860
MSW	4,000–8,000	1,715–3,430
Sewage/animal waste	1,164–1,863	500–800
Anthracite	31,952	13,720
Corn stover (dry)	17,470	7,502
Corn stover (30% moisture)	12,229	5,251

Source: El-Wakil, M. M., *Power Plant Technology*, McGraw-Hill, New York, 1984; Lizotte, P. L. et al., *Energies*, **8**, 4827–4838, 2015.

timber from forests causes deforestation, which is associated with adverse environmental and local climatic effects, such as lack of water, more frequent floods, and local warming. On the other hand, the high rate of utilization of municipal and agricultural waste is not only beneficial for the production of energy, but also offers an alternative pathway for the reduction of the volume of the waste and effectively removes the produced methane, which is a potent GHG.

Since biomass is used as fuel in combustion processes, the low heating values (LHVs) are useful figures of merit for the several types of biomass. Table 6.11 shows the LHV—sometimes called the energy density—of several common types of biomass [13,30]. The corresponding numbers for anthracite are also given for comparison.

As it becomes apparent from this table, the heating value of the biomass is affected by the moisture/water content. Water adds to the weight of the biomass, does not burn, and is released as vapor on combustion. Because water is trapped within the fibers of the biomass, even "naturally dried" biomass contains some moisture, typically 25–30% by weight. Figure 6.22 depicts the LHV of rice hulls and bagasse* as a function of their moisture content [3]. Moisture/water content significantly reduces the heating value of biomass, and water content more than about 80% renders the biomass useless. At such high water content, almost all the heat produced by the combustion of the organic materials is spent in vaporizing the water.

A glance at Tables 4.1 and 6.11 as well as at Figure 6.22 proves that the LHV (or energy density) of most biomass forms is significantly lower than that of the fossil fuels. The LHV of natural rice hull containing 30% moisture is 3.5 times less than that of anthracite and 5.5 times less than that of methane. Simply, the natural forms of biomass do not have the high-energy content of the fossil fuels. For this reason, unprocessed biomass is considered as low-quality fuel.

---

* Bagasse is the residual from sugar cane production, after the stalk is ground and pressed to release its recoverable sugar content.

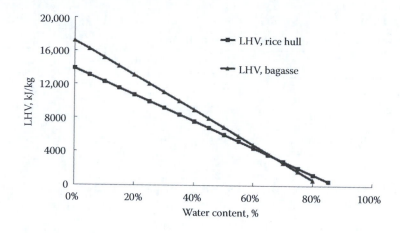

FIGURE 6.22
Heat content (LHV) of rice hulls and bagasse with moisture.

Example 6.11

It is proposed to construct a 150 MW power plant in an agricultural region that will be entirely fuelled by the waste products of corn stems and husks (stover), which contain 30% moisture. The estimated thermal efficiency of this power plant is 38%. (a) Determine the mass of stover required annually by this power plant. (b) If the production of stover is 9 t/ha (9,000 kg/10,000 m²) and there are two harvests per year, how much harvested area is necessary to produce the corn waste needed for this power plant?

Solution:

a. The rate of heat required by the proposed power plant is 150/0.38 = 395 MW. Therefore, during a year, this biomass power plant will need 395 × 60 × 60 × 24 × 365 = 12.46 × 10⁹ MJ or 12.46 × 10¹² kJ of heat. Since the heating value of corn stover is 12,229 kJ/kg (Table 6.11), this heat will be provided by 12.46 × 10¹²/12,229 = 1.02 × 10⁹ kg of corn stover.

b. With 9,000 kg produced by 10,000 m² and two harvests per year, the total area needed for this project is 10,000 × 1.02 × 10⁹/(2 × 9,000) = 566 × 10⁶ m² or 566 km² (23.8 × 23.8 km²).

It is observed in this example that the harvesting area required for this biomass power plant is enormous. With such a large area to produce the biomass, the transportation cost for bringing the fuel to the power plant is considerable, in both monetary and energy terms.

The potential of biomass to contribute to the world energy challenge may be measured in terms of annual production rates. The current production of renewable biomass includes agricultural byproducts, forestry byproducts, animal waste, and urban waste including sewage. These are the *recoverable biomass residuals*. For estimates of the full global potential of biomass, we may include *energy crops*, which are crops grown with the sole purpose to be used for energy production. Suitable energy crops include sorghum; energy cane; sugar cane; switchgrass; eucalyptus trees; miscanthus; giant reed; and *Leauceana lucacephala*, a small, fast-growing plant. Because part of the arable land on earth is not utilized, we may add to the total potential for biomass the energy crops to be produced by the unused land.

TABLE 6.12

Potential Annual Production of Biomass in Quads

Region/Country	Recoverable Residuals	Potential from Energy Crops	Total Potential
North America	5.6	33.0	38.6
Europe	3.6	10.8	14.4
Africa	2.5	49.5	52.0
China	3.2	15.5	18.7
Japan	0.3	0.9	1.2
Total earth[a]	29.6	253.1	282.7

[a] This is an estimate that does not take into account the availability of energy and water inputs for the growth of the crops.

Table 6.12 shows the annual energy equivalent of the recoverable residuals as well as the annual amount of energy that could be produced in the continents from land utilization for the production of energy [29,31]. The third column in the table *potential from energy crops* assumes that all arable areas, which are currently not used for the production of food, are fully utilized for the production of energy crops. The latter is an estimate based on land area and does not take into account the availability of water for the growth of energy crops.

It is observed that densely populated countries, such as Japan and the European Union (EU), have very small potential for energy production from biomass. Since the total global primary energy consumption in 2013 was approximately 537 Q [2], it appears that the extensive production and utilization of biomass has the potential to supply slightly more than 50% of the global primary energy needs. However, this is a misleading indication because of the following inherent assumptions in the table [3,29]:

1. The quantities of fresh water, which are required for the growth of the energy crops, may not be locally available.

2. The assumption that all the currently unused land may be harvested to continuously produce energy crops is doubtful. Some of the unused land is not arable, and the personnel to work on the farms in these regions may not be available.

3. The amount of energy, which is needed for the field preparation, seeding, fertilization, growth, harvesting, and processing of the energy crops, is not included in the entries of Table 6.12. As it will be seen in the next section, these energy needs are significant, and in the case of ethanol production from corn, the energy inputs amount to almost 100% of the LHV of the final fuel.

6.5.2 Biofuels: Ethanol Production from Corn

Biofuels are the liquid fuels derived from all forms of biomass including waste products. Of these, methanol (CH_3OH, often called *wood spirit*) has been produced for centuries from the distillation of wood products, natural gas, and coal. Ethanol fuel (C_2H_5OH or alcohol) is produced primarily from the fermentation of sugar cane. A mixture of 10% ethanol and 90% gasoline by volume, denoted as E10, is the *gasohol*, which is used as a transportation fuel in several countries including the United States. The E10 mixture may be used in all spark-ignition engines without any modification to the internal combustion (IC) engines. A mixture containing 20% ethanol, the E20, has been introduced in several US states and

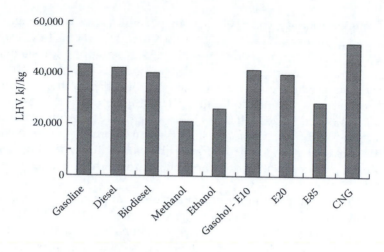

FIGURE 6.23
Heat content (LHV) of common fuels and biofuels.

countries, also without significant engine modification. Using a percentage of ethanol in the fuel mixture more than 25% requires modifications in the fuel supply system of the spark-ignition IC engines. A much richer in ethanol mixture, E85, is composed of 85% denatured* ethanol. This fuel has been used in Brazil since the 1990s and was introduced in the EU, primarily in Sweden. Only cars and trucks fitted with the so-called flex-fueled engines may use the E85 fuel. Because the LHV of ethanol is 60% of the LHV of gasoline, the addition of ethanol in gasoline reduces the LHV (specific energy) of the fuel mixture. Cars running on a gasoline–ethanol mixture have lower mileage than similar cars running on pure gasoline. Figure 6.23 depicts the LHV of several common biofuels as well as common liquid fossil fuels used in the transportation sector. The lower LHV of ethanol and its mixtures is apparent in this figure. Because the fuels E10, E20, and E85 have 10%, 20%, and 85% ethanol, respectively, the LHV of these fuels is proportionately lower than that of gasoline.

Example 6.12

A car has mileage of 35 miles per gallon (mpg) when it is fuelled with gasoline. What will be the mileage of this car when it is supplied with E10 and E15 fuel?

Solution: The LHV of gasoline is 43,000 kJ/kg, and the LHV of ethanol is 26,000 kJ/kg. Therefore, the LHV of E10 is $0.9 \times 43,000 + 0.1 \times 26,000 = 41,300$ kJ/kg, and the LHV of E85 is $0.85 \times 43,000 + 0.15 \times 26,000 = 40,450$ kJ/kg. The spark-ignition engine needs the same quantity of heat per mile, regardless of the fuel. Hence, the mileage with E10 will be $35 \times 41,300/43,000 = 33.6$ mpg, and the mileage with the E15 fuel will be $35 \times 40,450/43,000 = 32.9$ mpg.

While the addition of ethanol in petroleum products lowers the LHV of these products, an environmental benefit of using ethanol is that it is considered a cleaner fuel. Ethanol burns at lower temperatures and produces lower amounts of NO_x and almost zero SO_x.

---

* Denatured alcohol contains additives to make it undrinkable. If consumed by humans or animals, it is toxic.

Biodiesel is a generic name given to liquid transportation fuels with a mixture of heavier hydrocarbons and alcohols, such as $C_{10}H_{22}$, $C_5H_{11}OH$, and $C_8H_{17}OH$. Biodiesel is produced from oily seeds (primarily cotton seeds, olive oil residue, and pumpkin seeds), from industrial waste products or from spent cooking oil. The biodiesel has almost the same composition and very similar transport and combustion properties as the common diesel fuel. Hence, diesel engines may run on biodiesel without any modifications. Ironically, the first "diesel engine," produced by Rudolf Diesel was designed to run on what is now called "biodiesel," oil produced from vegetables. If this original practice were to continue today, there would not have been enough fuel for all the cars with diesel engines. Simply, the entire earth cannot produce enough vegetable oil to run all the cars and trucks with diesel engines.

Biofuels may also be directly derived from agricultural products—cotton, peanut, flax, soybean, sunflower, safflower, sesame, palm, jatropha, and Chinese tallow. These products grow in both warm and cold areas of the planet, which means that almost all the countries may produce one or more of these crops. However, because the crops have traditionally provided food for the local populations, their widespread use for the production of fuel may become problematic and lead to a rise in food prices and food shortages.

The conversion of biomass to usable transportation fuels may be very much energy consuming. How important are the energy inputs to this conversion is illustrated with the processes of ethanol production from corn, which are shown in the diagram of Figure 6.24. After all the processes for this conversion, one bushel (25.5 kg) of corn yields approximately 2.5 gal of ethanol, with an energy value 200,500 kJ (190,000 Btu). The figure also shows the processes where water, bacteria, yeast, heat (Q), and electric power (W) inputs are also required. In addition, the agricultural production of one bushel (25.5 kg) of corn entails energy-intensive processes: the plowing, seeding, water pumping, fertilizing, and harvesting of the corn fields; the transportation and drying of the agricultural products; and the distillation of the fermentation products consume high quantities of energy.

The net energy produced by the formation of ethanol has been the subject of several studies, which sometimes disagree on the analysis of the data [32–36]. Some studies include the energetic equivalent of the machinery replacement [32,33], while others neglect it. Several studies [34–36] also include the energetic value of the "coproducts," but invariably, the coproducts are wasted in practice and not used for energy production. Also, a few studies

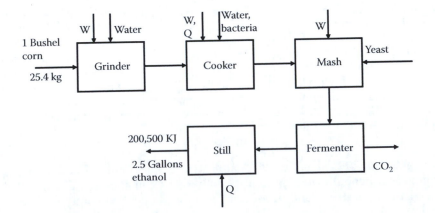

FIGURE 6.24
Processes for the production of ethanol from corn.

[36] use the higher heating value (HHV), instead of the LHV, for their conclusions. When the replacement of machinery is taken into account—and this is a method consistent with sustainability principles—all studies indicate that the total energy input for the production of ethanol exceeds the LHV of the product [32,33]. A study that ignored the energetic value of machinery replacement, but included the energetic value of the coproducts concluded that a very small (approximately 2% of the LHV) net energy generation is possible from the conversion of corn to ethanol fuel [35].

All the pertinent energetic studies agree that 1 ha (1 hectare is 10,000 m² or 2.47 acres) of land produces approximately 3176 L of ethanol. With the LHV value for ethanol at 21.1 MJ/L, 1 ha produces 67.0 GJ. The corresponding energy produced, when the HHV is used, is 74.0 GJ. We may construct two figures of merit (efficiencies) for the production of ethanol from corn using the ratios of the heating values of the fuel and the total energy spent for the production of ethanol, the *energy output-to-input ratio*. Table 6.13 summarizes the principal results of the studies [32–35] without taking into account the energetic value of coproducts that are invariably wasted. The values for the preprocessing include all the energy used for the production of the corn in the field. The values for the processing include the transportation energy for the corn and any other products.

A remarkable observation from Table 6.13 is that the results of three of the four studies indicate that the energy output-to-input ratio, based on the LHV, is less than 1, and the result of the fourth (1.02) is practically equal to 1. The energy output-to-input ratios, based on the HHV, are also very close to 1. This demonstrates that in the production of ethanol from corn (as well as from similar processes based on starch materials), we usually spend more energy than we actually obtain from the final product. After all the effort for the production and conversion of the agricultural product, at best, we get approximately the same amount of energy as the inputs of the crop growth and fuel conversion processes.

Corn, as well as other biomass crops, will not grow in arid regions. In addition to the energy input, the production of 1 L of ethanol necessitates the use of approximately 1000 L of freshwater, primarily for the growth of corn. One may conclude that the production of ethanol from corn for its use as a transportation fuel is highly questionable. Because this method of producing transportation fuels often consumes more energy than it delivers, it is not sustainable and should not be considered as a viable alternative to sustainable transportation fuel production.

TABLE 6.13

Total Energy Inputs (GJ/ha) and Energy Output-to-Input Ratio Based on the LHV of 67,000 MJ/ha and HHV of 74,000 MJ/ha

Study	Total Preprocessing	Total Processing	Total Energy Consumption	LHV Ratio	HHV Ratio
Pimentel, 2003 [32]	33.8	51.4	85.2	0.79	0.87
Patzek, 2004 [33]	29.2	51.4	80.6	0.83	0.92
Shapouri et al., 2002 [34]	22.3	46.7	69	0.97	1.07
Wang et al., 1997 [35]	19.8	45.9	65.7	1.02	1.13
Average	26.28	48.85	75.13	0.90	1.00
Percent standard deviation	24	6	12	12	12

Source: With kind permission from Springer Science+Business Media: *Alternative Energy Sources*, 2012, Michaelides, E. E.

While the production of ethanol from starch material is an unacceptable option for sustainable energy production, other methods for the production of ethanol, e.g., from cellulosic materials,* have more favorable energetic analysis and may be better alternatives in the production of biofuels [37]. However, even with cellulosic materials, the land and energy input for the large-scale production of ethanol are very significant. Small departures from the ideal (high efficiency) operation of the equipment and intermediate processes for the conversion to fuel ethanol may easily lead to net energy deficit.

6.5.3 Aquatic Biomass

The sea, the lakes, and the rivers cover almost 75% of the earth's surface. Plants that grow in aquatic areas include seaweed, kelp, and several types of algae. Because the amount of potable freshwater on the planet is only 0.3% of the total mass of water and because fresh water is becoming a scarce resource in several regions, it is not environmentally prudent to use large quantities of freshwater for energy plants or algae. The saltwater and brackish or marsh water areas, which comprise 97.3% of the total water on the planet, are readily available for the production of biomass and biofuels. Harvesting marine microorganisms for the production of energy is a way to produce liquid fuel for transportation or solid fuel for the production of electricity. The use of aquatic biomass has several advantages for the sustainable energy production over the land-grown biomass:

1. The aquatic biomass does not use land surface, which is a scarce resource primarily used for the production of food.
2. No irrigation is necessary for the production of aquatic biomass.
3. Nutrient trace elements, such as K and P, are abundant in the sea, and hence, it is not necessary to use fertilizers for the growth of biomass.
4. Typically, the food of aquatic bacteria and algae is sewage. This helps in the reduction of liquid waste; sewage reduction; and the possible elimination of liquid waste, which becomes "food" for energy-rich biomass.

Modern biology has produced a number of synthetic bacteria and has discovered algae, which thrive in aquatic environments and have high content of fatty oils. When these organisms are dried up, the fatty oils are processed to produce a biofuel, often called *aquatic biodiesel*. One of the advantages of these algae is that they may grow in very short times. For example, microalgae can regenerate in 48–72 hours and cyanobacteria can regenerate in 5–20 hours, in regions where the sunlight is plentiful. The short growth and harvesting times of algae produces more energy per unit area. The production of biodiesel from algae in gallons per hectare per year is 10–20 times higher than that of palm oil and jatropha; and it is 50 times higher than that of oilseed.

A significant technological problem with the use of aquatic biomass for the production of energy is the high water content or the produced biological organisms. Drying the aquatic biomass is necessary before it is processed to produce biofuels or to be burned in a boiler for the production of electricity. The high-energy requirement of the drying process often exceeds the energy obtained from the biomass itself. When sunshine is used for drying, the drying process is energy neutral, but solar drying is significantly slower and rather

---

* Cellulosic materials consist of the fibers that are commonly known as "wood." Their scientific name is lignocellulose, which is principally composed of cellulose, hemicellulose, and lignin.

unreliable, especially during the wet, rainy seasons. Because of this, the contribution of aquatic biomass to the global energy demand in the first two decades of the twenty-first century is insignificant. Technological breakthroughs in the drying of algae, the extraction of oils and lipids from the algae cells, and better-engineered biological organisms are needed for the aquatic biomass to make significant contributions to the supply of the global energy needs.

Whether aquatic or land produced, biomass for energy production has the inherent disadvantages of periodic production, uncertain yield and high volume per unit energy. There are one or two harvests of crops every year, primarily in the warmer seasons, and the yield of the land highly depends on the weather conditions of the growing season. Crops are produced during a 2–3 month period, at the end of which all the biomass material must be harvested and used. Similarly, and because of the seasonal variability of available insolation, there is significantly less alga production during the colder months. The volume of biomass is very high in comparison to fossil fuels (e.g., coal and petroleum products), and this makes the storage of biomass for long periods impractical. In addition, long-term storage will spoil the biomass and cause the production of methane, which is a potent GHG. If the biomass crops are to be used for the production of electricity in a power plant, another auxiliary supply of fuel would be needed for the continuous operation of the plant. The large-scale power source conversion from one fuel to another always causes operational problems to a power plant.

6.5.4 Environmental and Ecological Impacts of Biomass Use

Although the increased use of biomass has been advocated by a few as a viable method to reduce GHG emissions and national programs have been adopted for the widespread use of the E10 and higher ethanol-content fuels, the widespread use of energy crops worldwide is highly questionable when one considers the environmental and social effects of biomass utilization for energy production. The following examines the most important of environmental effects for the energy crops:

1. *Arable land use*: Arable land is a scarce resource on the planet. Agricultural crops and forests use a very large area and the development of energy crops is very demanding in the use of land. The widespread planting of switchgrass and eucalyptus trees for combustion in power plants, or corn and sugar cane for the production of ethanol fuel, entails the utilization of agricultural land, which is better to be used for the production of food. The use of agricultural waste—bagasse, tree clippings, rice husk, and wasted forestry products—does not use additional land and water resources and is a very welcome energy source. In particular, the cofiring of coal and agricultural waste saves a fraction of the coal currently used in power plants and prevent the associated carbon dioxide emissions [3].

2. *Freshwater requirements*: A significant environmental effect of the growth of biomass for energy is the very high requirements of freshwater for irrigation and processing. The production of a single gallon of ethanol from corn requires approximately 1000 gal of water (1000:1 volume ratio). If more of the available land were to be used for the growth of energy crops, this practice would require very high quantities of freshwater, which are not available everywhere. Unless the energy crops are planted close to large river systems with unused or underutilized agricultural land—e.g., the Amazon, the Mississippi, the Rhine, and a few Siberian rivers—the

competition for the existing water resources would result in significant strain to the local human and animal populations. Planting energy crops in the southwestern part of the United States is not feasible because of water scarcity. Planting additional crops to be used for energy production along the Rhine in Germany, the Nile in Africa, or the Mekong in Southeast Asia is not feasible, because most of the agricultural land along these river systems is currently utilized. Simply, the irrigation requirements of energy crops are very high and very few regions on the planet are suitable for the wider production of energy crops.

This does not apply to the use of the sea and seawater for the production of energy. There are vast areas of land close to the ocean, which are unsuitable for the production of food. A technological breakthrough (e.g., the discovery of plant species or organisms that thrive on seawater instead of freshwater) would allow the irrigation of crops with seawater and would make the production of energy crops in several coastal regions of the world feasible.

3. *Use of fertilizers and pesticides*: Energy crops are very fast-growing plants and need the input of large amounts of fertilizers, pesticides, and insecticide chemicals. These chemicals contain phosphorus, sulfur, nitrates, arsenic, and trace metals such as zinc, lead, and manganese. Many of these elements are toxic to humans and harmful to the environment. The widespread combustion of biomass is a pathway that introduces these toxic elements to the atmosphere and to the living organisms. In addition, the residue of all the chemical compounds in the fertilizers, pesticides, and insecticides invariably find their way to the hydrosphere—rivers, lakes, estuaries, and oceans—following the runoff waters from rainfall. The addition of these chemicals in the hydrosphere alters the chemical composition of the freshwater and seawater. This is especially significant for the fragile ecosystems in lakes, estuaries, and bays with low water flow or circulation. The periodically observed *hypoxia*—very low oxygen concentration in water that kills larger fish—in the northern Gulf of Mexico is caused by the high nutrient content in the waters of the Mississippi River, which carries fertilizers and other nutrients from 43% of the landmass in the United States. The water effluent locally alters the chemistry of the northern Gulf of Mexico, off the Louisiana and Texas coasts, and promotes the growth of smaller organisms that consume the dissolved oxygen. As a result, larger fish die because they do not have sufficient oxygen to thrive. Similarly, lake *eutrophication** in Europe and North America causes the depletion of oxygen by algae, and this endangers the existence of other more complex species, such as fish and shellfish. These are two examples of how the fertilizer effluent from the land disturbs the ecological balance in the sea and the freshwater lakes.

4. *Unintended production of methane and other GHGs*: If left stored and untreated for periods of a few weeks, biomass naturally decomposes and produces carbon dioxide, carbon monoxide, and methane, all potent GHGs. The anaerobic decomposition of biomass, e.g., when it is immersed in water and buried underground, always produces methane gas, which diffuses into the atmosphere and significantly contributes to global warming. If biomass is not promptly used and is left to decompose, either in the field or in storage facilities, it would cause significantly

---

* This is the uncontrolled growth of the algae population because of the high concentration of nutrients. The large population of algae consumes the dissolved oxygen at a fast rate and causes the death of fish and other large aquatic animals.

greater damage than the environmental benefit of removing some of the carbon dioxide from the atmosphere.

The Brazilian experience during the 2002–2008 period of uncontrolled and high demand for biomass products is a warning about what may happen to the land if economic incentives for the growth of biomass are offered. Because of the high demand for biofuels that are produced from sugar cane and the high price of sugar cane, Brazilian farmers converted large parcels of land along the River Amazon to biofuel-producing crops. In addition, they cleared very large tracts of tropical forest near the Amazon to establish farms for the production of sugar cane, corn, and other high energy-yielding crops. The result was that large trees were uprooted and were burned or, even worse, were left decomposing to produce methane. Bearing in mind that 30–40% of the mass of the tree is in the underground roots, this organized deforestation and conversion to arable land brought into the atmosphere more GHGs than what the subsequent biomass production removed.

5. *Other environmental effects*: Other less significant environmental effects of expanded biomass production include the following:

   a. Soil erosion and depletion of soil nutrients

   b. Loss of biodiversity

   c. Partial or total deforestation, which may lead to desertification and microclimate change

   d. Growth of monocultures, which are highly vulnerable to agricultural diseases and bacteria

   e. Higher river silt concentration and enhanced siltation rates that may lead to river, lake, or estuary eutrophication

   f. Changes in land use and irrigation patterns, which may change the microclimate of the region

   g. Dust production from plowing, harvesting, and transportation

Almost all the adverse environmental effects of extensive biomass production are associated with energy crops and not with the treatment of agricultural, human, and animal wastes. The expanded production of energy from any type of anthropogenic waste and the reduction of the volume of waste on a global scale has beneficial environmental and ecological effects, in addition to producing the needed energy.

6.5.5 Social, Economic, and Other Issues Related to Biomass

The widespread use of crops specifically grown for energy production is highly controversial. Unlike the other renewable energy sources, which do not have any other uses, biomass is intricately connected to food production. The excessive use of energy crops is associated with world poverty and malnutrition; high rate of freshwater use, needed by humans and animals; high rate of arable land use; and the deterioration of agricultural land. Increased reliance of the world population on biomass-derived energy will have adverse effects on the food and freshwater supplies, with all the economic and social consequences this would entail. For this reason, there are several key issues, which must be resolved, before national or regional policies are formulated that encourage and further promote the higher utilization of energy crops.

6.5.5.1 Food Production and Food Prices

The experience in the United States during the period of 2005–2008 underscores the food production and food price issue. During this period of increasing petroleum prices, and following the introduction of generous national and regional subsidies for the production of ethanol, a great deal of the domestic corn production was diverted to ethanol production. Simple demand and supply considerations dictate that the removal of corn from the marketplace for the production of ethanol will always increase its price as food. The immediate effect of the corn-for-ethanol diversion in the United States was a significant increase in the price of corn and similar products that are used as livestock feeds. As a result, the prices of milk and meat doubled in the period of 2006–2008 [38]. The prices of several other food items followed this trend.* In addition, because Mexican corn was imported at higher prices to satisfy the increased demand in the United States, the economic effect spilled into Mexico and resulted in the tripling of the price of corn tortillas, an indispensable foodstuff for the Mexican population. This led to considerable social unrest and obliged the Mexican government to impose price controls on tortillas.

Several countries, including most of east Asia and Africa, simply cannot afford to divert scarce agricultural land for energy crop production, because almost every parcel of land is currently needed to feed the regional populations. A wiser and much more effective use of land for energy would be the installation of PV cells to produce electricity. The produced energy may be stored in the form of hydrogen and then exported in exchange for foodstuff. This exchange may be called the *food-for-energy trade* and is based on the fact that PV energy conversion has much higher conversion efficiency (15–20%) of the solar energy than plants (0.1–1.5%).

One can foresee a situation in the future where countries such as the United States, Germany, France, and Brazil, with their vast agricultural and water resources, produce significantly more food than they do now. Countries in the Saharan West Africa (e. g., Chad, Mali, Niger, and Mauritania) and the Arabian Peninsula are located in the best regions for the production of solar power as may be seen in Figure 6.4. These countries have very small areas of arable land, but they may produce a great deal of electric energy from PVs, store it as hydrogen fuel, and transport it. The excess production of food in the first group of countries may be exchanged for the hydrogen fuel from the second group. Similarly, hydrogen fuel produced in the highlands of Ethiopia by the abundant solar energy may be traded for food grown in Mozambique or Tanzania. The *food for energy trade* would result in a much better use of arable and nonarable land areas than, simply, the growth of energy crops. Of course, this large-scale production of electricity and hydrogen would entail high capital cost and investment. This is an investment that will contribute in the future to the sustainable economic development of the region and the better employment of the local populations.

6.5.5.2 Food Scarcity

Related to the earlier statement, regional food scarcity may become consequences of the increased use of land for energy crops in regions where the food supply is not plentiful. The production of ethanol in the United States and Brazil tripled from 4.9 billion gal to almost 15.9 billion gal between 2001 and 2008. The rapid expansion of ethanol production during

---

* Other factors, such as higher energy and fertilizer costs as well as commodity speculation, contributed to food price increases. The use of corn for the production of ethanol was the primary variable that caused the rapid increase of food prices.

this period was accompanied by a significant rise in food commodity prices worldwide [3,38,39]. Even though the conclusions of economic studies on the effect of corn-for-ethanol production on the globally increased food prices significantly vary, all the pertinent studies attribute 15–60% of the globally increased food prices to the production of ethanol and other biofuels from agricultural products that could have been used for food [39]. The diversion of agricultural products for the production of energy drives the prices of food products higher worldwide.

Moral concerns about food availability, especially in the famine-prone regions of the world, have the potential to stifle any use of food products for energy and to cause the removal of all food-production subsidies [38]. The popular press has widely reported in 2009 that 1 billion people on earth—one out of seven inhabitants—suffer from malnutrition and that in every 6 s, one child dies of malnutrition or outright starvation.* The rising global food prices in 2001–2009 have affected citizens living at or near the poverty level even in the wealthier OECD nations. Under this light, the increased use of subsidized corn and other agricultural products to feed the engines of Hummers, BMWs, Cadillacs, and sport utility vehicle (SUVs) is not ethical and not socially justifiable. The United Nations has issued a warning against the use of food crops for energy, and several national governments stopped the expansion of energy crop programs. An international policy along the line of *"do not burn what you can eat"* may justifiably become the cornerstone of the future of biomass use for energy.

6.5.5.3 Economic Subsidies

Food production and several agricultural products enjoy generous subsidies from national and regional governments. Farming subsidies are among the few items that are exempted from the international treaties brokered by the World Trade Organization. Frequently, these subsidies are in the form of tax credits. For example, in the United States, there is a federal tax credit of $0.51/gal of ethanol that is used as a transportation fuel. Because biomass is essentially a farming product, it enjoys the same generous subsidies in most countries including tax credits, direct grants, low-interest loans for its production and processing, exemption from taxation of the fuels used for its production, high depreciation rates, and even tax credits for the equipment used for its production and processing.

In the late 1990s, the real cost for the production of 1 gal of ethanol in the United States and Europe was in the range of $8–$11. Yet, because of the several subsidies, it became economically justifiable for corporations to mix ethanol with gasoline and sell it at significantly lower prices, approximately $1 in the United States. While this practice is a prime example of economic inefficiency, several of the agricultural subsidies may be socially justified when directed to small farmers, who (in most countries) live close to poverty levels and produce other staple foods for the benefit of the society at large. However, since the late 1990s, large enterprises and several multinational conglomerates entered the business of biomass production for conversion to transportation fuels. They enjoy the generous subsidies that are intended for poorer farmers and have derived high profits from the subsidized biomass production. In the case of large corporations, the social, economic, and national interests in continuing and extending the agricultural subsidies intended for small farmers cannot be justified, and the practice of such subsidies is controversial.

Current and future subsidies for energy crops must be justified on social, economic, and environmental grounds: While economic subsidies to small farmers for the production of

---

* Associated Press, October 15, 2009.

corn, wheat, and sunflower seeds for food are justified, subsidies to corporations for the large-scale production of these staples for their conversion to biofuels are highly questionable. The regulatory institutions that govern the production of biomass should ensure that any economic subsidies benefit the small farmers and are not used to enrich large profitable corporations. They should also ensure that the diversion of foodstuff crops to produce energy does not make food prices unaffordable for the population and does not cause environmental deterioration.

6.5.5.4 Global Poverty Levels

Some of the concerns related to world poverty over the price of food are partially offset by a worldwide increase in the agricultural incomes in the first decade of the twenty-first century. Because a large percentage of the population living in poverty levels derives its income from agricultural employment, the increase of their incomes has been a welcome effect of biomass production and a rationale for several national and regional governments to continue with the agricultural subsidies. However, while considering the economic subsidies for the production of biomass, governments must ensure that large for-profit corporations do not take unfair advantage of subsidies intended for the poorer farmers.

6.5.5.5 Stability of Energy Prices

The interest in biomass and biofuels is not new. The *energy crisis* of the 1970s and the 1973 oil embargo in Europe and North America caused a dramatic rise in the production of biomass, especially biofuels, that continued in the early 1980s. Interest and investment in the conversion of biomass waned in the mid-1980s, when energy prices dropped globally. Several biomass conversion plans were abandoned at that time, and a great deal of investment capital was lost.

It has become apparent that high global energy prices are needed to stimulate and maintain the utilization of biomass for energy. The Organization of the Petroleum Exporting Countries production trends and the resumption of the higher oil demand in Asia in the twenty-first century are indications that the energy prices will be maintained at relatively high levels in the near future. This also indicates that biomass utilization, especially biomass derived from waste products, will increase.

6.5.5.6 GHG Policies and Regulations

GHG emission concerns are the primary reason for the increased use of all renewable energy sources, including biomass. Well-intentioned but injudicious regional and national regulations may require a higher use of biomass or biofuels. For example, the regulatory change from the E10 fuel (gasohol), which is used in most of the states in the United States, to E20 has the potential to double the demand for ethanol. A provision was included in the Energy Policy Act of 2005, which arbitrarily mandated that 7.5 billion gal of "renewable fuels" be used in gasoline by 2012. Similar regulations in the EU and other nations helped to artificially increase the production of ethanol and other biofuels. Given the high energy consumption and the unfavorable energy balance in the production of ethanol from starch materials (Section 6.5.2), such regulations will not always achieve their goal for the reduction of GHG emissions, unless it is also stipulated in the regulatory or legislative framework that ethanol and biofuels may not be derived via processes that consume high-energy quantities—e.g., starch and foodstuff—but from low-energy sources without other uses—e.g., from waste products and cellulosic plant materials [37].

6.5.5.7 Technological Advances

The socially responsible global use of biomass will benefit from technological advances, such as the following:

1. Bioengineering of algae and bacteria with higher fat content
2. Faster and more efficient drying of the aquatic biomass
3. Higher crop yields in the currently unused fallow land
4. Energy crops that may grow with saltwater
5. Ethanol from cellulosic processes
6. Catalytic pyrolysis of biomass at lower temperatures
7. Bacteria that process solid and liquid waste at a faster rate

6.5.5.8 Global and Regional Climate Change

It is rather ironic that climate change, the very cause for the increased use of biomass, will have an impact on the production of crops, biomass, and biofuels. Potential regional climatic changes are a key issue in the crop yield of the land and the production of biomass. Warmer weather would increase the crop yield and the available amount of biomass, if it were accompanied by sufficient rainfall. On the other hand, drier weather and lack of water regionally may spell the end of the use of biomass for energy in several regions, because sufficient food supply will always have priority over energy supply.

6.6 Sea/Ocean Energy

More than 70% of the earth's surface is covered by sea. The ocean currents, the energy of sea waves, the tides, and the temperature difference between cold and warmer waters may be harnessed for the production of power. These energy sources are diffuse and require the construction of large installations to produce power, and the power produced is at present very expensive. Because the sea, including bays and estuaries, is a hostile environment for electric power generation units, pertinent power installations are often subjected to destructive natural forces—storms, high winds, high waves, hurricanes, cyclones—that cause a great deal of damage. For this reason, there have been very few successfully operating facilities that utilize the energy of the seas. The following sections give a short description of the several forms of sea/ocean energy and their potential to contribute to the energy challenge of the world.

6.6.1 Ocean Currents

Permanent ocean currents flow in several parts of the ocean. They are generated by a combination of the Coriolis force, the predominant surface wind, temperature gradients on the surface of the sea, salinity gradients, and the predominant regional tides. Among the well-known ocean currents are the Gulf Stream, which passes between Florida and Cuba, hugs the southern coast of the United States, and dissipates in Europe, and the Kuroshio current that hugs the coast of the Philippines and brings warmer water to the shores of Japan.

An ocean current is in all respects similar to a wind current. As with the wind power, the available power of ocean currents is given by the expression

$$\dot{W}_{av} = \frac{1}{2} A \rho V^3 = \frac{\pi}{8} D^2 \rho V^3, \tag{6.29}$$

where V is the velocity of the sea current; D is the diameter of the turbine; and ρ is the seawater density. A significant difference between wind and ocean currents is that the density of the seawater is 1,025 kg/m³, more than 800 times higher than that of air. While this is a favorable characteristic for the available power, the high water density also causes higher drag on the blades, and this precludes the construction of turbines with very large diameters. The actual power produced by the turbine is the product of the available power and the efficiency of the turbine.

$$\dot{W} = \eta_T \dot{W}_{av} = \frac{1}{2} \eta_T A \rho V^3 = \frac{\pi}{8} \eta_T D^2 \rho V^3. \tag{6.30}$$

If suitably utilized, ocean currents may produce sufficient power to offset the demand of coastal communities. The following example gives an indication of the magnitude of power that may be produced by the ocean currents.

Example 6.13

The Gulf Stream flows from Florida to the English Channel (La Manche) within an area of 100 km (62 mi) wide and 800 m (2,600 ft) to 1,200 m (3,900 ft) deep. The average velocity of the water is 2 m/s. Calculate (a) the available power of the entire Gulf Stream and (b) the electric power that may be produced by 100 turbines, with diameter of 6 m and efficiency of 28%, anchored inside the Gulf Stream.

Solution:

a. The average depth of the Gulf Stream is approximately ½(800 + 1,200) m = 1000 m. Since the width of the stream is 100,000 m, the cross-sectional area of the Gulf Stream is approximately $A = 100 \times 10^6$ m². Since the density of the seawater is $\rho = 1,025$ kg/m³, and the velocity V is 2 m/s, Equation 6.29 gives for the available power for the entire Gulf Stream:

$$\dot{W}_{av} = \frac{1}{2} A \rho V^3 = \frac{1}{2} 100 \times 10^6 \times 1,025 \times 8 = 410,000 \text{ MW}$$

b. For a turbine with $D = 6$ m and $\eta = 0.28$, Equation 6.30 yields

$$\dot{W} = \frac{\pi}{8} \eta_T D^2 \rho V^3 = \frac{3.14}{8} \times 0.28 \times 36 \times 1,025 \times 8 = 32,442 \text{ W}$$

Therefore the 100 ocean turbines will produce approximately 3,244 kW. This is not a very significant amount of power, but may fulfill the demand of a small coastal town.

One may also note in example 6.13 that if the diameter of the turbine blades were 10 m (5 m long blades), the power produced would almost triple to 9,012 kW. It is apparent that tapping even a small fraction of the enormous potential of the Gulf Stream would provide a significant amount of electric energy and would replace the use of fossil fuels. An advantage of the ocean current energy is that because ocean currents are continuous and their velocity is almost steady, the power produced by this form of renewable energy is not intermittent or predictably variable. Ocean current turbines may be used for the continuous and dispatchable production of electric power.

While the use of ocean currents for the production of electricity is feasible and may prove to have significant payoffs, an ocean current power plant has not been built yet, even at the pilot scale. The principal impediments for such a project are as follows:

1. Ocean storms and high-velocity transients that damage underwater installations.
2. The electricity must be transmitted onshore and fed into the national grid. High-voltage electric lines seldom pass near the coasts.
3. Lack of experience and appropriate research on underwater turbines and large-scale underwater electricity generating systems.
4. Most strong water currents are in international waters. Lack of international treaties and governing laws increase the uncertainty and risk of investment.
5. Sabotage and terrorism concerns for systems that, by their very nature, are inherently built a large distance offshore.
6. Ecological effects related to the reproduction of fish, fish migration patterns, and other effects on marine life.

6.6.2 Wave Energy

Ocean waves are directly caused by wind shear and, indirectly, by solar energy. Because the oceans cover more than 70% of the earth's surface, wave power is abundant, and if harnessed on a large scale, it has the potential to produce a great deal of electric power. However, it is unrealistic to think that all the wave power on the sea may be utilized for the production of energy that is transmitted onshore. The machinery for the utilization of ocean waves and the transmission of high amounts of electric power may be built and maintained close to the shores only. Hence, only the wave power in narrow coastal strips is feasible to be harnessed.

Waves are formed on the surface of the ocean. They are characterized by the amplitude α, the wavelength λ, the phase velocity c, and their frequency f. The last two parameters are related to the wavelength by [3,13]

$$c = \sqrt{\frac{g\lambda}{2\pi}} \quad \text{and} \quad f = \sqrt{\frac{g}{2\pi\lambda}}, \tag{6.31}$$

where g is the gravitational acceleration, 9.81 m/s².

The total power that may be harnessed from the waves is

$$\dot{W} = \eta_E \frac{\rho L \alpha^2 g^2}{4\pi f}, \tag{6.32}$$

where η_E is the efficiency of the engine used for the harnessing of the waves; ρ is the density of saltwater, approximately 1,025 kg/m³; L is the length of the machinery that utilizes the waves; α is the amplitude (height) of the waves; and f is the wave frequency.

Example 6.14

A startup corporation claims to have developed an engine that produces 5 MW of power from waves of 3 m amplitude and 2 s⁻¹ frequency. The corporation claims that the engine is no longer than 100 m and that it will cost less than $320,000 to manufacture and install. Evaluate its claims.

Solution: From Equation 6.32, a wave engine that is 100% efficient, with the specifications given in this example would produce

$$\dot{W} = \frac{\rho L \alpha^2 g^2}{4\pi f} = \frac{1,025 \times 100 \times 3^2 \times 9.81^2}{4\pi \times 2} = 3.53 \text{ MW of power.}$$

This is significantly less than the claim of 5 MW, and therefore, the technical claim is impossible.

The economic part of the claim does not matter. Such an engine does not exist!

Since the 1960s, there have been several experimental engines constructed for the harnessing of wave power, but not a commercial plant that would transmit significant power to the electricity grid. Among the more successful engines are as follows [3]:

1. The *salter duck*, which is made of a central cylindrical spine, with a buoyant cam. The cam responds to the motion of the incident waves, absorbs the energy of the incoming waves, and converts it to circular motion that drives an electric generator.

2. The inverted cylinder and piston assembly where the wave action moves a buoyant piston up and down in a vertical motion. The piston movement is used to compress air that is directed to a duct and an onshore turbine, which produces electricity.

3. The splashing effect of the waves on shores raises the water level and floods narrow parts of the coastline. Splashing converts the vertical motion of the waves to a horizontal motion of the water on the shore. The horizontal water movement is converted to power by a waterwheel or a low-head water turbine. The velocity of the water may be amplified by the use of a tapered channel (TAPCHAN). The TAPCHAN (from "tapered channel") system that was developed in Norway causes a 3–5 m elevation of the water level onshore and sufficient horizontal velocity to operate a small Kaplan turbine.

4. The attenuating wave–energy converter. This engine, developed by the now defunct corporation *Pelamis*, responds to the shape, not the amplitude, of the waves and pumps pressurized oil to a hydraulic turbine. A few prototypes of the "Pelamis engine" have been tested, and one of them was the first to supply small amounts of electricity to the national grid, in Scotland. The bankruptcy of the corporation in 2014 has put in jeopardy the further development of this engine.

5. The *dam–atoll* system, named from the combination of the actions of dams and atolls (small islands of volcanic origin in the Pacific Ocean), uses the wave

movement to create a vortex with spiral motion of high angular momentum. The water is directed to the center of the dam–atoll system where its spiral motion drives a turbine, located at the center of the system [13]. This system has not been tried in practice, not even at the pilot scale.

In the early part of the twenty-first century, there are very few wave power engines in operation, almost all of them supported by government grants or subsidies. The economical and purely commercial development of wave energy engines appears to be a difficult task because of the following reasons:

1. Wave energy systems must operate and maintain their structural integrity under heavy and low seas, under storms, and calm weather. This imposes a constraint on the size, the strength of the materials, and the design of the energy conversion engines, which must withstand a very high range of forces. In addition, the corrosive saltwater of the sea makes the use of more expensive corrosive-resistant materials necessary. These factors add to the manufacturing cost of the wave engines.
2. Because the wave power density is low (e.g., in kW/m^2) and the conversion efficiency of the engines is low, wave energy conversion systems are massive and require large quantities of materials for their construction. This adds to their manufacturing and installation cost.
3. The power produced is highly intermittent and unpredictable.
4. The electricity-generating engines are in a hostile environment and are frequently damaged by storms.
5. Anchoring systems are not fail-safe. Anchor failure, especially during storms, causes the drift of the power system, possible damage to the system, and likely collisions and further damage to boats and ships in the area.

Wave energy is a form of energy, which is abundant in nature. Its harnessing does not have a significant environmental impact. Among the environmental and ecological problems presented by this form of energy are as follows:

1. Most wave power systems are lengthy and close to the coastline. Oftentimes, they obstruct the navigation of ships.
2. Wave power systems include moving parts, which would kill fish and other sea animals if they were trapped in the systems.
3. Wave power systems generate noise that disrupts the aquatic life.

6.6.3 Tidal Energy

Tides are generated because of the gravitational and kinematic effects due to the relative positions of the earth, the moon, and the sun as well as of the earth's rotation. The moon, even though it has much smaller mass than the sun, plays a more important role in the development of tides because it is closer than the sun. The mass of the moon "pulls" the ocean water masses in its direction and creates a "bulge" and a moving wave on the surface of the ocean. The effect of the moon's pull is modified by the effect of the sun's pull, as shown in Figure 6.25. When the earth, the moon, and the sun are aligned—a position known as *syzygy*—the tides are amplified, and they are called *spring tides*. When the three

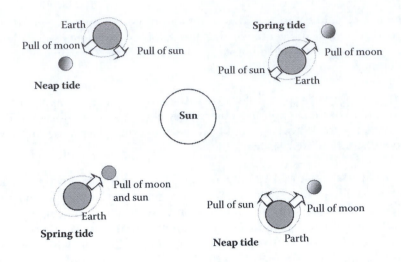

FIGURE 6.25
The positions of earth–moon–sun system for the development of spring tides and neap tides.

celestial bodies form a right-angle triangle, the tidal effect is modulated, and we have the *neap tides.* Because the tides depend on the relative position of the earth–moon–sun masses, tides occur periodically. During every moon cycle, which lasts for approximately 29.5 days, there are two spring tides and two neap tides. Between these high and low amplitude events, there are several "intermediate range" tides, which periodically occur because of the earth's rotation around its axis. These intermediate tides occur at approximately every 12.5 hours (45,000 s).

Although the globally averaged amplitude—called the "range"—of the tidal wave is less than 1 m, continental shelves, variable ocean depth, and coastline irregularities combine to cause distortions and resonances in several locations. These local characteristics create tidal waves with very high ranges, close to 10 m in some locations. Such tidal waves are generated in bays, gulfs, and estuaries and may be utilized to produce power.

A simple engineering system for utilizing tidal power is the *tidal barrage* or *single-pool tidal system*, schematically shown in Figure 6.26. The barrage is essentially a dam that separates the ocean from the basin, typically a gulf or an estuary. The function of the barrage is to allow for a height difference H to be temporarily created between the ocean and the basin. When H approaches its maximum value—the *range* of the site—several gates that

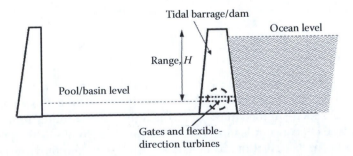

FIGURE 6.26
Schematic diagram of a single-pool tidal system for the production of power. The turbine generator systems produce power when the pool fills and when it empties.

are placed close to the bottom of the barrage open and direct the water to hydraulic turbines and generators that produce power. The turbines are of the *flexible-direction* type, designed to produce power when the water enters the basin and to return to the ocean side when the water leaves the basin. The total amount of energy produced during the filling of the basin from level 0 to H is

$$W = \frac{1}{2} A\eta\rho g H^2,$$ (6.33)

where η is the efficiency of the turbine; A is the cross sectional area of the pool; and ρ is the water density, approximately 1,025 kg/m³. At the end of the filling process, the basin/pool fills with water and the water level in the basin is approximately H. At this point, the gates of the barrage close and the plant does not produce power until the level of the ocean drops (low tide). At low tide, the level inside the pool is approximately at H, and that of the ocean, at 0. The gates open again and the flexible-direction turbines produce power for the second time within this half of the tidal wave cycle. The single-pool system produces twice the amount of energy of Equation 6.33 during each half of the tidal cycle, which lasts for approximately $T \approx 12.5$ hours (45,000 s). The average power produced during this period is

$$\dot{W}_{av} = \frac{A\eta\rho g H^2}{T},$$ (6.34)

where $T \approx 45,000$ s.

The single-pool system does not continuously produce power. Rather, its operation produces the potential energy calculated in Equation 6.33 in two "bursts," during the filling and during the emptying processes. The bursts last approximately for $T = 1$–2 hours (3,600–7,000 s), not the 12.5 hours of the tidal wave. Therefore, the actual power produced during the shorter duration of the operation of the power plant is significantly higher than that calculated from Equation 6.34. For example, if the discharge occurs within 1 hour (3,600 s), the actual power would be 12.5 times higher than the average power indicated by Equation 6.34. This type of periodically variable operation of a large power system entails several disadvantages:

1. The high power is produced during short periods, when the regional demand for power may not exist. For example, the power may be produced in the early morning hours, when electricity demand is at a minimum. In the absence of high electricity demand, the power level of other facilities will have to be reduced to accommodate the tidal power. This may involve significant and undesirable power fluctuations with the other power plants connected to the electric grid and, oftentimes, grid instability.

2. The system must employ high-power turbines to accommodate the bursts of power produced. The high-power turbines and generators are more expensive and add to the capital cost of the power plant and the cost of the electric energy produced by the tidal plant.

3. Switching the turbines, exciters, and generators on and off increases the wear of the equipment and shortens their useful lives.

The *two-pool tidal system* [3] has been conceptually developed to mitigate these effects: this system uses two basins instead of one, which are separated by a barrage. Water is exchanged between the two pools to lengthen the duration of the power producing process. The operation of the two-pool system may be arranged and optimized to achieve one of the following three objectives:

1. To produce the maximum amount of energy during a tidal cycle
2. To produce maximum power during periods of peak demand
3. To produce a fraction of the available power and store this energy in the tidal system to be used during peak demand

There are two medium-sized commercial tidal energy power plants in operation: The power plant at La Rance, France (near the town of Saint Malo), which has been operating since 1966 with capacity of 240 MW, and the one in Sihwa Lake, South Korea, which was commissioned in 2011 with capacity of 250 MW. Figure 6.27 shows an aerial view of the barrage and the surroundings of the tidal unit at La Rance. Smaller, mostly experimental units are in operation in several other countries. All these are of the single-pool type. Table 6.14 includes the locations on the planet, where tides with high ranges occur as well as the average power and total electric energy that may be produced annually [40]. It is apparent that only a very small fraction of the available tidal power is currently utilized, and this is primarily due to the construction cost of tidal systems.

FIGURE 6.27
Aerial view of the barrage and the tidal system in La Rance, France. (Courtesy of "Tswgb.")

TABLE 6.14

Locations with High Potential for Tidal Energy Conversion

Location	Country	Range, m	Potential Average Power,[a] MW	Annual Energy, GWh
Cobequid	Canada	10.7	20,000	175,200
Passamaquoddy	Canada	5.5	1,800	15,768
Severn River	England	9.8	1,680	14,717
Mont St. Michel	France	8.4	9,700	84,972
Kimberlay	Australia	6.4	630	5,519
Mesen (Mezen)	Russia	6.6	14,000	122,640
Khambat	India	7.0	7,000	61,320
Golfo Nuevo	Argentina	3.7	6,600	57,816

[a] Based on calculations of the one-pool system and Equation 6.34.

Example 6.15

A single-pool tidal water system is to be constructed in Khambat, India. It is desired that the system produces 120 MW average power to supply four nearby cities. It is anticipated that the turbo generators to be used for this project will have 78% overall efficiency. What is the area that needs to be enclosed for such an installation?

Solution: The expression for the average power may be rewritten as follows to yield the required area:

$$\dot{W}_{av} = \frac{A\eta\rho g H^2}{T} \Rightarrow A = \frac{\dot{W}_{av} T}{\eta\rho g H^2} = \frac{120 \times 10^6 \times 45,000}{0.78 \times 1,025 \times 9.81 \times 7^2} = 14.1 \times 10^6 \ m^2 = 14.1 \ km^2$$

It is observed that the enclosed area is very large and that the tidal basin will enclose more than 14 km². Tidal energy has very low energy density, and its area "footprint" can be very large.

Tidal power is renewable and clean. There are a few environmental effects of tidal systems, which are enumerated in the following:

1. Substantial construction is required for the barrage system, which uses a great deal of materials, especially cement.
2. Barrages obstruct the free passages of fish and other marine life.
3. Tidal power systems interfere with fish spawning and have an adverse effect on the local populations of the fish and other aquatic life.
4. Because of the large water flows involved, there are fluctuations in the turbidity (dissolved solid material) and chemical composition of the water. This may have adverse effects on the aquatic life of bays and estuaries.
5. Barrages affect the local navigation of boats and recreational watercraft.
6. Additional sedimentation in the artificial pools makes the frequent dredging and disposal of silt, which may be contaminated, necessary.

The careful construction of a tidal system may also serve multiple purposes to add to the quality of life of the surrounding communities. For example,

1. The upper surface of a barrage that connects two sides of an estuary may be paved and used as a road that connects communities.
2. Recreational activities—sailing, boating, and swimming—may be promoted within the enclosed estuary.
3. The barrage provides flood control to the coastal communities during severe storms and mitigates coastal erosion.

6.6.4 Ocean–Freshwater Salinity Gradient

When a membrane is exposed to saltwater on one side and freshwater on the other, the freshwater will flow to the other side until the pressure there becomes equal to the osmotic pressure. This creates a pressure differential—height differential—between the two sides of the membrane, which may be utilized to produce power. A prototype osmotic power plant was constructed in 2009 at the island of Tofte, Norway, by Statkraft, the Norwegian state energy corporation. The osmotic power plant was decommissioned in late 2013, and a few months later, Statkraft discontinued its investment in this energy source.

The process used in the Statkraft osmotic plant involved seawater that was pumped at 60–85% of the osmotic pressure against one side of a group of semipermeable membranes. The other side of the membranes was exposed to freshwater. Since the freshwater is compelled by osmosis to traverse the membranes and dilute the saltwater, the volume on the saltwater side increased and so did the static pressure in the chamber that held the saltwater. The increased pressure in the saltwater chamber was utilized to power a small turbine. The system produced 2–4 kW of electric power, a very low power production.

While a small amount of power may be produced from salinity gradients, and the technology for low-power production was demonstrated by the Statkraft prototype, the power produced is very low. Osmotic pressure is very diffuse and does not have the energy density to produce the high amounts of electric power demanded by our society and, thus, substantially contribute to the global energy challenge.

6.6.5 Ocean–Thermal Energy Conversion

The concept of OTEC is almost as old as the theory of heat engines and is included in most elementary thermodynamics textbooks as a demonstration of the consequences of the second law. In the tropical latitudes, the water at the surface of the sea may reach temperatures in the range of 27–32°C. Because the sunlight does not penetrate at depths below 30 m, water in the bottom of deep bays is colder, in the range of 5–8°C. Such conditions provide the two heat reservoirs that are necessary for the operation of the elementary heat engine, depicted in Figure 1.5: the warmer seawater at the surface and the cold water at the bottom become the two heat reservoirs that are necessary for the operation of a heat engine. All that is needed is the engineering system that would make this heat engine operational.

Because of the low-temperature differences, the efficiency of the OTEC heat engine would be very small. Even if we take the extreme temperatures of the last paragraph, 32°C and 5°C (305 and 278 K), the Carnot (maximum) efficiency of the OTEC cycle would be less than 9%, while the Carnot efficiencies of gas and steam cycles are in the range of 60–80%. The inherently low Carnot efficiency of the OTEC cycle implies that the efficiency of actual

OTEC systems would be even lower. The expected range of actual OTEC power plants is in the range 1–2%. However, since the heat extracted from the warm water of the oceans is almost inexhaustible and free, successfully designed OTEC power systems may become capable of providing inexpensive electric power in an environmentally friendly way.

Two practical systems have been designed to be used with OTEC resources:

1. The *Claude cycle* was developed by Georges Claude, who, in 1929, built an OTEC plant in Matanzas Bay, Cuba. Seawater in the Claude cycle is flashed to a lower pressure and produces a small amount of steam, which is directed to a low-pressure steam turbine and produces electric power. The power production system of the Claude cycle is similar to the *single-flash* units of geothermal power plants.

2. An ORC similar to the one used in the binary geothermal units. The organic fluid evaporates in a large heat exchanger, located at the warmer surface waters, and condenses in the colder waters, which are pumped from the bottom of the sea. While the ORC is simple in its concept, when the system is applied to the lower seawater temperatures, it requires the passage of very large volumes of seawater and the use of two large heat exchangers operating. Pumping the water must be done with significant power loss (parasitic loss) that reduces the net power produced by the unit. In addition, the heat exchangers are subjected to biological fouling because they handle natural seawater and not demineralized water. This causes the buildup of fouling materials in the heat exchanger equipment, a highly detrimental effect for the net power produced by the system.

Because of the vast amounts of available seawater, in principle, OTEC units may provide almost unlimited power [41]. At various times, several OTEC systems have been proposed or have been in the design stages. However, practical OTEC systems have not been successful in the past and typically closed after 1–2 years from the commencement of operations. In 2017, there are only two small OTEC pilot plants: the first on Kume Island, Okinawa, Japan, with 50 kW capacity, and the second in Hawaii with nominal capacity 105 kW.

OTEC is another form of clean renewable energy, which does not cause atmospheric pollution. There are very low-impact and rather insignificant environmental and ecological effects associated with the electricity production from OTEC: the local disturbance of aquatic life and minor impediments in the local ship navigation. Because the warm OTEC resources are predominantly located in smaller bays, all environmental and ecological effects are localized in the vicinity of the OTEC power plants.

6.7 Myths and Reality about Renewable Energy

Myth 1: There is enough solar power to provide cheap energy and to cover the energy needs of the entire world.

Reality: There is no doubt that the sun's energy is free and that the entire energy that falls on the earth is sufficient to cover the energy needs of the world (Section 6.2). However, the systems that harness this energy are neither free nor inexpensive. The cost of building and maintaining the solar systems is amortized into the price of energy they produce. Because of this, electric energy produced from PV and solar thermal units is currently significantly more expensive than the equivalent energy from nuclear and fossil fuel production units, in dollars per kilowatt hour. The conversion of solar energy to syngas and transportation fuels is another very expensive transformation. In addition, the periodic variability of solar energy makes the use of energy storage facilities necessary, which significantly increase the total amount of energy that needs to be produced as well as the cost of the energy to the consumers. At present, solar energy is abundant but far from inexpensive!

Myth 2: Solar energy is very expensive; we will never be able to produce as much energy as we need at a cost we can afford.

Reality: Environmental concerns and the several governmental subsidies make this type of energy more affordable at present, but still electric energy produced by PV cells is more expensive than energy produced by other power plants. One of the main reasons for this is that PV technology is still very young—almost at its infancy—while fossil fuel technology is more than 100 years old, and nuclear, more than 70 years old. One should not take current costs and project them into the future. Our experience with other systems that are massively produced and have become household appliances dictates that the cost of such systems decreases as they become massively produced, and often, their performance is improved in parallel. The personal computer, the cellular telephone, and the refrigerator are three systems whose price dropped by orders of magnitude since they were first introduced in the market, and their performance significantly improved. If massively produced from accessible materials, PV systems (perhaps with energy storage too) will follow the same pricing pattern and solar energy will become affordable to most households, as refrigeration has become since the 1960s.

Myth 3: The manufacture of solar cells requires a great deal of energy that cannot be recovered during the life cycle of the cell.

Reality: This was correct in the 1970s, when it would have taken more than 30 years to recover the energy consumed for the manufacture of a PV cell. Continuous research resulted in several technological advances that have significantly decreased the time it takes for a PV cell to produce the energy used during its manufacturing process, the energy payback time (EPBT). The polymer-based and cadmium–tellurium PV cells have EPBTs of less than 1 year. The EPBTs of the more common (and cheaper) silicon-based cells are in the range of 2–5 years. Given that the life cycle of new PV cells exceeds 15 years, it is clear that a PV cell in its lifetime produces more energy than it takes to manufacture it.

Myth 4: There is enough wind energy to provide the entire energy needs of the United States and the entire world.

Reality: As mentioned at the beginning of Section 6.3, the total available wind power may produce 218 times more energy than the total primary energy globally consumed in 2015. With typical wind turbine conversion efficiencies of 25%, we need to harness approximately 2% (1/(0.25/218)) of the total available wind power to satisfy the global energy demand. However, this is impractical with the current technology, because a great deal of the wind energy is at very high altitudes and at great distances from the coasts offshore, where wind towers cannot be built with the current technology. Practical, technologically feasible wind turbine systems may only be built onshore or at short distances from the coastlines. Consequently, a smaller fraction of the globally available wind energy may be harnessed. An additional impediment is that because wind power is intermittent, energy storage is necessary to meet the instantaneous power demand of the population. Storage increases the required amount of energy and significantly adds to the cost of the renewable energy system [42]. In 2017, there are no systems with sufficient capacity to store the necessary large quantities of energy in the United States or other countries of the world.

Myth 5: As in several other countries, e.g., Iceland and Costa Rica, geothermal energy can supply more than 50% of the electric energy in the United States.

Reality: The countries, where geothermal energy supplies a high fraction of the generated electricity, have a small population, and their electricity need is very low in comparison to that of the United States. For example, the 2013 electricity generation in Iceland and Costa Rica was 17.74 and 9.20 TWh, respectively, while in the United States, it was 4,110 TWh [2]. In addition, Iceland and Costa Rica lie entirely on geological formations where geothermal resources are abundant, while most of the United States does not have this advantage. As a consequence, the entire United States does not have the high density and high quality of geothermal resources these two smaller countries have. For this reason, the fraction of electricity that may be derived from geothermal units in the United States is significantly lower than that of other, smaller countries.

Myth 6: Since ethanol is a cheap, clean, and domestically produced fuel, we must produce more of it and use more of it.

Reality: This was the argument that resulted in the increased production of ethanol in the United States in the period of 2004–2008. While it is correct that ethanol is domestically produced, it is not produced cheaply, and it is not as "clean," as it has been advocated. The price of ethanol has been competitive because of the several subsidies—including tax breaks and direct subsidies—to farmers, agricultural corporations, and ethanol producers. If one accounts for these subsidies, the cost of production of ethanol is significantly higher than the cost of gasoline, and the difference is paid by the taxpayers. As it becomes clear from Table 6.13, the production processes of ethanol from corn actually requires more energy than it delivers and this makes questionable any argument for the increased use of ethanol as a transportation fuel. The production of more ethanol from corn will also necessitate significant amounts of water (almost 1000 L of water are needed for the growth of corn and the production of 1 L of ethanol), which will be taken away from agricultural and industrial uses.

Myth 7: Ethanol is a clean fuel.

Reality: Section 6.5.2 on the production of ethanol proves that there is a great deal of materials and energy that goes into the production of ethanol, which must be accounted in the carbon footprint of this fuel [32–35]. As shown in Table 6.13, the carbon footprint of ethanol must include the footprint for field preparation, seeding, harvesting, transporting, and processing. In addition, pesticides and fertilizers, which are invariably used for the production of ethanol, are occasionally carried by rain runoff water in rivers and lakes to pollute these water bodies.

Myth 8: Biofuels derived from kitchen oil waste can be widely used to save imported petroleum and gasoline.

Reality: While kitchen oil waste may be refined and used as transportation fuel (e.g., biodiesel), there are not sufficient quantities of this waste product to be widely used. Using the entire quantity of the oil waste from all the restaurants will hardly make a dent on the imported quantities of petroleum. For example, the United States—one of the most wasteful societies in the world—used approximately 13 million kg of cooking oils in 2015, primarily soybean oil, canola oil, palm oil, coconut oil, and olive oil. In the same year, the petroleum consumption of the United States was 700 billion kg. Assuming that 50% of the edible oil waste (6.5 million kg) becomes waste and is converted to biofuel, this quantity is less than 1×10^{-5} (one thousandth of one percent) of the petroleum consumption in the country. When one accounts for the additional fuel and energy used for the collection, transportation, and conversion of the kitchen oil waste to biofuels—which is of the order of magnitude of the energy content of the generated biofuel itself—one realizes that there is very little benefit in this practice.

Myth 9: Planting trees and other biomass will offset our carbon footprint.

Reality: Trees absorb a great deal of CO_2 from the atmosphere and convert it to carbo-hydrates. However, trees also use a lot of land area, which is a scarce resource. One fully grown pine tree (50' tall and 12" trunk diameter) weights approximately 2,000 lb (909 kg) with its root system, occupies an area between 25 and 36 m^2, and contains 800 lb (364 kg) of carbon C. It takes about 25 years for a tree to grow to this size in the south (e.g., in the states of Mississippi and Alabama) and 85–90 years in the north (e.g., in Vermont and Canada). Since one person in the United States produces approximately 38,500 lb of CO_2 annually, which correspond to 10,363 lb (4,710 kg) of C, each person will need to plant 13 trees/year to offset his/her share of the CO_2 production. The total offset will be realized when the trees are fully grown, after at least 25 years.

Now, let us assume a scenario where 50% of the United States citizens, approximately 160,000,000 people, decide to plant pine trees to offset their carbon footprint. The area that will be needed for this forest is 52,000–75,000 km^2 annually, and this is about the size of the entire state of West Virginia, with lakes, rivers, and mountains included. If this practice is continued, in approximately 12 years, an area equal to the territory of the entire deep south states—from South Carolina to the border of Texas—will be covered by this giant forest the "carbon offset" practice has created. This is clearly not a sustainable option because soon we will run out of land. If the inhabitants of Belgium, who produce only 23,540 lb (10,690 kg) of CO_2 per capita annually, decided to follow this offsetting practice, the area of their entire country will be totally covered in approximately 9–10 years! Clearly, the noble environmental goal of planting trees as "carbon offset" is unsustainable. The available fertile land should better be used to feed the growing population of the planet.

Myth 10: Trees absorb CO_2 from the atmosphere. Burning wood instead of natural gas to heat our homes is cheaper and CO_2 neutral.

Reality: 4.6 trillion scf of natural gas were used in the residential sector of the United States in 2015 [43]. The heating value of this quantity of fuel is $4.6 \times 10^{12} \times 1.07 \times 10^6$ J $= 4.92 \times 10^{18}$ J. From Table 6.11, the heating value of dry wood is 18,500,000 J/kg. Assuming that 50% of the residential demand for natural gas is shifted to wood, the 2.46×10^{18} J needed for this substitution would be provided by 133×10^9 kg (133 million t) of dry wood. A 50 ft (16 m) grown pine tree with 12 in. (30 cm) diameter trunk weighs approximately 900 kg, with its roots included. One concludes that for the supply of such an enormous quantity of heat, we will need to burn the equivalent of 148 million grown up pine trees and their roots annually. This is not a sustainable practice because the country will run out of trees and wood very soon. In addition, we will have the effects of increased pollution from the ash and metallic trace elements that are accumulated in the trees. While it may be carbon neutral, burning trees for heating and cooking is not a good solution to sustainable living.

Myth 11: A city of 1,000,000 inhabitants in the United States wastes so much that the MSW pro-duced is sufficient to supply a city in China with 1,350,000 inhabitants with energy.

It is correct that the MSW produced per person in the United States is the highest in the world and that it has grown at an unsustainable rate and should be reduced. However, statements such as this are grossly misleading, because they are based solely on the heat-ing value (HHV or LHV) of the MSW and not on the actual energy forms that are needed and consumed by the corresponding city in the PRC (or any other country). In addition, for the recovery of the heating value of the MSW, the waste must be burned. This practice is not allowed because MSW contains a plethora of toxic and pollutant materials that would be released in the environment. The anaerobic decomposition of MSW, which produces methane and is currently used for the production of small amounts of electric power, uses only a very small fraction of the heating value of the MSW and would be entirely insufficient to power a city in the PRC. In general, MSW is *waste*, and very little useful energy can be extracted from it. Finally, this statement was made in 2006, based on 2003 data. Since then, the energy consumed in a typical city in the PRC has more than quadrupled, and hence, the two numbers (notwithstanding what they represent) are incorrect.

Myth 12: All of Hawaii's electric needs could be satisfied by about 12 commercial-scale OTEC plants.

Reality: While, in theory, this may be correct, our experience with OTEC units does not support the statement. Actual OTEC units are small (about 100 kW) and produce approximately 0.8 GWh/year. The state of Hawaii produced and consumed 9.26 TWh of electricity in 2015. Therefore, Hawaii would have needed 11,600 typical OTEC power plants to produce the needed amount of electricity. In addition, our engineering experience with OTEC power plants is that they have low reliability in the long term because of heat exchanger biofouling and damage from storms. While OTEC may supply a small fraction of the electric energy currently used in Hawaii (and other islands), the state cannot depend solely on OTEC power for its electricity needs.

PROBLEMS

A. Hydroelectric Energy

1. A small river is to be dammed to create a hydroelectric power plant. After the completion of the dam, the river level will rise by 28 m. The river has a cross-sectional area of 340 m² and the average water velocity is 0.6 m/s. If the combined turbine-generator efficiency of the power plant is 0.75, what is the power this river will produce?

2. The water-level difference in the Three Gorges dam is approximately 110 m. How much water flow rate (in kg/s and m³/s) is required from the river for the power plant to produce its rated power, in Table 6.2, if the combined turbine-generator efficiency of the power plant is 0.76?

3. How much annual CO_2 emissions (in tons per year) would be avoided if the United States doubles its hydroelectric capacity and reduces the coal-fired power plants by the same amount? The average overall thermal efficiency of the coal-fired power plants is 36% and the heating value of carbon is 32,700 kJ/kg.

4. Small waterfalls of 2–5 m exist in several locations on the planet. Typical water flow rates in these waterfalls are 2–100 m³/s. If you were to divert 50% of the water for electricity production, what would be the upper and lower ranges of the power that would be produced from such waterfalls? A typical efficiency of the smaller turbine-generator pair is 75%.

B. Solar Energy

5. The solar collectors of a spacecraft that orbits the earth have 42 m² area, and they are always facing the sun. What is the incident solar energy on these collectors during 3 hours?

6. Using Figure 6.4, estimate the annually averaged energy, in kilowatt hours, absorbed by a 10 m² solar collectors in the following locations: (a) Hannover, Germany; (b) London, England; (c) Timbuktu, Mali; (d) Calcutta, India; and (e) Buenos Aires, Argentina.

7. PV cells with an overall average conversion efficiency of 10% and total area of 5 m² are placed in the following locations: (a) Hannover, Germany; (b) London, England; (c) Timbuktu, Mali; (d) Calcutta, India; (e) Buenos Aires, Argentina; and (f) New York, United States. Determine the total amount of electricity, in kilowatt hours, produced annually in each location.

8. At a location where the average insolation on a flat plate collector is 7,200 kJ/(m² day), a solar collector is used to continuously provide 38,000 Btu/hour of process heat. If the collector efficiency is 42%, what should be the area of the collector?

9. It has been suggested that refrigerant R-134a be used instead of water for a new type of thermal solar collector. What is your opinion on this? Explain in detail your reasoning.

10. A thermal solar power plant uses 1,300 heliostats, each one with 25 m² area. The annually averaged peak insolation in the region is 700 W/m², and the average insolation is 210 W/m². The thermal solar energy collection efficiency of the plant is 56% and the efficiency of the cycle is 38%. Determine the annually averaged peak power produced by this power plant and the total energy it produces during a year.

11. The drag force on a surface due to the wind flow is given by the expression

$$F_D = \frac{1}{2} A\rho C_D V^2, \quad \text{with } C_D = \frac{24}{Re}(1+0.32Re^{0.687}) \text{ and } Re = \frac{\rho L V}{\mu},$$

where V is the velocity normal to the surface; C_D is the drag coefficient; Re is the Reynolds number; and L is a characteristic dimension, which is equal to the square root of the area of the collector. A is the area that is perpendicular to the wind direction. When $Re > 10^5$, $C_D = 0.3$. A circular solar collector is placed at an angle 30° to the horizontal and has a diameter of $d = 7$ m. The wind flows in the horizontal direction. Draw a diagram of the drag force on the collector versus the wind velocity, when the latter varies in the range of 0–72 mph. What do you observe?

12. A 20 MW (peak) thermal solar power plant is to be built in a region where the annually averaged peak insolation is 640 W. Because of high winds in the region, the area of the heliostats is not to exceed 12 m². If the overall efficiency of the power plant is 18%, calculate the number of heliostats required for the operation of this plant.

13. A 10 MW (peak) PV solar power plant is proposed for the Dallas, Texas region. The peak insolation in this area is 1.065 kW/m² and the average insolation is 210 W/m². The overall efficiency of the plant is estimated to be 21%. Estimate the area of the heliostats that must be used for this plant and the total energy it will produce during a year.

14. Use values for the peak and average insolation in your area to calculate the area of the heliostats needed for a 5 MW PV power plant with an overall efficiency 20%. What is the annual amount of energy this plant will produce?

15. Use typical values in your locality to calculate the solar collector efficiencies. Also, calculate what would be the maximum temperature the fluid in the collector may reach and suggest methods to keep the internal temperature below 90°C.

16. "We have the know-how and the capability to produce a 60% efficient solar cell, but we lack the investment to produce it commercially." Comment on this statement by writing a 250–300 word essay.

17. "The sun is producing abundant and free energy that may be used to economically meet the entire energy need of this country, but the petroleum cartel undermines all efforts to do that." Comment on this statement by writing a 250–300 word essay.

C. Wind Power

18. What is the available power per unit area of a 10 knot uniform wind (in W/m²)?

19. The drag force on a flat surface due to the wind flow is given by the expression

$$F_D = \frac{1}{2} A \rho C_D V^2, \quad \text{with } C_D = 0.3 \text{ when } Re = \frac{\rho L V}{\mu} > 10^5,$$

where C_D is the drag coefficient; Re is the Reynolds number; and L is a characteristic dimension, which is equal to the square root of the area. The area A is the area that is perpendicular to the wind direction. A sail with 16.5 m² area faces the wind perpendicularly. What are the forces on the sail when the wind velocity is 6 knots, 10 knots, 25 knots, and 40 knots? What do you observe in these calculations?

20. A pitot tube at 3 m height measures wind velocity of 5.3 m/s. What is the wind velocity at a height of 50 m? If the center of a 45 m diameter wind turbine is placed at 50 m, what is the maximum power this turbine may produce?

21. The wind velocity at a height 3 m is measured to be 2.3 m/s. What is the wind velocity at 20 m? If the air density is 1.21 kg/m³, what is the mass flow rate of the air that passes through a rectangular surface of 10 m width that extends from 2–30 m?

22. Starting from first principles, analytically derive *Betz's law*. Justify every step in your analysis.

23. Determine the maximum power a wind turbine with $D = 60$ m may produce at the following wind speeds: 1 m/s, 5 m/s, and 10 m/s.

24. The cut-in velocity of a 30 m diameter wind turbine is 1.5 m/s, the rated velocity is 8 m/s, and the cut-out velocity 22 m/s. The turbine has 40% efficiency and is placed at a location where the following wind velocity data are known:

　　0 m/s < V < 1.5 m/s 22% of the time

　　1.5 m/s < V < 3 m/s 12% of the time

　　3 m/s < V < 5 m/s 12% of the time

　　5 m/s < V < 8 m/s 28% of the time

　　8 m/s < V < 15 m/s 12% of the time

15 m/s $< V <$ 22 m/s 10% of the time

22 m/s $< V$ m/s 4% of the time

What is the total energy (in kWh) you expect this turbine to produce annually?

25. Modifications are performed in the turbine of problem 24 and its rated velocity is extended to 15 m/s. What is the annual total energy the turbine will produce?

26. During a 6-hour period, the wind velocity linearly varies from 1.2 to 12 m/s. A 40 m diameter wind turbine is subjected to this wind. The turbine has cut-in velocity of 2 m/s, rated velocity of 7 m/s, and cut-out velocity of 15 m/s. The conversion efficiency of the turbine is 70% of the maximum efficiency, which is given by Betz's law. Determine the amount of energy this turbine produces during the 6-hour period.

27. During a particular 24-hour period, the wind velocity in Abilene, Texas, is approximated with a half-sine function, $V = V_0 \sin(\pi t/24)$, where t is measured in hours. The amplitude V_0 is 16.5 m/s during the day. A 38 m diameter wind turbine with cut-in velocity of 1.5 m/s, rated velocity of 9 m/s, and cut-out velocity of 21 m/s is placed perpendicular to the wind. If the overall efficiency of the turbine-generator system is constant and equal to 0.32, what is the energy produced during that 24-hour period?

28. It is apparent that a desirable attribute of wind turbines is to have high rated velocities. Under the conditions described in problem 27, what would have been the total energy produced if the rated velocity for the turbine were 12 m/s? How about if the rated velocity were 17 m/s?

29. A meteorological station measures the wind velocity at height of 30 ft (9.1 m) to be 12 m/s. Calculate the power density of the wind at this height (in W/m^2). A large wind turbine with 48 m diameter is placed at this location. The hub of the turbine is 50 m from the ground. Determine the power density of the wind at the hub. Also determine the maximum power that can be delivered by this turbine.

30. "We can extract so much power from the wind that it would be possible to satisfy the entire electricity demand of humanity using wind power alone." Comment in a 250–300 word essay.

D. Geothermal Energy*

31*. A geothermal well produces 60 kg/s of saturated liquid water at 210°C. What is the maximum power (in kW) this well may produce? The ambient temperature is 27°C.

32. A geothermal well produces 11.2 kg/s of dry steam, which is fed to a turbine. The enthalpy difference at the inlet and exit of the turbine is measured to be 420 kJ/kg. Determine the power the turbine will produce, in kilowatts.

33. A geothermal well produces 22 lb/s of dry steam, which is fed to a turbine. The enthalpy difference at the inlet and exit of the turbine is measured to be 180 Btu/lb. Determine the power the turbine will produce, in horsepowers.

34*. A geothermal well has a diameter of 25 cm and produces dry steam at an average velocity of 9 m/s. The steam is at 5 bar pressure and 200°C. The steam

---

* Use of steam tables or refrigerant tables is needed for the problems marked with an asterisk (*).

is supplied to a turbine with an isentropic efficiency of $\eta = 0.78$ and exhausts in a condenser at an average temperature of 36°C. Calculate (a) the mass flow rate of steam; (b) the power produced by the turbine; (c) the total amount of kilowatt hours the plant produces annually; and (d) the annual revenue to the operator, if the average sale price of energy is $0.087/kWh.

35*. Typical dry steam wells are approximately 25 cm (10″) in diameter and continuously produce at a typical average steam velocity of 9 m/s at the wellhead. The wellhead steam conditions at a particular location are $T = 200$°C and $P = 7$ bar. It is proposed to build a 60 MW geothermal power plant in this location. The turbines to be used have an isentropic efficiency of $\eta = 0.80$, and enough cooling water is available in this location to maintain the condenser of a power plant at 38°C. What is the power that a single well may produce and how many wells must be drilled to supply sufficient steam to this geothermal power plant?

36*. Five geothermal wells produce together 310 kg/s of a mixture of water and steam at 180°C with a dryness fraction 12%. What is the maximum power a power plant may produce from the water of the five wells? The atmospheric temperature is 27°C.

37*. Make a graph showing the amount of steam produced in the flashing chamber from 1 kg of saturated liquid water at temperatures in the range of 220–140°C. The temperature of the flashing chamber is the average of the supplied water temperature and the condenser temperature $T_C = 40$°C. What do you observe?

38*. A geothermal well produces 68 kg/s of saturated liquid water at 230°C. The water is used in a single-flash unit, where the turbine efficiency is 80%. What is the total power produced? Assume a condenser temperature of 40°C.

39*. A geothermal well in Warakei, New Zealand, produces an average of 56 kg/s of a two-phase mixture of steam and water at 190°C. The dryness fraction at the wellhead is $x = 0.15$. The steam is separated at the wellhead and fed to a turbine with isentropic efficiency of $\eta = 0.82$. The rest, which is saturated liquid water, is flashed, and the steam produced is fed to a low-pressure turbine with $\eta = 0.80$. The back pressure of both turbines is 8 kPa. Determine the amount of power produced by the two turbines.

40*. A geothermal well produces 60 kg/s of liquid water at 20 bar and 160°C. What is the maximum power you can get from this resource? Design a binary cycle with R-134a or another refrigerant to produce power from this resource. You may assume a condenser temperature of 40°C.

41. The average apartment in Paris, France, needs 1.2 kW of heat power during the winter months. It is proposed that geothermal district heating be used in the 15th District (15th Arrondisement) for a number of buildings with a total of 1,260 apartments. How much heat power is required? If the geothermal water enters the district heating plant at 70°C and is reinjected at 42°C, how much water volumetric flow rate, in cubic meters per second, is required for this plant?

42. A geothermal district project in Iceland has a continuous water supply of 7200 t of water per hour. The water enters the district at 73°C and is reinjected at 40°C. This geothermal heat district operates for an average 285 days/year. Determine (a) the heat rate in kilowatts, this project delivers; (b) if the only

other heating alternative is propane, the annual amount of propane that is saved because of this type of heating; and (c) the amount of CO_2 emission avoidance annually because of this project.

43. "Because the interior of the earth is cooling fast, geothermal energy will not be available in the future. Therefore, there is absolutely no reason to invest in geothermal projects." Comment in a short essay of 250–300 words.

44. "44×10^{12} W of power is continuously produced from geothermal resources and this represents three times the entire energy consumption by humans. If we concentrate on the development of our geothermal resources, we will not need any other form of energy." Comment in a short essay of 250–300 words.

E. Biomass Energy

45. In your opinion, what are the main reasons for the large disparity of the MSW production per capita in the United States and in Europe?

46. The use of biomass from fast-growing trees, such as sycamore, is often advocated as an alternative to burning coal. After 3 years, the average grown sycamore tree weighs 430 kg and the heating value of their wood is 15,300 kJ/kg. Find out how much heat power is consumed annually by a 400 MW coal power plant with an overall efficiency 40%. If the entire amount of this heat input were to be supplied from harvested sycamore trees, how many grown sycamore trees would the power plant need to consume annually? You may assume that the energy for the planting, harvesting, and transportation of the sycamore trees to the power plant is negligible.

47. The country of Estonia consumed 8.42 TWh of electricity in 2014 at an average overall thermal efficiency 32%. It has been suggested that Estonia produces all its electricity from locally grown pine trees. The average pine tree there requires 22 m^2 to grow and weighs 600 kg in 18 years, and its wood has heating value 13,700 kJ/kg. How much area is required for this solution to Estonia's electric energy problem? Is this alternative feasible? Assume that the energy for the planting, harvesting, and transportation of the pine trees to the power plants is negligible.

48. How much electric energy (in TJ) was produced in your country? If this energy was produced by power plants with an average overall efficiency 32%, how much heat energy (in TJ) was consumed? Assume that part of the area in your country were planted with native trees that grow to 500 kg in 10 years have a heating value of 14,000 kJ/kg, and each requires an area 20 m^2 to grow. How much area for tree planting would your country need to produce its electricity demand from trees? Is this a feasible solution for the country?

49. Rice hull residue with 30% humidity is used in the boiler of a small thermal power plant. The plant produces 60 MW of electricity and has an average efficiency of 34%. How many tons of rice hull does the power plant consume every year? If the average distance for the transportation of the fuel is 58 mi and the transportation trucks carry 12 t/trip and consume 5 mpg of diesel on average, how many gallons of diesel are needed annually for this plant to operate?

50. A 600 MW coal electric power plant with 32% thermal efficiency substitutes 8% of its fuel with "tree clippings," which are tree branches that are cut to clear roads as well as the access lanes (rights of way) of power lines. The tree clippings are fed to the boiler together with the coal. The coal used by

the power plant has 83% carbon and a heating value of 26,560 kJ/kg. How much coal is saved annually by this practice and how much CO_2 emissions are prevented?

51. The New York metropolitan area has a population of approximately 10,000,000. If the New York population behaves like the average US person, calculate the following:

    a. How many tons of solid waste they produce every year?

    b. The range of the heating value of the solid waste for the city per year.

    c. If it were possible to use this waste in thermal power plants with 35% overall efficiency, how much power (in MW) would this waste produce?

52. Cuba produces 38% of its electric power from bagasse. In 2014, Cuba produced 16.41 TWh of electricity in power plants with overall efficiency 32%. How much bagasse was consumed?

53. What do you think will be the main parts and processes of a biodiesel plant that uses marine algae? Make a conceptual design of a plant that produces biofuels from marine organisms.

54. "We have plenty of land in this great country to produce enough corn and corn-based biofuels for us to become completely independent of foreign oil. We only need the will and a small amount of investment." Comment on this statement by writing a 250–300 word essay.

55. Calculate how many gallons of ethanol may be produced annually from a farm of 100 ha (250 acres). Then calculate the equivalent amount of gasoline of this quantity of fuel. If the entire quantity of this fuel were to be used in a SUV with 11.5 mpg, how many miles may be driven in this SUV?

56. Calculate how many liters of ethanol may be produced annually from a farm of 100 ha (250 acres). Then calculate the equivalent amount of gasoline for this quantity of fuel. If the entire quantity of this fuel were to be used in a car with 15 km/L consumption, how many kilometers may be driven in this car?

57. "At a time when 50% of the children in the world are undernourished, it is unconscionable of the leaders of the richer nations to divert food resources for the production of fuel for their cars." Comment on this statement by writing a 250–300 word essay.

F. Sea/Ocean Energy

58. If the proposed tidal power plant in Cobequid, Canada, is to produce 2,000 MW of average power what would be the basin area that must be enclosed by the dam? Assume that the turbine-generator efficiency is 75%.

59. The range of the tides in the Gulf of Mexico is 0.6 m. A small, tidal power plant of 15 kW is proposed to provide process power to a shrimp factory. What is the basin area required for this power plant if the efficiency of its turbine is expected to be 70%.

60. In several places, the Gulf Stream current near Cape Hatteras, North Carolina, is equal to 4 knots. If a small water turbine with blade diameter 3.2 m were placed there, what would be the maximum power this turbine would produce? How many of these water turbines are required to produce the equivalent of one typical nuclear power unit (1000 MW)?

61. It is proposed to place 5,000 small water turbines with blade diameters of 3.5 m at the bottom of the ocean and inside the Gulf Stream, where the average velocity is 2 m/s. What is the maximum power these turbines would produce? What type of engineering and technological difficulties do you expect this scheme may have in the short- and the long runs?

62. The waves in the North Sea have average amplitude of 3 m and wavelengths of 30 m. What is the power that may be harnessed from these waves by a 60 m long engine with 70% efficiency?

63. Waves of 30 m height and 60 m wavelength have been observed during adverse weather conditions in the Gulf of Mexico (e.g., during hurricanes and tropical storms). What are the total power and the power per unit length for these waves? What are the engineering problems that would be encountered in harnessing these waves?

64. What is the maximum power that may be produced from an OTEC cycle when the upper temperature is 34°C, the lower temperature is 6°C, and the cycle receives 180 kW of heat? How much heat would be rejected from this cycle?

65*. Refrigerant 134a is used as the working fluid in an OTEC power plant, which utilizes a simple Rankine cycle without superheat. The condenser is at 12°C, and the boiling of R-134a occurs at 28°C. The efficiency of the turbine is 76%. What is the thermal efficiency of this cycle and how much heat input is required for the cycle to produce 0.5 MW?

66. What is the maximum power that may be produced from an OTEC when the high water temperature is 34°C, the lower water temperature is 10°C, and the cycle receives 200 kW of heat? Take 4°C difference for the boiler and condenser heat transfer.

67. "The oceans comprise more than 70% of the surface of the earth and may provide an unlimited amount of power from ocean current and thermal–ocean energy. We simply have to invest in this vast potential, and soon we will have energy independence." Comment with an essay of 250–300 words.

References

1. REN21, *Renewables 2016 Global Status Report*, REN21 Secretariat, Paris, 2016.
2. International Energy Agency, *Key World Statistics*, IEA-Chirat, Paris, 2016.
3. Michaelides, E.E., *Alternative Energy Sources*, Springer, Berlin, 2012.
4. Osnos, E., Faust, China and nuclear power, *The New Yorker*, October 12, 2011.
5. Ahmed, A.T., Elsanabary, M.H., Hydrological and environmental impacts of Grand Ethiopian Renaissance Dam on the Nile River, *18th Intern. Water Technology Conference*, Sharm El-Sheikh, March 2015.
6. Wilcox, S., *National Solar Radiation Database 1991–2010 Update: User's Manual* NREL/TP-5500-54824, National Renewable Energy Laboratory, Golden, CO, 2012.
7. NREL (National Renewable Energy Laboratory), *NREL—National Solar Radiation Data Base: 1991–2010 Update*, http://rredc.nrel.gov/solar/old_data/nsrdb/1991-2010/ (website), last visited August 16, 2016.
8. Egger C., ed., *How Upper Austria Became the World's Leading Solar Thermal Market*, O.O. Energiesparverband, Linz, 2009.

9. Howell, J.R., Bannerot, R.B., Vliet, G.C., *Solar-Thermal Energy Systems*, McGraw-Hill Book, New York, 1982.

10. Goswami, D.Y., Kreith, F., Kreider, J.F., *Principles of Solar Engineering*, 2nd edition, Taylor & Francis, Philadelphia, PA, pp. 39–51, 1999.

11. Hull, R.H., Nielsen, C.E., Golding, P., *Salinity-Gradient Solar Ponds*, CRC Press, Boca Raton, FL, 1989.

12. Ugras, E., Simulation of a Solar Pond for Production of Power in San Antonio, Texas, MS Thesis, University of Texas at San Antonio, San Antonio, TX, 2010.

13. El-Wakil, M.M., *Power Plant Technology*, McGraw-Hill, New York, 1984.

14. Green, M.A., Emery, K., Hishikawa, Y., Warta, W., Dunlop, E.D., Solar cell efficiency tables (version 47), *Prog. Photovolt: Res. Appl.*, **24**, 3–11, 2016.

15. Kurtz, S., Levi, D., 2017, *Best Research-Cell Efficiencies* at https://www.nrel.gov/pv/assets /images/efficiency-chart.png (website), last visited July 2017.

16. Dubey, S., Sarvaiya, N.J., Sheshadri, B., Temperature dependent photovoltaic (PV) efficiency and its effect on PV production in the world—A review, *Energy Procedia*, **33**, 311–321, 2013.

17. Vestas, https://www.vestas.com/en/products/turbines/v90-3_0_mw# (website), last visited July 2017.

18. GE, *1.5 MW Wind Turbine*, https://geosci.uchicago.edu/~moyer/GEOS24705/Readings /GEA14954C15-MW-Broch.pdf (website), last visited July 2017.

19. Siemens, https://www.siemens.com/wind (website), last visited July 2017.

20. Geothermal Energy Association, *2015 Annual US & Global Geothermal Power Production Report*, Geothermal Energy Association, Washington, DC, February 2015.

21. Michaelides, E.E., Thermodynamic properties of geothermal fluids, *Trans. Geoth. Resources Council*, **5**, 361, 1981.

22. Kestin, J. ed., *Sourcebook of Geothermal Energy*, US Department of Energy, Washington, DC, 1980.

23. Davis, A., Michaelides, E.E., Geothermal power production from abandoned oil wells, *Energy*, **34**, 866–872, 2009.

24. Bestec, http://www.bestec-for-nature.com/j2510m/index.php/projects-en (website), last accessed on September 29, 2016.

25. Michaelides, E.E., Future directions and cycles for electricity production from geothermal resources, *Energy Convers. Manag.*, **107**, 3, 2016.

26. DiPippo, R., *Geothermal Power Plants: Principles, Applications, Case Studies and Environmental Impact*, 2nd edition, Butterworth-Heinemann, Elsevier, London, 2008.

27. Edrisi, B., Michaelides, E.E., Effect of the working fluid on the optimum work of binary-flashing geothermal power plants, *Energy*, **38**, 389–394, 2013.

28. Michaelides, E.E., Separation of non-condensables in geothermal installations by means of primary flashing, *Trans. Geotherm. Resour. Counc.*, **4**, 515–518, 1980.

29. Tester, J.W., Drake, E.M., Driscoll, M.J., Golay, M.W., Peters, W.A., *Sustainable Energy*, MIT Press, Cambridge, MA, 2005.

30. Lizotte, P.L., Savoie, P., De Champlain, A., Ash content and calorific energy of corn stover components in Eastern Canada, *Energies*, **8**, 4827–4838, 2015.

31. Larson, E.D., Technology for electricity and fuels from biomass, *Ann. Rev. Energy Environ.*, **21**, 403–465, 1993.

32. Pimentel, D., Ethanol fuels: Energy balance, economics, and environmental impacts are negative, *Nat. Resour. Res.* **12**, 127–134, 2003.

33. Patzek, T., Thermodynamics of the corn-ethanol biofuel cycle, *Crit. Rev. Plant Sci.* **23**, 519–567, 2004.

34. Shapouri, H., Duffield, J.A., Wang, M., *The Energy Balance of Corn Ethanol: An Update*, US Department of Agriculture, Washington, DC, 2002.

35. Wang, M., Saricks, C., Wu, M., *Fuel-Cycle Fossil Energy Use and Greenhouse Gas Emissions of Fuel Ethanol Produced from US Midwest Corn*, Argonne National Laboratory, DuPage County, IL, December 1997.

36. Farrell, A.E., Plevin, R.J., Turner, B.T., Jones, A.D., O'Hare, M., Kammen, D.M., Ethanol can contribute to energy and environmental goals, *Science*, **311**, 506–508, 2006.

37. Lynd, L.R., Cushman, J.H., Nichols, R.J., Wyman, C.E., Fuel ethanol, from cellulosic biomass, *Science*, **251**, 1318–1323, 1991.
38. Tenenbaum, D.J., Food vs. fuel: Diversion of crops could cause more hunger, *Env. Health Perspect.*, **116**, A254–A257, 2008.
39. World Bank, *World Development Report 2008: Agriculture for Development*, World Bank, Washington, DC, 2008.
40. Charlier, R., *Tidal Energy*, Van Nostrand Reinhold, New York, 1992.
41. Bruch, V.L., *An Assessment of Research and Development Leadership in Ocean Energy Technologies*, SAND93-3946, Sandia National Laboratories, Albuquerque, NM, 1994.
42. Leonard, M.D., Michaelides, E.E., Grid-independent residential buildings with renewable energy sources, *Energy*, **148C**, 448–460, 2018.
43. US Energy Information Administration, *Natural Gas Explained*, US Energy Information Administration, Washington, DC, 2016, from https://www.eia.gov/Energyexplained/index.cfm?page=natural_gas_use (website), last visited July 2017.

7

Energy Storage

Since the beginning of the industrial revolution, our society has regularly used energy that was accumulated and stored for millennia: Fossil fuels are substances with very high quantities of stored chemical energy, which we now use to satisfy our primary energy needs. Nuclear energy is the energy that has been stored in the nuclei of atoms since the beginnings of the universe. It is apparent from the recent trends of energy consumption that fossil fuels are depleting at an alarming rate and will likely be exhausted by the end of the twenty-first century [1]. Nuclear fuels may last longer, but they also exist in finite quantities and will be exhausted at some point in the future. Environmental factors, such as the production of greenhouse gases (GHGs), may result in international treaties and regulations that would curtail the use of fossil fuels, even before their finite quantities are exhausted. All indications for the future are that humans will have to rely more on renewable energy sources, chiefly wind and solar power.

Solar and wind powers do not follow the patterns of the human demand for power and exhibit high variability. Solar is a periodically variable source: while we know that at noon on September 20, there will be solar power—the exact amount depends on the cloudiness of the day—we definitely know that at midnight on the same day, the amount of available solar power vanishes. Wind is an intermittent energy source: we know that a wind turbine will produce a number of kilowatt hours every year, but we do not know for certainty the precise periods during the year when this energy will be available to satisfy the consumer demand. Can anyone make an accurate prediction on what will be the wind velocity at your location 6 months from the present time, or even 1 month from the present? Meteorology is capable of predicting, with good accuracy, the available wind velocity and the associated wind power in the next 24 hours. The accuracy in predicting these variables diminishes with the length of time, and predictions after 7 days lack reliability. Other, lesser used forms of renewable energy, e.g., wave and tidal power, exhibit similar variability. Geothermal energy is the only renewable energy source that is available all the time, but this source of energy cannot satisfy the entire global energy demand.

The societal demand for power follows different patterns that do not match the supply of renewable energy. Human activities must take place regardless of when power (from renewable or non-renewable sources) is available. For example, during the night hours of July 17, when there is high demand for air-conditioning and electricity in the northern hemisphere, there is no solar energy, and the wind power is significantly lower than average. On that night, the human society cannot rely on the immediate availability of solar and wind power for a comfortable environment in buildings. Similarly, transportation that relies on these renewable sources becomes impossible, with all the inconvenience this entails. Unless sufficient energy is stored, during the hours when energy is readily available to be used during the periods of little or no energy availability, solar and wind—the two abundant and most promising forms of renewable energy—will be inadequate to supply the energy needs of our society. Thus, energy storage becomes a necessity for a society that relies on renewable energy sources. As our society transitions toward alternative energy sources, the need for storage to satisfy the fluctuating energy demand of a diverse population, becomes apparent and more acute.

The processes of energy storage and recovery inevitably involve thermodynamic irreversibilities. Energy dissipation occurs during the storage and discharge processes that significantly reduce the amount of available energy after storage. The storage and discharge processes involve energy transformations (e.g., from electricity, to chemical or mechanical energy, and back to electricity) and consume power that must be accounted in the overall energy balance of the systems that are considered. The *round-trip efficiency* of an energy storage system is the ratio of the energy available at the end of the discharge process to the energy that was consumed by the storage system at the beginning of the storage process. As it will be seen in the following subsections, round-trip efficiencies of practical energy storage systems are often close to 50%, which implies that half of the energy produced is dissipated in the storage–discharge processes. Clearly, the improvement of round-trip efficiencies with practical energy systems must become a goal, if renewable energy is to supply a high fraction of the global electric power demand.

7.1 Demand for Electricity: The Need to Store Energy

Our demand for electricity is subject to diurnal, weekly, and seasonal fluctuations: Electricity consumption is higher during the day and falls off during the night, when the majority of the population retires. On a given day, there are demand peaks that coincide with the several daily activities of the population, such as the preparation of the daily meals, air-conditioning during the hot season, and entertainment during the evening hours. In a typical week, in Organisation for Economic Co-operation and Development (OECD) countries, there is higher demand during the working days and lower during Saturdays, when many of the offices and businesses close. The lowest demand is on Sundays, when most of the businesses close, the majority of the population does not work, and many families engage in outdoor activities that do not use electricity. The widespread use of air-conditioning has significantly increased the use of electricity during the hot summer season. This has caused a daily demand peak for electricity in the early afternoon hours, when the ambient temperature is higher, and the need for air-conditioning is higher. Figures 7.1 and 7.2

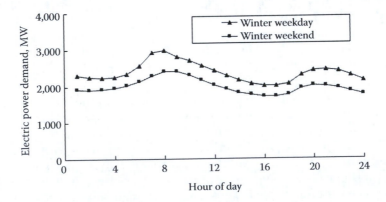

FIGURE 7.1
Typical electric power demand during winter days in San Antonio, Texas. (Data courtesy of CPS Energy Corp., San Antonio, TX.)

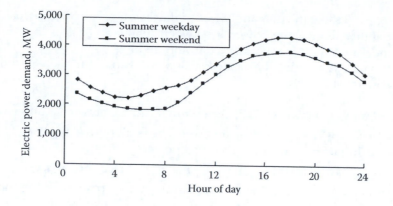

FIGURE 7.2
Typical electric power demand during summer days in San Antonio, Texas. (Data courtesy of CPS Energy Corp., San Antonio, TX.)

depict the hourly electric power demand during two typical winter and summer days in San Antonio, Texas. The figures include weekdays and weekends, and the data for the figures have been provided by the CPS Energy Corp., which supplies power to the entire San Antonio area [1].

A close look at the trends in the two figures leads to the following conclusions:

1. On a given working day, for both winter and summer, the peak daily electric power demand is approximately twice the minimum demand, which occurs in the early morning hours.

2. During the summer, the peak workday demand is approximately three times more than the minimum demand for the weekend.

3. The summer power demand is more than 60% higher than the winter demand on a similar day. Air-conditioning is the reason for this.

4. The power demand on summer weekdays almost doubles within a short period, representing a 2,200 MW shift. During the night, the electric power demand drops by 50% within a 4–6 hour period, representing a similar shift in power.

With the rapidly increasing use of air-conditioning in all countries, these trends are becoming universal. Because electricity must be produced instantaneously, such trends imply that the power systems, which supply electricity to the cities, must be flexible enough to accommodate the diurnal fluctuations.

7.1.1 Electricity Supply by Types of Power Plants

Since electricity cannot be stored, the entire electric power demand must be met instantaneously on the supply side by the group of the electric power plants, which are connected to the regional electricity grid. Because of the frequent and wide fluctuations in the demand for power, the group of power plants in the region must be able to satisfactorily meet the demand fluctuations and to provide the needed power to the consumers at any time. For this reason, not all the power plants continuously operate, and several power plants operate at a fraction of their stated power. Gas turbines may start when there is excess demand and stop when the demand decreases. Hydroelectric stations may be shut at nights to partly follow the demand. Most of the steam power plants (coal and nuclear) have the capability to generate power at between 105% and 85% of their rated capacity. These units operate almost continuously, day and night.

Parameters that characterize the capability of a power plant to meet part of the electric power demand are [1,2]:

1. The *rated power* or *plate capacity* is the nominal electric power a plant may produce.

2. The *peak power* is close to the rated power. The peak power may be a few percentage points higher or lower than the rated power.

3. The *average power* is the average of the power produced during a year.

4. The *availability factor* (AF) is the fraction of time that the plant is online and may contribute to the electric power demand, regardless of the power level it contributes.

5. The *power operating factor* (POF) or *plant capacity factor* (CF) is the ratio of the total electric energy generated by the plant during a year, to the total energy the plant would have produced if it continuously operated at its rated power. The POF takes into account the fluctuations of the energy produced during the year.

Figure 7.3 is a schematic diagram of the operation of a hypothetical power plant during the time interval $[0, T]$, which is typically 1 year, that is, $T = 8,760$ hours. The power plant is online and generates electric power during the time intervals $[0, t_1]$, $[t_2, t_3]$, $[t_4, t_5]$, and $[t_6, T]$, and it is off-line (e.g., for maintenance, or because power is not needed) during the time intervals $[t_1, t_2]$, $[t_3, t_4]$, and $[t_5, t_6]$. The rated, peak, and average powers of this power plant are shown in the figure. The AF of this power plant AF is

$$AF = \frac{t_1 + (t_3 - t_2) + (t_5 - t_4) + (T - t_6)}{T}. \tag{7.1}$$

The AF is a measure of the fraction of time during a year, when the power plant is available for the production of electricity and contributes to the supply of the electric grid. A glance at Figure 7.3 proves that the areas inside the polygons with the symbols E_1, E_2, E_3, and E_4 represent the total energy produced by the plant during the time intervals $[0, t_1]$,

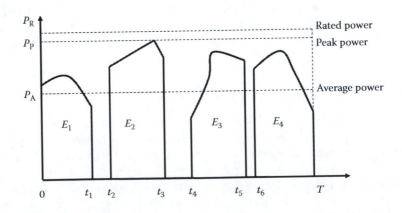

FIGURE 7.3
Operational parameters of a power plant with variable power production.

$[t_2, t_3]$, $[t_4, t_5]$, and $[t_6, T]$, respectively. The POF or CF is a measure of the average power produced by the plant throughout the year and is defined in terms of the rated power $\dot{W}_{rat}$ as

$$POF = \frac{E_1 + E_2 + E_3 + E_4}{\dot{W}_{rat}T}. \tag{7.2}$$

The POF in the last equation may be used to calculate the total energy produced by the power plant during an entire year. It is apparent from Figure 7.3 that the total energy produced by this power plant is significantly different from the total energy the plant *could have produced* if it continuously operated at its rated power during the time interval $[0, T]$.

The AF of the power plants significantly depends on their type and design: Most of the existing large steam power plants operating with steam (Rankine) cycles, such as nuclear pressurized water reactor units in the United States and coal power plants, do not have the capability and, actually, may be damaged if they follow the daily fluctuations of the demand for power. These units remain in continuous operation for long periods of time, months, or years. Several of the steam power plants may operate at slightly reduced power (80% of rated power), rather than at peak power, for several hours. For example, throttling the steam and bypassing part of the steam before the entrance to the turbine will reduce the generated power to 80–85% of the rated power in large coal and nuclear power plants. This reduction of the power produced also decreases the overall efficiency and increases the cost of power for the power plant. Nuclear power plants in France, and a few nuclear plants in Germany use control systems with "gray rods" that allow the units to operate in the range of 30–105% of their rated power. These power plants approximately follow the daily electricity demand, but do not follow any rapid transients of the demand. Smaller power plants, primarily those operating with gas cycles—the so-called gas turbines—may be taken in and out of line at will and are best suited to meet the daily fluctuations of the electric power demand.

Example 7.1

A gas turbine with 60 MW plate capacity is used for 2 hours/day during the months of November, December, January, February (28 days), and March at the reduced power of 45 MW; for 4 hours per day during October, April, and May at 55 MW; and for 8 hours per day during June, July, August, and September at the rated capacity. Determine the AF and POF of this gas turbine.

Solution: The gas turbine operates for a total of 60, 62, 62, 56, and 62 hours respectively, during November, December, January, February, and March. The total time of operation during this period is 302 hours, and the energy production is $45 \times 302 = 13{,}590$ MWh. During October, April and May the turbine operates for $4 \times (31 + 30 + 31) = 368$ hours and produces $55 \times 368 = 20{,}240$ MWh. During June, July, August, and September the hours of operation are $8 \times (30 + 31 + 31 + 30) = 976$ hours, and the energy production is $58{,}560$ MWh. The total hours of this year are 8,760.

$$AF = (302 + 368 + 976) / 8{,}760 = 0.1879 \ (18.79\%)$$

$$POF = (13{,}590 + 20{,}240 + 58{,}560) / (60 \times 8{,}760) = 0.1758 \ (17.58\%)$$

Example 7.2

Determine the AF and POF for the wind turbine of Example 6.7.

Solution: From the solution of that example, the rated power of this turbine is 1.58 MW. The other operational characteristics of the turbine are given by the following table:

Wind Velocity, m/s	Average Power, MW	Total Hours	Energy Produced, MWh
$0 < V < 3.5$	0	613	0
$3.5 < V < 8$	0.26	2,015	523
$8 < V < 12$	1.03	2,628	2,707
$12 < V < 16$	1.58	1,402	2,215
$16 < V < 20$	1.58	1,051	1,661
$20 < V < 25$	1.58	876	1,384
$25 < V$	0	175	0

The total hours in the year this turbine produces power are calculated by the sum in the third column, when the power is nonzero: 2,015 + 2,628 + 1,402 + 1,051 + 876 = 7,972 hours.

Similarly, the total energy produced by this turbine annually is obtained by summing the amounts in the fourth column to obtain 8490 MWh annually.

The wind turbine produces power in 7972 hours/year. The AF of the turbine is AF = 7,972/8,760 = 91.0%.

The POF is $8,490/(1.58 \times 8,760) = 61.3\%$.

A significant observation in this example is that wind turbines have very high AFs, because they produce some power during most of the hours of the year. Because during most of the hours in a year, the generated energy is lower than the rated power of the turbines, and their plant operating factors are substantially lower.

The availability of solar, wind, and wave power plants depends on the availability of their respective resources: a solar unit does not produce any power during the night, a wind turbine is motionless when there is no wind, and an engine powered by waves does not produce power during calm seas. Other factors that affect the availability of a power plant are as follows:

- The *operational cost* of the plant, which primarily depends on the cost of its fuel. Other things being equal, units that are expensive to produce electricity are taken off-line more often, while units that produce power cheaper operate for longer periods. In terms of energy generated per kilogram of fuel (e.g., kWh/kg of fuel), nuclear fuel and coal are typically less expensive than other fuels.

- The strain to switch the system on and off. Steam power plants (including, geothermal, coal and nuclear) may be damaged if they are frequently switched on and off. Steam power plants must operate almost continuously. Hydroelectric power plants, which produce electricity inexpensively, may start and stop without any impairment, and for this reason, they are often used to meet the fluctuating electricity demand.

- The offered market price in "deregulated markets," where the price of electric energy for short-term contracts fluctuates, typically every 15 minutes.

Based on these considerations, the electric power plants are broadly divided in the following categories:

1. *Base-load plants*: These are large (200–1,300 MW), very efficient units that operate with the Rankine steam cycle. Typically, the base-load units are nuclear and coal power plants that operate almost continuously throughout the year, except for periods of maintenance. During periods of lower demand (e.g., nights), the power generation of base-load plants is slightly reduced. However, these plants do not stop completely, and they are not taken off the electricity grid. Their AFs are usually higher than 90% and their POFs are in the range of 70–100%.

2. *Intermediate load or cycling plants*: These are usually older, less efficient steam units, hydroelectric power plants, and the more efficient gas turbines. They are specifically designed to be taken online and off-line relatively often. The AF of these units is close to 60%, and their POFs are in the range of 25–70%. The cycling plants typically supply power during the daytime and early evening hours (when the demand is high) and shut down during the late night hours (when the electricity demand is lower).

3. *Peak load plants or peaking plants*: These units come online and off-line often, sometimes two or three times during a single day. They are smaller units, typically in the power range of 5–120 MW, and use an air-cycle, Brayton or Diesel cycle. Their fuel may be natural gas, synthetic gas, diesel, or pulverized coal. Depending on the cost of the fuel, the electric energy produced from peaking plants is usually more expensive than that of other larger units, but this is often compensated by the higher price for electricity at peak hours. The peaking units produce a relatively small amount of electric energy annually, but they fulfill a very important function: they enable the electricity corporations (in all the countries of the globe) to meet the peak electric demand of their customers. Peaking units may be in operation for as little as 2 hours during a week (as in Example 7.1 earlier), and their POF is in the range of 5–25%.

In a given electricity grid, the base-load plants produce a very high fraction of the annual electric energy because they operate for long periods. Figure 7.4 depicts the hourly electric

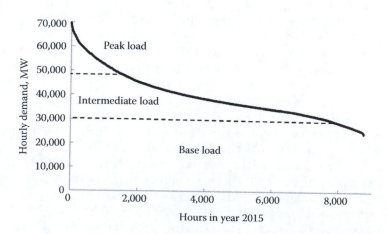

FIGURE 7.4
Hourly demand of the ERCOT electricity grid during the year 2015.

power supply during the year 2015 in the Electric Reliability Council of Texas (ERCOT) power grid that supplies with electric power almost 27 million people in the state of Texas, United States [3]. The maximum power supply during a single hour of the year is 69,620 MW and the minimum power is 24,337 MW. The areas shown in the figure represent the total energy supplied by the three types of power plants during the year 2015. It is observed in this figure that the base-load units provided a high fraction of the total electric energy during the year, while the peak power units contributed a very small fraction of the total energy demand. The base-load plants contributed 74.9% of the total electric energy to the electricity grid, the intermediate units 21.6%, and the peak power units 3.5% of the annual electric energy [3]. It is also observed from the shape of the figure that a great deal of the electric power capacity is underutilized for most of the year: 10,000 MW of the installed electric power capacity were used during only 360 hours of the year, and 15,000 MW were used for 765 hours—less than 10% of the total hours in the year. Clearly, the current patterns of electric energy demand result in the underutilization of the installed electricity production capacity, and this represents an additional cost to the electricity production corporations that must build and maintain all the peak power plants.

The realization of the adverse effects of carbon dioxide accumulation in the atmosphere has put pressure on governments, corporations, and individuals to *become greener,* by using a higher percentage of nonpolluting renewable energy sources in their mix for power production. Solar and wind are the most abundant and most popular renewable energy sources. However, during a windless summer night, neither a solar nor a wind power plant is capable of providing the necessary energy to run essential electric systems and appliances that are necessary for households and industry.

Let us contemplate a society in the future, where a high percentage of its energy is supplied by solar and wind power units. How will the energy demand of the society be satisfied during the hot, windless night of July 17? Without energy storage, air-conditioners will not run; entertainment systems will become silent; and industrial activities will stop. Total or even highly increased reliance on renewable sources with high variability (such as solar and wind) would result in prolonged power shortages and adverse economic effects that are unacceptable to the modern society. A significantly increased solar and wind power capacity for the production of electricity would create major economic and societal problems, unless it were supplemented by the development of substantial energy storage capacity. With sufficient energy storage capacity, a fraction of the energy produced during windy evenings in the spring and during sunny summer days will be stored to be used during the windless summer nights.

A second reason for energy storage is related to the operational details of the base-load power plants: These power plants are large nuclear and coal units that operate with Rankine steam cycles and cannot be switched on and off at will. Their production capacity may be occasionally reduced down to 80% of the rated capacity, but the power plants must be continuously in operation, day and night. Frequent power generation stoppages or further power reduction would damage the machinery of the steam units. If a region decided to "become solar" and produce a great deal of its electric power demand by solar energy, which is only available at daylight hours, there will be a significant reduction of the electric energy demand from nonsolar units during daylight. The minimum demand for the rest of the power plants will shift to the morning hours and will become less than the current minimum. This interferes with the operation of the base-load units for the parts of the day, when the insolation is high, but the power demand is not. Figure 7.5 depicts the expected modification of the summertime weekday power demand in San Antonio, Texas, which was depicted in Figure 7.2, when 10%, 20%, and 25% of the buildings in the city become zero-energy buildings (ZEBs)

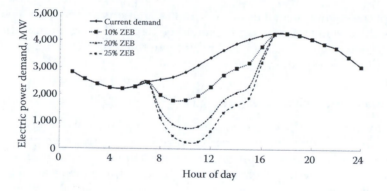

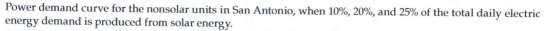

FIGURE 7.5

Power demand curve for the nonsolar units in San Antonio, when 10%, 20%, and 25% of the total daily electric energy demand is produced from solar energy.

and produce as much energy as they consume, while still connected to the electricity grid. The ZEBs draw electricity during the nighttime and deliver an equal amount of electricity to the grid during the daylight hours. Because of the heavy air-conditioning demand during the afternoon hours, most of the energy is delivered back to the electric grid during the morning hours, when the insolation is high and the air-conditioning demand is not. As a result, the power demand from the nonsolar power plants shifts from the solid line (which is the same as the weekday line of Figure 7.2) to the broken lines that are labeled by the fraction of the total energy generated by solar during the 24 hours. The area enclosed by the broken lines and the solid line represents the fraction of electricity generated by solar energy during the day. The demand curves for the nonsolar units exhibit a large dip during the early daylight hours and have been called *U-shaped demand* curves or "duck curves" [4,5]. The sharp dip implies that the power production from the nonsolar power plants must be reduced accordingly in order to accommodate the energy production of the solar units. In the case of San Antonio, if 28% of the energy is supplied by the solar units, all the other power plants will have to shut down between 9:00 and 11:00 am.

Similar shifts in the daily demand curves will occur in all the regional electric grids that have peak demand in the afternoon hours, when a significant fraction of the energy demanded is produced by solar units, thermal, and photovoltaic (PV). Because large base-load steam units cannot adjust their power production as frequently as the insolation changes, the regions served by the electricity grids will have to pursue a combination of the following:

1. Reduce the number or completely eliminate the current base-load power plants and substitute them with others that follow the demand. The substitution requires substantial investment that will make electricity more expensive.

2. Adopt incentives to adjust the electricity demand of the consumers of the region in a way that increases electric power demand during the early daylight hours.

3. Invest in significant, utility-level storage capacity that would store the excess power produced during the early daylight hours and enable the base-load plants to operate in conjunction with the solar units.

Energy storage systems are essential in a contemporary society that contemplates a "carbon-free future" with increased utilization of wind, solar, tidal, and wave energies.

Even without the environmental incentives and regulations for the reduction of CO_2 in the atmosphere, the certainty that fossil fuels will be depleted in the future dictates that the utilization of solar and wind power will increase in the future, and this must be accompanied by significant energy storage capacity.

7.1.2 Wholesale Electricity Prices: Deregulation

Early in the twentieth century, when electrification around the world grew at phenomenal rates, the electricity supply became the responsibility of private or public corporations, which had monopoly powers within their own geographical regions and were subjected to regulations by local or national authorities. These corporations (called *public utilities* in the United States) produced, transmitted, and distributed electric energy in the homes, businesses, and factories. With the approval of their regulatory authorities, the electricity corporations also set the electricity price for their consumers.

The public monopoly model does not encourage competition, increased efficiency in the production of power, and higher use of renewable energy sources. Starting in the 1990s, several countries and states adopted legislation to *deregulate* the energy market by opening the supply of electricity to competition. While the several national processes of *deregulation* were tailored to particular countries and regions and differ from each other, the several deregulated electricity markets that have been created from 1990 to 2017 have two common features. The first feature of deregulation is that the older electricity monopolies, either private or public, were split into several corporations that have the following distinct functions:

1. The *production* of electricity. These are the corporations that own the power plants, including nuclear, coal, hydroelectric, solar, and wind. The production businesses are in competition to produce electric energy at lower price and sell it at the market price. This invariably includes higher efficiency for the existing and new power plants.

2. The *transmission* of electricity. Transmission corporations own the regional electric distribution grids, the transformers that step up and step down the voltage, the distribution lines, and support towers. In general, they transmit electric power from the power plant to the electric meter of the consumer and are responsible for the maintenance and efficient operation of the transmission system.

3. The *distribution* corporations buy the electric energy from several producers and market it to their consumers. Their main function is to market and sell the electricity in retail to their consumers. Because of price competition, the distribution corporations typically offer several pricing options/plans to their customers, who may choose from a number of competing distribution corporations and plans. Options for the consumers include the purchase of *green energy*, which is electricity produced from renewable sources—primarily wind, geothermal, and solar. This option, which, in many countries, is augmented by grants, subsidies, and tax benefits to the producers of green energy, has been a very important contributor to the recent significant growth of renewable energy that is apparent in Table 6.1. Another option offers a very low price per kilowatt hour during nights and weekends with a significantly higher price during the rest of the week. With this option, consumers shift their demand for power at periods of off-peak demand.

The second feature of deregulation is the creation and development of national and regional energy wholesale markets. The energy distributors in a region will contract with

the energy producers for the delivery of the needed quantities of electric energy at specified days and times of the year. Since the energy demand in every region is variable, the
quantities of energy significantly fluctuate depending on the season, the time of the day,
and the expected weather conditions—e.g., more electric energy is demanded in the south
part of the United States during hot summer days, while communities in the north consume higher quantities of energy during the very cold winter days. Two types of contracts
are common in the electric power sector:

1. The long-term contracts that cover almost the entire energy supply of the base-
   load and intermediate-load power plants. There is a minimum of electric power
   demand at every hour of the year in all geographic regions. In the region supplied
   by ERCOT, the minimum electric power demand during 2015 was 24,337 MW, as
   depicted in Figure 7.4. Energy distributors may contract with producers to *dispatch** the minimum demand at an agreed price. Similarly, most of the intermediate demand may also be contracted because it is expected to occur at known hours
   during the year. For example, an electricity producer that operates a hydroelectric
   power plant may contract to provide 100 MW of electric power from 9:00 am to
   8:00 pm for the summer months and 60 MW of power from 7:00 am to 9:00 pm
   during the winter months. The long-term contracts are set at specified prices (e.g.,
   close to $30/MWh) and cover the minimum base demand and most of the typical
   and expected diurnal demand variations during a year. Typically, these contracts
   are agreed several months (up to 18 months for ERCOT) before the due day for the
   delivery of the electric energy.

2. The short-term contracts pertain to short-term variations and primarily cover the
   peak demand for electric power. These contracts are agreed upon a few hours
   before the delivery of power is needed; they reflect previously unanticipated
   peaks in electricity demand and usually fetch higher prices, the *spot prices* (e.g.,
   $150/MWh). An example of an unanticipated demand surge that must be met by
   a short-term contract is the following: On August 6, 2015 meteorologists predicted
   higher than anticipated temperatures for the next day in the Dallas/Fort Worth
   area of Texas, with the implication of higher than expected air-conditioning usage
   and, as a consequence, higher than anticipated electric energy demand in the
   entire region for August 7. Two local distribution corporations that anticipated
   electric energy shortage contracted with producers for the delivery of an additional 800 MWh between the hours of 3:00 and 6:00 pm at the price $175/MWh
   (approximately six times the typical prices of long-term contracts and significantly
   more than the retail price of electricity in the region). The contracts were fulfilled
   by the production corporations, and enough power became available for the consumers in the region who did not experience the inconvenience of power shortage.

It must be noted that if multiple contracts are established on a given day, the agreed
energy prices may be very different, depending on the energy market conditions at the
time the contracts are agreed. Figure 7.6 depicts the minimum and maximum prices of
short-term energy contracts (spot prices) and the number of contracts that were executed
during the days of the year 2015 by the several producers and distributors at the *PJM
Interconnection* [6]. The latter is a regional transmission limited partnership, with producers

---

* *Dispatch* and *dispatchable* are terms used for quantities of electric energy that are certain to be produced and
  may be sold to the customers.

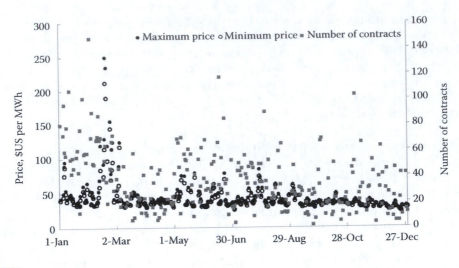

FIGURE 7.6
Spot prices and number of short-term contracts in the *PJM Interconnect* grid during 2015.

of approximately 184,000 MW generating capacity, serving 61 million customers from the mid-Atlantic region of the United States and extending to Ohio, Kentucky, and Tennessee. It is observed in this figure that the spot price of electricity in the region varied from a maximum of $250/MWh to a minimum $23/MWh. It is also noteworthy that the maximum prices of $250 and $235/MWh occurred as a result of unseasonally cold weather in parts of the region served on February 18 and 19 of that year.

The electricity distribution corporations have two principal objectives:

1. To satisfy the electric power demand of their customers, whenever it occurs. This is a contractual obligation of these corporations and, in several countries, an integral part of their charters.

2. To realize profits by purchasing energy at lower prices and selling at higher prices, while adhering to the environmental and statutory regulations of the region they serve. This helps improve the efficiency of power production.

Electricity corporations—both public and privately owned—use a variety of electric energy demand management methods to realize these objectives. Among the methods used for the management of the power demand and supply from producers and distributors are as follows [1]:

1. Shift the power production to the base- and intermediate-load units. This minimizes the production of power from the more expensive peaking units and minimizes the number of the contracts at the (typically higher) spot prices.

2. Incorporate into their electric grid several geographic areas where the peak demand occurs at different times.

3. Charge the consumers higher prices for peak power. This is usually accomplished by supplying the consumers with *smart meters*, which show the price charged at any time. Naturally, the majority of the consumers curtail their demand for power during periods when the price of energy is high, by conserving energy or

by switching off appliances that may be used at another time. During a high-price interval, an informed consumer may raise the temperature of the air-conditioning thermostat and postpone the use of the dishwasher, the washing machine, and the dryer for the late evening, when the demand is less and the electricity price is lower. A few distribution corporations offer plans for zero-cost electric power during the nighttime and the entire weekends. They compensate for this by charging a higher average price during the rest of the week.

4. Shift the heavy electric demand by 10–15 minutes, by delaying the starting times of the operation of equipment that consume high amounts of power. This is achieved when the electric power production corporation remotely controls high-powered units (e.g., air-conditioning units and industrial refrigeration units) and arranges for their staggered operation in a way that power demand is smoothened across their entire area. When the shifting period is short (10–15 minutes), the effect of this practice to the consumer is almost imperceptible.

5. Use power plants that have lower operational cost as base load units and allow more costly units to run only when necessary.

6. Construct smaller low capital-cost units as peaking units, even though the fuel cost of these power plants may be more expensive.

7. Use large energy storage systems to produce energy at low cost and then sell the stored energy, when the demand peaks.

7.1.3 Energy Storage Applications and Figures of Merit

Energy storage is becoming very important for solar and wind power producers, who do not necessarily produce power when the consumers demand it. With the impending depletion of fossil fuels and the apparent realization of the climatic effects of CO_2 and other GHG emissions, national regulations and international treaties dictate the increased use of renewable and highly variable energy sources, primarily wind and solar. As a result, the number of the highly variable solar and wind power units is expected to continuously increase in the entire world; and the energy storage systems will multiply and will become ubiquitous in our society. In general, the size of the energy storage systems depends on their applications, which may be broadly classified in the following categories:

1. *Utility level*: This requires large storage capacity systems to be used when the renewable energy sources do not produce sufficient power, as for example, in the nighttime supply of energy in a region where most of the energy is produced by solar systems.

2. *Power quality*: This is also at the utility level, but involves lower capacity and faster response times, to ensure uninterrupted power supply. An example of this type of storage is energy storage in flywheels to provide uninterrupted power supply from a PV unit, when there is temporary cloudiness and the power production of the PV panels drops for short time periods.

3. *Distributed grid*: This type of storage is at the location of the consumers, who may use electric or thermal storage systems to satisfy their energy demand. Because they supply relatively low power, the distributed grid systems have significantly lower capacity than utility-level storage.

4. *Automotive*: This includes the battery-operated and fuel cell-operated vehicles as well as the hybrid vehicles. In the early twenty-first century, there is a great deal of research activity and several innovations in this type of technology, as higher numbers of consumers use electric vehicles and hybrids.

5. *Electronics*: Several types of battery systems are used to supply power to smaller personal electronic devices (from watches to laptop computers) as well as to large computer and data storage systems that need backup power.

Desirable attributes of energy storage systems are reliability, low energy losses during the storage cycle, low mass and volume, and longevity. The following are useful parameters that characterize the different energy storage systems and help in the choice of the most suitable systems for the desired applications.

1. *Efficiency* or *round-trip efficiency* is the ratio of the energy output of the system to the energy input. Input and output must be the same type of energy and measured in the same units.

2. *Specific energy* is the energy stored per unit mass of the system. High specific energy is important in automotive systems, but almost insignificant for utility-level systems.

3. *Energy density* is the energy stored per unit volume. This parameter measures how large is the size of the energy storage system.

4. *Depth of discharge* is the fraction of the total energy stored that has been discharged, e.g., 70% discharge implies that the energy storage device contains only 30% of its maximum charge and has already discharged 70% of its maximum capacity.

5. *Self-discharge time* is the time it takes for a fully charged energy system to discharge down to a predefined depth of discharge, when it does not supply useful energy to an outside system. The self-discharge time for flywheels to 20% of their depth of discharge is on the order of minutes; that of solid-state batteries, on the order of weeks; and the self-discharge time of hydrogen storage systems is almost infinite.

6. *The cycle life* is the parameter that indicates how many times the energy storage system may be charged and discharged before its storage capacity and round-trip efficiency deteriorate. This parameter is particularly important for battery systems, which have a finite number of charge–discharge cycles.

The following sections explain the operation and capabilities of the various energy storage systems, with particular emphasis on systems that may store a significant amount of electric energy for utility-level and distributed grid usage.

7.2 Electromechanical Storage

Electromechanical systems store energy as potential energy, kinetic energy, and static electrical energy. The most commonly used systems in this category are above- and below-ground water pumping, compressed air, mechanical springs, flywheels, superconducting coils, and ultracapacitors.

7.2.1 Pumped Water

Pumped water systems—also referred to as *pumped hydro systems* (PHSs)—lift water from a lower-level source and store it in a natural or artificial lake at a higher elevation. When water is returned to its original lower elevation, it produces electric power. Water is abundant, inexpensive, easily transportable in pipelines, and has relatively high density. Since the potential energy is $E = mgZ$, 1 m³ of water, which has a mass of approximately 1 t (1000 kg), at a height of 360 m, possesses potential energy approximately equal to 3.6 million J, the equivalent of 1 kWh. A natural or artificial lake may store millions of cubic meters of water for long periods. These characteristics make the transformation of the potential energy of water an ideal medium for energy storage.

The pumped water system starts with a conventional base-load power plant located at a low elevation—typically a valley with a river to supply the water—and is close to high hills or mountains, the higher elevation. If a natural lake does not exist at the higher elevation, an artificial lake is constructed. During periods of low electricity demand (e.g., at nights), pumps that are powered by the electricity produced by the power plant transport water to the lake at the higher elevation. In several pumped water systems, the pump–motor combination used for the pumping of water is reversed and operates as a turbine-generator system to produce the electric power. The reversal of the pumping system to a turbine-generator system not only entails lesser round-trip efficiencies but also reduces significantly the capital cost of energy storage and utilization.

A schematic diagram of a pumped water system is shown in Figure 7.7. Excess energy produced in the power plant is transmitted to the pumping/generating station. Water from the nearby river or lake is pumped through a common pipeline to the artificial lake at the higher elevation. The pumped water at the higher elevation has high potential energy, which may be converted to electric power in the generating station at any time by reversing the flow in the pipeline.

Example 7.3

A base-load nuclear power plant has been constructed in a low valley and is supplemented with a pumped water system to store energy during low demand. A 200 × 200 m² artificial reservoir with an average depth of 1.5 m (approximately the size of 40 Olympic swimming pools) has been constructed at an elevation of 800 m higher than

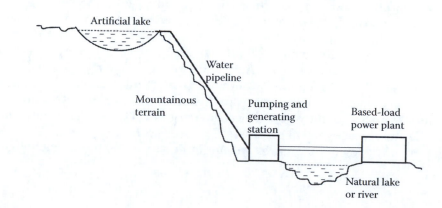

FIGURE 7.7
Above-ground pumped-hydro system.

the valley. (a) How much is the maximum amount of water that may be stored in the artificial lake; (b) what is the maximum stored energy; and (c) if all this energy is converted to electric power within 3 hours at a 75% overall efficiency, what is the additional power produced by the power plant?

Solution:

a. When the entire volume of the reservoir is completely filled, the reservoir contains the maximum amount of water: $200 \times 200 \times 1.5 = 60,000$ m^3 of water and since the density of water is 1000 kg/m^3, the maximum mass of the water in the reservoir is 60×10^6 kg (60,0000 t).

b. From $E = mgZ$, $E = 60 \times 10^6 \times 9.81 \times 800 = 471 \times 10^9$ J or 130,833 kWh.

c. If all this energy is converted at 75% efficiency within 3 hours (10,800 s), the additional power that will be produced is $471 \times 10^9 \times 0.75/10,800 = 32.7 \times 10^6$ W (32.7 MW).

This example highlights a very important fact about PHS energy storage and utility-level energy storage in general: The electric energy our society uses—several kilowatt hours for a single household—is very significant and is measured in units that represent a very high amount of potential energy (1 kWh = 3.6×10^6 J). In this example, a reservoir with very large surface area (200×200 m^2) stores just enough energy to produce 32.7 MW of power for 3 hours. The mass of water required for utility-level energy storage is very high, because the specific potential energy of water is low in comparison to the energy of nuclear fuels and the chemical energy of fossil fuels that have stored this energy for millennia. Typical coal power plants continuously produce 400 MW; gas turbines, 100 MW, etc. PHSs that would produce power of this magnitude for prolonged periods must have enormous water-holding capacities.

A second problem that is inherent in all energy storage methods is related to the thermodynamic irreversibilities of the storage process and the subsequent conversion of the stored energy to electricity. Following standard engineering practice, the irreversibilities for an entire process may be lumped together and expressed as the efficiency of the process. All the pumping and transportation of the water to the reservoir in the preceding example may be expressed as the efficiencies of the pumping system η_P and the transportation system η_{Tr1}. The latter consists mainly of friction energy losses in the long pipeline that transports the water to the reservoir. On the water return and conversion side, we will have two similar efficiencies η_{Tr2} and η_T, with the latter representing the efficiency of the turbine generator system that converts the stored water energy to electric energy. In the case of pumped water, typical efficiencies of the four processes are $\eta_P \approx 70\%$, $\eta_{Tr1} \approx 95\%$, $\eta_{Tr2} \approx 92\%$, and $\eta_T \approx 75\%$. If a quantity of electric energy E is produced in excess and becomes available for storage, a quantity $E \times \eta_P \times \eta_{Tr1} \approx 0.665E$ is stored as potential energy in the reservoir (66.5% of the original energy), and $E \times \eta_P \times \eta_{Tr1} \times \eta_{Tr2} \times \eta_T \approx 0.459E$ is the amount of energy recovered from the storage process. In this typical example, more than half of the original electric energy (54.1%) is wasted in the irreversibilities of the storage–conversion processes. The product $\eta_P \times \eta_{Tr1} \times \eta_{Tr2} \times \eta_T$ is equal to the round-trip efficiency of the storage system. In this case, the round trip efficiency of storage is 45.9%.

Despite the rather low round-trip storage efficiency, large pumped water systems present a viable alternative to energy storage for the following reasons:

1. In comparison to other systems, a PHS may store a significant amount of energy (millions of kWh) at relatively low capital cost.
2. The energy in a PHS may be stored for long periods. If a large reservoir is available—e.g., a large natural lake—the energy produced from wind farms may be stored during the spring to be used in the summer months, when the electricity demand is higher.
3. Except for a small amount of water that may be lost to evaporation,* there are no other losses. Oftentimes, the evaporated water is recouped and even enhanced from rainfall water.

An alternative system to storing water above ground is to store water in underground caverns. In underground storage, the pumping/generating station is located inside the cavern and water flows to the cavern from the surface by gravity. When the power plant produces electric power in excess, the pumping station retrieves water from the cavern and stores it at the earth's surface, close to the power plant. Underground energy storage has the same advantages and the same storage round-trip efficiency as the aboveground storage. The disadvantages of underground storage are as follows:

1. Underground caverns oftentimes leak through ground fissures, thus wasting some of the water.
2. The storage capacity is limited to the size of the cavern.
3. Power must be transmitted from the power plant to the underground pumping station.
4. Access and maintenance of the underground pumping/generating equipment is difficult and more expensive.

In 2017, there are not any known operational underground PHS units, while there are several utility-level aboveground PHS units.

7.2.2 Compressed Air

Compressed air energy storage (CAES) operates with a Brayton power cycle and typically uses a water column to maintain constant pressure, as shown in Figure 7.8. When there is low demand for power, the gas turbine drives the compressor, which pressurizes air that is diverted to an underground cavern or an artificial compressed air reservoir. Constant pressure in the cavern is maintained by the addition or displacement of water. When air is added to the cavern, the water is displaced to the aboveground water supply reservoir, which may be a river, a lake, or an artificial pond. When pressurized air is removed,

---

* Evaporation losses may be minimized by covering the lake surface with a thin plastic film.

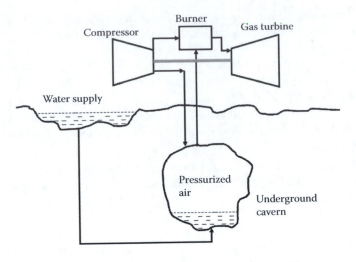

FIGURE 7.8
Schematic diagram of a CAES.

the water rushes inside the cavern to displace the air and maintains a constant pressure above atmospheric, $P = \rho g H$. During periods of high power demand, the pressurized air is extracted from the cavern; it is fed to the burner of the gas turbine system to increase its temperature; and subsequently, it is directed to the turbine. The additional quantity of the compressed air augments the power produced by the Brayton cycle. A variation of this method of energy storage—but of lesser efficiency—is to feed the compressed air directly to the turbine without heating it in the burner.

Salt caverns, hard-rock caverns, aquifers, and artificial reservoirs have been suggested for the air storage. An ideal air storage medium would have the following characteristics:

1. Low thermal conductivity to maintain the higher air temperature of the compressed air
2. Air tight—absence of fissures—to contain the pressurized air without air mass loss
3. To the extent possible, high wall elasticity to avoid crack generation under frequent loading and unloading.

Salt caverns and hard-rock caverns exhibit several of these characteristics and appear to be promising structures for compressed air storage. Of the CAES power plants that have been constructed around the world, the first one was a 290 MW unit in Huntorf, Germany, which was built in 1978. The second is the 110 MW unit of the Alabama Electric Corporation in McIntosh, Alabama, United States, which was commissioned in 1991. Both power plants use large underground caverns to store the compressed air. The two caverns of the Huntorf unit have a total capacity of 310,000 m^3 and operate between 43 and 70 bar pressure. At this range of pressures, air may still be considered an ideal gas ($PV = mRT$, where $R = 0.287$ kJ/(kg·K)).

The two power plants have demonstrated the technical soundness and reliability of CAES technology in storing and supplying significant power to the electric grid. However, both plants have also demonstrated that CAES energy storage is expensive and significantly adds to the cost of electric energy. As with the PHS units, the additional expense is primarily due to the irreversibilities in the production–storage–production processes, which consume a great deal of the energy produced to start with. The equipment efficiencies during these processes keep the round-trip efficiency of CAES units to less than 50%. Since the capital cost associated with the construction of the CAES systems is high (significantly higher than PHS) and the energy losses are substantial, the cost of the produced electric power by CAES is high [7,8]. Because of this, there are no more large commercial CAES plants that are in operation in 2017. A few demonstration projects and experimental or pilot plants around the globe have been built, using primarily government funds, to further study this type of technology.

Example 7.4

Pressurized air at 38 bar and 240°C is stored in a large facility with volume of 1,200 m³. The air is supplied to a turbine, which produces electric power. What is the maximum power this turbine may provide if it operates for 2 hours? The environment is at 30°C.

Solution: At first, air is an ideal gas, and the mass contained in the pressurized vessel is obtained from the ideal gas equation ($T = 513$ K, $R = 287$ J/(kg·K)):

$$PV = mRT \Rightarrow m = \frac{PV}{RT} = \frac{38 \times 10^5 \times 1,200}{287 \times 513} = 30,972 \text{ kg.}$$

In order to calculate the maximum power, one may assume that the turbine is isentropic (100% efficient) and determine the maximum work this quantity of air may produce in 2 hours (7,200 s). The maximum work that can be obtained from this mass of air is given by the exergy difference. With $c_p = 1.005$ kJ/(kg·K) and $R = 0.287$ kJ/(kg·K), the maximum work is

$$W_{max} = m[h_1 - h_0 - T_0(s_1 - s_0)] = m\left[c_p(T_1 - T_0) - T_0\left(c_p \ln\frac{T_1}{T_0} - R\ln\frac{P_1}{P_0}\right)\right]$$

$$= 30,972 \times 367 \text{ kJ} = 11.367 \times 10^6 \text{ kJ (3158 kWh).}$$

In 2 hours (7200 s), this amount of stored energy will produce power equal to 1579 kW (1.579 MW).

Examples 7.3 and 7.4 underline the difficulty of storing enough energy to produce utility-level quantities of power for long periods. The power produced by typical power plants is measured in hundreds of megawatts. The last two examples utilize systems of very high volumes but are only capable of producing power of 32.7 and 1.6 MW, respectively. Storage systems that would reliably produce hundreds of megawatts of power must be hundreds of times larger (and much more expensive!).

7.2.3 Flywheels, Springs, and Torsion Bars

Flywheels have the shape of bicycle tires with a heavy metal rim and rotate around a fixed axis. They have been extensively used with reciprocating engines, attached to the crankshaft. During the power impulse, the flywheel absorbs some of the energy from the crankshaft and releases the energy in the latter part of the stroke, when the power of the engine is reduced. The function of the flywheel is to make the power delivered by the reciprocating engines uniform and smoothen the operation of the engine. Flywheels are occasionally used with the electric motors of elevators to store small amounts of mechanical energy during the descent of the elevator. They have also been used with subway cars on rails in urban transportation systems to store energy during the braking process and release it during the acceleration process. All these applications pertain to relatively small quantities of stored energy with short storage–release times.

A high fraction of the mass of the flywheel is at the metal rim. The energy stored in a flywheel is rotational energy given by

$$E = \frac{4\pi^2}{2} mR^2 n^2, \tag{7.3}$$

where m is the mass; R is the radius; and n is the number of revolutions per second (the angular speed of the flywheel is $\omega = 2\pi n$). Accordingly, a metal flywheel, with 100 kg of mass uniformly distributed at a radius of 1 m and rotating with $n = 10$ revolutions per second, stores 197.4 kJ, or 0.055 kWh of energy. A similar flywheel, weighting 1 t (1000 kg) rotating at 30 revolutions per second (a very high rate of rotation for this mass), would store approximately 5 kWh. Flywheels do not have the capability to store energy at the utility-level for a long time. They are occasionally used to smoothen the high-frequency fluctuations of energy produced (e.g., from the passage of clouds over a PV electric production unit). Because of the continuously incurred friction losses, flywheels dissipate a high fraction of their energy, and have short self-discharge times. All the energy stored in flywheels must be used within a very short time.

The use of smaller flywheels has been advocated as an energy storage device in automobiles. A flywheel may absorb a fraction of the engine power during braking and release this power during the subsequent acceleration of the automobile. Such a flywheel system would improve the efficiency and the fuel consumption of a typical automobile. However, the weight of the flywheel and its mechanism would significantly add to the weight of the automobile. In addition, a fast rotating flywheel would reduce safety and reliability: gyroscopic forces on the axis of the flywheel are developed during changes in the direction of motion. The effect of the gyroscopic forces would be to decrease the horizontal stability of the automobile, a significant safety risk that might increase traffic accidents.

Springs and torsion bars typically store significantly less energy than flywheels. The energy stored in a spring is

$$E = \frac{1}{2} k(x - x_0)^2, \tag{7.4}$$

where k is the spring constant, a property of the material of the spring, and $x - x_0$ is the displacement of the end of the spring from its equilibrium position. Typical values of spring constants for large steel springs are on the order of 1 MN/m. With displacements on the order of 0.2 m, the energy that may be stored in such springs is on the order of 20 kJ (0.0056 kWh). This is a very small amount of electric energy.

Torsion rods are similar energy storage devices. The energy stored in a cylindrical torsion rod is

$$E = \frac{1}{2} (\pi R^2 L) G(\Omega - \Omega_0)^2, \tag{7.5}$$

where L and R are the length and radius of the torsion bar, respectively; G is the shear modulus, a property of the material of the rod; and $\Omega - \Omega_0$ is the angular displacement from its equilibrium position. Steel has a shear modulus $G = 81 \times 10^7$ N/m^2. A 2 m long cylinder of steel with radius of $R = 0.1$ m and subjected to an angular displacement of $\pi/12$ (15°) will store 1.7 MJ or 0.48 kWh. This stored energy is very low in comparison to typical electric energy needs. As with the springs, the torsion bars are not good candidates for the storage of significant amounts of energy that would provide electric power. The two mechanical devices may store mechanical energy for long periods and are typically used in highly specialized low-energy applications, such as mechanical clocks and weight balancing devices.

7.2.4 Capacitors, Ultracapacitors, and Superconducting Coils

Capacitors and ultracapacitors store static electric energy, while superconducting induction coils store magnetic energy. When magnetic and electric energy are released in the appropriate systems, they are converted directly and almost instantaneously to electric power with high efficiencies, in the range of 80–95%. The storage of energy in these devices does not suffer from significant round trip losses. The electric energy stored in a capacitor is given by the expression

$$E = \frac{1}{2}CV^2,$$ (7.6)

where C is the capacitance, a property of the material, which greatly depends on the size of the capacitor, and V is the applied voltage. When the capacitance is measured in farad (F) and the voltage in volts (V), the calculated energy is in joules (J).

Conventional capacitors are made of two metal plates, or a sphere and a surrounding spherical shell. The two plates or spheres acquire equal and opposite charges, which create the voltage difference across the capacitor. Dielectric materials are inserted between the charged plates or spheres to counteract the attractive forces and maintain the gap that separate the charges. Optimizing the dielectric properties of these materials creates higher energy densities for any given volume of the capacitor. Conventional capacitors store modest amounts of energy in a given volume, primarily because the dielectric material between the plates is bulky. The *ultracapacitors*, developed in the first decade of the twenty-first century, do not use as high volume for the dielectric material. They utilize the concept of the double electric layer and make use of the polarization of an electrolytic solution to store energy electrostatically. The principle of operation and the geometry of ultracapacitors allow the packing of a much larger electrostatic surface area into a small volume, resulting in extremely high capacitances C. Therefore, ultracapacitors store significantly higher amounts of energy than conventional capacitors. However, and because the voltage across a double layer in an electrolyte is limited to a few volts, even ultracapacitors cannot store the high amounts of energy required by the electric power industry.

An advantage of capacitors and ultracapacitors is that they may be charged and discharged in fractions of a second. Despite their low-energy storage capacity, these devices deliver high power from their relatively low volume, but for a very short duration. For this reason, they are used when high amounts of power are required at short time periods, as for example, during the startup of heavy equipment, including compressors and pumps.

The energy stored in an electric coil (solenoid) is given by an equation similar to Equation 7.6:

$$E = \frac{1}{2}LI^2, \tag{7.7}$$

where I is the current flowing in the coil and L is the inductance of the coil. The latter is proportional to the radius of the coil r and to the square of the number of turns of the coil. When the inductance is measured in henrys (H), and the current, in amperes (A), the stored energy is calculated in joules (J).

When made with the usual metal conductors, an electric coil has finite electrical resistance R and exhibits significant ohmic energy losses I^2R. A conventional coil is not a good medium for energy storage, because the stored energy is dissipated fast. Superconducting materials have almost vanishing resistance ($R \rightarrow 0$) and, therefore, extremely low ohmic energy losses. Electric current fluxes on the order of 10^8 A/m^2 and currents on the order of 10,000 A are now routinely achieved with superconducting coils. With typical inductance values of 500 H (1 H = 1 volt × second/ampere), Equation 7.7 yields that a typical superconducting coil has the capacity to store 2.5×10^{10} J, or approximately 7,000 kWh, a significant amount of energy for utility-level storage. Superconducting coils may be charged and discharged at very short times, and therefore, they may deliver extremely high amounts of power in a well-controlled process. What prevents their more widespread use is that the current generation of superconductors is expensive to manufacture and operates at very low (cryogenic) temperatures that must be artificially maintained at the expense of additional power.

7.3 Thermal Storage

Thermal storage is primarily used for three purposes:

1. To store energy at low temperature (below 100°C) for later use as hot water or for the heating of buildings. Residential water heaters are effectively thermal energy storage devices.

2. To store energy at higher temperatures, usually as latent heat of steam or molten salts, to produce electric power.

3. To store "coolness," usually water chilled to very low temperature or ice, for refrigeration or air-conditioning at a later time.

7.3.1 Sensible and Latent Heat Storage

When thermal energy is stored as *sensible heat*, the temperature of the storage material (pressurized water, organic fluid, and solids, including solid beds of pebbles) increases. When the thermal energy is extracted, the temperature of the material decreases. The thermal energy storage material is well insulated, so that it does not lose a great deal of heat. Typical sensible energy storage units are space and water heaters that have been extensively used for the heating of buildings and hot water for domestic use.

The sensible heat space heaters, which have been in use for decades in several European countries, are made of packed bricks or similar high-density materials with high specific volumetric heat capacity. Electrical resistances and air channels run between the energy storage materials. The electric power is dissipated in the resistances, and the temperature of the storage materials rises during the night and early morning hours, when the electricity demand is low and the base-load power plants have spare capacity. During the day, when the base-load power plants do not have spare capacity, the heating units gradually release the stored thermal energy by circulating a controlled stream of air in their interior channels. The air cools the storage material and transfers the heat to the interior of buildings, where the temperature is kept at the desired levels.

One may achieve better temperature control and heat flow with *latent heat storage*. This is accomplished with pressurized steam, another pressurized vapor, or molten solids such as metals and salts. The energy stored in the vapor or the molten solids may be used for (a) space heating, (b) process heating, and (c) production of electric power. The main advantage of latent heat storage methods is that the energy is released at a constant and controlled temperature. This is very important for several industrial processes as well as for power production, where constant temperatures for the supply of heat are desirable. All other factors being equal, the stored latent heat is significantly higher than sensible stored heat. For example, 1 kg of water vapor (steam) at 1 atm will supply approximately 2580 kJ at 100°C for space or process heating,* while 1 kg of liquid water at the same temperature may only supply approximately 300 kJ for space heating. Latent heat storage is ideal for storing large quantities of thermal energy for heating and for the production of electricity.

Latent heat storage has been proposed for extending the use of base-load steam power plants to peaking units, by storing high amounts of energy in steam or molten solids: When a steam power plant does not run at full load, some of the steam produced is diverted from the turbine to be stored at high pressure in large vessels, called *steam accumulators*. At periods of high electric demand, this steam is fed to auxiliary turbines, to produce additional power. Alternatively, the excess thermal energy may be stored in a molten solid—metal or salt—to be used for the production of additional steam during the peak demand periods. Latent heat values of salts used for thermal energy storage are on the order of 1000 kJ/kg and are higher than the latent heat values of most common metals [9].

Gas-cooled nuclear power plants working with high-temperature cycles, such as the advanced gas-cooled reactor, the high-temperature gas-cooled reactor (HTGR), and packed bed reactor (PBR) power plants, are excellent candidates for this type of operation and for supplying higher power during peaking hours through the following operations:

1. During the off-peak hours, all the heat produced by the reactor is transferred by the circulating high-temperature helium to a pool of the salt, which heats up and melts. The pool temperature may rise to 650–800°C, depending on the capacity of the storage material and its container.

2. The stored energy in the molten salt is removed by circulating high-pressure water, which becomes steam at high pressure and temperature.

3. The produced steam is directed to a turbine to produce electric power. During this operation, the molten salt may cool to a lower temperature, and depending on the design of the unit, some or all of it may solidify.

---

* The steam releases 2,257 kJ/kg when it condenses and becomes liquid water at 100°C. An additional 322 kJ is released as sensible heat when the temperature of the condensed water decreases to 27°C.

Because of the high temperatures achieved in the heat storage media, the temperature of the steam may be in the range of 550–600°C. This is in the higher range of temperatures achieved by modern fossil fuel plants and implies that the steam component of the combined power production unit will be as efficient as any modern fossil fuel power plant.

These modifications to their operation will enable the HTGR and PBR nuclear power plants to produce variable, controlled power and to contribute to the base-load as well as to peak demand. One of the disadvantages of this modification is the very high mass of the storage material required, as the following example illustrates:

Example 7.5

A molten salt mixture of lithium, calcium, and potassium nitrates is considered to be used as thermal storage medium for a nuclear HTGR. The sensible and latent heat to be stored and extracted from the mixture of salts is 1,300 kJ/kg. The melting salts are used to raise steam that will provide an additional 100 MW of electric power for 3 hours in a steam cycle with 38% overall efficiency. Determine the mass of salts to be used for this HTGR cycle modification.

Solution: During the 3 hours of operation of the steam cycle, 3×100 MWh or 1.08×10^9 kJ of electric energy is produced. Since the steam cycle has 38% efficiency, this electric energy is produced from 2.84×10^9 kJ of thermal energy, which must be stored in the molten salts. Because the total energy stored and recovered in the mixture of salts is 1300 kJ/kg, the mass of salts required is 2.186×10^6 kg (2186 t).

This is a very large amount of salts, but by no means unobtainable. The modification of this HTGR nuclear power plant is feasible, but may come with the expense of additional capital. A more serious disadvantage of this storage method is that several of the nitrate salts [e.g., the $Ca(NO_3)_2$ salt] are chemically unstable at high temperatures and decompose after a few thermal cycles. Research is currently conducted to produce materials with desirable thermal properties that are also stable at high temperatures and have long life cycles.

A significant disadvantage of all thermal storage systems is the heat loss. Common experience tells us that if a hot object is left for a long time at ambient conditions, its temperature will gradually decline, even if it is well insulated. Given enough storage time, the stored energy dissipates in the environment and is eventually lost. Thermal storage systems are designed to hold heat for relatively short time periods, on the order of several hours. This is sufficient for utility-level energy storage that follows the diurnal fluctuations of the demand for power.

The rate of heat loss from a thermal storage vessel of area A at an elevated temperature T is

$$\dot{Q} = UA(T - T_{amb}), \tag{7.8}$$

where T is the temperature of the energy storage material, which is assumed to be uniform inside the storage vessel; T_{amb} is the ambient temperature; and U is the overall heat transfer coefficient of the vessel. The latter represents the heat conduction or convection inside and outside of the vessel as well as the conduction through the insulation of the vessel.

If the initial temperature of the thermal storage material is $T(0)$, after a time period t, the temperature becomes [1]

$$T = T(0) + [T_{amb} - T(0)][1 - \exp(-t / \tau_{st})], \tag{7.9}$$

where τ_{st} is the characteristic time for the thermal storage material. For a material of density ρ and specific heat capacity c, contained in a vessel with volume V and outside area A, the thermal characteristic time τ_{st} is given by the expression

$$\tau_{st} = \frac{V\rho c}{AU}. \tag{7.10}$$

The thermal energy loss, or thermal discharge, when the temperature of the storage system drops from $T(0)$ to T is

$$E(0) - E(t) = V\rho c[T(0) - T] = V\rho c[T(0) - T_{amb}][1 - \exp(-t / \tau_{st})]. \tag{7.11}$$

Equation 7.11 demonstrates that the temperature of a sensible heat storage system drops exponentially, and eventually—at very high values of t—it becomes equal to the ambient temperature. A well-insulated vessel, which has very low U, has lower heat loss to the ambient, and its temperature drop is delayed. Thermal storage systems are best used for short to intermediate energy storage situations. In the long run, all the thermal energy is lost to the environment.

Insulating materials that are currently used with thermal energy storage systems have low enough overall heat transfer coefficients U to enable the storage of large quantities of heat for periods of 10–12 hours without significant temperature drop and energy losses. Thermal storage systems may be used with diurnal heating cycles (e.g., storage of heat during the night and use during the day) but not with seasonal cycles (e.g., storage of heat during the spring and summer months to be used during winter).

7.3.2 Storage of "Coolness"

As it is apparent from Figures 7.1 and 7.2, the continual need for air-conditioning in the United States has shifted the peak demand for electric power from the winter to the summer months. In the southern part of the United States, the peak demand for electric power during the summer is 35% higher than the peak power demanded during the winter, for both weekends and weekdays. Similar shifts in the electric power demand—albeit with lower difference between winter and summer—have also been observed in the countries of southern Europe and Japan. As the global use of air-conditioning is continuously expanding, it is expected that similar electric power demand trends will soon be observed in Latin America, India, Indonesia, and the People's Republic of China.

Because the peak power demand is primarily spent on the cooling of buildings, it makes sense to store coolness that is to cool energy storage materials at temperatures lower than the internal temperatures of buildings. Coolness may be stored for several hours as latent heat, as sensible heat, or combination of the two. Chilled water at low temperatures, 5–10°C; ice; a mixture of water and ice; and frozen salts are some of the materials used for coolness storage. Materials primarily working with latent storage have an advantage because they store significantly higher amounts of coolness per unit mass, but liquid chilled water is cheaper and much more convenient to transport in pipelines.

Chilled water is used in most coolness storage systems as the working fluid for air conditioning: Water is chilled during the off-peak hours in the refrigerator of the air conditioning system; it is stored in a large tank for a few hours until it is needed and is then circulated in the buildings to cool them. The air-conditioning system uses electric power to produce chilled water during the off-peak hours, and the stored chilled water cools the buildings. With such coolness storage, the air-conditioning system avoids using a great deal of electric power during the peak demand hours.* A chilled water storage system has two main advantages:

1. It shifts the electric power demand from peak hours to off-peak hours, when cheaper electricity is supplied by base- and intermediate-load power plants. Electricity production corporations do not need to have a large spare capacity to meet the high peak power demand.

2. Because base-load and intermediate-load power plants are more efficient than peaking units, the electric power used in the air-conditioning is produced at higher average plant efficiency. Because of this, lesser primary energy resources are used, and this entails lesser emissions and lesser environmental impacts.

Since air-conditioning uses a great deal of energy, coolness storage systems have high holding capacity. Figure 7.9 shows the building used for chilled water storage at the Dallas–Fort Worth (DFW) International Airport, with other airport features shown for scale comparison [10]. The DFW airport is one of the major airports in the United States and consists of five large terminal buildings that need a great deal of air-conditioning during the summer months, when ambient temperatures routinely reach 100°F (38°C). The chilled water storage building, which is shown in the figure, contains a well-insulated tank with 6,000,000 gal (22,712 m^3) capacity of chilled water, stored in the building at an average temperature 47°F (8.3°C) [10]. The chilled water has 90,000 t hours of refrigeration capacity (1.08 × 10^9 Btu or 1.14 × 10^{12} J). The refrigeration system of the airport chills the water during the late night and early morning hours, and the water is stored in the building until the late morning, when it is gradually used to provide some or all the air-conditioning needs of the airport. A typical daily operation of the chilled water system is shown in Figure 7.10, with the tank charging process in negative numbers [10]. The hourly power units are in tons of refrigeration; 1 t of refrigeration is 12,000 Btu/hour or 3.52 kW. It is observed in this figure that the spare capacity of the water refrigeration system is used to charge the tank during the night from 21:00 hours to 7:00 hours of the morning. Small amounts of the chilled water are used to assist the air-conditioning system of the airport from 8:00 to 13:00, while from 14:00 to 20:00 hours, the stored chilled water provides most of the needed air-conditioning to the five terminals of the airport. The system operates so that no electric power is used for the refrigeration part of airport air-conditioning system during the peak demand hours 15:00 to 18:00, when the electric grid is strained. Because electricity is more expensive for the airport operator during peak hours, the chilled water storage system represents significant monetary savings.

---

* A low amount of electric power is still needed for the circulation of air.

FIGURE 7.9
Chilled water storage building in the DFW airport. Cars and airplanes are shown for scale comparison. (Courtesy of Rusty T. Hodapp.)

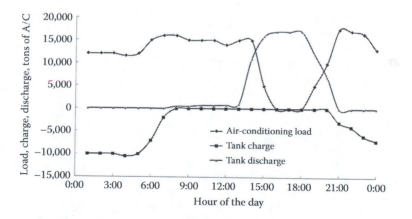

FIGURE 7.10
Hourly air-conditioning load, chilled water tank charge, and discharge in the DFW airport; 1 ton of air conditioning represents 12,000 Btu/hour (12,660 kJ/hour or 3.52 kW).

7.3.3 Phase-Change Materials: Eutectic Salts

A very simple and efficient system to maintain comfortable temperatures in buildings for both summer and winter is to use phase-change materials (PCMs) with the capacity to store high amounts of energy that also exhibit hysteresis in their melting–freezing phase change behavior. Ideal PCMs melt at lower temperatures and solidify at higher temperatures. The temperature–enthalpy diagram of a desired PCM for building cooling and heating is depicted in Figure 7.11. The melting of the PCM occurs at 10°C (50°F), and solidification, at 45°C (11°F), while the temperature of the building T_{bld} is kept at an intermediate value, e.g., at 25°C. Materials that exhibit this type of thermal hysteresis are the *eutectic salts*.

In the summer months, the PCM stores coolness for air-conditioning. The melting at 10°C, occurs during off-peak hours, and the temperature of the PCM drops to values that are lower than the melting/solidification range, e.g., to 5°C. During the peak demand hours of the summer, the PCM temperature increases to 10°C when melting occurs. Throughout the melting process, the PCM absorbs heat from the interior of the building and keeps its temperature at a desired level, e.g., at $T_{bld} = 25°C$. In the winter months, the operation of the air-conditioning system is reversed to a heat pump and provides heat to the PCM. In this mode of operation, the PCM absorbs heat during the off-peak hours, and its temperature rises higher than 45°C. During the peak hours, the warmer PCM supplies the building with heat; its temperature first lowers to 45°C; and the PCM solidifies at this temperature as shown in the solidification curve in the figure. Since, the solidification temperature is higher than the desired internal temperature of the building, the PCM supplies heat to the building throughout the cooling and solidification processes and maintains the building temperature close to the desired temperature T_{bld}.

While this is an ideal operation of a PCM system for heat and coolness storage, operations with actual PCM systems are different, because most of these materials have different and unsuitable melting and solidification temperatures. Commonly used PCMs, e.g., the salt sodium sulfite decahydrate ($Na_2SO_4 \cdot 10H_2O$) also known as Glauber's salt, is unstable and decomposes after several heating and cooling cycles. The solidification temperature of Glauber's salt is 32.2°C [11], and this limits the temperature difference of the heat exchanger that transfers heat to the interior of the building. Lower temperature variations result in more expensive heat exchangers that have larger areas. More research in this area and the discovery/development of new PCMs with properties close to the ideal operation of Figure 7.11 will significantly assist with the shifting of the peak electric power demand in both summer and winter.

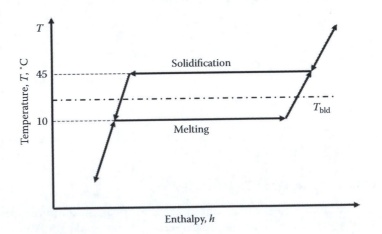

FIGURE 7.11
Melting and solidification curves of a desirable PCM system for heating and cooling.

7.4 Chemical Storage: Batteries

Most of the primary energy sources we use today and apply them to transportation, space heating, and electricity production are fossil fuels, with the energy stored in their molecules as chemical energy. Coal, oil, and natural gas consist of chemicals that have been formed millions of years ago and are now burned at an alarmingly fast rate. As shown in Table 4.1, these chemicals have very high specific energy (in kJ/kg or kJ/kmol). The chemical energy released from the combustion of solid graphite ($-\Delta H^0$) is 32,770 kJ/kg; for octane (the main component of gasoline), the heat of combustion is 49,010 kJ/kg; and for methane (the main component of natural gas), 55,510 kJ/kg. Fossil fuels are stable; the energy in them may be stored for long periods (petroleum and coal have been stored for millennia) and are very easy and convenient to be used by combustion in boilers, burners, and internal combustion (IC) engines.

Because the specific energy (in kJ of stored energy per kg of material) of most stable chemical bonds is significantly high, chemical storage is a very desirable method for energy storage. Apart from high specific energy and high storage density, an energy storage device with desirable properties should charge and discharge quickly, to provide not only large quantities of energy, but also high power when it is needed.

7.4.1 Wet and Dry Cell Batteries

Acids, bases, and metallic salts have much shorter formation times than fossil fuels and are universally used as *electric accumulators*. These are energy storage devices that are commonly referred to as "batteries" or "electrochemical cells." The *Volta cell*, invented by Alessandro Volta in the late eighteenth century, was the first electrochemical cell to be used. It was made of copper and zinc electrodes with copper sulfate solution as the electrolyte between the electrodes. The *Daniel cell*, which also uses copper and zinc in an aqueous medium, was invented in 1836 and makes use of a semipermeable membrane that prevents the deposition of copper on the zinc anode. The Daniel cell, also called a wet cell because it uses an aqueous solution, is made with solid copper and zinc *electrodes*, the *cathode* and the *anode*, respectively. The two electrodes are surrounded by solutions of copper sulfate ($CuSO_4$) and zinc sulfate ($ZnSO_4$), respectively. The aqueous solutions constitute the *electrolyte*. A porous membrane separates the two electrolyte solutions; it allows the SO_4^{2-} ions to pass through its pores; but it inhibits the passage of the Cu^{2+} and Zn^{2+} ions and effectively keeps them in the vicinity of their respective electrodes. At the anode, a solid zinc atom deposits two electrons and enters the solution as a positive ion according to the reaction

$$Zn \rightarrow Zn^{2+} + 2e^-. \tag{7.12}$$

At the cathode, a copper ion (Cu^{2+}) absorbs two electrons and is converted to a solid copper atom, which is deposited on the cathode,

$$Cu^{2+} + 2e^- \rightarrow Cu. \tag{7.13}$$

The overall reaction that takes place in the wet electrochemical cell is

$$Zn + Cu^{2+} \rightarrow Cu + Zn^{2+} \tag{7.14}$$

Since the anode of the Daniel cell has a surplus of electrons, while the cathode has a deficiency of electrons, an electromotive force (emf or voltage difference) is created between the anode and the cathode. This voltage is used in an external circuit to produce current I and electric power: $\dot{W} = VI$. The current is maintained within the electrolyte solution by the movement of the SO_4^{2-}, ions, which may permeate the membrane within the cell. Common batteries consist of several individual cells, connected in series to generate higher voltage.

Household batteries are made of *dry cells* and use solid chemicals or pastes. Of these, the *zinc–carbon* cell is one of the most common types and consists of a zinc container (anode) with a carbon rod at the center (cathode), which remains inert. Manganese oxide (MgO_2) with a paste of ammonium nitrate (NH_4Cl) is the "dry" material in the interior of the cell. During the operation of this electrochemical cell, the following reaction takes place:

$$Zn + 2MnO_2 + 2NH_4Cl \rightarrow Mn_2O_3 + Zn(NH_3)_2Cl_2 + H_2O. \tag{7.15}$$

A voltage difference (emf) of approximately 1.25 V is produced from the electrochemical cell reaction between the anode and the cathode. If the ammonium chloride in the last reaction is substituted with zinc chloride ($ZnCl_2$), the overall reaction in the electrochemical cell becomes

$$4Zn + 8MnO_2 + ZnCl_2 + 5H_2O \rightarrow 8MnO(OH) + 2Zn(OH)Cl + 3ZnO, \tag{7.16}$$

and the produced emf is approximately 1.5 V.

In contrast to the chemical energy in fossil fuels, which is used by combustion and transformation to thermal energy that is subjected to the Carnot limitations of Equation 1.9, the electrochemical cells convert the energy stored in the chemical bonds of their materials directly into electricity, without the intermediate transformation to thermal energy. Since electrochemical cells are not subjected to the Carnot limitations, in principle, the entire chemical energy of the materials in the cell may be converted to electrical energy. This process is called *direct energy conversion* (DEC).

The total electric energy that may be produced by the reaction of the electrochemical cell is equal to the change of the Gibbs free energy $-\Delta G^0$ of the overall chemical reaction in the cell. In the absence of thermodynamic irreversibilities, the Gibbs free energy is converted entirely to electrical energy, which is equal to the product of the voltage developed and the electric charge transferred. When 1 kmol of a substance reacts—e.g., 1 kmol of Zn in the reaction of Equation 7.12—the charge transferred is equal to ζF, where F is the *Faraday constant* (= 96,500,000 C/kmol) and ζ is the number of valence electrons, the number of electrons transferred in the anode per reaction. In the case of zinc, $\zeta = 2$. The maximum electromotive force produced by an electrochemical cell is obtained when the operation of the cell is reversible and the entire Gibbs free energy of the reacting materials $-\Delta G^0$ is converted to electric energy:

$$-\Delta G^0 = \zeta F V_{max} \Rightarrow V_{max} = \frac{-\Delta G^0}{\zeta F}. \tag{7.17}$$

In the Zn–Cu reaction of Equation 7.14, $\Delta G^0 = 216,160$ kJ/kmol and $\zeta = 2$. Hence, the maximum voltage that the Daniel cell may produce is approximately 1.1 V, an order of magnitude that is typical for most electrochemical cells. Several electrochemical cells are connected in series to develop the higher voltage in the range of 10–20 V, which is needed in several applications. In actual situations, internal irreversibilities related to the flow of ions in the electrolyte and external irreversibilities by the current, would reduce the voltage developed even further by 10–25%.

Example 7.6

The Gibbs free energy of water is –237,000 kJ/kmol. Calculate the maximum voltage that may be produced from the combination of hydrogen and oxygen in an electrochemical cell.

Solution: Two electrons are transferred for the formation of a water molecule ($\zeta = 2$). From Equation 7.17, the maximum voltage is 237,000/(2 × 96,500) = 1.23 V. This is the maximum voltage produced in a hydrogen *fuel cell*.

7.4.2 Lead Batteries

One of the most common types of cells for batteries is the lead cell, which is invariably used in the automotive and marine industry. The lead cell consists of an anode, made of solid lead (Pb). The cathode is made of solid lead oxide (PbO_2). The electrolyte, between the two electrodes, is a dilute sulfuric acid solution (H_2SO_4). The acid in the electrolyte solution separates into the hydrogen (H^+) and sulfuric $\left(SO_4^{2-}\right)$ ions. The chemical activity and mobility of the ions in the electrolyte causes chemical reactions at the surface of the electrodes, which release or absorb electrons. The predominant chemical reactions are as follows:

1. At the anode:

$$Pb + SO_4^{2-} \rightarrow PbSO_4 + 2e^- \tag{7.18}$$

2. At the cathode:

$$PbO_2 + 4H^+ + 2e^- \rightarrow 2H_2O + Pb^{2+} \tag{7.19}$$

3. In the electrolyte:

$$2H_2SO_4 \rightarrow 4H^+ + 2SO_4^{2-}; \quad Pb^{2+} + SO_4^{2-} \rightarrow PbSO_4 \tag{7.20}$$

The solid lead in the anode and the lead oxide in the cathode are eroded by the hydrogen cations and the sulfuric anions. The excess electrons in the anode and the deficit of electrons in the cathode generate the voltage difference (emf) of the cell. The overall reaction in the cell is obtained by combining Equations 7.18 through 7.20 to yield

$$Pb + PbO_2 + 2H_2SO_4 \rightarrow 2PbSO_4 + 2H_2O. \tag{7.21}$$

This overall reaction produces an electromotive force of approximately 2 V. Six cells in series constitute the typical automotive and marine batteries that produce approximately 12 V.

Example 7.7

The owners of a household have decided to install deep cycle marine batteries to guard against power failures in their area. The batteries have a nominal voltage 12 V, they are rated at 35 Ah (ampere hours), and they should not be discharged below 20% of their capacity. The voltage inverter to be used for this system has an efficiency of 92%. From their electricity bills, the owners estimate that during the hot summer nights, the heating, ventilation, and air-conditioning (HVAC) unit of the household consumes almost continuously 8.5 kW of electric power. How many batteries should the homeowners buy and install to ensure that the HVAC unit will operate for 6 hours on the power supplied by the battery system?

Solution: From the household demand, the energy required for 6 hours of operation is $W = 8.5 \times 6 = 51$ kWh. With an inverter efficiency of 92%, the system of batteries must be able to supply $51/0.92 = 55.4$ kWh. This amount must be provided by the batteries when they discharge from 100% to 20% of their capacity. Therefore, the total capacity (stored energy) of the batteries must be $55.4/0.8 = 69.3$ kWh.

From the given battery specifications, one deep cycle battery has a capacity of $12 \times 35 = 420$ Wh $= 0.42$ kWh. Hence, this household will need to install $69.3/0.42 = 165$ deep cycle marine batteries.

One realizes from Example 7.7 that a very large number of conventional batteries are needed to supply the energy needs of a household with high HVAC power demand. The use of conventional battery storage is not the best option for households that use a great deal of HVAC power.

7.4.3 Lithium Batteries

The lithium batteries were developed in the 1990s to be used in electronic devices and light vehicles. The term *lithium battery*, or *lithium-ion battery* (Li-ion), refers to several different types of batteries with a variety of cathodes, anodes, and electrolytes. The most common type of lithium batteries uses metallic lithium as the anode and manganese dioxide as the cathode. The solvent is an organic solvent with a dissolved lithium salt. A second type of commercial Li-ion battery is the lithium–thionyl chloride cell, which also uses metallic lithium as the anode. A liquid solution of thionyl chloride ($SOCl_2$), which is dissolved in a porous carbon material, is the cathode of the cell. Lithium tetrachloroaluminate ($LiAlCl_4$) solution is the electrolyte in this cell. Lithium batteries are well suited for very low-current applications, where reliability and long life is desirable.

Table 7.1 shows the properties of four types of Li-ion cells, the typical voltage they produce, the electric charge capacity they have, in ampere hours per kilogram of weight and the specific energy that may be stored in such cells, in kilojoules per kilogram [1].

TABLE 7.1

Voltage, Charge Capacity, and Specific Energy (in kJ/kg and kWh/kg) for Several Types of Li-Ion Cells

Cathode	Voltage, V	Capacity, A Hour/kg	Specific Energy, kJ/kg, kWh/kg	
$LiCoO_2$	3.7	140	1,865,	0.518
$LiMnO_2$	4.0	100	1,440,	0.400
$LiFePO_4$	3.3	120	1,426,	0.396
Li_2FePO_4F	3.6	115	1490,	0.414

Example 7.8

For each of the lithium batteries listed in Table 7.1, determine the battery weight needed (in kg) if the batteries are to supply the energy needs of the household of Example 7.7. You may assume that the batteries can be safely discharged to 10% of their nominal capacity.

Solution: With the 10% threshold for discharge, the system of lithium batteries need to supply $51/(0.92 \times 0.90) = 61.6$ kWh, or approximately 222,000 kJ. Based on the energy values from Table 7.1, the systems of the four types of batteries listed must have masses 119, 154, 156, and 149 kg, respectively.

For energy-to-weight comparison, the specific energy of octane from Table 4.1 is 44,430 kJ/kg. Although the Li-ion batteries may store a significant amount of electric energy per unit mass and be used in electric vehicles, the recoverable specific energy (kJ/kg) stored in fossil fuels, such as gasoline and diesel fuel, is approximately 30 times more than the specific energy of the Li-ion cells. This implies that electric cars must carry considerable weight in their battery systems, as the following example illustrates.

Example 7.9

An electric car is designed with $LiMnO_2$ batteries to replace the IC engine of a gasoline-powered car with a 20 gal (74 L) tank. The efficiency of the gasoline engine is estimated at 22%, and the efficiency of the new electric motor system is 95%. Determine the weight of the Li-ion batteries needed if the electric car is to have the same distance range as the gasoline car. The batteries may be discharged to 15% of their nominal capacity.

Solution: What is of interest here is the actual work that becomes available as the motive power of the car. Since the density of gasoline is 0.719 kg/L, the mass of the gasoline in the tank is 53.2 kg, and the heat of combustion (lower heating value) of gasoline is 44,430 kJ/kg, the IC engine vehicle has available with full tank $74 \times 0.719 \times 44,430 = 2.36 \times 10^6$ kJ of thermal energy.

This amount of gasoline, when converted at 22% thermal efficiency, provides a total of 0.52×10^6 kJ of mechanical work, which is used for the propulsion of the car.

In order for the electric car battery to provide the same amount of mechanical work at the 95% efficiency of its motor system, it must have $0.52 \times 10^6/0.95 = 0.55 \times 10^6$ kJ of stored electrical energy in the battery. From Table 7.1, the specific energy of the $LiMnO_2$ battery is 1,440 kJ/kg. Hence, 380 kg mass of the $LiMnO_2$ batteries is needed for the two cars to have the same range, if the batteries might be discharged completely. At discharge to 15%, the actual mass of the batteries is $380/0.85 = 447$ kg. The ratio of the battery to gasoline masses is 8.4.

It is apparent in Example 7.9 that electric cars carry significantly more weight with their battery systems than the gasoline and diesel fuel used by IC engines. This weight disadvantage is offset because they operate with electric motors that are much lighter than the IC engines and the other auxiliary equipment of conventional automobiles. A major drawback of all electric cars is the recharge time for the conventional as well as the Li-ion batteries, which are on the order of tens of minutes to hours.* Recharging an electric car takes much longer than filling a gasoline tank.

---

* The fast-charging of a battery (e.g., tens of minutes) is highly irreversible and involves significantly higher electric energy dissipation. It also shortens the cycle life of the batteries.

7.4.4 Advantages and Disadvantages of Batteries

Batteries are easy to manufacture and to use; they provide power readily, at demand; they are clean; and they are based on relatively known technology that may be adapted by all nations. Their charging process may be long, but it is accomplished efficiently. The round-trip storage efficiency of most types of batteries is close to 80%, when charged slowly. Batteries have been used for the supply of relatively small quantities of energy to household appliances, computers, and portable telephones and for the starting of automobile and boat engines.

Conventional batteries based on chemicals, are not ideal for the storage of large quantities of energy. A glance at Table 7.1 proves that the storage of 1 kWh of electricity (3.6×10^3 kJ) requires the equivalent of 2–2.5 kg of Li-ion batteries. In order to develop the utility-level electric storage capacity of Example 7.3 (130,833 kWh), one would need approximately 300 t of Li-ion batteries, a very large quantity of mass. The weight requirement would be much higher for typical (and less expensive) automobile or marine batteries that are based on lead, because a typical marine or automobile battery stores approximately 0.4 kWh and weighs 17 kg. An electric storage system, based on marine or automobile batteries, that has the same storage capacity of 130,833 kWh, requires the equivalent of 5,560 t of battery mass, even if the batteries were allowed to completely discharge. Marine and automobile batteries are not practical for utility-level electricity storage.

A second disadvantage of all batteries is the energy loss (discharge) during long periods of energy storage. The voltage difference within the battery cells causes a *current drift*, which is a weak internal current, even when the poles of the battery are not connected to an external circuit. It is common knowledge that a car battery that remains idle and is not charged for a few weeks is "drained" (completely discharged) and becomes a "dead" battery. This is caused by the weak current drift, which gradually dissipates the stored energy. Typical self-discharge rates for common rechargeable cells are, for lead acid, 4–6% per month; for nickel cadmium, 15–20% per month; for nickel metal hydride, 30% per month; and for lithium, 2–3% per month. The self-discharge implies that conventional batteries are not reliable energy storage systems for long periods, such as the seasonal storage of energy from the high winds of the spring to the high electricity demand of the summer.

A third disadvantage of conventional batteries is their finite life and limit of charging–discharging cycles and the environmental threat posed by their regeneration or disposal. Almost all the solid battery types, including lithium batteries, contain heavy metals (Pb, Ni, Co, Mn, Cd, etc.) that have been proven to be harmful to living organisms, including humans. In particular, lead is an environmental pollutant and very harmful to humans as explained in Chapter 3. After a few thousand charge–recharge cycles (with the current technology on the order of 1000 cycles), batteries fail to recharge and "die." Dead batteries are recycled to be chemically regenerated in specialized facilities. Let us assume that our economic system uses solid batteries in residential buildings for the storage of significant amounts of energy, which is produced from renewable sources. With the current electrical energy demand, this practice would necessitate the use of the equivalent of 160 heavy-duty marine batteries in a typical household of the south and southwest United States, where the use of air-conditioning is widespread and necessitates large quantities of stored energy in the buildings [12]. The batteries will need to be regenerated every 24–48 months. As a result, every household in this region must have the equivalent of 40–80 batteries per year sent to a central facility for refurbishment or replacement. The volume of this battery traffic is simply very high for the environment to sustain, because there will be many environmental and public health problems that will be created from misplaced, discarded, and badly disposed batteries. The use of lithium and Li-ion batteries, which contain small quantities of heavy metals (Cd, Ni, Cr, Pb), presents the same problem, albeit at a smaller scale. Our contemporary society, simply, cannot rely solely on solid chemical batteries for utility-level or building-level energy storage without risking environmental damage and future public health problems. Even if the life cycle of the solid batteries improves by a factor of two, a rather unlikely event if the cost of batteries is to be kept at affordable levels, the environmental risk posed by misplaced and discarded batteries would still be enormous. In an economy heavily dependent on batteries for energy storage, the extensive transportation and chemical regeneration of batteries will cause significant environmental and public health problems.

Since the 1990s, there has been progress in the manufacturing of durable, reliable, and high-capacity batteries for electric vehicles. With the current state of technology, the conventional solid batteries cannot readily replace the gasoline tanks in all the vehicles, primarily because of the self-discharge problems, the time to recharge, and the energy losses during faster recharges. For batteries to play an important role in transportation and seasonal electric energy storage, a significant amount of research and technological advances must occur with goals to

1. Increase the specific energy and energy density of the batteries
2. Reduce the recharge time and associated energy losses
3. Lengthen the cycle life of batteries to more than 10,000 charge–discharge cycles
4. Eliminate the internal current drift to enable seasonal energy storage

---

7.5 Hydrogen Storage

Hydrogen may substitute all fossil fuels, in power plants, IC engines, and jet engines. It may also be used in *fuel cells* that directly convert its chemical energy to electricity. Because all water in the sea, lakes, and rivers is composed of hydrogen and oxygen, this element is abundant and almost inexhaustible. Hydrogen, as a chemical element by itself, is not a naturally occurring energy source and must be artificially produced by methods such as electrolysis or chemical reactions at the expense of other energy forms. Hydrogen storage has been advocated as an effective, environmentally friendly, and physically powerful energy storage medium for the following reasons:

1. Hydrogen is the lightest element, and the specific chemical energy stored in this element at ambient conditions is higher than that of most other materials ($\Delta H = -142,700$ kJ/kg in hydrogen vs. $-49,500$ kJ/kg for octane). The energy density of hydrogen, on a volumetric basis, is also high but not as high as the energy density of liquid fuels. Compared to compressed gases, its energy density is much higher than that of compressed air. At 20 bar pressure, 1 m^3 of H_2 stores 229 MJ of energy (in chemical form), while 1 m^3 of air at the same pressure would store merely 6.0 MJ of energy (as the exergy of compressed air). At 700 bar, the energy density of hydrogen is 4800 MJ/m^3, while that of compressed air is 459 MJ/m^3.

2. Because hydrogen is a stable compound and does not change form, it may be stored and used after a long time, e.g., for the seasonal storage of wind energy from the spring to the summer season.

3. It readily combines with oxygen to form water. Emissions from hydrogen systems are harmless and nonpolluting.

4. It is abundant on the surface of the earth and may be relatively easily produced by the electrolysis of water, a well-known technology and available to all. As with all energy conversion processes, a fraction of the energy input is lost in the electrolytic conversion, and this is reflected in the efficiency of the process, currently in the range of 65–78% [13].

5. It is not harmful to the environment if released and does not pose any health threats to humans.

6. It may be readily used in fuel cells with potentially high conversion efficiency.

7. It may be used as transportation fuel in vehicles that are powered by either IC engines or by fuel cell motors.

8. It has very low viscosity and may be transported in pipelines with relatively low frictional losses and cost.

9. It may be used as combustion fuel, e.g., in conventional gas turbines, as well as for the direct conversion to electricity in fuel cells.

10. There is significant expertise in storing and handling hydrogen, since for a long time, hydrogen has been used as an industrial material and as propulsion fuel in the space shuttle and several types of rockets.

Hydrogen has several disadvantages and shortcomings as an energy storage medium, with the most important of them as follows:

1. Because the hydrogen molecule is very small and light, hydrogen readily diffuses through the matrix of metals including steel. When hydrogen diffuses through metal matrices, it causes *hydrogen embrittlement* and *decarburization*. In order to avoid these processes that weaken the materials, special polymer coatings are used for the interior of storage vessels and the hydrogen-handling pipelines. In addition, tanks made of carbon–fiber composite materials have been developed that may withstand pressures up to 700 bar [14].

2. Hydrogen is flammable and explosive. It must be kept in special containers and transported under controlled conditions. Because of its high specific energy and high diffusivity, hydrogen may form a powerful explosive mixture with air. For this reason, hydrogen containers or pipelines must be well designed to avoid leakage and mixing with air.

3. Since hydrogen is the lightest element, its material density is very low under ambient conditions (0.08 kg/m^3 vs. 1.2 kg/m^3 for air). Hydrogen gas does not favorably compare with fossil fuels on a volumetric basis (energy density): 1 m^3 of hydrogen at standard conditions would carry 11.8 MJ, while the same volume of gasoline carries 34,800 MJ and that of natural gas carries close to 28 MJ. Such considerations make it necessary for the use and transportation of compressed hydrogen.

If hydrogen is to be used as energy storage medium, it must be stored under high pressure or as a liquid. The higher pressure for gaseous containments implies stronger vessels with thicker walls that make the vessels massive and costly. In addition, the compression process consumes energy that is not recovered in a typical fuel cell or in a burner. Hydrogen tanks have been manufactured for prototype vehicles that operate in the range of 300–700 atm. The material density of hydrogen at the upper limit of 700 atm is approximately 37 kg/m^3, and its energy density is 5,280 MJ/m^3, still lower than that of gasoline, but very significant. Because of the much higher pressure, the tanks of the hydrogen vehicles must be larger and heavier than gasoline or diesel tanks, which operate at atmospheric pressure.

Liquid hydrogen is also a possibility for energy storage: Liquid hydrogen has a density close to 71 kg/m^3, which corresponds to an energy density of 10,132 MJ/m^3. However, liquid hydrogen must be kept at very low temperatures, because the boiling point of hydrogen at 1 atm is approximately 20 K, or −253°C. Such cryogenic temperatures are impractical and costly to maintain for everyday applications.

Another method is to store hydrogen in metal hydrides, such as NaH, MgH_2, and $LiAlH_4$, which have the inherent disadvantages of very high weight for the energy storage medium—the mass of the metal in the hydride by far outweighs the mass of the hydrogen. Hydrides have very low specific energy in megajoules per kilogram. Using hydrides for vehicular transport significantly adds to the weight of the vehicle and lowers its overall energy efficiency. Chemical stability after several energy storage cycles is another problem, because metal hydrides are rather unstable complex compounds and decompose.

7.5.1 Fuel Cells

Fuel cells are essentially batteries that operate as open thermodynamic systems with fluid fuels—gases or liquids. Fuel cells are DEC devices where the conversion of the chemical energy of the fuel to electricity occurs directly without the intermediate production of heat and the associated Carnot limitations on the conversion efficiency. DEC processes are convenient and clean and typically have higher efficiency than thermal power conversion. Fuel cells are open thermodynamic systems supplied with a fuel and an oxidant, e.g., hydrogen and oxygen or air. Their starting time is very short; their operation may be interrupted or ended at will; and they operate continuously. In short, fuel cells may operate at the will of the operator continuously or intermittently to produce electric power from their fuels.

Fuel cells were invented in 1802 by Sir Humphrey Davy, almost one century before the IC engine, and were used for the operation of tractors as early as 1839 by Sir William Grove. However, they have not been widely used in practical applications of transportation because of the convenience and attractiveness of the IC engine. This happened despite their inherent and distinct thermodynamic and environmental advantages for the conversion of energy. The principal disadvantage of fuel cells is that the voltage they generate is very low, on the order of 1 V for the direct conversion of chemical energy. High current has to compensate for the production of the required power, and high current in engines is associated with high power dissipation (generation of heat).

A schematic diagram of a fuel cell operating with hydrogen as the fuel and oxygen as the oxidant is shown in Figure 7.12. Hydrogen is supplied to the cell on the side of the cathode, the negative terminal, and oxygen is supplied on the side of the anode, the positive

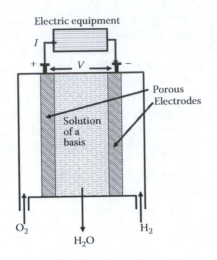

FIGURE 7.12
Schematic diagram of the hydrogen–oxygen fuel cell.

terminal. The two electrodes are composed of a porous material and enclose a basis solution of potassium hydroxide (KOH), which dissociates to produce K^+ and OH^- ions. The hydrogen gas diffuses through the porous electrode in the cathode and combines with the hydroxyl ions (OH^-) there to form water molecules. In this process, the hydrogen releases two electrons according to the reaction

$$H_2 + 2HO^- \rightarrow 2H_2O + 2e^-. \tag{7.22}$$

The cathode of the fuel cell has a surplus of electrons, and this creates a negative charge. On the side of the anode, oxygen gas also diffuses into the porous electrode of the anode and combines with water molecules and electrons to form hydroxyl ions (OH^-):

$$\frac{1}{2}O_2 + H_2O + 2e^- \rightarrow 2HO^-. \tag{7.23}$$

As in the case of the solid batteries, a deficiency of electrons is created in the anode, and this induces a positive electric charge. The potential difference created between the anode and the cathode induces an electric current to pass through an external circuit. It is easy to conclude (by adding the last two chemical reactions) that the overall reaction in the hydrogen–oxygen fuel cell is the formation of water:

$$H_2 + \frac{1}{2}O_2 \rightarrow H_2O. \tag{7.24}$$

The water-forming reaction from oxygen and hydrogen takes place in the fuel cell, and the ions in the solution of the basis are the catalyst for the reaction. Hydrogen and oxygen may be continuously supplied to their respective chambers, from storage tanks, and the produced water may also be continuously removed from the fuel cell through a drain. This allows the fuel cell to continuously operate for long periods. In practical fuel cells, the

TABLE 7.2

Fuel Cell Reactions, Gibbs Free Energy, and Maximum Voltage at 298 K

Fuel	Reaction	$-\Delta G^0$, kJ/kmol	V_{max}, V
H_2	$H_2 + 1/2 O_2 \rightarrow H_2O$	236,100	1.22
CO	$CO + 1/2 O_2 \rightarrow CO_2$	275,100	1.43
CH_4	$CH_4 + 2O_2 \rightarrow CO_2 + 2H_2O$	831,650	1.08
CH_3OH	$CH_3OH + 3/2 O_2 \rightarrow CO_2 + 2H_2O$	718,000	1.24
C_2H_5OH	$C_2H_5OH + 3O_2 \rightarrow 2CO_2 + 3H_2O$	1,357,700	1.17

production of water dilutes the basic KOH solution. Part of the diluted solution is drained from the cell for operators to periodically supply the solution with concentrated KOH solution and restore the desired concentration in the fuel cell.

The maximum voltage developed in the hydrogen–oxygen fuel cell from Equation 7.17 is

$$V_{max} = \frac{-\Delta G^0}{\zeta F} = \frac{236,100}{2 \times 96,500} = 1.223 \text{ V}. \tag{7.25}$$

This is a very low voltage, which is typical of fuel cells. Several types of fuels, other than hydrogen–oxygen, have been developed. Table 7.2 lists a few of these fuels with the corresponding overall oxidation reactions, the Gibbs free energy of these reactions [15], and the electromotive force, or maximum voltage, developed according to Equation 7.25.

Of the fuels listed in Table 7.2, methanol (CH_3OH) and ethanol (C_2H_5OH) are liquids with significantly higher material densities than hydrogen and other gases. While the specific energy of these liquid fuels is lower than that of hydrogen, their high material density and high energy density is an advantage because fuel cells (and storage systems based on liquids) do not need to be voluminous or under high pressure.

Because of their high energy density, a great deal of research has been conducted in the development of fuel cells that would directly convert methanol and ethanol* to electric power. A problem encountered in the development of hydrocarbon fuel cells is that liquid hydrocarbon molecules that are water soluble are "dragged" with the H^+ ions through the pores of the membranes to the cathode by electroosmosis. In the cathode, the hydrocarbon molecules combine with the oxygen ions and burn, but this combustion process does not contribute to the production of electric energy, and a fraction of the liquid hydrocarbon fuel is wasted. This is sometimes called the *crossover effect* that reduces the voltage of the fuel cell and degrades the fuel cell performance. A few polymeric membranes have shown promise to have lower fuel crossover [16]. The development of membranes that minimize fuel crossover and the manufacture of commercial-grade liquid hydrocarbon fuel cells is an ongoing area of scientific investigation.

It is also apparent in Table 7.2 that all the practical fuel cells produce a maximum voltage, on the order of 1 V. The low voltage implies that for practical applications, several fuel cells must be connected in series, and these are referred to as "stacks" of fuel cells. This practice increases the thermodynamic irreversibility of the system; it reduces the fuel cell efficiency and reduces the quantity of electrical energy that becomes available. Typically, fuel cells are in stacks of 20–30 units, which provide operational voltages close to 30 V. The fuel cells

---

* It will be ironic if a small dose of vodka (or another type of spirit that contains ethanol) may be used as the fuel that powers our laptops via a fuel cell.

are joined by the *interconnect*, a metallic or ceramic layer that connects the different cells. Because the interconnect is exposed to both the oxidizing and the reducing sides of the cell at high temperatures, it must be made of a very stable material.

The Gibbs free energy and, by extent, the voltage developed in fuel cells are strong functions of the temperature. Several reactions are known for which the maximum voltage at elevated temperatures is significantly higher than the voltage at ambient conditions. In addition, the rates of all chemical reactions are faster at elevated temperatures, and hence, the electric current and the power produced by high-temperature cells are significantly higher than the power produced by the same cells at ambient temperatures. High-temperature fuel cells (HTFCs) have been developed with operating temperatures in the range of 600–1000°C. The maximum voltages achieved are in the range of 3–7 V. Because water evaporates at such high temperatures, other more stable materials are used to facilitate the ionization reactions at the two electrodes. Several solid oxides withstand the high temperatures and are used in the solid oxide fuel cells (SOFCs).

Because of the high temperatures, a SOFC does not use water as the electrolyte, but a dense layer of ceramic materials that conducts the ions of the oxygen and the fuel. Both the cathode and the anode in an SOFC are good electric conductors with high porosity, to allow the diffusion and transport of ions. Lanthanum–strontium–manganite (LSM) is a material that is frequently used for the cathode of HTFC, and the yttria-stabilized zirconia mixed with nickel metal for high conductivity is a material commonly used as the anode. Recent advances in materials science have produced a variety of materials with desirable properties that may be adopted for the development of better, more powerful, and more efficient fuel cells.

In comparison to the batteries, the fuel cells have the distinct advantage that they may continuously operate. Fuel cells do not need to be recharged, because the fuel and the oxidant are continuously supplied from storage tanks. Chemical energy stored in a hydrogen tank (or another suitable fuel) is continuously fed to stacks of fuel cells, which, in turn, supply vehicle engines and buildings with electric power. When massively produced, fuel cells have the potential to become the future power plants in a society that produces large quantities of hydrogen from electrolysis using renewable energy sources.

As DEC devices, fuel cells convert the chemical energy directly into electric energy without the intermediate stage of combustion and the production of thermal energy. Because of this, fuel cells are not subject to the Carnot limitations of the heat engines, given by Equation 1.7. In principle, the entire chemical energy of a fuel cell $-\Delta G^0$ may be converted to electric energy ($-\Delta G^0 = VIt$) in an ideal fuel cell that operates reversibly. In practice, thermodynamic losses occur, the practical fuel cells do not operate reversibly, and a fraction of their chemical energy is dissipated as heat. The result of the irreversibilities in the fuel cells is that the actual voltage obtained is lower than the theoretical maximum voltage given by Equation 7.17.

All the irreversibilities associated with the practical operation of fuel cells are lumped together in the concept of efficiency. The *voltage efficiency* and the *energetic efficiency* (or simply, the *efficiency* of the fuel cell) are two figures of merit commonly used for fuel cells. The two are defined as follows:

$$\eta_V = \frac{V}{V_{max}} \quad \text{and} \quad \eta = \frac{E}{-\Delta G^0}, \tag{7.26}$$

where V is the actual voltage produced and V_{max} is the maximum voltage obtained from Equation 7.17; E is the actual energy produced by the fuel cell per kilomole of fuel and $-\Delta G^0$ is the molar Gibbs free energy. Typical energetic efficiencies for fuel cells η are in the range of 60–75%.

The fraction of the chemical energy $-\Delta G^0$ that is not converted to electricity is dissipated as heat in the fuel cell stack. Because of this, fuel cell systems generate heat fluxes that must be removed by a cooling system, typically by a cooling water loop. In order to harness the dissipated heat, fuel cells may be used as cogeneration devices to supply both heat and electric power. A large fraction of the heat generated by the HTFCs, which operate at 600–1000°C, may be used to raise steam and produce electric power with a conventional Rankine cycle. The combination of electric energy produced by the HTFC and a smaller Rankine cycle plant may achieve overall efficiencies in the range of 80–85%.

Example 7.10

A cylindrical tank with 1.2 m diameter and 5 m height contains hydrogen at 300 bar at 300 K. The hydrogen in the tank may be used to produce electric energy by combustion in a gas turbine with 36% overall efficiency or by DEC in a fuel cell with 75% efficiency. Determine the amount of electric energy that may be produced using the two methods.

Solution: The volume of the cylinder is $V = \pi D^2 H/4 = 5.65$ m^3. At 300 bar, hydrogen may be approximated as an ideal gas, and hence, the mass contained in the cylinder is $m = PV/RT = 300 \times 10^5 \times 5.65/(4{,}157 \times 300) = 135.9$ kg.

From Table 4.1, the heat of combustion of hydrogen is 119,950 kJ/kg. The combustion of this quantity of hydrogen produces $135.9 \times 142{,}700 = 16.3 \times 10^6$ kJ of heat. When this amount of heat is converted with 36% efficiency, it produces 5.86×10^6 kJ or 1,630 kWh.

If a fuel cell is used, its efficiency, according to Equation 7.26, is based on $-\Delta G^0$, which is $236{,}100/2 = 118{,}050$ kJ/kg. With 75% efficiency, the fuel cell will produce $118{,}050 \times 135.9 \times 0.75 = 12.03 \times 10^6$ kJ $= 3{,}342$ kWh.

The DEC conversion method produces more than twice as much electrical energy and is by far preferable.

Example 7.11

The gasoline-operated IC engine of Example 7.9 is substituted with a hydrogen–oxygen fuel cell that has a 65% efficiency. The motor of the new car is 95% efficient. Determine the mass of hydrogen required in a hydrogen tank for this car to have the same distance range. If the hydrogen is kept at 350 bar, assuming it is an ideal gas, how much must be the volume of this tank if the car operates at ambient temperatures close to 300 K?

Solution: It was determined in Example 7.9 that 0.52×10^6 kJ of mechanical work is needed for the propulsion of this car in the desired distance range. With 95% electric motor efficiency and 65% fuel cell efficiency, there must be enough hydrogen in the tank to supply $0.52 \times 10^6/(0.95 \times 0.65) = 0.84 \times 10^6$ kJ of work.

From Table 7.2, for hydrogen, $-\Delta G^0 = 236{,}100$ kJ/kmol, and hence, $0.84 \times 10^6/236{,}100 = 3.56$ kmol (7.12 kg) of hydrogen need to be stored in the hydrogen tank at 350 bar (350×10^5 Pa). At this pressure, the ideal gas approximation is valid for hydrogen, and from $PV = nRT$, $V = 3.56 \times 8{,}314 \times 300/350 \times 10^5 = 0.254$ m$^3 = 254$ L.

A comparison of the data from Examples 7.9 and 7.11 is as follows:

Type of Car	Work Needed, kJ	Mass of Fuel/ Battery, kg	Tank Volume, L
IC gasoline engine	0.52×10^6	53.2	74
Electric LiMnO$_2$ battery	0.55×10^6	380	n/a
H$_2$ fuel cell	0.84×10^6	7.12	254

7.5.2 Practical Types of Fuel Cells

With fossil fuels depleting at a fast rate and the emphasis on renewable energy sources, fuel cells operating with hydrogen and synthetic fuels have become critical devices to resolve the energy challenge of the future. At the beginning of the twenty-first century, there is a great deal of research on practical fuel cells that will not only have high efficiency, but will also be durable and will reliably produce power. Several types of fuels cells have been developed to power different engineering systems—e.g., vehicles, space crafts, domestic power, and industrial power. The practical fuel cells are characterized by the type of their electrolytes and their membranes. Most of the practical fuel cells use hydrogen as fuel and oxygen or air as the oxidant. The most common types of them are as follows [17,18]:

1. The *polymer electrolyte* fuel cells use polymer membranes (Nafion™ 117 is the most common) and operate at lower temperatures, below 100°C. Their electrodes are coated with gold, silver, palladium, or platinum, which act as catalysts for the oxygen–hydrogen reaction.

2. *Alkaline fuel cells* (AFCs) use KOH solution as an electrolyte and operate in the temperature range of 60–250°C. Platinum, chromium, and nickel have been used as catalysts for these fuel cells that have powered the Apollo mission to the moon and the space shuttle.

3. *Phosphoric acid fuel cells* (PAFCs) use this acid (H_3PO_4) as their electrolyte and operate in the range of 160–210°C. Porous silicon carbide is used for the separation of the oxidant and fuel and for the diffusion of ions. Their electrons are made of noble metals (ruthenium, rhodium, palladium, silver, osmium, iridium, platinum, and gold), which are not eroded by the acid.

4. *Molten carbonate fuels cells* (MCFCs) utilize molten alkali metals (Li, K, Na, or alloys of them) in a porous matrix of lithium–aluminum oxide, with nickel electrodes. The range of their operating temperatures is 600–800°C, and their fuel can be hydrogen or carbon monoxide.

5. SOFCs use phosphoric acid (H_3PO_4) in a porous matrix of silicon carbide, with electrodes made of noble metals that act as catalysts. Their operating temperatures are very high, in the range of 800–1000°C.

Table 7.3 shows several state-of-the-art types of fuel cells and their characteristics.

It is apparent from Table 7.3 that the several types of fuel cells have the capability to produce a high amount of power for consumption in a cluster of buildings or a small factory. However, the currently achieved efficiencies of the fuel cells are well below the limit of 100% for DEC. More research and development efforts are required for practical

TABLE 7.3

Characteristics of Practical Fuel Cells

Type	Temperature, °C	Power	Efficiency, %	Electrolyte
PEM	up to 100	1–100 kW	40–50	Sulfonic acid
AFC	50–90	10–100 kW	50–60	KOH solution
PAFS	150–200	0.2–10 MW	~50	Phosphoric acid
MCFC	~700	0.2–3 MW	60–70	Carbonate solution
SOFC	up to 1000	1–2 MW	60–70	Yt or Zr oxide

fuel cells to become highly efficient, reliable, long-lasting, and economically viable. The development of energy storage systems with highly efficient electrolysis processes for the production of hydrogen and cost-effective fuel cells to convert the stored energy to electricity (at the time when electricity is needed) will open the "flood gates" for the utilization of solar and wind energies that are not dispatchable and need to be stored seasonally and diurnally.

7.5.3 Hydrogen Economy

The high specific energy of hydrogen in combination with its abundance on the surface of the earth (as water) and the relative easy way to produce it (by electrolysis or chemical reactions) make this gas a very viable and desirable energy storage medium. The chemical stability and energy storage characteristics and properties of hydrogen make it ideal for the long-term storage of energy. When the fossil fuels are depleted and renewable energy is to be produced in large quantities and stored between day and night (e.g., solar) or between days and seasons (e.g., wind energy from the spring to be used in the summer or from the autumn to be used in the winter), the only viable options for storage are very large-scale PHS and hydrogen. Hydrogen has a significant advantage because of its very high specific energy and relatively high energy density. Hydrogen storage is expected to become universal and will be adopted as fuel for power plants in municipalities and households. It is expected that improved materials and metal coatings for the handling and storage of hydrogen will be developed, in combination with improved methods for its production, storage, and transportation. Several scientists have suggested that the widespread use of hydrogen as energy storage fuel will transform the future economy of the planet into the *hydrogen economy* [18,19]. In a futuristic hydrogen economy,

- Renewable energy sources are utilized to produce hydrogen, primarily by electrolysis.
- Hydrogen is supplied everywhere by a system of pipelines, similar to the pipeline network that delivers currently natural gas.
- Automobiles, trucks, and trains use hydrogen as a fuel for their IC engines or they use fuel cells.
- Instead of centralized power plants for the production and distribution of electricity, electric power is locally produced by fuel cells, in buildings that are supplied with hydrogen from the pipeline network.
- Industrial processes, where high temperatures are necessary, use hydrogen combustion or electric power to achieve the desired temperatures.
- There are almost zero CO_2 emissions in a hydrogen economy, and the pollution is minimal.

A drawback to materializing these expectations is that hydrogen does not naturally occur in large quantities and must be artificially produced using other primary energy sources at significant energy cost. At present, the large-scale hydrogen production is by far more expensive than the use of fossil fuels.

The term *hydrogen economy* was coined in the 1970s to describe the economy after the depletion of fossil fuels [17–19]. Hydrogen is not a primary energy source but an energy carrier, similar to electricity. Hydrogen will be artificially generated from water, e.g., by electrolysis using the energy harnessed by the abundant solar or wind power, or by a chemical method using the heat generated in nuclear power plants. With suitable materials for the storage and transportation of the produced hydrogen, this chemically stable element will be transported to the final users. The low viscosity of hydrogen mitigates its low density because higher volumetric rates of hydrogen than those of natural gas may be transported in pipelines with the same pumping power. Hydrogen gas pipelines will deliver similar energy density (e.g., in kJ/m^3) to the current network of natural gas pipelines. Unlike electricity, hydrogen may be stored for long times and can be used when the demand arises. The production of hydrogen from renewable energy sources and the use of the stored and transported hydrogen would virtually eliminate carbon dioxide emissions. A hydrogen economy would have de facto eliminated the most plentiful of the greenhouse gases and would have alleviated global climate change. In addition, the hydrogen economy may become a panacea for developing countries that do not have fossil fuel reserves: since renewable energy sources, such as wind and solar, are widely and uniformly distributed on the planet, all nations will be capable of achieving energy independence and avoid expensive fossil fuel imports, by using a combination of renewable energy and hydrogen storage.*

Whether or not a global hydrogen economy will evolve in the near or far future greatly depends on the technological advances related to the storage and transportation of this gas, other methods for energy storage, the cost of alternative energy storage systems, and the future implementations of national and international regulations for the curtailment of fossil fuel combustion. Proponents of a world-scale use of hydrogen argue that this gas is the cleanest source of energy known to end users, particularly in transportation applications, where it does not release any particulate matter and greenhouse gases. Critics of the transition to a hydrogen economy contend that the cost of switching to a national or a global hydrogen distribution system may be prohibitive, and an intermediate step may become economically more viable: for example, synthetic fuels from locally produced hydrogen and atmospheric CO_2, such as ethanol and methanol, might accomplish the same goals of a hydrogen economy at significantly lower investment. Since the CO_2 input will be from the atmosphere, this arrangement will not contribute to the growth of the GHG emissions. While this is also a viable solution to the energy challenge of the world, in the second decade of the twenty-first century, one may clearly see small but persuasive signs that point to more hydrogen production and a slow transition to a global hydrogen economy [1]:

1. Several hospitals, schools, universities, and governmental buildings have installed combined units that accomplish the electrolysis of water and storage of hydrogen to be used with fuel cells for emergency power, whenever there is an electric power disruption. These energy storage systems are excellent for emergency use, because they have very low maintenance requirements, may instantly produce power, and are nonpollutant. Fuel cells may be located at any place within the building, as opposed to combustion-driven generators that must have adequate space for the ventilation of exhaust gases.

2. Hydrogen and fuel cell pilot programs for commercial applications have started in all OECD countries as well as in Russia, China, India, Brazil, and several countries of the Middle East.

---

* This will require significant capital investment for equipment for renewable energy production, hydrogen production, storage, and transportation.

3. Major automobile manufacturers have started to market car models that operate with hydrogen fuel cells. The Honda FCX Clarity; the Kia Borrego; the Hyundai ix35 FCEV; the Ford Focus FCV; and the Toyota Mirai, which is depicted in Figure 7.13, have models available with engines powered by hydrogen fuel cells. Most of these models also include solid-state batteries to improve their driving range.

4. Several European communities from Iceland to Greece have adopted public buses that make use of hydrogen and use fuel cells for propulsion.

5. In order to serve the hydrogen-powered vehicles, countries such as Portugal, Iceland, Norway, Denmark, Germany, Japan, and Canada, as well as several states in the United States, such as California, Oregon, Minnesota, and Texas, have started investing in hydrogen distribution network systems. Even though these systems have proven to be initially costly, technological breakthroughs and improved methods of hydrogen transport will become very lucrative in a society dominated by the production and use of hydrogen.

The concept of the hydrogen economy has become a futuristic concept that has drawn a great deal of criticism and debate, primarily stemming from the expense of the hydrogen fuel cells and the expense of a hydrogen distribution infrastructure. Our society is currently dominated by fossil fuels to the point that it is almost impossible to think of a mode of transportation outside the framework of liquid fossil fuels—gasoline and diesel. However, one must not forget that both of these fuels were entirely novel and very little used until the end of the nineteenth century. At that time, the use of petroleum and natural gas was at an embryonic stage. The vast pipeline distribution systems for the two now vital

FIGURE 7.13
Toyota Mirai is powered by fuel cells and is marketed in Japan, several countries of the European Union, and California.

fossil fuels did not exist and, perhaps, were beyond the imagination of the nineteenth-century humans.

Hydrogen is to the humans of 2017 what gasoline was in 1880, and one may draw parallels in their development and uses: During the twentieth century, a century that is characterized by the widespread use, the extensive exploration, and the rapid depletion of fossil fuels, a planet-wide infrastructure was developed for the mining, transportation, and distribution of liquid and gaseous fossil fuels. The petrol/gasoline station, which is ubiquitous in the present-day landscape, was not always present in almost every corner of urban developments. To the citizens of 1880s, today's infrastructure for the production, transportation, and distribution of fossil fuels would have appeared prohibitively expensive and out of reach. Similarly, a future with an extensive hydrogen infrastructure appears the same way to the citizens of the early twenty-first century. The universal high demand for petroleum and natural gas was instrumental in building the expensive infrastructure for fossil fuels (oil and gas wells, tankers, refineries, pipelines, gasoline stations, etc.) in both the market and the centrally planned economies. A similar demand for hydrogen in the future will also help build the required infrastructure for the production, storage, and transportation of hydrogen worldwide. Profitability and the market forces will take care of the needed investment for the production and distribution infrastructure of a hydrogen economy.

It is undisputable that gasoline and diesel will be exhausted at some point in the near or far future. There will be a point in the future, when the use of fossil fuels will be significantly curtailed and will finally cease. Another energy source will inevitably take the place of the currently used fossil fuels, and hydrogen, produced from renewable energy sources, is a good candidate to become the means to store and distribute this energy. This inevitable technological and societal evolution will be accomplished with private and public investments in the production, transportation, and distribution of hydrogen. The hydrogen filling station may become in the future as ubiquitous as the gasoline/petrol station is in the beginning of the twenty-first century. The difference between the two fuels is that hydrogen may be continuously produced by renewable or other reliable and environmentally friendly energy sources in all the countries and by all human communities, instead of being created at finite quantities eons ago in only a few privileged regions in the globe. The hydrogen distribution station and the hydrogen-based economy are sustainable and will last much longer than the fossil fuels.

7.5.4 Case Study of Hydrogen Energy Storage for Buildings

A study was undertaken [12] for the conversion of two existing buildings to become grid-independent buildings (GIBs) using hydrogen storage. GIBs are not connected to the electric grid and do not contribute to the U-shaped electricity demand pattern (or duck curve pattern), which is the focus of Figure 7.5. The U-shaped demand will become a serious issue with the proliferation of solar ZNEBs (which supply the grid with electric power during the early daylight hours and draw a great deal of power in the evening and nighttime) because it will require the shutting of large base-load power plants. The two buildings are located (a) in Fort Worth, Texas, at the latitude $31°$, where a great deal of energy is used for air-conditioning in the summer months and little energy as heat in the winter, and (b) in Duluth, Minnesota, at the latitude $47°$, where there is no need for air-conditioning in the summer, but a great deal of heat is needed in the winter months. The buildings have two stories of total area 3,750 ft^2 (348 m^2) that is serviced by the HVAC system. Both buildings correspond to large residences in the two cities—their living space places them in the upper 20% of residential buildings.

The hourly demand for thermal energy and electricity was obtained for the two buildings for the 8,760 hours of the year. Currently, the buildings have a natural gas heating system. When the two buildings become GIBs and are totally dependent on the electricity generated from the PV systems, the heating is provided by an efficient ground source heat pump (GSHP) with an average coefficient of performance (COP) of 4.5. Hydrogen, which is produced by electrolysis, is the energy storage medium. The hydrogen is compressed and stored in tanks that may withstand high pressures, up to 500 bar. To satisfy the hourly demand for electric power and thermal energy, the buildings use either the electric power of the PV cells directly—during the daytime—or the hydrogen-stored energy in the tank—during the nighttime and hours of low insolation. Table 7.4 gives the operational parameters of the components in the energy production, storage, and delivery for the two buildings.

Figure 7.14a shows the monthly energy demand and energy production for the building in Fort Worth as well as the maximum hydrogen storage during the month. One may see the high electricity demand for air-conditioning and the almost zero heat demand in the summer. The heat (thermal energy) units were converted from standard cubic feet of natural gas and kilojoules to kilowatt hours, to fit the graph. Because a GSHP with COP of 4.5 is used in both buildings, these numbers are divided by 4.5 to yield the total electricity demand of the building. Figure 7.14b has the same information for the building in Duluth. It is observed in both figures that the total energy produced by the PV cells is significantly higher than the energy demand of the buildings. This happens because a high fraction of the demand is supplied by the storage system and the round-trip efficiency of the system (electrolysis, fuel cell and inverter) is close to 41%. The calculations revealed that converting the two buildings to GIBs (without any conservation measures) would require approximately 99 m^2 of PV panels with rated power of 18.2 kW for the Fort Worth building and 161 m^2 of PV panels with rated power 29.1 kW for the building in Duluth. These are very high PV cell areas and very high rated powers, which may be reduced by energy conservation measures in the buildings and improved round-trip efficiency of the energy storage systems [12].

Figure 7.15a depicts the storage capacity of the hydrogen tank, in kilomoles and in kilowatt hours, corresponding to the building in Fort Worth. The ordinate is the number of hours since the beginning of the year (1:00 am of January 1 is hour 1). It is observed that the maximum energy is stored in the beginning of June and is spent at a fast rate during the hot summer months. The minimum storage level occurs in the middle of October, when the demand for air-conditioning in the region subsides. Figure 7.15b depicts the same information for the building in Duluth. It is observed that the maximum storage capacity for this building occurs in October, when the cold season starts. The minimum occurs in March, when the weather improves, lesser heating is needed in the building,

TABLE 7.4

Operational Parameters of the Energy Production, Storage, and Delivery in the Two Buildings, at Fort Worth and Duluth

PV cell efficiency at 25°C	18%
Electrolysis efficiency	70%
Fuel cell efficiency	65%
COP of the GSHP	4.5
Voltage inverter efficiency	90%
Hydrogen tank maximum pressure	500 atm

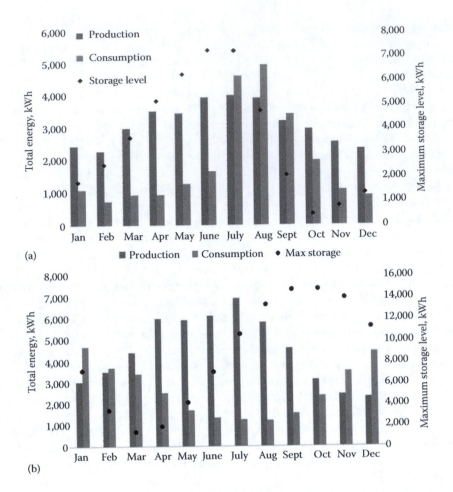

FIGURE 7.14
(a) Monthly energy demand and production as well as maximum level of storage for the GIB in Fort Worth;
(b) monthly energy demand and production as well as maximum level of storage for the GIB in Duluth.

and in addition, there is higher insolation to produce electric energy. It is apparent in the two figures that (apart from the diurnal cycle energy storage) the two buildings store energy for several seasons to use it when the energy demand is high. The timescale for efficient building energy storage spans several seasons.

Another observation in the last two figures is that the energy storage capacity at the two locations has very strong negative correlation. If the buildings did not have their own storage systems and were supplied with hydrogen from a centralized location (e.g., in Arizona), the monthly hydrogen demand in the centralized facility would not have high variability.

The computations also showed that both the need for PV cell ratings and area and the hydrogen tank capacity would significantly decrease with (a) better building insulation and (b) higher efficiency for the fuel cells and for the electrolysis process [12].

Example 7.12

An existing building is to be converted to a GIB. The building will be fully powered by solar energy with hydrogen storage for the low- and zero-insolation hours. The annually averaged power consumed by the building is 6.8 kW. The insolation will directly supply 38%

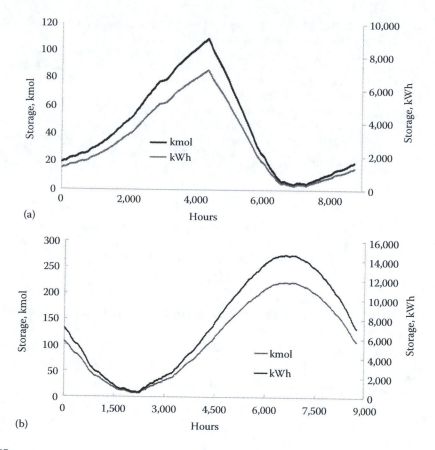

(a)

(b)

FIGURE 7.15

(a) Amount of hydrogen stored and equivalent electric energy in kilowatt hours for the GIB in Fort Worth; (b) amount of hydrogen stored and equivalent electric energy in kilowatt hours for the GIB in Duluth.

of the annually consumed energy of the building with the remainder of 62% coming from hydrogen fuel cells. Hydrogen will be produced by electrolysis with efficiency of 72%, and the fuels will have 65% efficiency. The DC-to-AC inverter will have 95% efficiency.

Determine for an entire year (a) the total energy supplied to the building directly by the solar panels (b) the total energy supplied by the fuel cells, (c) the total energy the solar panels need to produce and (d) the total mass of hydrogen produced and consumed in the building.

Solution:

a. With 6.8 kW average power consumption, the building consumes 6.8 × 24 × 365 = 59,568 kWh annually. Since the solar panels will directly supply 38% of the total energy, 59,568 × 0.38 = 22,636 kWh of energy will be directly supplied by the solar panels.

b. The rest of the annually consumed energy 59,568 − 22,636 = 36,932 kWh will be supplied by the fuel cells.

c. For the energy supplied via the fuel cells, the solar panels must produce a higher amount to compensate for the losses in the process of electrolysis, the conversion to DC in the fuel cells, and the inversion to AC. Since the efficiencies of the last three processes are 70%, 65%, and 95%, the solar panels must produce $36{,}932/(0.7 \times 0.65 \times 0.95) = 85{,}441$ kWh (this is more than two times the actual energy supplied to the building by the fuel cells!).

When the energy is directly supplied by the solar panels, it passes through the inverter with 95% efficiency. Hence, the solar cells must produce $22{,}636/095 = 23{,}827$ kWh. The total amount of energy the solar panels must produce annually is $85{,}441 + 23{,}827 = 109{,}268$ kWh.

   d. The fuel cell–invertor systems supply 36,932 kWh (132.9×10^6 kJ) to the household with 65% conversion efficiency and with 95% inversion efficiency. Since $-\Delta G^0 = 236{,}100$ kJ/kmol for hydrogen, the amount of hydrogen to be produced and consumed throughout the year is $132.9 \times 10^6/(0.65 \times 0.95 \times 236{,}100) = 911.6$ kmol or 1823.2 kg of hydrogen.

7.6 Characteristics, Timescales, and Cost of Energy Storage

Desirable characteristics of actual energy storage systems are as follows:

1. High round-trip efficiency. This ensures that only a small amount of energy is dissipated and wasted during storage.

2. High specific energy (kJ/kg) and energy density (kJ/m^3). This warrants low weight and/or low volume for the storage of a given quantity of energy.

3. Low capital cost to build and maintain the energy storage system.

4. Low energy loss during the storage period. This is particularly important for energy storage systems designed to store energy seasonally, e.g., from spring to summer.

5. Immediate availability of the energy and needed power. This implies short response times for storage and discharge. When the characteristic time of discharge is low, a high amount of power may be produced.

6. Reliability. The stored energy must become available when it is needed, not hours later. Also, the storage system must satisfy the entire demand for power.

7. Durability and long cycle life. This also reduces the lifetime costs of the system.

From the descriptions of the energy storage systems, it becomes apparent that mechanical springs and torsion bars are not adequate to store large quantities of energy for the production of electric power. Flywheels, capacitors, and ultracapacitors may store moderate quantities of energy for short times and may be used to improve the power quality, but not for long-term storage. Solid batteries may be used for the storage of smaller quantities of energy, but they are not suitable for utility-level storage. PHS, CAES, and hydrogen are the most suitable candidates for utility-level energy storage. Table 7.5 summarizes the salient storage capability characteristics of several systems.

Table 7.5 shows that most of the energy storage systems are suitable for storage during the diurnal cycles of energy demand. Batteries, CAES, PHS, and hydrogen storage systems may be used in electric power plants or households to store energy during the hours of low demand and deliver it later on the same day. For seasonal and large-quantity storage of energy, e.g., storage of wind-generated electricity in the spring for use in the summer,

TABLE 7.5

Characteristics of Energy Storage Systems

System	Power Range, MW	Specific Energy, kWh/kg	Storage Losses per Day	Storage Timescale
PHS	10–400	0.001	Almost 0%	Hours–months
CAES	10–400	0.03–0.06	Very low	Hours–days
Flywheel	0–0.3	0.01–0.1	100%	Seconds–minutes
Lead battery	0–10	0.03–0.05	~1%	Hours–days
Li-ion battery	0–0.01	0.15–0.4	~1%	Hours–days
Ni-Cd battery	0–40	0.01–0.3	~1%	Hours–days
Capacitors	0.05	0.005	~50%	Seconds–minutes
Superconducting coils	0.1–5	0.5–2	~12%	Minutes–hours
Hydrogen	0–1000	40	0%	Seconds–years

Source: Michaelides, E.E., *Alternative Energy Sources*, Springer, Berlin, 2012; Zakeri, B., and Syri, S., *Renew. Sustain. Energy Rev.*, **42**, 569–596; Corrigendum, **53**, 1634–1635, 2015.

the PHS and the hydrogen systems are the most suitable systems to store high quantities of energy, to be used with seasonal timescales.

The current cost of building the energy storage system and energy losses in the storage process are also of interest to electricity corporations and to consumers. Table 7.6 gives the total capital cost (TCC) per unit power (€/kW) and the total capital cost per unit energy stored (€/kWh) for the same energy storage systems. For both columns of the total capital costs, the minimum values of the systems from Zakeri and Syri [8] are used, because they represent current technology and best practices. Realistic values of the round-trip efficiencies of these energy storage systems are also shown, based on the actual operation systems and the known efficiency values for the energy storage and recovery processes [1,8,15].

It should be emphasized that the TCC values are subjected to a high degree of uncertainty and that they are mostly estimates, subject to several assumptions [8]. Our experience shows that the capital cost of all technical devices and systems significantly drops when the systems are massively produced and, especially, when they become household items. A relatively recent example of this economic phenomenon is the dramatic drop of the prices of electronic calculators, personal computers, and cellular telephones. The price

TABLE 7.6

Cost and Round-Trip Efficiencies of Energy Storage Systems

System	TCC, €/kW	TCC, €/kWh	Round-Trip Efficiency, %
PHS	1,030	96	~50
CAES	774	48	~50
Flywheel	590	1,850	70–90
Lead battery	1,388	346	70–90
Li-ion battery	874	973	80–95
Ni-Cd battery	927	1,071	60–80
Capacitors	214	691	60–70
Superconducting boils	212	399	92–95
Hydrogen (fuel cells)	2,395	227	~50

of refrigerators (a bulky, household item with dimensions similar to the energy storage systems) dropped 100-fold in the first part of the twentieth century, when these appliances became household items and were massively produced. One reasonably expects that if energy storage systems are widely applied to store renewable energy and massively produced, their cost will significantly drop to the prices of similar household appliances. Other factors that will contribute to lower the energy storage costs are technological breakthroughs, new storage methods, new materials, and improved round-trip efficiency. One can only expect that the cost numbers in Table 7.6 will significantly drop when renewable energy is commonly used and storage becomes a necessity.

7.7 Myths and Reality on Energy Storage

Myth 1: By 2022, our great state will produce 35% of the electricity needed from cheap, clean solar energy.

Reality: While this is a lofty and admirable goal, the production of electricity from solar energy has proven to be more expensive than that produced by nuclear and fossil fuel power units. Judging from current trends and prices, firstly, the electricity produced from insolation will not be so cheap in 2022. Secondly, a major issue with this proposition is the drastic shift of the daily electricity demand pattern to the U-shaped demand curve demonstrated in Figure 7.5 and what this change will imply for the base-load power plants, which currently produce a very high fraction of the electricity for the great state at low cost. Because the significant contribution of the solar installations must be instantly accommodated, the power generated from the rest of the electric grid must be substantially reduced.

When 35% of the total annual electricity demand in the great state is produced by solar installations, the power demand shift will become pronounced. If the demand for the nonsolar units drops to zero, all the nonsolar plants must be shut.

The shift of the demand to a U-shaped curve implies that several or all base-load power plants in the great state will have to shut down during part of the daylight hours, when the electric load is very low. The base-load plants will have to start again before dusk, every day. Large power units operating with steam cycles—nuclear and coal-fired units—cannot be shut down and start again within a short time period. Of the thermal power plants, only gas turbines and diesel engines can operate in this manner. In order to achieve this lofty goal, the great state will have to substitute all or most of its base-load power plants with gas turbines and diesel engines or will have to install substantial utility-level storage systems. This dramatic change of the electricity generation infrastructure is costly for the consumers and needs to be considered before significant changes to the current power production mix are made.

Myth 2: Wind energy is abundant in our state and has now produced cheaper than any other type of energy. We should therefore strive to produce all the energy we need from the wind.

Reality: Even if the cost of energy produced by wind becomes cheaper—a doubtful statement in the absence of governmental incentives and subsidies—wind power is not readily available when the demand arises. The question "how do we run the air-conditioners and lights in our homes on a windless night?" has to be answered before a state or a region commits to produce a high percentage of their power from wind (as it is with solar). In the absence of significant energy storage capacity, which will increase the cost of energy delivered, it will be impossible to satisfy a high fraction of the energy demand of a state.

Myth 3: Car and marine batteries are very efficient and the technology has been with us for decades. We can use systems of a few car batteries in our homes to go 100% green.

Reality: While it is correct that a large system of car (or better, marine) batteries will store enough electricity for a building, this storage system is very large and weighs a lot. A typical car or marine battery with 35 Ah rating may store $12 \times 35 = 420$ Wh or 0.42 kWh. If the batteries are discharged to 20% of their capacity (as recommended by their manufacturers), the energy that becomes available to the household per battery is 0.336 kWh. A typical home in the United States uses 3 kW of daily average power during the summer season, when the air-conditioning demand is high. If the battery system is to provide 12 hours of operation for this home, the storage of 36 kWh is needed, which may be provided by a minimum of $36/0.336 = 107$ batteries. A higher number (e.g., 140) may be needed to overcome the losses in the inverter and to provide for a safety margin. This is by all means a very large system of lead-based batteries. At the average weight 39 lbs (17.7 kg) per battery, this storage system of 140 batteries will add 5,460 lb (2,482 kg) to the weight of the house. In addition, the transportation (for maintenance and periodic regeneration) of such a large number of lead-based batteries will add to environmental deterioration.

Myth 4: Hydrogen will never substitute our energy sources. It is too expensive to even contemplate it.

Reality: The fossil fuels are depleting at a fast rate, and at some point in the future, they will be exhausted and will very likely need to be substituted with the renewable wind and solar energies. While it is correct that the extensive use of hydrogen as an energy carrier will need a new and expensive infrastructure, our society simply cannot function without energy and will have to pay the price. One must remember that the infrastructure we now have for the production, transportation, and distribution of fossil fuels (oil rigs, tankers, refineries, pipelines, gas stations, etc.) did not exist in the middle of the nineteenth century. Developing the fossil fuels infrastructure was also very expensive at the end of the nineteenth century, but it proved to be profitable, and one cannot even contemplate our society without it. The fossil fuel infrastructure was built because energy production and distribution has been necessary for human progress and a highly profitable economic sector. In a market-oriented society, the profitability of the new paradigms for energy supply will attract the necessary investment to build the indispensable energy infrastructure for the future.

PROBLEMS

1. A small 40 MW peak-power gas turbine operates every day from 10:00 am to 6:00 pm during the months of May, June July, and August. The gas turbine is shut down during the rest of the time. Determine the AF and the POF of this power plant.

2. A 1000 MW nuclear power plant operates at full power during the day and at 80% from 11:00 pm to 6:00 am every day of the year. The plant is recharged with fuel and maintained every 18 months. During recharging and maintenance, the plant does not produce any power for 20 days. Determine the AF and the POF of this unit.

3. An 800 MW nuclear power plant produces electricity at full capacity. All the power produced during the day is fed to the power grid. During the nighttime, 40% of the electric capacity is diverted to pump water from a river to a mountainous lake at 740 m higher elevation. The pumped water is used for the production of

additional power during the daytime. The same conduits are used for the return of the water and the pump–motor assemblies double as turbine–generator pairs when the flow is reversed. It is estimated that the frictional and other losses during the pumping and returning operation amount to 15% of the stored energy. The efficiency of the turbomachinery is 70% when they operate in the pump–motor mode and 75% when they operate in turbine–generator mode. What is the peak power this combination of nuclear/hydroelectric power plant produces and how much energy is lost annually by the storage–generation part of this power plant?

4. A base-load nuclear power plant has been constructed in a low valley and is supplemented with a pumped water system to store energy during low demand. A lake with approximate dimensions of $1,250 \times 500 \times 4$ m^3 is located 5 km from the power plant at 630 m elevation. The electric power corporation has received permits to use 33% of the water mass in the lake for a PHS. The water transportation losses are 5% each way, the efficiency of the turbomachinery is 72% when they operate in the pump–motor mode and 76% when they operate in the turbine–generator mode. Calculate the following: (a) What is the stored energy available to the power plant; (b) if all this energy is converted to electric power within 5 hours, what is the additional power produced by the power plant; and (c) the round-trip efficiency of this storage system.

5. A 120 m^3 air vessel has been designed to store pressurized air. A compressor fills this vessel with 420 K and 100 atm air. During the storage period, the temperature of the air drops to 370 K. Determine the following:

   a. The mass of air stored in the vessel

   b. The total exergy of the air when the filling process stops

   c. The final pressure of the air and the total exergy of the air

   d. The exergy loss during the storage period

6. A steel torsion bar is to be used for the storage of energy. The bar is cylindrical ($R = 5$ cm) and has a length of 1.2 m. What is the energy stored in the bar when it is rotated by 15°? What are the friction forces at the grips of the two ends of the bar that would maintain this shear?

7. A flywheel is constructed with thin spokes and resembles a bicycle wheel. Assuming that the mass of the flywheel is 500 kg and that all the mass is concentrated at a radius of 2 m, what would be the revolutions per minute of the flywheel if it were to provide 0.1 MW power for 1 minute?

8. Rock has often been suggested as a medium for the thermal storage of energy. A cylindrical piece of rock with diameter of 10 m and height of 20 m is used for the storage of thermal energy. The rock is insulated and has an average heat transfer coefficient of $U = 0.28$ kW/m^2 K. The density of the rock material is 2,650 kg/m^3, and its specific heat capacity is 0.72 kJ/kg K. The temperature of the rock is raised to 500°C, and the ambient temperature is 25°C. Determine the following:

   a. The total energy stored in the rock in megawatt hours

   b. The temperature of the rock 12 hours after the heating process stops

   c. The heat that has escaped from the rock during this 12-hour period

9. The power demand during a typical summer day in San Antonio, Texas, is depicted in Figure 7.2. This power demand is typical of urban environments in OECD countries during the hot summer season. Approximately 60% of the demand between the

hours of noon to 10:00 pm is due to air-conditioning. In order to reduce peak power demand, it is suggested that eutectic salts be used in conjunction with air conditioners to provide "coolness." The eutectic salts will be frozen from midnight to 10:00 am and will provide cooling from noon to 10:00 pm. Draw a diagram of the power demand during a typical summer day if 20% of the consumers in San Antonio adopt this air-conditioning method. What is the percent reduction of the peak power?

10. The Gibbs free energy change of the Pb–PbO_2 reaction with sulfuric acid is $\Delta G^0 = 286{,}160$ kJ/kmol. Determine what is the maximum voltage an electrochemical cell based on this reaction will produce.

11. Because of frequent power outages in an area, the owners of a small house wish to develop an energy storage system with marine batteries to provide electric energy to their house for 12 hours. They estimate that the average power they will need from the batteries is 1.5 kW. The kind of marine batteries they will buy develops 12 V voltage and has capacity of 65 Ah. The batteries may be safely discharged to 25% of their full capacity. The inverter efficiency of the household is 92%. How many batteries do they need to purchase? If every marine battery weighs 39 lb (17.7 kg), what is the total weight of the system?

12. The owners of the home of problem 11 decide to use $LiCoO_2$ batteries. How much is the weight of this system of batteries?

13. The combustion of methane produces $\Delta H^\circ = 800{,}320$ kJ/kmol (50,020 kJ/kg) of heat, which is typically used in a Rankine or Brayton cycle for the production of electricity. If the overall efficiency of the thermal cycle is 43%, what is the mass flow rate of methane that would power a 10 MW small power plant? If the same quantity of methane were to be used in a system of fuel cells with 70% efficiency, what is the power these fuel cells would produce?

14. A steel vessel of 0.5 m³ may be used for the storage of pressurized air or pressurized hydrogen. The maximum pressure for air is 100 bar, and the maximum pressure for hydrogen is 70 bar. What is the maximum electric work that may be obtained from the storage of the two substances? What is the ratio of the two and what do you conclude about the capability of the two gases for energy storage?

15. A system of hydrogen–oxygen fuel cells is designed to produce 32 MW of electric power for a small town. The efficiency of the fuel cell is 72% and the efficiency of the inverter is 88%. For the rated power of 32 MW, calculate the following:

a. The flow rates (kmol/s) of hydrogen and oxygen consumed by the fuel cells

b. The waste heat generated in the fuel cells and in the inverter

Suggest ways to remove the waste heat from the fuel cell and the inverter.

References

1. Michaelides, E.E., *Alternative Energy Sources*, Springer, Berlin, 2012.
2. El-Wakil, M.M., *Power Plant Technology*, McGraw-Hill, New York, 1984.
3. ERCOT (Electric Reliability Council of Texas), *2015 ERCOT Hourly Load Data*, Hourly Data Archives, ERCOT, Austin, TX.

4. Freeman, E., Occello, D., Barnes, F., Energy storage for electrical systems in the United States, *AIMS Energy*, **4**, 856–875, 2016.

5. Weber, M.E., Making renewables work, *ME Magazine—ASME*, **138**(12), 12, 2016.

6. US EIA (US Energy Information Administration), 2016 *Wholesale Electricity and Natural Gas Market Data*, US EIA, Washington, DC, October 2016.

7. Luo, X., Wang, J., Dooner, M., Overview of current development in electrical energy storage technologies and the application potential in power system operation. *Appl. Energy*, **137**, 511–553, 2015.

8. Zakeri, B., Syri, S., Electrical energy storage systems: A comparative life cycle cost analysis. *Renew. Sustain. Energy Rev.*, **42**, 569–596; Corrigendum, **53**, 1634–1635, 2015.

9. Siegel, N.P., Bradshaw, R.W., Cordaro, J.B., Kruizenga, A.M., *Thermophysical Property Measurement of Nitrate Salt Heat Transfer Fluids*, 5th International Conference on Energy Sustainability, American Society of Mechanical Engineers, Washington, DC, August 2011.

10. Hodapp, R.T., Dallas/Fort Worth International Airport District Energy Plant Upgrades Project—Making More with Less, *Sustainable Communities Conference*, Dallas, TX, March 2009.

11. Keller, L., Phase-change—A new type of thermal storage, *Solar Energy*, **21**, 449, 1978.

12. Leonard, M.D, Michaelides, E.E., Grid-independent residential buildings with renewable energy sources, *Energy*, **148C**, 448–460, 2018.

13. Mazloomi, K., Sulaiman, N., Moayedi, H., Review—Electrical efficiency of electrolytic hydrogen production, *Int. J. Electrochem. Sci.*, 7, 3314–3326, 2012.

14. Riis, T., Hagen, E.F., Vie, P.J.S., Ulleberg, Ø., *Hydrogen Production and Storage*, International Energy Agency, Paris, France, 2006.

15. Moran, M.J., Shapiro, H.N., *Fundamentals of Engineering Thermodynamics*, 6th edition, Wiley, New York, 2004.

16. Neburchilov, V., Martin, J., Wang, H., Zhang, J., A review of polymer electrolyte membranes for direct methanol fuel cells, *J. Power Sources*, **169**, 221–238, 2007.

17. Bockris, J.O'M., Srinivasan, S., *Fuel Cells: Their Electrochemistry*, McGraw-Hill, New York, 1969.

18. Bockris, J.O'M., The origin of ideas on a hydrogen economy and its solution to the decay of the environment, *Int. J. Hydrogen Energy*, **27**, 731–740, 2002.

19. Rifkin, J., *The Hydrogen Economy: The Creation of the Worldwide Energy Web and the Redistribution of Power on Earth*, Tarcher/Penguin, New York, 2003.

8

Energy Conservation and Higher Efficiency

"Energy conservation" is a misnomer because the first law of thermodynamics dictates that energy is conserved. What is usually meant by this colloquial term is that lesser primary energy is used for the performance of a desired action or a process that consumes energy. When one examines the origin of the energy supplied to the process that fulfills the desired action, a lesser quantity of total primary energy source (TPES) is used and a part of the available TPES is "conserved" for future use. The colloquial term *energy conservation* may be expressed more accurately in thermodynamics as *exergy conservation, minimum exergy destruction*, or *minimum entropy production*. The applications of the exergy concept and exergetic calculations ultimately lead to the minimum consumption of primary energy resources, and by extent, to the conservation of natural resources. The upgraded efficiency of machinery and processes also leads to lesser energy consumption and the ultimate conservation of part of TPES for future use.

While most of the other chapters in this book pertain to the supply side of energy, conservation and improved efficiency are directly related to the demand side. By conserving energy and forgoing the use of a fraction of primary energy sources for the performance of desired energy-consuming operations and processes, the global society demands less TPES that supply the demand for energy. As a result, the *natural resources* that supply this energy are conserved for future use. In this chapter, we distinguish between *conservation* and *higher efficiency*, and we present several examples of methods and systems that lead to the lesser consumption of natural resources in all the economic sectors.

It must be noted from the beginning that energy conservation and higher efficiency activities are highly influenced by governmental policies, regulations, and guidelines. Such policies are formulated by taking into account some or all the following parameters:

1. The energy needs of the population
2. Plans for the growth of the population and the economy; larger population and higher gross domestic product (GDP) lead to higher TPES consumption
3. The availability of domestic energy resources
4. The availability and reliability of energy imports
5. Environmental effects of each energy source

The governmental policies and incentives may have any form, including the following:

1. Public campaigns for the promotion of energy conservation measures and energy efficient appliances and products
2. Monetary incentives: rebates, subsidies, high depreciation of investment, tax credits, etc.
3. Voluntary guidelines the industry should meet

4. Governmental regulations that must be met, e.g., the corporate average fuel economy (CAFE) standards for automobiles

5. Product discontinuation policies, e.g., discontinuing incandescent electric bulbs in favor of fluorescent bulbs and light-emitting diodes (LEDs)

6. Environmental regulations on effluents from industrial processes that may favor more energy efficient processes and products

Several countries—among them the European Union (EU), the United States, Canada, Mexico, Australia, Brazil, the People's Republic of China, India, Russia, Japan, South Africa, and South Korea—have recently formed the autonomous *International Partnership for Energy Efficiency Cooperation* (IPEEC), which aims to promote international cooperation among governmental and regulatory agencies that promote energy conservation and higher efficiency. The organization assesses and promotes energy efficiency in transportation, industrial processes, electricity production, buildings, etc. One part of the IPEEC facilitates the diffusion of energy efficiency technology and policies to developing nations. The directives of IPEEC are simply recommendations; they are not binding for the member states until they are adopted by the respective national governments.

8.1 Desired Actions, Energy Consumption, Conservation, and Higher Efficiency

Energy conservation and *energy efficiency* appear to be synonymous concepts and are often used interchangeably. However, there are subtle differences between the two: The best way to describe these differences is to start with the concept of a *desired action* or *societal task* that needs to be fulfilled with the consumption of energy. The action or task in this case is any function, operation, or process performed by individuals, groups, or the entire society that is normally accomplished using some form of energy. Examples of such tasks are as follows [1]:

- Maintaining a warm and comfortable temperature for the residents of a building in Berlin, Germany, during the winter months
- Maintaining a cool and comfortable temperature for the residents of the households of Miami, Florida, during the summer months
- Providing adequate lighting in the classrooms of a high school in Myanmar, so that students may read their books comfortably
- Cooking 0.4 kg of pasta to produce a meal
- Producing 1000 gal of gasoline from crude oil
- Producing 1 t of cement
- Transporting 300 passengers from London to Oxford
- Manufacturing 1,500 m of 8 gauge copper wire

It is important to realize that in the performance of these actions, the groups of individuals who will perform these actions/processes do not demand energy per se. They will use energy forms in order to accomplish one or more of these actions, but their goal is to accomplish the action, not to use a certain form or a certain quantity of energy. If the action/task can be performed reliably and conveniently using lesser energy or another cheaper form of energy, they do not care at all, and actually, they may welcome the outcome.

When one looks at these energy-consuming desired actions our society is undertaking, one quickly realizes that energy use is not an end by itself, but the means to accomplish these and other similar actions. If the desired actions, processes, and tasks are accomplished, those who undertake them are satisfied, the economy functions well, and the human society progresses. As long as the desired actions are accomplished well, how much energy is consumed for their accomplishment is entirely irrelevant to the human society and to the human comfort, satisfaction, and happiness. Simply put, the actual amount of energy consumed is immaterial to everyone as long as "business is done." On the contrary, when humans are unable to accomplish such desired actions, processes, and tasks—either because of lack of sufficient energy supply or because energy has become too costly—discontent and, on several occasions, frustration and rage become noticeable in groups of people and entire communities. Such discontent and public rage was confirmed with several examples of demonstrations and violence during the energy crisis of the 1970s in the United States and the European countries; in 2007–2008, in India, Indonesia, Pakistan, the Philippines and several other developing nations as a result of the higher energy prices that caused food prices to dramatically increase; and in January 2017, in Mexico as a result of a rise in the price of gasoline.

The global energy demand is generated from the aggregate performance of desired energy-consuming actions and processes by the human population. When it is said that the energy demand increases or will increase in the future, what is actually meant is that either humans will perform a larger number of energy-consuming desired actions and processes or a higher number of humans—because of the continuously increasing global population—will perform the same number of actions individually, and this will require the use of more energy. Based on the statistical information of Chapter 2 on energy demand and supply, higher affluence in a nation typically implies that more energy-consuming actions will be desired to be accomplished. When the standard of living—measured by the GDP—of a nation rises, the citizens of this nation will drive more cars; will install more air-conditioning units for the summer months, will buy more energy-consuming products, and will take more vehicle and airline trips. The addition of these energy-consuming actions by the citizens of this nation significantly increases the aggregate national energy demand. Similarly, when the population of a nation increases, the energy use also increases because there are more citizens to desire energy-consuming actions. Therefore, if one denotes by E the total aggregate energy consumption in a nation, the total number of the desired actions or processes by DA, and the average energy consumed per action/process by E_{DA}, one may obtain the simple expression [1]:

$$E = (DA)E_{DA}. \tag{8.1}$$

Then, the aggregate rate of increase or decrease in the energy demand for this nation may be written as

$$\frac{dE}{dt} = \left(\frac{\partial E}{\partial (DA)} \right) \frac{d(DA)}{dt} + \frac{\partial E}{\partial E_{DA}} \frac{dE_{DA}}{dt}. \tag{8.2}$$

This simple and very general equation reveals that the energy consumption in a nation or in the entire globe will decrease when (a) the total number of desired actions/processes decreases or (b) the energy consumed per desired action E_{DA} decreases over time. The first outcome is directly related to the population and GDP growth. The second implies using lesser average TPES per desired action and is directly related to what is commonly called "energy conservation" as well as the increased efficiency of energy-consuming actions and processes. Public policies that aim at "energy conservation" and "energy efficiency" essentially aim at reducing the last term dE_{DA}/dt.

The energy-consuming actions that are desired by the human society may be accomplished in more than one ways, methods, or processes, which use different forms and different amounts of energy. Let us consider the first task in the previous list: to maintain a comfortable temperature for the residents of a building in Berlin during the winter months. This task may be accomplished in a variety of ways, including the following:

1. Use the existing natural gas burner of the building and maintain the interior temperature at 24°C, as it was done in the last 20 years.

2. Use the existing natural gas burner of the building, maintain the temperature at 20°C and ask the residents to wear an extra sweater to keep warmer and comfortable. Less energy will be used for the heating of the building.

3. Replace the existing natural gas burner with an efficient heat pump, which consumes electricity instead of natural gas and maintains the temperature at 24°C. Because of the higher coefficient of performance (COP) of the heat pump, even if we account for the natural gas energy conversion to electric energy, less overall energy will be consumed for the fulfillment of the desired outcome, which is to maintain a comfortable temperature for the inhabitants of a building.

The first alternative is the status quo of accomplishing the desired action: do nothing and continue with the same method the task was accomplished in the past. The second alternative necessitates that the residents of the building will have to change their normal behavior by putting on an extra sweater or another piece of clothing. This alternative implies that the residents will have to cooperate or to assist in the accomplishment of the task. In this case, there is *conservation of energy* because an action of cooperation is required by the residents for less energy to be used without the installation of new machinery and capital expense. The third alternative is typical of an increased energy efficiency project: With the replacement of the hot-water burner by an efficient heat pump, the temperature of the building is maintained at the same level of 24°C, and no cooperation is required of the residents of the building. Lesser primary energy is used with the higher efficiency of the heat pump without any discomfort to the residents. In most of the cases, energy efficiency projects require new or modified/improved machinery and the expense of capital.

In summary, energy conservation implies an action, the cooperation, or a "sacrifice" from the community, such as to switch off lights, to drive less miles, to use more bicycles rather than cars for transportation, and to reduce the building temperature in the winter and raise the temperature in the summer. Energy efficiency is typically accomplished by the modification or replacement of machinery, equipment, or processes, with all other conditions remaining the same. Energy efficiency does not require any cooperation or "sacrifice" from those affected.

The lesser consumption of natural primary energy resources may be accomplished by either energy conservation or improved energy efficiency or a combination of both.

TABLE 8.1

Several Activities Leading to Energy Conservation and Energy Efficiency

Energy Conservation Action	Energy Efficiency Action
Switch off lights when out of a room	Replace incandescent bulbs with fluorescent
Use carpooling for the daily commute	Replace the BMW-381i with a Ford Escort for the daily commute
Increase the thermostat temperature from 21°C to 24°C in the summer. Do the opposite in the winter.	Use a GSHP to replace the old air-conditioning and heating systems
Implement the daylight energy savings time	Install thermal insulation in the roof of buildings
Mandate a maximum vehicle speed of 55 miles/hour on the highways	Install an additional feed-water heater in the electric power plant

Oftentimes, the two terms are used interchangeably in informal conversations. For the better and optimum design of energy systems, it is best that the two are distinguished and differentiated. Table 8.1 gives several examples of energy conservation and energy efficiency activities as used in everyday life. All activities ultimately result in natural resource conservation that is the minimization of natural resource consumption.

The common characteristic of both energy conservation and energy efficiency efforts is that the desired actions and processes are performed using a lesser quantity of primary energy resources. In the case of the second and third alternatives for the heating of the building in Berlin, the net effect of either alternative method—to use a heat pump or to reduce the internal temperature—will be a lesser use of heating fuel, which implies lesser consumption of primary energy sources.

A related concept is that of *energy substitution* where another form of energy, usually a renewable form, substitutes for another form to accomplish a desired action. For example, the substitution of a gas water heater with a solar collector and the production of electric power from wind turbines may be claimed by nonprofessionals as "energy conservation measures." For the correct accounting of energy resources and the achievement of optimized processes, scientists and engineers must be able to differentiate between energy substitution, efficiency, and conservation.

When engineers or the entire society strives to consume less primary energy sources for the accomplishment of the societal tasks, it is useful to know what the *minimum amount of energy* for the accomplishment of a certain desired outcome is. For example, what is the minimum amount of energy one will have to use for the production of one ton of cement? Or what is the minimum quantity of energy that will cook 0.4 kg of pasta? The minimum energy for the accomplishment of a given task or a desired action becomes the *benchmark* for the energy consumption process. Then the task of the engineers and scientists is to design processes and equipment that would more closely approach this benchmark. As explained in Section 1.5, the minimum energy benchmark is calculated using the concept of exergy.

8.2 Use of the Exergy Concept to Reduce Energy Resource Consumption

Exergy calculations are very helpful in the decision-making process for the choice of practices and equipment that best utilize the available primary energy resources. In the following sections, it will be demonstrated with practical examples how the concept of exergy may be used in choosing or improving the engineering systems that perform a desired outcome using the minimum amount of energy.

8.2.1 Utilization of Fossil Fuel Resources

Let us assume that we have a given amount of a natural resource, e.g., a mass of methane (CH_4) equal to 1 kmol (16 kg), and we wish to design an energy conversion system that would maximize the amount of electricity produced. Since methane is a fossil fuel, a common way to convert its chemical energy into work is to combine it with oxygen from the atmosphere, burn it to release its chemical energy in the form of heat, and subsequently convert the heat in a turbine generator system to electricity. This is the series of processes used in a gas turbine system. At first, methane and oxygen from the air undergo a chemical reaction and produce water and carbon dioxide:

$$CH_4 + 2O_2 \rightarrow CO_2 + 2H_2O. \qquad (8.3)$$

The reaction also releases $-\Delta H^0 = 800,320$ kJ of heat. Thus, the methane is fed to the combustion chamber of a gas turbine, which utilizes a Brayton cycle and releases the heat of reaction to the working fluid. The gas turbine produces power and electric work. In a well-designed gas turbine system, an overall thermal efficiency of approximately 40% may be achieved, which implies that our original 1 kmol of methane will produce $W = 320,128$ kJ of electricity. The amount of work may vary a little, depending on the efficiency of the gas cycle, which for most conventional gas turbine power plants is in the range of 45–30%.

A moment's reflection will prove that methane is a chemical substance, a hydrocarbon, where the available energy is stored in the form of chemical energy. The maximum work that may be extracted from 1 kmol of any substance is given by the negative of the Gibbs free energy change during the reaction $-\Delta G^0$ as expressed in Equation 1.28. A glance at thermodynamic tables shows that for methane $\Delta G^0 = -816,650$ kJ/kmol, and, hence, the maximum amount of work, which may be extracted from the available 1 kmol of methane, would be $W_{max} = 816,650$ kJ. This is more than 2.5 times the electric work produced by the gas turbine cycle. This quantity, 816,650 kJ of electricity/kmol of methane, is the benchmark engineers have to strive for. This process of thought elicits the question "What is the engineering system or the series of processes that would enable us to reach this benchmark?"

It is apparent that the combustion of methane in a conventional work-producing cycle—Rankine, Brayton, Diesel, etc.—would supply only a fraction of the benchmark. The combustion of all fossil fuels yields significantly less work than the maximum work, which might be obtained from these resources, primarily because the chemical energy of the fuels is first converted to heat. The subsequent conversion of the heat into work is subjected to the "Carnot limitation," which was explained in Section 1.3.3. The large difference between the actual work obtained from burning the 1 kmol of methane and the benchmark/maximum work will induce the scientists and engineers to devise alternative methods—systems, processes, or a combination of processes—that would produce a higher quantity of electric work, closer to the benchmark $-\Delta G^0$.

When searching for an alternative method for the combustion of fossil fuels, it becomes apparent that the benchmark may be better approached using fuel cells, which were described in more detail in Section 7.5.1. Fuel cells directly convert the chemical energy to electricity; they do not use heat as the intermediate form of energy and are not subjected to the Carnot limitation. Fuel cells may potentially convert the full amount of $-\Delta G^0$ into electric work. Practical fuel cells, which are now in a development stage, have overall efficiencies in the range of 65–80%. Therefore, one may produce electric energy in the range of 408,325–612,488 kJ/kmol using fuel cells, a quantity that is significantly higher than the electric energy produced by a thermal power plant.

This applies to all liquid and gaseous fossil fuels: instead of using a burner/boiler for the combustion of the fossil fuels, these fuel resources would produce significantly more electric work if they were used in a direct energy conversion device, such as a fuel cell.

In this example, the simple use of the concept of exergy and simple calculations that establish the benchmark/maximum work from a primary energy source guide scientists and engineers to choose an energy conversion system, which better utilizes the primary energy sources. It must be noted that the use of the exergy concept points to the use of an entirely different system, for the conversion of the chemical energy of fossil fuels into electric work, than is currently used. This is a new paradigm for energy conversion. The application of the exergy concept does not merely lead to small-scale efficiency improvements of systems and methods that have been used for centuries, but helps in the development of new systems and methods that offer a better energy-conversion alternative.

By determining the benchmark—the maximum work that may be produced from a natural resource—the exergy method does not reveal the method, the system, or the equipment that will produce this maximum work. This is the mission of the knowledgeable engineers and scientists who are tasked with the production of the maximum work. They must interpret the exergy calculations, and based on their knowledge and experience, they design the appropriate processes and equipment that would approach the benchmark. Exergy will point to a direction, but experienced scientists and engineers are needed to lead in this direction and solve the energy problem.

The indication of a certain type of equipment, as with the fuel cell, does not necessarily imply that the maximum work is produced, because the efficiency of all equipment is less than 1. The invaluable contribution of the exergy concept to engineering is to provide the amount of the maximum work that can be obtained from a resource and to supply the benchmark engineers may strive to achieve. The knowledgeable engineers and scientists determine the method, design the system, and choose the equipment that achieves a performance as close as it is feasible to the calculated benchmark, within the economic, social, and environmental constraints.

8.2.2 Minimization of Energy or Power Used for Desired Actions

The concept of exergy may also be used to determine the "minimum work/power" for activities that consume power and work. We use work or power in our everyday activities in order to accomplish certain tasks and desired actions: Electric power is used in refrigerators to keep food at a low temperature, typically below 5°C, and gasoline is used in automobiles for transportation. The desired actions are the preservation of food and our transportation from one place to another. The consumption of electricity (for the refrigerator) and gasoline (for the car) are not the primary purpose of our actions, but the enablers for the accomplishment of the desired actions. Because both electricity and gasoline are derived from primary energy sources and, in addition, are costly, rational consumers will try to accomplish the activities by using the minimum possible amount of the two energy forms.

It must be recalled that according to thermodynamic convention, which is depicted in Figure 1.3, work produced by a system is positive, while work consumed by a system is negative. Therefore, in thermodynamics, the work consumed by a refrigerator for the preservation of food is a negative number. Let us assume that we have three types of refrigerators, which fulfill the task of keeping the food below 5°C during a year, that consume 1,356, 1,672, and 2,198 kWh, respectively.* The refrigerators have the same capacity and are otherwise similar. A rational, environmentally conscious consumer would choose the refrigerator that consumes 1,356 kWh/year, the appliance that consumes the "minimum" amount of electricity annually. A moment's reflection, however, proves that according to the established thermodynamic convention, the actual numerical/scientific values of the work required to run the three refrigerators are negative: −1,356, −1,672, and −2,198 kWh. Of these numbers, −1,356 is, actually, the algebraic maximum of the three. Therefore, in order to conserve energy resources, the rational choice is to choose the algebraic maximum value for the work, and this is what the use of the exergy concept specifies.

What is colloquially referred to as *the minimum amount of work consumed* is an algebraic maximum according to thermodynamic convention. One may interpret this as "the maximum work produced." Since the exergy analysis always determines the maximum value of work involved with equipment or processes and does not make a distinction between positive and negative values of this work, the concept of exergy may be used for the determination of the best way to utilize the energy resources, not only in work-producing processes, where the value of the work is positive, but also in work-consuming processes, where the value of the work is negative. The following example of air compression will help illustrate how one may use the concept of exergy in decision-making processes related to the consumption of lesser work or power.

Let us assume that the desired action is to compress 1 kg of air at atmospheric pressure and temperature of 300 K, to 20 atm. This process could be part of a gas power cycle or of energy storage in a compressed air energy storage. A typical compressor, with 80% isentropic efficiency, would consume approximately 506 kJ to accomplish this desired action (actually, −506 kJ, according to the thermodynamic convention). This compressor increases the pressure of the air to the required 20 atm and simultaneously increases its temperature from 300 K to approximately 780 K. The temperature increase, which occurs with the choice of this compressor, is unnecessary for the task at hand and, in addition, consumes a good amount of work.[†]

Now, let us employ the concept of exergy to determine if the task of pressurizing the air from 1 to 20 atm may be accomplished using a lesser amount of work. Since air is a compressible substance, we may determine the "maximum work" required by the physical principles (the laws of thermodynamics) to accomplish the process of increasing the air pressure from 1 to 20 atm. According to Section 1.6, the initial state of this process is the environmental state, denoted by the subscript 0. The final state will be denoted by the subscript 1. Thus, $P_0 = 1$ atm, $T_0 = 300$ K, and $P_1 = 20$ atm. Rewriting Equation 1.23, which applies to open systems, for the compression process 0-1 and using the ideal gas with constant specific heats model for the air [2,3], we obtain the maximum work (per kg) for this process as follows:

---

* In several markets, refrigerators and other household appliances are sold by the manufacturer with an estimate of their annual consumption of energy.
† The significant temperature increase is a consequence of the type of equipment used for the accomplishment of the desired activity, the almost isentropic compressor. Even if we had a perfectly isentropic compressor (and such equipment do not exist), the amount of work spent would have been 405 kJ, and the exit temperature, approximately 693 K, which is, still, high.

$$w_{max} = e_0 - e_1 = h_0 - h_1 - T_0(s_0 - s_1) = c_P(T_0 - T_1) - c_P T_0 \ln \frac{T_0}{T_1} + RT_0 \ln \frac{P_0}{P_1}. \tag{8.4}$$

Of the three terms in the last part of this equation, only the term $RT_0 \ln(P_0/P_1)$ is directly relevant to the desired action, that is the pressurization of air. As expected, this term is always negative since $P_0 < P_1$, signifying that work must be spent for the pressurization. The other two parts of Equation 8.4, which pertain to the initial and final temperature of the system, $c_P(T_0 - T_1) - c_P T_0 \ln(T_0/T_1)$, are irrelevant to the original task, which is to provide pressurized air. The desired action does not necessarily entail the increased temperature for the air.

It may be easily proven by elementary calculus that, for all values of T_0 and T_1,

$$c_P(T_0 - T_1) - c_P T_0 \ln(T_0 / T_1) \leq 0. \tag{8.5}$$

The fact that this expression is negative (or zero for $T_0 = T_1$) implies that additional work must be spent for the air temperature to increase during the pressurization process. The two terms of Equation 8.4 that pertain to the temperature always add to the work required for the completion of the process, even though the final effect—the temperature increase—does not directly pertain to the desired activity. Since this part of Equation 8.5 becomes equal to zero if $T_0 = T_1$, it follows that if the compression occurs at $T_0 = T_1$ (that is, we have an isothermal compression of the gas), then the absolute value of the specific work required for the compression process, as given by Equation 8.4, would always be less than the work required by an isentropic compressor [1,4]. The "minimum work" required for this isothermal compression is actually an algebraic (and thermodynamic) maximum, and in this case, it is given by the expression

$$w_{max} = RT_0 \ln \frac{P_0}{P_1} = -258 \text{ kJ/kg}. \tag{8.6}$$

The result may be interpreted that *at least* −258 kJ of work must be performed for the pressurization of 1 kg of air or that *at least* 258 kJ/kg of work must be consumed for the task to be performed. This is the benchmark of the desired activity engineers must strive to achieve. It is seen that the exergy analysis of the air pressurization process points to the best alternative process, the isothermal compression. The isothermal amount of work is significantly lower than that required for even the idealized isentropic compression of air, which is 405 kJ/kg (in absolute value).

It must be emphasized that although the value −258 kJ/kg is an algebraic maximum, in practice, we call this "the minimum work," because in everyday parlance, we consider only the absolute value of the work consumed, not its algebraic value. It is apparent in this example that the correct use of thermodynamic theory, and in particular of the exergy concept, provides only the numerical value of the minimum work, which becomes the benchmark for the air compression process. The concept also gives an indication for the process to be used for the minimum work to be achieved: this is an isothermal process, or as close to isothermal as practically feasible [1,3,4]. Because isothermal processes require a significant amount of heat transfer from the compressed gas, such processes are very slow and rather difficult to achieve in practice, if a significant quantity of the gas is to be compressed at a reasonably fast rate.

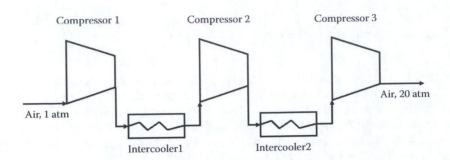

FIGURE 8.1
Minimization of compression work using a series of three compressors with intercoolers.

In order to achieve a practical process, which is as close as possible to isothermal, several smaller, isentropic compressors with intercoolers are used in engineering practice. The intercoolers are essentially heat exchangers that admit the gas from the compressor, cool the gas to a lower temperature—as close as possible to the atmospheric temperature—and feed it to the next compressor for further pressurization. The process is repeated until the fluid is pressurized to the specified pressure. This compression process is schematically shown in Figure 8.1, where three compressors and two intercoolers are used for the compression of air from 1 to 20 atm.

In the example of the pressurization of 1 kg of air from 1 to 20 atm, when only one intercooler is used with two 80% efficient compressors, the work consumed is 400 kJ. When two intercoolers are used, the work consumed is reduced to 371. An almost isothermal process—which consumes only the 258 kJ/kg indicated by the exergy calculation—may be achieved by using a very large number of intercoolers with isentropic compressors. If a very large number of compressors and intercoolers are used with 80% efficient compressors, then the amount of required work is 258/0.8 = 323 kJ/kg. Table 8.2 shows the absolute value of the work required for the compression of atmospheric air from 1 to 20 atm using several methods with different compressors.

In the case of compression with intercoolers, the application of the exergy concept will also prove that for optimum performance, the pressure ratio across all the compressors must be equal. This implies that the pressure increase in each compressor is $(P_1/P_0)^{1/n}$, where n is the number of compressors (and $n - 1$ is the number of intercoolers).

It is observed again that the application of the exergy concept leads to the work/power minimization for processes that consume power and work. An experienced engineer will use the exergy method and will design the compression process with one or more

TABLE 8.2

Work (kJ/kg) for the Compression Process of Air from 1 to 20 atm

	Work, kJ/kg	Energy Savings,[a] %
Isentropic compression, no intercooler, $\eta = 80\%$	506	0
Isentropic compression, no intercooler, $\eta = 100\%$	405	20
Isentropic compressions, one intercooler, $\eta = 80\%$	400	21
Isentropic compressions, two intercoolers, $\eta = 80\%$	371	27
Isentropic compressions, many intercoolers, $\eta = 80\%$	323	36
Isothermal compression, $\eta = 100\%$	258	49

[a] Energy savings with respect to the currently prevalent compression method.

intercooling stages to achieve significant energy savings for the performance of the given task—the compression of atmospheric air to 20 atm.

8.3 Improved Efficiency in Electric Power Generation

The basic vapor and gas power cycles that are presented in Section 1.5 may be significantly improved by optimizing the pressures and temperatures and by adding components and processes that improve the thermal efficiency of the cycle.

8.3.1 For Rankine Vapor Cycles

The following methods may be implemented to increase the thermal efficiency of Rankine cycles. The numbered states in this section refer to Figures 1.7 and 1.8.

1. Increase the upper cycle temperature T_3 by superheating the steam in the boiler and superheat the produced vapor to higher temperatures. There is an upper temperature limit for superheating, the *metallurgical limit*, where the temperature is too high for the turbine materials to handle and damage may occur. With the materials currently used in boilers and turbines, this limit is approximately 640°C.

2. Increase the upper pressure of the cycle P_3. The materials used in the superheaters and the turbine inlets impose an upper pressure limit, which is approximately 35 MPa (350 atm).

3. Reduce the lower temperature and pressure of the cycle T_4, which is also equal to T_1. The limit here is imposed by the ambient (environmental) temperature, where finally all the heat from the condenser must be rejected. When water is available, from a river or a lake, the temperature of the water is the limit of the lower temperature, the condenser temperature is designed to be lower and the thermal efficiency of the cycle is higher, because water is colder than the ambient air. Most of the modern, large power plants are located close to lakes or rivers to make use of the ambient water for cooling.

4. Use a reheat process: steam is partly expanded in the turbine to a lower pressure (and produces power in the process) and is then fed back to the superheater, where it receives additional heat, and its temperature is increased again to approximately T_3. The partial expansion and reheating processes may be repeated two or three times and are facilitated when the pressure P_3 is high. Supercritical cycles—with upper pressures higher than 230 atm—are ideal cycles for the addition of several superheat processes.

5. Use the *regeneration* or *bleeding* process: A fraction of the steam (10–20%) is extracted from the turbine at moderately high temperature and is used to heat up the water effluent of the pump at state 2 in a separate heat exchanger, which is called the *feedwater heater*. This process may also be repeated two or three times by extracting several streams of steam at two to three points from the high-pressure turbines. The effect of regeneration is to use heat at a lower temperature from the expanding steam to increase the liquid water temperature in the low-temperature parts of the

cycle. The high temperature heat in the boiler/superheater is saved to be used only at the higher temperatures of the cycle, and this improves the thermal efficiency of the Rankine cycle.

Optimized vapor cycles that utilize the preceding methods have upper cycle temperatures in the range of 550–580°C and may reach thermal efficiencies in the range of 42–44%.

Example 8.1

A 400 MW coal power plant has thermal efficiency 36% and 92% capacity factor. It is proposed to add a regenerator to this power plant that would increase its efficiency to 37.8%. (a) Determine how much additional energy, in kilowatt hours, this power plant will produce annually for the same heat input. (b) If the owner of the plant sells the energy produced at an average value of $35/MWh, what is the additional annual revenue this improvement will bring?

Solution: The rate of heat input of this power plant is 400/0.36 = 1111 MW. For the same heat input and fuel cost, and with the increased efficiency of 37.8%, the power plant will produce 420 MW of electric power, an additional 20 MW.

  a. The additional electric energy produced during a year will be 20,000 × 8,760 × 0.92 = 161.2 × 10^6 kWh or 161.2 × 10^3 MWh.
  b. The additional annual revenue is 161.2 × 10^3 × 35 = $5,641,440.

8.3.2 For Brayton Gas Cycles

The efficiency of the basic Brayton (gas) cycle may be improved using additional equipment and processes. The numbered states in this section refer to Figures 1.9 and 1.10.

1. Use a reheat process, similar to the vapor cycle: The gas expands in the first turbine to lower pressure and is then introduced to the combustor, or a second combustor, where the combustion of additional fuel again increases its temperature to a value close to T_3. The gas is then directed to a second gas turbine, where it produces additional power.

2. The temperature at the last turbine exhaust is relatively high, typically at 400–500°C. This is oftentimes higher than the temperature T_2 at the exit of the compressor. In such cases, one may use a *regenerator,* a heat exchanger that preheats the compressor output by using the turbine exhaust mixture. The result is that lesser heat is required in the combustor and the thermal efficiency improves.

3. Use intercoolers in the compression process as explained in Section 8.2.2. In this case, a single large compressor is replaced by a series of smaller compressors that operate in series and consume lesser work.

4. Use a *bottoming cycle* by producing vapor [2,5]. The high-temperature combustion products at the turbine exhaust are used in a heat exchanger to generate steam (or another vapor) at high pressure, which produces power using the turbine–condenser–pump combination of the Rankine cycles. Effectively, the heat exchanger that uses the heat of the exhaust gases is the boiler of the bottoming cycle. The additional vapor cycle does not require external heat input, and this significantly

enhances the thermal efficiency of the combined system. The thermal efficiency of practical combined cycle systems may reach values close to 60%. More details on this combination of the two cycles are given in Section 8.4.2.

Example 8.2

A 37 MW gas turbine operates with 38% efficiency. A study conducted by a young engineer determined that 21% of the waste heat from the exhaust of the gas turbine may be used with a regenerator to reduce the heat supplied to the combustion chamber and, of course, reduce the fuel supplied. For the same power output, determine what will be the efficiency of the gas turbine after the addition of the regenerator and the percentage of fuel that will be saved.

Solution: The rate of heat input to this gas turbine is 37/0.38 = 97.4 MW, and the rate of waste heat is 60.4 MW. With the regenerator, 21% or 12.7 MW of the waste heat is diverted to the regenerator inlet air, and this helps offset 12.7 MW of the heat input to the cycle. Therefore, the new rate of heat input will be 84.7 MW, and the new cycle efficiency is 43.7%.

Since the same type of fuel will be used before and after the improvement to the system, the fraction of the fuel saved is 12.7/97.4 = 0.13.0 (13%).

8.3.3 Combination of Processes and Desired Actions: Cogeneration

Several industrial, domestic, and commercial applications are performed using a number of processes that have the need of more than one energy forms: A steel mill needs high-temperature heat for the production of steel from iron ore and coal, as well as electricity for the forming of steel plates; a refinery uses a great deal of heat at moderate temperatures (110–130°C) as well as a significant amount of electric power for the operation of its turbomachinery; a large supermarket uses natural gas for heating in the winter months as well as electricity for lights and the several refrigerators and freezers; and many households use electricity for lighting and air-conditioning as well as natural gas for cooking and hot water supply. The desired actions in all these cases are the production of the needed electric power $\dot{W}$ and the needed rate of heat $\dot{Q}$. When these desired actions are performed separately, they consume a great deal of primary energy sources.

It is possible to combine two or more of the needed processes and desired actions to produce the same outcomes using lesser primary energy. In such cases, two or more desired actions are accomplished by the same system and produce two or more desired outcomes. When the desired outcomes are the production of electric power and heat and the outcomes are combined in a single system, the result is the *cogeneration* system, which generates both electric power and heat. The cogeneration of electricity and heat simultaneously produces the required amounts of heat and electric power using a single system. In the case of the refinery that needs heat at temperatures in the range of 110–130°C for its evaporator as well as electric power, a system may be designed that utilizes a steam cycle to produce the electric power. The needed rate of heat comes from this cycle by one of the following methods:

1. Using a condenser at 130°C or higher temperature. Effectively, the condenser becomes the evaporator for the crude oil.
2. Extracting a fraction of the steam from the turbine (bleeding) at 130°C or at higher temperature and feeding the steam to the evaporator of the crude oil.

Figure 8.2 is a schematic diagram of the first method, where the steam discharged by the turbine is at high enough temperature and pressure. The steam enters the condenser and condenses at a temperature higher than 130°C. The heat produced from condensation is transferred to the crude oil input to produce the petroleum fractions. The cycle for this method is identical to the Rankine steam cycle with the condensate being extracted at higher than usual pressure and temperature. The cycle produces the needed electric power as well as the heat for the petroleum evaporation process. The second method is depicted in Figure 8.3 with the steam extracted from the turbine and diverted to the heat exchanger. The latter delivers the heat of condensation to the crude oil for the fulfillment of the oil heating/evaporation process. The rest of the cycle produces the needed electric power.

The cogeneration of electric power and heat may also be achieved with a gas cycle, using the turbine exhaust gas. The latter is always at high temperature, and hence, the waste heat of the gas cycle may be used in a process that requires heat input. The cogeneration of heat and electric power in a single system, as depicted in Figures 8.2 and 8.3, always requires less primary energy/fuel input than two separate systems, which would perform the two desired

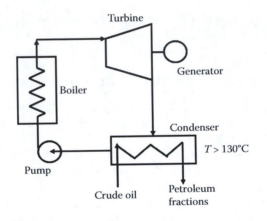

FIGURE 8.2
Cogeneration with heat supplied by steam condensation.

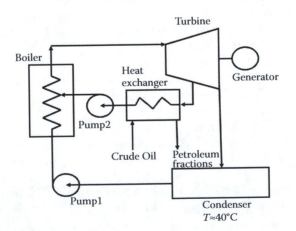

FIGURE 8.3
Cogeneration with heat supplied by steam extraction (bleeding) from the turbine.

actions, electricity production, and heat production [5]. Because heat and power cogeneration systems save a great deal of primary energy, have very high efficiencies, and reduce the cost associated with fuel consumption, they have become very popular systems since the 1980s and are now widely used with commercial establishments, large residential complexes, and office buildings. One of the regulatory measures that assisted the widespread use of cogeneration is the deregulation of the electric power industry in most Organisation for Economic Co-operation and Development (OECD) countries and the provision that requires electricity transmission and sale corporations to purchase any excess electric power produced by smaller producers (i.e., through cogeneration) at the prevailing wholesale price.

Example 8.3

A 50 MW gas turbine in a university campus operates with efficiency of 42% and capacity factor of 68%; 60% of the waste heat from the gas turbine is used to supply the dormitories and university offices with space heating and hot water. Determine the electric energy and the quantity of heat supplied annually by this gas turbine.

Solution: The electric energy produced annually by this gas turbine is 50,000 × 0.68 × 8,760 = 297.8 × 10^6 kWh/year (8,760 is the number of hours per year).

The average rate of waste heat produced by this gas turbine is 50 × 0.68 × (1/0.42 − 1) = 46.95 MW.

The heat supplied annually to the dormitories of the university is 0.6 × 46,950 × 60 × 60 × 8,760 = 888 × 10^9 kJ.

Example 8.4

A large milk pasteurization plant uses hot water at 85°C at a rate 75.2 m³/minute. The water is supplied at an average temperature of 18°C and is currently heated by natural gas in an old burner with 80% efficiency. It is proposed to buy a 50 MW gas turbine, which has 32% efficiency; use the waste heat of the turbine for the supply of hot water; and sell the produced electricity. (a) Determine the amount of natural gas that will be saved daily. (b) Will the system operate as proposed? If not, what modification(s) do you suggest?

Solution:

a. The volumetric flow rate of the water is 75.2 m³/60 = 1.253 m³/s, and the mass flow rate is approximately 1,253 kg/s. The rate of heat required for the water temperature to increase from 18°C to 85°C is 1,253 × 4.184 × (85 − 18) = 351,251 kW. Since the heat supplied by 1 scf of natural gas is 1.07 × 10^6 J (1,070 kJ/scf), and the burner has 80% efficiency, the pasteurization plant currently consumes 351,251/(0.8 × 1,070) = 410 scf/s of natural gas. The daily amount consumed is 410 × 60 × 60 × 24 = 35.42 × 10^6 scf.

b. The 50 MW gas turbine has heat input 50/0.32 = 156.3 MW, and its waste heat is 106.3 MW. This rate of heat is not sufficient to supply the demand of 351,251 kW (351.3 MW) for the pasteurization process.

Suggestions:

1. Install three additional gas turbines of 50 MW (for a total of 200 MW installed), which will produce 425.2 MW of waste heat, more than enough to supply the demand of 351.3 MW.
2. Install a lesser number of gas turbines and a more efficient burner to supply the difference. Both suggestions will lower the daily consumption of natural gas.

8.4 Waste Heat Utilization

All the power plants reject very large quantities of heat to the environment, the so-called waste heat. In the year 2014, 16,977 TWh (61.1×10^{21} J or 57.6 Q) of electric energy were produced in the world from thermal power plants [6]. At an average thermal efficiency of 34%, these power plants rejected approximately 118.6×10^{21} J (or 120 Q) as waste heat to the environment. The latter is 22% of the TPES consumed by the entire world. While this is a very high amount of energy, because it is inherently produced at low temperatures, the practical uses of the waste heat are very much limited. A few processes and systems that are capable to utilize part of the low-temperature waste heat are presented in the following two sections.

8.4.1 From Rankine (Steam) Cycles

The waste heat from steam power plants is typically removed by a cooling water system, which becomes available to transfer this heat to be utilized in another process. A typical 400 MW coal-fired power plant has overall thermal efficiency approximately 40% and rejects 600 MW to its surroundings. Even though this is an enormous amount of heat power, most of it is not used in practice because it is available at the low condenser temperature, typically close to 40°C. Very few practical applications may utilize such low-temperature heat, because most of the industrial heat is required at significantly higher temperature. The waste heat may find practical applications, and substantial energy savings may be generated if users of low-temperature heat would collocate with power plants. Among the low-temperature applications, which may use the waste heat from Rankine cycles are as follows:

1. *Seawater desalination*, which is very important in arid regions that are close to the sea, such as the countries of the Persian Gulf, the countries that border the Red Sea, and the Texan part of the Gulf of Mexico. Water desalination is commonly achieved by the reverse osmosis of seawater in a series of membranes. Since the efficiency of the reverse osmosis process improves at higher temperatures, heating the seawater from the ambient temperature to 30–35°C will significantly increase the efficiency of the process and the yield of the desalination plant. A second method that entails the evaporation and condensation of seawater is schematically depicted in Figure 8.4: The colder seawater on the left enters a vessel where it is heated by the waste heat water to a higher temperature (35–42°C). A constant stream of air from the top partly evaporates the warmer water and produces an air stream with higher humidity. The air is directed to another vessel, the condenser, where seawater is fed at ambient temperature and part of the high-humidity air–stream condenses. This condensate is fresh water, which is collected at the bottom of this vessel and sent for consumption. For the production of more fresh water condensate, the warmer saline water output of the condenser may be mixed with the seawater input of the evaporator at the left.

2. *Agricultural soil heating*. A small increase of the soil temperature results in significantly higher crop yields and, oftentimes, multiple crops per season in regions where multiple crops are otherwise not feasible. In temperate climates, the yields of several agricultural crops double when the soil temperature is maintained

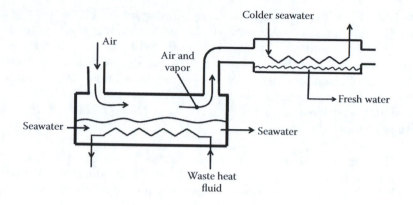

FIGURE 8.4
Water desalination by evaporation and condensation.

between 30°C and 34°C, and the use of warmer greenhouses has resulted in continuous vegetable and fruit production throughout the year [7]. Higher soil temperatures may be achieved when the waste heat from the power plants is transferred
by warm water to neighboring fields and greenhouses by underground pipes. The
shortcomings of such systems are the cost of the construction of underground
(close to the surface) piping systems under the large fields for the warming of the
fields and the power for the pumping requirements of the system.

3. *Aquaculture.* Similar to agricultural crops, heated ponds of fresh and salt water
   have significantly higher yields of fish and seafood. Shrimp growth improves by
   80% when the water temperature increases from 21°C to 27°C. Catfish grow three
   times faster at 28°C than at 24°C [7]. The optimum temperatures for the growth
   of both species and of similar aquaculture products are in the range of 30–35°C.
   The water of the ponds for aquaculture may be controlled and, when necessary,
   heated up by the waste heat from neighboring power plants to the desired temperatures, where not only the fish protein yield is higher per unit surface of the
   pond, but also different more desirable aquatic species may grow to be harvested.
   The hydraulic systems that would facilitate this temperature control of aquaculture ponds are very simple and rather inexpensive if the power plant is close by.
   Figure 8.5 depicts such a schematic diagram for the transfer of the waste heat from
   a nearby power plant to an aquaculture pond.

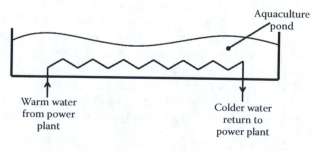

FIGURE 8.5
Waste heat utilization for aquaculture.

4. *District heating* is the heating of a large number of buildings by a single heat source. Geothermal water may be used for district heating as explained in Section 6.4.4. The large quantities of waste heat produced by power plants may be used in a similar manner for the heating of nearby buildings. When the electric power plant is located close to a district of a city, this district may satisfy a great deal of its space heating needs using the waste heat of the plant. Pumped water is the medium that transfers the waste heat from the condenser of the power plant to the building interior.*

When the distance of the steam power plant, which is the source of heat, to the consumption location of the waste heat is long, e.g., more than a few kilometers, the further use of the waste heat becomes almost impossible, primarily for the following reasons:

1. The temperature of the heat-carrying medium, usually water, drops because the surrounding environment is colder.
2. The pumping power for the transport of the water becomes significant and costly.
3. The capital cost in piping and pumps required for the transfer of heat becomes high enough to render the project uneconomical.

It must be noted that the steam power plant rejects waste heat continuously, and therefore, this heat must be continuously removed. If the waste heat system operates for part of the year, e.g., when district heating is needed during the colder months, a backup heat removal system must become available in the power plant to remove the unused heat. The construction and maintenance of backup systems always adds to the capital cost of the waste heat utilization system.

8.4.2 From Brayton (Gas) Cycles: Combined Cycles

Gas turbines operate with the open Brayton cycle and reject the waste heat via the exhaust of the combustion gases, which are at high temperatures, typically in the range of 300–500°C. This range of temperatures is high enough for heat to be transferred to a secondary fluid, evaporate this fluid, and produce additional electric power with a vapor turbine/generator system. The whole scheme is depicted in Figure 8.6, which shows the Brayton (gas turbine) cycle at the top. The exhaust gases at the exit of the gas turbine are directed to a heat exchanger (the vapor generator), where the secondary fluid evaporates at high pressure and is then directed to the vapor turbine. The vapor at the exit of the turbine is condensed, pressurized by the pump, and fed back to the vapor generator to complete a simple Rankine cycle. In most systems, water/steam is used in the Rankine cycle, but any other suitable fluid—e.g., a refrigerant, ammonia, or a hydrocarbon—may be used. As the figure shows, the final system is the combination of a Brayton cycle and a Rankine cycle, with the latter cycle at the bottom receiving all its heat from the waste heat of the top cycle. For this reason, the system of Figure 8.6 is referred to as *combined cycle* [2,5] or *bottoming cycle*. The combined cycle produces two streams of power, from the gas turbine and the vapor turbine, and uses fuel only for the gas combustion of the upper cycle. The result is

---

* Apart from waste heat utilization, other sources of hot water, such as water from hot springs and aquifers, have been used for district heating. Oftentimes such district heating systems are classified as *geothermal district heating*.

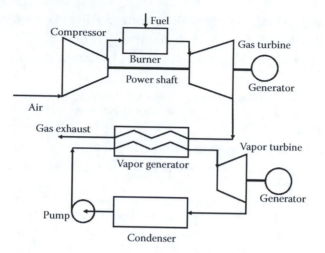

FIGURE 8.6
Gas turbine with a bottoming vapor cycle.

an advanced system with significantly higher efficiency than conventional electricity generation units. Overall, thermal efficiencies close to 60% have been achieved with combined cycles.

It must be noted that the combined cycle still produces waste heat via the condenser and the exhaust gases. This heat is approximately equal to the difference of the heat the Brayton (top) cycle would have produced in the absence of the bottoming cycle and the net electric work produced by the bottom cycle. Some of the waste heat from the combined cycle may still be used in low-temperature applications.

8.5 Conservation and Efficiency Improvement in Buildings

Approximately one-third of the energy consumption in OECD countries is spent in private and public buildings. The temperature of the buildings is maintained within a narrow range throughout the year for the comfort of their occupants. Buildings are heated during the colder months of the year and air-conditioned (cooled) during the hotter months. The buildings must also have electric lighting and be supplied with hot water throughout the year. Heating, cooling, lighting, and hot water supply consume most of the energy required by residential and commercial buildings in the world. The rest of the needed energy is consumed by appliances, such as refrigerators, cooking stoves, microwave ovens, television systems, computers, and communication systems, which usually operate with electric power.

The construction of energy-efficient buildings is easy to regulate because all building construction is governed by regional and municipal codes for safety. The 2009 and 2012 *directives* on energy efficiency in the countries of the EU are examples of more centralized regulations that must be followed by all member states. Most municipalities have incorporated in their building codes energy-efficient construction methods and materials. The EU directives include the following:

1. All new buildings to meet minimum energy efficiency standards
2. Energy certifications before large, older buildings are sold
3. Large residential and industrial buildings must comply with newly adopted energy standards
4. Regular inspections of large energy-consuming systems—such as furnaces, boilers, and air-conditioning units—to ensure environmental regulation compliance and energy efficiency
5. Voluntary installation of real-time electricity monitors
6. Plans to increase the fraction of zero net energy buildings (ZNEBs)

In the United States, several municipal and state authorities have enacted provisions in their building codes for the mandatory construction of 15% highly energy-efficient buildings. It is expected that by 2030, all new buildings constructed for the federal government will be ZNEB—although there is not an official definition of a ZNEB. A voluntary system for the construction of new buildings and the refitting of older buildings in the United States and Canada is the Leadership in Energy and Environmental Design (LEED) system, which issues third-party certifications to buildings (*platinum, gold, silver,* and *certificate*). Although this is a voluntary certification program, it has been met with success, because it has become a matter of civic pride for individuals, private and state corporations, educational establishments, and local or state governments to construct new buildings that meet one of the levels of the LEED certification. The result of this program is a plethora of new very energy-efficient buildings in both countries.

Any reduction of the electricity and the heat required for the operation of the heating, cooling, and lighting systems in the buildings as well of the appliances that provide homely comfort to the inhabitants of the buildings would result in the conservation of energy resources. The following describe some of the common methods and actions that may be implemented or are currently used for the minimization of the energy use in buildings.

8.5.1 Use of Fluorescent Bulbs or Light-Emitting Diodes

Incandescent bulbs—an invention of the early twentieth century—have been fixtures of buildings since the early 1900s. The incandescent bulb has very low efficiency in converting electric power to light and provides very low amount of visible light energy in comparison to the electric energy input. The efficiency of lighting devices is defined as the visible radiation energy emitted divided by the electric energy input to the bulb. Typical efficiencies of incandescent bulbs are in the range of 2–4%, which implies that a great deal of the energy supplied is wasted.

Florescent bulbs have higher efficiencies, typically in the range of 10–12% [8], and sodium lamps, which emit a yellowish light and are used on roads and highways, have efficiencies that approach 20%. The more modern LEDs are the most efficient lighting devices, and their efficiency is close to 60% [8]. It is apparent that the substitution of an incandescent bulb with a fluorescent bulb or, better, with a LED would significantly increase the lighting efficiency of a building. As a result, the electric energy required by the building for lighting would significantly decrease: The same amount of lighting (in lumens) produced by a 100 W incandescent bulb may be provided by a 25 W fluorescent bulb, or by a LED that consumes only 5 W. In a typical multiuse commercial building, where lighting is provided for 50% of the time during the year—that is, for approximately 4,380 hours/year—the mere substitution of a single incandescent bulb with its fluorescent equivalent would save 60 × 60 × 4,380 × (100 − 25) J = 1183 MJ of electricity, or 328.5 kWh/year. Substitution of that incandescent bulb with a LED would save 60 × 60 × 4,380 × (100 − 5) = 1,498 MJ (416.1 kWh) per year.

The energy savings from the reduced lighting power are not the final quantity of energy savings for the building. The energy used by all the lighting devices dissipates in the building and heats up the interior of the building. Since a building requires cooling during the hotter season and heating during the colder season, the lighting energy that is dissipated does not need to be removed by the air-conditioning system during the hotter months. During the colder months, an equivalent amount of heat must be supplied by the heating system of the building, usually by natural gas. For this reason, in the calculation of the total energy savings, the location of the building makes a significant difference as will be demonstrated in Example 8.5.

Example 8.5

Calculate the annual amount of electricity saved from the substitution of a single 100 W incandescent bulb with a fluorescent bulb of 25 W and a LED bulb of 5 W in two buildings: (a) in Fort Worth, Texas, United States, where air-conditioning is required 65% of the days during a year and heating 15% of the year, and (b) in Berlin, Germany, where heating is required for 60% of the year and cooling for 20% of the year. Assume that the buildings are occupied for an average of 12 hours/day and that heating and cooling are accomplished with similar heating, ventilation, and air-conditioning (HVAC) systems that have coefficients of performance of 2.8 for cooling and 3.8 for heating.

Solution: The substitution of the incandescent bulb with fluorescent saves in both buildings (100 − 25) × 60 × 60 × 12 × 365 = 1,183 MJ (329 kWh) of electricity per year and with a LED (100 − 5) × 60 × 60 × 12 × 365 = 1,498 MJ (416 kWh) per year.

The energy saved does not need to be removed by air conditioning but needs to be supplied as heat during the heating season.

a. For the building in Fort Worth, at the 50% utilization rate (the fraction of time lights are on), if fluorescent lighting is used, then less cooling is required for 0.65 × 0.5 × 8,760 = 2,847 hours and heating for 0.15 × 0.5 × 8,760 = 657 hours/year. With a fluorescent bulb, the building will save an additional 2,847 × 60 × 60 × (100 − 25)/2.8 J = 275 MJ (COP for cooling is 2.8) or 76.3 kWh of electricity from the cooling requirements of the building. An additional 657 × 60 × 60 × (100 − 25)/3.8 MJ = 46.7 MJ (COP for heating is 3.8) or 13.0 kWh must be supplied for the heating requirements of the building during the winter months. With the use of the fluorescent bulb, the total annual electricity savings for this building would be (328.5 + 76.3 − 13.0) = 391.8 kWh.

With the LED, the electricity savings from cooling are 2,847 × 60 × 60 × (100 − 5)/2.8 J = 347.7 MJ or 96.6 kWh. The additional heat input during the winter month is 657 × 60 × 60 × (100 − 5)/3.8 MJ = 60.7 MJ or 16.9 kWh. The total savings are 416.1 + 96.6 − 16.9= 495.8 kWh.

b. For the building in Berlin, the electricity savings from the lighting are the same. This building is cooled for 0.2 × 0.5 × 8,760 = 876 hours/year and is heated for 0.6 × 0.5 × 8,760 = 2,628 hours/year at the 50% utilization rate.

If a fluorescent bulb is used instead of the incandescent, the cooling savings amount to 23.5 kWh/year and the heating supplement by the heat pump system is 51.9 kWh/year for the total savings of (328.5 + 23.5 − 51.9) = 300.1 kWh/year.

TABLE 8.3

Annual Energy Savings (kWh) in a Large Building Where 100,000 W of Incandescent Light Bulbs Are Substituted with Fluorescent Lights That Consume 25,000 W or with LEDs That Consume Only 5,000 W and Produce the Same Luminescence

Source of Savings	Location: Fort Worth, Texas		Location: Berlin, Germany	
	Fluorescent	LED	Fluorescent	LED
Electricity to lights	328,500	416,100	328,500	416,100
From air-conditioning	76,300	96,600	23,500	29,800
From heating	(13,000)	(16,900)	(51,900)	(65,700)
Total Annual Savings	391,800	495,800	300,100	380,200

Note: Numbers in parentheses indicate a negative number.

If a 5 W LED substitutes for the 100 W incandescent bulb, the cooling savings are 29.8 kWh and the heating supplement becomes 65.7 kWh for total yearly savings of 416.1 + 29.8 − 65.7 = 380.2 kWh.

It is observed that the electricity savings in both locations are significant, especially with the substitution to LED lighting. The electricity savings in the warmer climate are augmented by the lesser need for cooling.

Based on this simple example for the substitution of a single incandescent bulb, Table 8.3 summarizes the energy savings in a large office building resulting from the substitution of the equivalent of 100 kW of incandescent bulbs with fluorescent ones that consume 25 kW and with LEDs that consume 5 kW and provide the same amount of luminescence in lumens. All numbers are in kilowatt hours per year. The buildings are used on average for 12 hours per day and the HVAC systems in the two buildings operate with COP of 2.8 for cooling and 3.8 for heating.

Table 8.3 proves that using fluorescent lamps and LEDs for the lighting needs of large buildings significantly reduces the annual electric energy consumption, and this cuts down on the annual electricity cost. It is also apparent that buildings in hot climates, with high air-conditioning usage, have additional energy and cost savings from the reduced load of the air-conditioning units. Because commercial and residential buildings in large urban areas utilize several megawatts of electric power for lighting, substantial energy savings are realized with the mere substitution of traditional lighting devices by more efficient ones.

While the substitution of incandescent bulbs with more efficient lighting devices is a measure that falls under the energy efficiency category, energy conservation measures pertinent to lighting may also be implemented in residential and commercial buildings. Among these measures are switching off lights when they are not needed and switching off light-emitting appliances, such as televisions and computer screens, when they are not in use.

8.5.2 Use of Heat Pump Cycles for Heating and Cooling

The interior of buildings is maintained within a narrow range of temperatures to ensure the living comfort of the inhabitants. When the ambient temperature T_{amb} is higher than that of the building interior T_{in}, heat enters the building and must be removed by the air-conditioning unit. When the ambient temperature is lower than that of the building interior, heat leaves the building and must be supplied by a heat pump or a fossil fuel burner. The rate of heat entering or leaving a building to the environment is given by the expression

$$\dot{Q} = UA(T_{amb} - T_{in}) = \frac{A(T_{amb} - T_{in})}{R},$$ (8.7)

where A is the area of the building and U is the overall heat transfer coefficient (similar to that of the heat exchangers). The latter is usually represented by its inverse R (= $1/U$), which is called the *R-value* of the building. Equation 8.7 applies to entire buildings or parts of buildings, such as roofs, windows, and walls.

Example 8.6

The aggregate R-value of an entire building is 15 $(m^2 \cdot K)/W$ and the outside area of the building is 108 m^2. The interior of the building is kept at 24°C during the summer and at 22°C during the winter. Determine: a) the heat that must be supplied to this building by a burner during a winter day when the average outside temperature is 0°C; and b) the daily amount of heat to be removed from the building during a summer day when the ambient temperature is 36°C.

Solution:

a. During the winter day, the rate of heat lost is $\dot{Q} = -108 \times (0 - 22)/15$ W or $\dot{Q} = -158.4$ W (the negative sign indicates that heat is leaving the building/system according to thermodynamic convention). During the entire winter day, the burner must supply the building with heat to replace the heat lost. Therefore, the heat supplied by the burner is $Q = 158.4 \times 60 \times 60 \times 24 = 13.69 \times 10^6$ J.

b. In the summer day, the rate of heat entering the building from the surroundings is $\dot{Q} = 108 \times (36 - 24)/15 = 86.4$ W. During the entire day, the air-conditioning unit must remove $Q = -86.4 \times 60 \times 60 \times 24 = -7.46 \times 10^6$ J (the negative sign indicates that heat is removed from the building).

Heat pumps are used in most of the newly constructed large buildings throughout the world, because they are clean, they have lower operational cost, and they are energetically more efficient than burners and boilers. As explained in Section 1.4.2, a heat pump consumes electric work and delivers heat to a building equal to a multiple of the electric work it receives. Heat pumps with a COP of 4 (seasonal energy efficiency ratio [SEER] 13.6) deliver four times the energy they receive as heat. A system that operates as a heat pump during the cold season may be reversed to become an air-conditioning system during the hot months.

The refrigeration/heat pump cycle, which is schematically depicted in Figure 1.6, essentially maintains two heat reservoirs: the first at a low temperature T_L and the second at a higher temperature T_H. When the system is in the heat pump mode, a quantity of heat Q_H is transferred to the interior of a building at temperature T_H. At the same time, the cycle absorbs heat Q_L from the ambient air, which is at lower temperature T_L, and consumes work W. The COP of the heat pump $\beta_{hp} = Q_H/W$ is a figure of merit for this system that represents the benefit (the heat to the building)-to-cost (the electric energy consumed) ratio. The COP of well-designed heat pumps may exceed 5 (16 SEER), and those of ground source heat pumps (GSHPs) may approach the value 8 (27 SEER). Because they deliver heat that is several times more than the electric energy they consume, heat pumps are very efficient systems for space heating, even if one considers that the electric energy is produced in a power plant at the expense of a larger quantity of heat, as the following example illustrates:

Example 8.7

A large building consumes $Q_H = 23 \times 10^6$ kJ of heat during a cold winter day. The building is heated by a central system with a burner that uses natural gas with heating value of 43,000 kJ/kg. The combustion efficiency of the burner is 92%. This heating system is replaced by a heat pump with COP of 4.8. The electric energy W needed for the operation of the heat pump is supplied by a gas turbine which has 35% overall thermal efficiency. Determine the amount of the primary energy (the natural gas) that is saved per day by the new heating system.

Solution: With $Q_H = 23 \times 10^6$ per day and the heat obtained from the combustion of 1 kg of natural gas being $43,000 \times 0.92$ kg, the burner of the building consumes $23 \times 10^6/$ $(43,000 \times 0.92) = 581$ kg (23,375 scf) of natural gas per day. With the heat pump system (COP = 4.8), the heating requirement of 23×10^6 kJ is supplied to the building with the consumption of $W = 23 \times 10^6/4.8$ kJ $= 4.79 \times 10^6$ kJ of electric energy. This electric energy is produced in a power plant of 35% efficiency, and hence, it is produced by the consumption of $Q_{in} = 4.79 \times 10^6/0.35$ kJ $= 13.69 \times 10^6$ kJ of heat at the power plant.

With natural gas used for the production of electricity, the heat at the power plant would be produced from the combustion of $13.69 \times 10^6/43,000$ kg = 318 kg natural gas (12,794 scf). Therefore, the substitution of the old burner with a heat pump system saves $581 - 318 = 263$ kg (10,581 scf) of natural gas per day.

It must be noted in the last example that the primary energy savings are the consequence of the significantly higher COP β_{hp} of the heat pump. If the overall thermal efficiency of the power plant is η_t, and the transmission efficiency for the electric energy is η_{tr}, the substitution of the burner with a heat pump system results in primary energy savings if the following inequality is satisfied:

$$\beta_{hp}\eta_t\eta_{tr} > 1 \qquad\qquad (8.8)$$

In Example 8.7, the transmission efficiency of the electric energy was assumed to be equal to 100%. In general, large electric power plants are far from the consumers, and transmission efficiencies are in the range of 86–98%.

An additional advantage of utilizing a heat pump cycle for buildings is that during the hotter season, the operation of the heat pump system may be reversed to become an air-conditioning system. Since the refrigeration cycle maintains two heat sources/sinks, one cold at temperature T_L and one hot (warm) at T_H, the colder heat source is utilized in the hotter season to supply the same building with cooler air. In practice, this is achieved by reversing the air streams in the building using the equipment of the same refrigeration cycle. In the hotter season, the air stream from the building is cooled by the evaporator of the refrigeration cycle. In the colder season, the air stream from the building is heated by removing heat from the condenser of the refrigeration cycle. Because the same cycle and equipment are utilized for both heating and cooling, the use of heat pumps/air-conditioning systems has become widespread in all types of buildings. Because of the different definitions of the COP for the heat pump and for the air-conditioning systems, when the same system is used as heat pump and as air-conditioner, it has a higher COP (and SEER) as heat pump: $COP_{hp} = COP_{ref} + 1$.

8.5.3 Ground Source Heat Pumps

One of the drawbacks for the use of conventional heat pumps for the heating or cooling of buildings is that they absorb heat from the atmospheric air, whose temperature is variable. The evaporator of the refrigeration cycle removes heat Q_L from the atmospheric air and the condenser of the cycle dissipates heat Q_H to the atmosphere. However, the atmospheric temperature is highly variable and ranges in some regions from −40°C during the colder months to more than 40°C during the hotter months. The ambient temperature variability makes the design of the refrigeration cycle difficult to optimize.

The COP of a heat pump significantly drops when the outside air temperature is very low. Similarly, the COP of the air-conditioning cycle drops when the atmospheric air temperature is at its extreme highs. Thus, the COP (and the SEER) of an air-conditioner is significantly lower when the outside temperature is 42°C than when the outside temperature is 30°C. As a consequence, the thermodynamic performance of heat pumps and air-conditioning systems deteriorates when the system is needed and utilized most, that is, during the coldest days (as a heat pump system) and during the hottest days (as an air-conditioning system). Other systems for the cooling or heating of the building are required when the performance of the HVAC systems deteriorates.

The reason for this performance deterioration is the variability of the ambient air temperature, which adds or removes heat from the HVAC system. If another heat reservoir with almost constant temperature were available, where large quantities of heat could be absorbed during the heating season and dissipated during the cooling season, then refrigeration cycles for the heating and cooling of buildings would operate with higher and almost constant COPs throughout the year and would not be affected by seasonal temperature extremes. Moreover, this system would operate more reliably and its components may be better optimized, on the basis of the almost constant heat source/sink temperature. The underground environment is such a heat reservoir: in most locations of the planet, while the surface of the ground is variable and is affected by the atmospheric air temperature, the temperature deeper than 6 m (20 ft) is almost constant, all the way down to depths of 100–200 m (330–660 ft).

The surface temperature of the ground is almost equal to the air temperature and has high annual variability. The annual variability of the underground temperature is reduced by a factor of 2 at a depth of 1.2 m (4 ft) and by a factor of 4 at 2.4 m. The temperature is almost constant throughout the year at depths below 10 m (30 ft) and slightly depends on the latitude of the location. The underground temperature is approximately 6°C (42°F) in Scandinavia throughout the year; 10°C (5°F) in Germany and England; 14–15°C in France*; 16°C (59°F) in southern Italy; 18°C (65°F) in Dallas, Texas; and close to 20°C (69°F) in San Antonio, Texas. Figure 8.7 shows the ground temperature profile for Giessen, Germany, during the months of July and January [9]. It is observed in the figure that the ground temperature below 8 m is almost constant in both winter and summer, at 9.5°C. Also, it is observed that even close to the surface of the ground, e.g., at 4 m, the temperature variability is much lower than the variability on the surface.

A GSHP, sometimes called geothermal heat pump, makes use of a refrigeration cycle with the ground as the heat dissipation medium from the condenser during the summer. The same cycle absorbs heat from the ground through its evaporator during the winter months. The heat exchange to the ground is achieved by a system of tubes or coils that

---

* This is also the recommended temperature for storing wine. Its origin is that it corresponds to the temperature of underground cellars in the Burgundy and the Bordeaux regions of France.

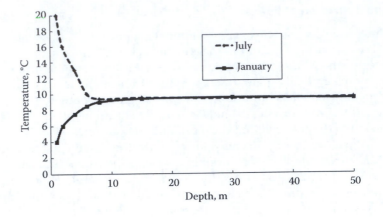

FIGURE 8.7
Ground temperatures for Giessen, Germany, in the months of July and January.

carry the cooling or heating fluid. Three main types of heat exchangers are used in the design of GSHPs. The first two use a horizontal configuration and the last uses a vertical configuration [9].

1. The horizontal ground piping system, which consists of a series of horizontal trenches with pipes at depths between 2 and 8 m. The heat exchanger fluid, typically water, is forced by a pump to circulate in the system of shallow underground tubes and dissipates the heat to the ground during the air-conditioning operation of the system, while it absorbs heat from the ground during the heating operation. Given that the depth of the trenches is shallow, the heat is dissipated close to the ground, where, nevertheless, there is a low annual temperature fluctuation.

2. The trench spiral collector system, for which the trenches dug are at a similar depth of 2–8 m, but wider. The heat exchanger tubing is placed in the trenches in circular loops, in a configuration that is similar to the toy *slinky*. A large number of loops in the "slinky" facilitate the heat transfer from the circulating fluid to the ground.

3. The vertical loop system, which essentially consists of several long vertical U-tubes in wells that are dug deep in the ground. This system requires the least amount of surface area, but extends to depths between 50 and 100 m. The piping of the tubes is typically of plastic material and a special bonding agent (bentonite or another heat-conducting gel) is used to fill the entire well and to provide structural stability for the U-tube.

The first two types of GSHP systems exchange heat close to the surface of the ground and are suited for regions where more heating than cooling is required during the year. The shallow, horizontal systems are not suitable for climates where the cooling season is longer than the heating season, such as the southern part of United States, where significant amounts of heat power need to be dissipated in the ground and for long periods. The third system, with wells extending to 100–120 m underground, is more suitable to be used in hot and arid climates. This vertical loop system has higher heat exchanger capacity. Typical wells for this system have diameters on the order of 0.3 m. The deep wells dissipate the heat deeper into the ground without appreciably raising the ground temperature. It has been observed that even with the deep wells, the ground temperature rises sometimes by 3–6°C. This happens during the hot summer months when air-conditioning is heavily used and very large quantities of heat are continuously dissipated underground [10].

Because of the almost constant ground temperatures, the substitution of a conventional air-conditioning system with a GSHP is usually accompanied by a very significant increase in the COP and SEER of the equipment. The COP of an optimized GSHP cycle may become as high as 8 (27.3 SEER), and that of an air-conditioner, as high as 7 (23.9 SEER). The heating and air-conditioning systems of buildings perform the same cooling task while consuming lesser electric power. This implies lower cost of cooling for the owners of the buildings, fewer emissions, and lesser environmental pollution associated with the electricity production.

The substitution to GSHPs also has a rather unexpected advantage for power generation corporations: In the warmer regions that need a great deal of air-conditioning, the peak power demand occurs in the hot season. For regions, such as the southwest of the United States, the summer power demand by far exceeds the winter demand as it is shown in Figures 7.1 and 7.2. As a consequence, the peak power demand occurs in the summer and highly depends on the need for cooling. The high population growth in these regions—Texas and Arizona in the United States are prime examples—imposes a corresponding growth of the regional peak electric power production. This necessitates the construction of new power plants and significant capital expenditure for power generation corporations. Oftentimes, it is financially advantageous for these corporations to invest in energy efficiency measures at the consumer level, such as higher SEER for air-conditioning systems, than in new power plants. For this reason, several power generation corporations have offered incentives (rebates and special contracts) to consumers that install GSHPs. This trend is more apparent with the not-for-profit publicly owned utilities, which do not derive a significant financial advantage from selling more power to consumers: Since 2009, the San Antonio and Austin, Texas, utilities (CPS, and Austin Power, which are owned by the two cities) have offered financial incentives to consumers for the installation of GSHPs with higher SEER than conventional air-conditioning systems. This substitution has proven to be cost effective for both the consumers and the publicly owned utilities that do not have to invest in additional power-producing capacity.

8.5.4 Hot Water Supply

In addition to space heating and cooling, hot water usage in buildings at temperatures of 40–50°C consumes a great deal of thermal energy, and this is typically provided by electric or gas heaters. Because of the energy losses at the electricity generation station, gas hot water heaters are by far better to be used than electric hot water heaters. The electricity used in an electric heater is produced by a power plant with efficiency typically in the range of 30–40%. Therefore, a significantly higher quantity of heat—since $Q = W/\eta_t$, this quantity of heat is from 3.3 to 2.5 times higher than the electricity needed—is consumed at the power plant for the production of the electric energy that is converted back into thermal energy in the electric water heater. In contrast, a well-maintained gas burner has 90–99% thermal energy conversion efficiency and uses a lesser amount of gas. The series of energy transformations, chemical energy → heat → electricity → heat, in the electric heater consumes 2.3–1.5 times more primary energy than the simpler transformation, chemical energy → heat in the gas heater. The following example demonstrates this effect.

Example 8.8

Determine the standard cubic foot of natural gas used for the heating of 200 kg of water (approximately 56 gal) from 20°C to 50°C if (a) a gas burner is used with 95% combustion efficiency and (b) an electric heater utilizing electricity generated in a gas turbine with 35% thermal efficiency is used.

Solution: The desired action to increase the water temperature requires

$$Q = m \times c_P(T_2 - T_1) = 25.14 \text{ MJ of heat.}$$

a. From Table 1.1, 1 scf of natural gas provides 1.072 MJ of heat when burned at 100% efficiency. At 95% efficiency, 25.14/(1.072 × 0.95) = 24.7 scf of natural gas is needed.
b. When an electric heater is used (their efficiency is close to 100%), the required 25.14 MJ of heat are produced from electricity, which was generated using 24.15/0.35 = 69 MJ of heat. This amount of heat is supplied to the electricity-generating station by 67.0 scf of natural gas.

It is apparent in this example that the mere substitution of an electric heater with a gas heater saves 42.3 scf of natural gas (primary energy) for the accomplishment of this desired action. The substitution of an electric stove (cooker) with one that uses gas also saves significant amounts of a primary energy source. As a rule, and because electricity is typically generated from other primary energy sources in power plants with 30–40% thermal efficiency, using electricity for heating always wastes primary energy resources. Using natural gas, synthetic gas, or liquid petroleum products for heating and cooking are much better options for primary energy source conservation and, in most cases, more economical for the consumers.

Residential gas and electric water heaters maintain a large volume of water (30–80 gal in the United States) in a tank at a higher temperature, typically in the range of 45–50°C. The heaters are typically located in attics or basements, where the air temperature is significantly lower. As a result, the water in the tank cools with the heat loss being higher during the night hours, when the ambient temperature is lower. Ironically, hot water is not needed during the night. Better insulation of the water heater is a good conservation measure that reduces the average heat losses of the heater. A water heater design that supplies the household with hot water only when this is needed is another way to reduce the energy spent. This type of heater is smaller and is located close to the hot water demand point, the bathroom or the kitchen sink. The heater is essentially a small burner, equipped with a sensor that ignites the gas when water flows in the pipes. When water flows through the pipelines, the gas combustion in the heater supplies heat to the coils inside the heater and causes the increase of the water temperature as it exits the heater to the desired temperature. When the water supply to the heater is shut, the gas flow is switched off by the flow sensor and the combustion stops. This type of self-igniting heater supplies hot water on demand and does not store any hot water, and therefore, lesser heat is lost to the surroundings. The technology for such heaters is well known, because the heaters have been used in several European countries since the 1950s. Water heaters in the United States consume on average 12% of the annual household energy consumption of residential buildings. It is estimated that the installation of such small, self-igniting heaters in a household would reduce the energy spent for hot water by 56% and would thus reduce the energy needs of the average United States household by 6.7% [11].

Hot water may also be supplied to residential and commercial buildings from the condensers of HVAC systems. All heat pump and air-conditioning systems use refrigeration cycles with condensers at moderate temperatures that have the capability to transfer heat and supply hot water to buildings during the cooling season at no additional consumption of fuel. A glance at the heat pump/refrigeration cycle of Figure 1.6 proves that the quantity of heat Q_H is removed from the cycle at moderately high temperatures. When the system operates in the air-conditioning mode, this heat is dissipated in the environment. Part of this waste heat may be recovered to raise the temperature of water for the hot water supply of buildings. When the HVAC system operates in the heat pump mode, part of the heat from the condenser may be diverted to increase the water temperature. Typical condenser temperatures for refrigeration cycles are in the range of 55–65°C, and typical hot water temperatures are in the range of 45–50°C. Therefore, all or some of the waste heat of the refrigeration cycle may be transferred from the condenser of the refrigeration cycle to supply all or part of the hot water needs of buildings. In practice, this is accomplished with heat exchanger coils, which become part of the condenser of the air-conditioning cycles and are inserted in the hot water tanks. The condensing refrigerant passes through the coils that transfer heat to the hot water supply of the buildings. As a result of this waste heat utilization process, the hot water heaters in buildings consume significantly less primary energy (gas or heating oil).

A schematic diagram of the temperatures during this heat transfer process is shown in Figure 8.8. The refrigerant desuperheats and condenses at constant temperature inside the coil, while the temperature of the water, outside the coil, increases to the desired hot water temperature. When the HVAC system (the refrigeration cycle) works in the cooling mode, all the heat in the refrigerant condensation process needs to be dissipated. Using this heat for hot water does not add to the energy consumption of the building. When the refrigeration cycle operates in the heating mode and the HVAC system is essentially a heat pump, the heat released in the condenser of the cycle is used for the heating of the building and the supply of hot water is an additional heat load. In this case, the HVAC system needs to operate longer to supply the additional heat to the water heater. The application of this method for domestic water heating is particularly advantageous in hotter climates, where the refrigeration cycle operates in the cooling mode for extended periods of the year and the heat used for the hot water supply is the

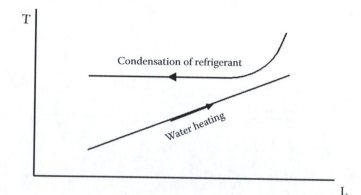

FIGURE 8.8
Hot water supply from the condenser of a refrigeration (air-conditioning) cycle.

waste heat of the cycle. Experience shows that heat pump/air-conditioning cycles, especially those using the GSHPs, will provide 100% of the hot water needs of households in the southern part of the United States during the months of April to October without any additional use (and cost) of energy. These systems will also provide hot water to the building during the winter months at 40–70% cost of the gas for the hot water needs of the building.

8.5.5 Adiabatic Evaporation

Adiabatic evaporation in hot and arid climates significantly reduces the cooling needs of buildings. When a stream of warm and dry air passes on top of a water body, a fraction of the water evaporates. The warm air stream provides the latent heat for the vaporization of the liquid water, and as a result, the temperature of the air stream decreases, oftentimes significantly. Simultaneously, the specific humidity of the air stream increases. Figure 8.9 depicts a schematic diagram of the evaporative cooling process as a thermodynamic process. The mass balance in the control volume denoted by the dashed-line rectangle is

$$\dot{m}_w = \dot{m}_a(\omega_2 - \omega_1),\tag{8.9}$$

where $\dot{m}_a$ is the mass flow rate of the air; $\dot{m}_w$ is the rate of the water mass that evaporates and enters the air stream; and ω is the specific humidity of the air. Similarly, the energy balance—first law of thermodynamics—for this control volume is

$$\dot{m}_a(h_{a1} + \omega_1 h_{v1}) + \dot{m}_w h_w = \dot{m}_a(h_{a2} + \omega_2 h_{v2}).\tag{8.10}$$

The subscripts v and a denote the water vapor and air, respectively. Because the temperature difference $(T_2 - T_1)$ is small, the enthalpy difference of air is equal to the product of the specific heat of air and the temperature difference. The cooling effect, or temperature reduction, $(T_2 - T_1)$, from the evaporative cooling is

$$T_2 - T_1 = \frac{1}{c_P}(\omega_1 h_{v1} - \omega_2 h_{v2} + \omega_2 h_w - \omega_1 h_w).\tag{8.11}$$

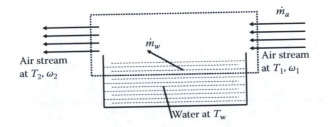

FIGURE 8.9
Schematic diagram of evaporative cooling.

TABLE 8.4

Inlet and Outlet Temperatures in the Evaporative Cooling of Ambient Air

T_1, °C	φ_1, %	φ_2, %	T_2, °C
35	10	60	21
35	20	70	23
40	20	70	26
40	20	50	30
35	30	60	23

The last expression may be solved using thermodynamic tables for the enthalpies. Alternatively, one may derive an approximate expression in terms of the latent heat of evaporation of water h_{fg}:

$$T_2 - T_1 \approx \frac{h_{fg}}{c_P}(\omega_1 - \omega_2). \tag{8.12}$$

With $c_p = 4.184$ kJ/(kg·K) and $h_{fg} \approx 2,400$ kJ/kg, the increase of the specific humidity of air by 0.01 reduces the air temperature by 5.7°C (10°F). The evaporative cooling effect can be significant in dry weather and may be used to partially or totally cool the air of a building. In addition, evaporative cooling may be achieved with rudimentary and inexpensive equipment, and this makes it very attractive financially. For example, blowing air with a fan over a small open water container or injecting small water droplets in front of an air fan will appreciably cool the air. Locating residences at the shores of the sea or a lake, where local natural breezes bring cooler air, has the same cooling effect and has been used for centuries for domestic comfort during the summertime.

Table 8.4 includes several cases that show the evaporative cooling effect, in terms of the relative humidity φ for several inlet and outlet conditions. T_1 is the inlet temperature of the ambient air and T_2 is the outlet temperature, after the cooling effect. It is observed that 10–18°C cooling of the ambient air may be achieved by merely increasing the relative humidity of the air from the range of 10–30% to 50–70%. In most cases, this temperature drop sufficiently cools the ambient air, and no additional air-conditioning is needed. This type of natural residential cooling has been used for centuries in Mediterranean countries, especially in the islands, where artificial air-conditioning was almost unknown until the beginning of the twenty-first century.

8.5.6 District Cooling

The concept of *district cooling* is similar to that of *district heating*. It is based on the higher efficiency of larger and better maintained air-conditioning installations in comparison to smaller and less efficient installations, not based on the use of waste heat. The typical household air-conditioning unit is composed of a smaller 0.5–10 kW engine. Affordability and low price (not high efficiency) are the primary design considerations of these smaller units. As a result, the COPs of typical small building air-conditioners are in the range of 2.0–3.0. A larger air-conditioning/refrigeration unit, which might supply chilled water to a district of several buildings, could be designed with a significantly higher COP. Cooling the condenser of the system by water rather than air would increase the COP of the HVAC system by 1 to 1.5 units. A larger and more efficient compressor, frequent maintenance,

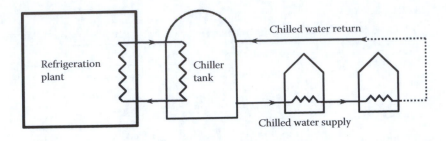

FIGURE 8.10
District cooling concept.

and the continuously controlled and optimized operation of the larger HVAC system would further increase its overall COP.

A schematic diagram of district cooling is shown in Figure 8.10. A large, highly efficient refrigeration plant cools water and maintains it at a low (8–12°C) temperature in the *chiller*—a large, well-insulated tank. The cold water is pumped in series or in parallel by well-insulated underground ducts and delivered to nearby buildings, where it is used to cool their interior. The returned, warmer water is cooled in the chiller by the central refrigeration system. Another advantage of district cooling is that the large chiller tank may be used for the "storage of coolness" as mentioned in Section 7.3.2. The refrigeration plant may operate to chill the water during the night hours when the demand and the cost of electricity are low. The chiller tank pictured in Figure 7.9 is essentially part of a large district cooling system for all the Dallas/Fort Worth airport buildings.

District cooling operates at best when the buildings are close enough for the water of the chiller not to be significantly heated from the ambient air or the ground. For this reason, district cooling works best with larger buildings in high population density areas, such as high-rise buildings, commercial centers, large hotels, and municipal centers. Several districts in city centers of the United States and southern Europe have adopted this concept of cooling with significant energy savings: Eleven large hotels in the downtown district of San Antonio, Texas, use a central highly efficient cooling plant to provide chilled water for their air-conditioning needs. The use of the district cooling in suburban areas, where the building density is low, the distances between buildings are long, and there is significant heat transfer to the chilled water from the environment, does not result in appreciable energy savings, and sometimes may even add to the aggregate energy consumption.

8.5.7 Fenestration (Windows) Improvement

Windows are thinner than walls; they have much higher thermal conductivity; they are transparent to radiation heat transfer; and they do not possess the additional insulation the interior of a wall may have. Simple glass with 1/8 in. thickness (3.2 mm) has 500 times higher conductivity than an insulated wall or a roof. Even though the area of the windows is much less than that of walls and roofs, most of the heat enters and leaves residential and commercial buildings through the windows. Up to 90% of the heat transfer to or from typical buildings may occur through the fenestration system, which includes frames and seals, if this system is not "weatherized."

Improvements of the energy effectiveness of fenestration will significantly cut the energy used for the air-conditioning of buildings during the hot season as well as for the heating during the colder season. The following modifications may be used to improve the energy efficiency and reduce the heat transfer rate through the fenestration of buildings:

1. Installation of double-pane (double-glazed) or triple-pane windows with a thin gap between the two panes. The gas in the thin gap provides an additional layer of insulation. The inert argon gas has better insulating properties—higher R-value—than air.
2. Installation of vacuum-glazed windows, where the gap between the panes is evacuated. However, the vacuum seals in fenestration degrade after several thermal cycles and air may leak in to fill the gaps and reduce the *R*-value.
3. Covering the outside glass surface with radiation coatings that reflect part of the interior heat back into the building during the cold season and reflect part of the solar radiation out of the building during the hot season.

A natural method that reduces the heating and cooling needs of buildings is the use of awnings outside of windows that predominantly face the sun. This is the southern side of buildings in the northern hemisphere and the northern side of buildings in the southern hemisphere. Well-designed awnings significantly reduce the sunlight that enters a window during the summer, when the sun is at the highest. The same awnings allow sunrays to enter the building during the winter, when the sun is lower in the horizon. More solar energy enters the building during the heating season and less during the cooling season, a combination that reduces both the heating and the cooling requirements of buildings. The use of simple awnings in the south-facing windows reduces by 15–25% the cooling needs of a household in the southern part of the United States and of most Mediterranean countries.

8.5.8 Improved Efficiency of Appliances

While most of the energy consumption in residential buildings is used by the HVAC systems for heating and cooling, large appliances—among them refrigerators, cooking stoves, hotplates, dishwashers, washers, and dryers—consume approximately 35% of the total energy used in the United States households [11]. The percentage of the energy used in large appliances in the households of the EU and other OECD countries is in the range of 30–45%. It is axiomatic that improving the efficiency of these appliances, which primarily consume electricity, would significantly reduce the national and global energy consumptions. Since the 1970s, national guidelines have been adopted by several governments to improve the efficiency of household appliances. With the improved efficiency, the same desired action is accomplished, e.g., the washing of 3 kg of clothes, with the consumption of lesser energy. Because most of the large household appliances operate with electricity, the savings in primary energy are even higher. In the United States, several "standards," adopted by the federal government and individual states (California is a pioneer among them), have resulted in significant reductions of the energy used by large household appliances. Figure 8.11 shows that the average energy used by refrigerators in the United States dropped by a factor of 4 in the period of 1972–2014. It is also of interest to note in the figure that the higher efficiency and energy savings did not come with an increase of the refrigerator prices, which actually dropped by 53% in constant 2010 US dollars, or with a reduced refrigeration capacity—the latter increased by 30% [12]. Similar standards

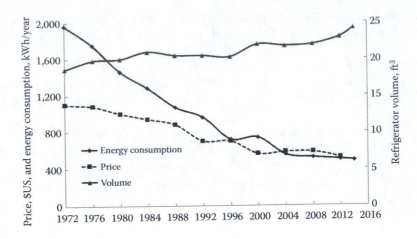

FIGURE 8.11
Average annual energy consumption, average volume, and average price in constant US dollars, for refrigerators sold in the United States.

adopted for other large household appliances have also resulted in noteworthy efficiency improvements and reduction of the total energy consumption in residential buildings.

Most nations have governmental regulatory agencies that promote higher efficiency for appliances and processes. In the United States, the Department of Energy established the *National Appliance Energy Conservation Act* to specify minimum efficiencies for appliances in commercial and residential buildings. This has resulted in the *Energy Star* system of appliance assessment and the mandatory labels for the expected annual energy consumption of the major appliances. In the EU, the European Commission issues frequent *energy efficiency directives*, for the efficiency of household appliances. In Japan, energy efficiency has become part of the mission of *Ministry of Economy, Trade and Industry*, while in the People's Republic of China, this subject has become the purview of the state-owned *China Energy Conservation and Environmental Protection Group Corporation*. These and similar agencies in other countries regularly issue reports on energy conservation and efficiency of household appliances and adopt directives to improve appliance efficiencies.

8.5.9 Other Energy Conservation Measures for Buildings

Other conservation methods that decrease the energy consumption in buildings are as follows:

1. Use better-fitting doors and windows that reduce the air draft in buildings. Similarly, in older buildings, use insulating adhesive strips to better seal doors and windows.
2. Use fans for the circulation of air. This improves the human comfort and reduces the need for air-conditioning.
3. Place the heating and air-conditioning vents on the floor rather than the ceiling. This measure keeps the supply of hot and cold air where it is mostly needed by the occupants and is especially effective in high-ceiling spaces.

4. Use diffuse natural lighting rather than artificial lighting. In addition to conserving the electric energy for the lights, this measure also reduces the cooling needs of the buildings.

5. At the time of planning and construction, optimize the orientation of the buildings and their fenestration placement for less total annual energy consumption.

6. Use screens and thin films on windows to reduce the insolation during the summer.

7. Insulate better the attics, basements, and other exposed parts of the buildings.

8. Use programmable thermostats that allow the internal temperature to be adjusted outside the human comfort zone, when the buildings are not occupied.

9. Maintain well the heaters and burners for complete combustion and replace old units with modern more efficient units.

The reduction of annual energy consumption also reduces the maintenance cost of commercial and residential buildings. Older buildings, which were built in an era of cheaper energy and are very costly to maintain may be *retrofitted* to reduce their annual energy consumption. In all cases, retrofitting generates significant monetary savings for the owners, who are able to amortize the cost of retrofitting in a short time, typically 2–6 years. Buildings that were constructed before the 1970s—especially the ones where air-conditioning has been added after the original construction—are prime candidates for significant energy and monetary savings from retrofits. In several countries, local governments as well as local electricity production corporations provide loans and rebates to assist with the capital expenditures of building retrofitting.

It must be noted that for-profit energy production corporations, which derive their profits from the sale of energy at prices determined by contracts, do not have incentives to encourage energy efficiency or energy conservation for their customers. At the end, if these corporations sell less energy, they will gain lesser profits. Public electricity corporations that do not operate on the principle of profit maximization and include in their mission energy conservation and environmental stewardship are more likely to promote and assist with conservation and higher efficiency efforts. For the for-profit electricity corporations, a new paradigm is needed, through which they will accrue revenue and profit for "satisfying the energy needs of their customers," rather than for "selling as much energy as possible to their customers." Such a paradigm in the mission, administration, and regulation of regional electricity corporations would very much assist with regional and national goals for energy conservation, energy efficiency, and better environmental stewardship.

Example 8.9

Two multiapartment buildings are heated by heating oil and require 4.80×10^9 kJ of heat per year. The LHV of the heating oil is 140,000 Btu/gal, and the fuel costs $2.3/gal. The "boiler" of the buildings is only 62% efficient, and the owner decides to substitute it with a newer, 95% efficient unit. Determine the amount of heating oil saved per year, and the annual monetary savings for the owner of the building.

Solution: With 62% efficiency, the old boiler uses the equivalent of $4.80 \times 10^9/0.62 = 7.74 \times 10^9$ kJ or 7.34×10^9 Btu, and this corresponds to $7.34 \times 10^9/140,000 = 52,417$ gal of the fuel. The annual cost of this of fuel is $120,551.

With the new boiler, the equivalent amount of fuel is $4.80 \times 10^9/0.95 = 5.05 \times 10^9$ kJ = 4.79×10^9 Btu, which may be supplied by $4.79 \times 10^9/140,000 = 34,209$ gal at a cost of $78,680. Hence, the annual savings are 18,208 gal of heating oil and $41,847.

8.6 Conservation and Improved Efficiency in Transportation

The transportation sector accounts for 25–35% of the primary energy demand in most OECD countries; between 30–55% in the developing nations and encompasses the transportation of persons as well as goods. The significant increase of global petroleum consumption, shown in Figure 2.7, is primarily due to the transport of commercial goods and the wider recreational use of the personal automobile. The globally strong correlation between the number of automobiles and aircraft and petroleum product—gasoline, diesel oil, kerosene, and similar fuels—consumption in a country is corroborated with the data from the United States, which are depicted in Figures 2.11 and 2.12. In the early twenty-first century, the transportation of individuals and goods—domestic and international trades—have become essential parts of everyday life. Developing better and more efficient methods for the transportation of goods and persons will significantly reduce the total global energy consumption.

While trying to devise methods for the reduction of the energy consumption in the transportation sector, one must always keep in mind what is the desired action that needs to be fulfilled: the comfortable transportation of persons from one location to another or the transportation of goods from the production to the marketplace. In most cases, several alternative methods are available to accomplish these actions. The desired action may be 100 t of potatoes need to be transported from Idaho to New York City or 135 passengers need to travel from Marseille, France, to Barcelona, Spain. The means of transportation are not necessarily defined and may become a matter of choice. The transportation of potatoes is fulfilled when the potatoes are transported by truck, by train, by airplane, or by a fleet of passenger cars. The passenger transportation may be accomplished by ship, airplane, train, bus, or, by individual automobiles. It becomes immediately apparent that of the available methods to accomplish the transportation of persons and goods, one would use the least amount of energy, in this case, transportation fuel. An analysis of the methods proposed would reveal that the transportation of potatoes from Idaho to New York by train consumes the least amount of fuel. Similarly, the travel of the 135 passengers from Marseille to Barcelona by ship also consumes the least amount of fuel. Typically, rail and ship transportation are the methods that consume the least amount of fuel, while airplanes and automobile/truck transportation use higher amounts of fuel to fulfill the desired action.

Figure 8.12 depicts the data of the energy consumed per passenger mile in the United States for several commonly used modes of transportation [13] in the period of 1960–2014. It is apparent in this figure that rail transport—although not the most popular means of passenger transport in the United States—requires the least amount of energy per passenger mile. It is also apparent that the average energy spent per passenger mile has decreased for all modes of transportation since 1960, reflecting the improved efficiency of the engines. This is particularly apparent in the air transport data, where the introduction of new and more efficient jet engines has made this mode of transportation less energy demanding than private automobiles and buses.

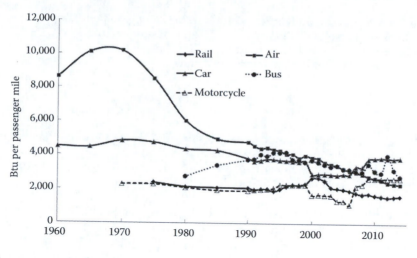

FIGURE 8.12
Historical data of the energy consumed per passenger mile in the United States for several commonly used modes of transportation.

Public transportation in urban environments with high population density, and especially the metro/train transportation, is the least energy consuming method of all the available means of transportation, when it is used by a high percentage of the citizenry. However, personal convenience, societal values, perceived status, and long-time habits play important roles in the way our society values and utilizes the public transportation systems. Residents of several cities in the United States consider it more convenient and status enhancing to drive their automobiles to work rather than use public transport. On the contrary, the vast majority of Parisians, Londoners, and Berliners would rather take the bus/metro combination (where they also read their daily paper, a book, or safely check their e-mails).

Passenger transportation is different from the transportation of merchandise, because safety and convenience play more important roles than energy savings. While safety is paramount and should not be compromised, a certain amount of convenience may often be sacrificed for higher energy efficiency. For example, commuting by subway or urban buses saves a significant amount of fuel, but may be less convenient than individual automobile travel, and may add to the time of commute. Additional benefits to the community are that public transport contributes to the reduction of urban traffic congestion, lesser harmful emissions, better environmental quality, and more reading time for the passengers (or more time to connect with friends through social media without the risk of a car accident).

Example 8.10

A new bus line transports on average 25 passengers per trip to an urban center from a suburban area and back. The one-way trip is 43 km. The average mileage of the cars in the area is 8 km/L, while the bus consumes 3.2 km/L. Determine the amount of fuel saved per trip of the bus. If 10 round trips are made per day, what are the gasoline savings per year and the annual CO_2 avoidance?

Solution: If the 25 passengers used their own car with average mileage of 8 km/L, they would have consumed a total $25 \times 2 \times 43/8 = 268.8$ L of gasoline. Instead, the bus consumes $2 \times 43/3.2 = 26.9$ L per round trip. The savings per round trip are 241.9 L of gasoline.

The savings per year are $10 \times 365 \times 241.9 = 882,935$ L of gasoline.

The density of gasoline is 720 kg/m^3 or 0.72 kg/L. Therefore, $882,935 \times 0.72 = 635,713$ kg of gasoline is saved with the switch to public transportation.

From the equation for the combustion of octane,

$$C_8H_{18} + 25/2O_2 \rightarrow 8CO_2 + 9H_2O,$$

we deduce that 114 kg of octane produces $8 \times 44 = 352$ kg of CO_2. Therefore, the annual CO_2 avoidance is $352 \times 635,713/114 = 1963,000$ kg.

A useful number to remember in this example is that the CO_2 emission avoidance is approximately 2.22 kg of CO_2/L of gasoline saved.

Regarding the transportation of commercial goods, transportation by ship or train is the most energy-efficient method. Petroleum is transported by tanker ships; grain from the Midwest United States reaches South America by ships; and vegetables from Spain reach Britain by train. The few and very specialized goods with very short life—e.g., food items and flowers—necessitate faster transportation at the expense of higher amounts of energy. For example, tulips from Holland delivered to flower shops in Philadelphia are transported fast by airplane instead of the more economical, less energy-consuming, but slower ship, and the perishable blue tuna fish from the Gulf of Mexico is also transported fast by airplane to Japan to supply the sushi restaurants.

A great deal of energy savings may be accomplished in the transportation sector, both for the transport of goods and for individuals: The bulk transport of goods in trains and ships is by far less energy consuming (in kJ consumed per kg of goods transported) than transport by cars, trucks, or airplanes. Therefore, whenever there is no *a priori* reason to do otherwise—e.g., the goods have a short life and may be damaged—transportation by train or ship should be preferred. At the same time, it makes sense both economically and from the energy point of view to pool resources and fill the entire transportation medium during a trip. A train, truck, airplane, or ship that is filled to capacity transports goods cheaper and with less specific energy consumption (kJ/kg of goods transported) than a partially filled vehicle. For a long time, nautical companies have been using several methods, including visits to a number of ports, to fill their ships with cargo before they crossed the oceans. Commercial airlines are increasingly using similar strategies to fill the aircrafts in most trips by using differential pricing, shared codes, and offering flights during periods of high demand.

Because of the importance of the transportation sector to the national economies, several governments have introduced regulations on the expected energy consumption of cars and trucks. The regulations provide the standards for the mileage of the new vehicles introduced in the market and are usually measured in miles per gallon (mpg) or hundreds of kilometers per liter (100 km/L) of gasoline (or the equivalent to gasoline fuel the vehicle consumes). In the United States, CAFE standards were introduced after the energy crisis of the 1970s. CAFE standards require vehicle manufacturers to comply with the gasoline mileage (the fuel economy standards) that are set by the US Department of Transportation (US-DOT). The values for these standards are obtained using the city and highway test results and the weighted average of vehicle sales in the country. Because CAFE standards apply to new vehicles, the efficiency of the entire fleet of vehicles continuously improves as older cars are replaced by newer more efficient models. Figure 8.13 shows, for the years 1980–2015, CAFE standards adopted by the US-DOT, the fuel efficiency (mileage in mpg) for the new automobiles that were introduced during the year, and the average mileage of all the automobiles—new and older—that were registered in the United States [13].

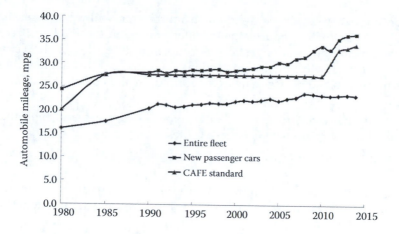

FIGURE 8.13

CAFE standard adopted by the US-DOT in the years 1980–2015 and their effect on the entire fleet of cars in the country.

It is apparent that the cars introduced during each year met or slightly exceed the standard. As a result, the mileage of the entire automobile fleet in the country, while lower than that of the new automobiles, has continuously improved since 1980. The effect of hybrid cars that were introduced after 2005 is apparent in both the increased mileage of the new cars and the adoption of higher CAFE standards in 2010–2013. Increasing the numbers of new automobiles in the national fleet improves the average efficiency of transportation and results in lesser energy consumption.

Example 8.11

A saleswoman uses her private car to make calls on customers. On the average, she travels 600 mi/week for 48 weeks/year. She recently decided to trade her older car that achieves 22 mpg with a hybrid that averages 47 mpg. Determine how many gallons of gasoline she saves annually. Assuming that the gasoline is solely composed of octane, determine the CO_2 avoidance from this vehicle substitution.

Solution: During a year, the saleswoman travels 48 × 600 = 28,800 mi and consumes 1309 gal of gasoline. If the same distance is travelled by a hybrid with 47 mpg performance, then only 613 gal of gasoline are needed. The annual gasoline savings are 696 gal.

The density of gasoline is 720 kg/m³ or 2.727 kg/gal. Therefore, 1898 kg of gasoline is saved with the vehicle substitution.

From the equation for the combustion of octane,

$$C_8H_{18} + 25/2 O_2 \rightarrow 8CO_2 + 9H_2O,$$

we deduce that 114 kg of octane produce 8 × 44 = 352 kg of CO_2. The savings of 1,898 kg of gasoline also alleviate the emission of 1,898 × 352/114 = 5,860 kg (5.86 t) of CO_2 in the atmosphere.

A useful number to remember is that the emissions avoidance is approximately 8.42 kg of CO_2 per gallon of gasoline saved.

Other than buying new cars with higher mileage, among the ways individuals and communities may reduce their energy consumption for transportation are as follows:

1. Carpooling, where several individuals from the same suburban area share a ride to the workplace. This reduces the number of cars on the road and reduces gasoline consumption and congestion. Several municipalities in the United States encourage this practice by designating fast-moving high-occupancy vehicle lanes on the highways and by assisting in the formation and coordination of carpooling groups.

2. Establishing more bus routes in densely populated areas and light rail systems and encouraging citizens to use these routes.

3. Encouraging the wider use of smaller cars and electric cars for commuting to work.

4. Developing close-to-work communities with minimal daily commute. This is advocated by several urban planners but is not agreeable to urban employees who wish to live in the country.

5. Traffic light synchronization. Frequent stops, which entail decelerations and accelerations, significantly increase the energy consumption of a vehicle. Optimized traffic patterns with average constant vehicle speeds between 40 and 45 mpg are appropriate in urban and suburban environments.

6. Seamless connection of highways by the use of dedicated entrances and exits without the delays that slow down or stop the traffic and activities that waste fuel.

7. Avoidance of peak traffic hours by adopting staggered work hours in large urban areas. The working day for most urban businesses is rigid: it starts between 8:00 am and 9:00 am and ends between 5:00 and 6:00 pm. This creates peak hour "traffic jams" in urban centers, which slow down the traffic, increase the fuel consumption in the region, and significantly contribute to air pollution. The adoption of more flexible working hours (e.g., working days that start between 6:30 am and 10:00 am and end between 3:30 and 7:00 pm) would alleviate the peak hour "traffic jams," and would reduce the congestion, the pollution, and the time of commute for the urban workers.

8.6.1 Electric Vehicles with Batteries

Vehicles in motion require power to counteract the ground and air friction forces. Since power is the product of a force and a velocity (Equation 1.2), vehicles need additional power to overcome the gravity force and move up a hill at the desired speed.* The needed power for a vehicle is produced by its engine. In the vast majority of vehicles, the engine is an internal combustion (IC) engine. The rest are electric engines.

The IC engine operates on a thermodynamic cycle and has a thermal efficiency η_{tv} typically between 12% and 30%. The power produced by the engine cycle is transmitted to the wheels by the transmission system with a small loss. This brings the overall efficiency of a vehicle—sometimes called the *well-to-wheels efficiency*—within the range of 10–30%.

The electric vehicle concept is not new. It has been successfully applied to rail transport in underground (e.g., metro, subway) and aboveground railways (electric intercity trains) for more than a century. The electric railway movement is confined to the rails; while in

---

* Because of friction in the engine and transmission, only a fraction of the work spent during the ascent of a hill is recovered during the descent.

operation, the railway engines are connected to the electricity grid; they are directly powered from the electricity transmission lines; and their well-to-wheels efficiency is higher than that of diesel and gasoline locomotives. Electric cars and buses have higher freedom of movement and cannot be continually connected to the electric grid. The electricity that propels them must come from a source within these vehicles, a battery or a fuel cell. Batteries need to be charged with electricity, while fuel cells may operate for long periods with a tank that supplies hydrogen or a liquid light hydrocarbon (methane, alcohol, etc.).

Battery-powered electric cars were briefly introduced at the turn of the nineteenth century and, again, in the late 1950s with no commercial success. The availability of cheaper fuels (gasoline and diesel) in combination with the reliability of the IC engine drove the producers of the early electric vehicles to bankruptcy. In the early twenty-first century, environmental concerns of CO_2 and other combustion product emissions as well as improved battery technology ushered a new era of new electric battery-powered vehicles that are produced by almost all the major car manufacturers. Governmental incentives—tax credits and rebates—ranging from $4,000 in Sweden to $8,000 in the United States to $20,000 in Norway* have also helped the recent market of electric cars [14]. Without being ubiquitous, small electric vehicles, similar to the one shown in Figure 8.14, have become a feature in urban settings. Battery charging stations in garages as well as on several urban streets are also a feature of the modern urban setting. At the end of 2015, there were approximately 740,000 electric cars in the world. The histogram in Figure 8.15 shows in which countries most of these cars are registered [14].

The engines of the electric cars are essentially battery-operated motors, and their efficiency η_M is high, in the range of 90–95%. The charging efficiency of the car batteries η_B is also high, in the range of 85–90%. For this reason, oftentimes, electric cars are touted as very efficient. However, electricity is not a primary energy source and is produced by another fuel. Typical thermal efficiencies η_t of thermal power plants (coal, natural gas, nuclear) are in the range of 30–40%. In addition, losses of electric energy occur during the transmission of electricity from the production unit to the charging station. If the transmission losses are 10%, then the transmission efficiency of the electric power is η_{tr} is 90%. When all these processes are taken into account, the well-to-wheels efficiency for an electric car is

$$\eta_{wwel} = \eta_t \eta_{tr} \eta_B \eta_M. \tag{8.13}$$

With the typical efficiency values given earlier, the well-to-wheels efficiency of electric cars is in the range of 21–31%, which is comparable to the efficiency of current automobiles with an IC engine. The energy savings for most of the electric cars stem from the fact that most electric cars are by far smaller and lighter than typical IC engine cars, as seen in the car depicted in Figure 8.14. The substitution of a midsize sedan with a much smaller electric car, similar to the one in the figure, would result in energy savings. Hybrid cars, whose well-to-wheels efficiency may reach and sometimes exceed 45%, conserve significantly more primary energy than purely electric cars.

Example 8.12

A commuter drives 13,000 km/year in a midsize car that consumes 9 L of gasoline per 100 km. The commuter decides to substitute this car with a smaller electric car, operated

---

* Norway possesses a great deal of unused hydroelectric capacity.

FIGURE 8.14
An electric vehicle is charging on a street in Milan, Italy, and the publicly available charging system.

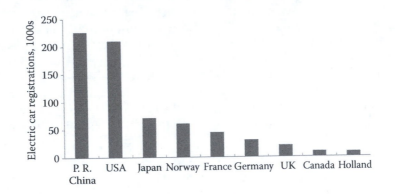

FIGURE 8.15
Countries with high numbers of electric cars registrations.

with a battery, which needs 15 kWh/100 km. This car has $\eta_M = 90\%$, the batteries charge with efficiency of $\eta_B = 88\%$, the electricity is produced at a coal power plant with $\eta_t = 38\%$, and the transmission efficiency of electricity is $\eta_{tr} = 90\%$. The power plant consumes anthracite with 90% carbon content and LHV = 31,000 kJ/kg. Determine (a) the annual amount of gasoline saved per year; (b) the annual amount of electricity used by the car; (c) the additional amount of anthracite needed to produce the energy for this car; and (d) the annual amounts of CO_2 emitted by the operation of the two cars.

Solution:

    a. In order to drive 13,000 km, the commuter uses $13,000 \times 9/100 = 1170$ L of gasoline annually. Assuming all the gasoline is composed of octane (density 0.720 kg/L), this car uses 842 kg octane per year.

b. At 15 kWh/100 km, the commuter uses 1,950 kWh/year to drive the same commuting distance. With the transmission, battery charging, and motor efficiencies given, the additional amount of electricity produced by the power plan is $1,950/(0.9 \times 0.88 \times 0.9) = 2,736$ kWh annually.

c. The additional 2,736 kWh (9,850 MJ) are produced in the power station using an additional $9,850/0.38 = 25,920$ MJ of heat. With the coal LHV of 31,000 kJ/kg (31 MJ/kg), the power plant consumes an additional 836 kg of anthracite annually.

d. The additional 836 kg of anthracite contains 753 kg of C. When burned, this amount of C produces $753 \times 44/12 = 2,761$ kg of CO_2.

From Example 8.10, 1 L of gasoline (octane) produces 2.22 kg of CO_2. Hence, the IC engine car that was substituted produced $1,170 \times 2.22 = 2,597$ kg of CO_2.

It is apparent in Example 8.12 that the substitution of a larger car with a smaller electric car has saved $37,428 - 25,920 = 11,508$ MJ of thermal energy annually. Paradoxically, this substitution resulted in the emission of an additional 164 kg of CO_2 annually. This happens because all the electricity consumed in this example is generated by a coal power plant. If the electricity were to be generated by a noncarbon source, e.g., by hydroelectric power, there would have been 2,597 kg of CO_2 emission avoidance. This example demonstrates that substitution to electric cars does not necessarily imply energy savings, CO_2 emission avoidance, and other pollution avoidance. How much electricity the car consumes and how this electricity is produced play important roles in this matter.

While the wider use of smaller electric vehicles has advantages for energy efficiency and, possibly, because of pollution reduction in the urban environments, the electric vehicles are still a novelty and have not been widely adopted, because of the following disadvantages:

1. Their cost is significantly higher than comparable cars with IC engines.

2. The distance traveled with one charge—the range of the electric vehicle—is limited to approximately 300 mi (500 km).*

3. The typical battery life is currently limited to several hundred recharges, beyond which the performance of the battery deteriorates. New materials and technological advances in this area may alleviate this problem in the future, when it is anticipated that battery lives will extend to thousands of recharges.

4. Convenience: The typical time for the recharge of the batteries is significant, more than 3 hours for slow-charging batteries. This implies that a road trip from Dallas to New Orleans or from Frankfurt to Paris becomes inconvenient, because batteries need to be charged after 500 km and the trip must be interrupted for a long time. The few *fast-recharge batteries* in service, which claim recharge cycles of the order of 10 minutes, are costly, require significantly higher voltage to charge, have lower charging efficiencies, and have the potential to create imbalance problems with the local electricity grid because 50–100 kWh must be stored in the battery during the shorter recharging period. In addition, fast-charging batteries have not been convincingly tested for reliability and long-term endurance and performance.

---

* The daily range of several city buses is less than this figure. Several municipalities in the United States and Europe have adopted electric buses for urban routes. In addition to the significant energy savings, electric buses do not emit combustion gases and do not directly contribute to environmental and noise pollution in the cities they serve.

5. Reliability on the road: A car that runs out of gasoline or diesel may be easily supplied with fuel from a tank carried by road assistance. A disabled electric car must be towed to a charging station.

If the charging of the electric car batteries is accomplished during the night hours, when electricity-producing corporations have ample capacity to produce more power, the additional electric power consumed by the electric vehicles will not impose a strain on the peak power demand of the electric grid. Contrary to opinions that there is not sufficient electric energy to power the fleet of vehicles in the United States [15], there is a great deal of spare electric power capacity during the evening and night hours to slowly (over a 4–6 hours of charging period) charge the entire fleet of vehicles in the United States and in most of the other nations. The wider use of electric vehicles that are charged during the nighttime would make the daily electric power demand fluctuations smoother (examples are shown in Figures 7.1 and 7.2) and will save significant amounts of primary liquid hydrocarbon resources, mainly crude oil. Electric cars will cause capacity and stability problems for the electricity grid if they are charged during daytime hours (when they would increase the peak demand) and when a large number of fast-charging (e.g., 10 minute charging period) large batteries are simultaneously charged.

Example 8.13

One hundred electric vehicles with average battery capacity of 85 kWh are charged over a period of 10 minutes with 80% charging efficiency.

a. Determine the additional power the regional electricity grid must provide for this charging process.
b. Determine the additional power the electric grid must provide if the same number of cars are charged over 8 hours (during the night) and, because of the slower charging process, the charging efficiency is 92%

Solution: The total energy stored in the car batteries is 8,500 kWh.

a. Because the charging efficiency is 80%, the regional grid must provide 8,500/0.8 = 10,625 kWh in the 10-minute period. Since this energy is consumed within 10 minutes (1/6 of an hour), the additional power of the electric grid during this 10-minute period is 10,625 × 6 = 63,750 kW (63.75 MW). Note that this is a significant amount of power to be provided by the electricity grid at short notice, especially during daytime hours.
b. With 92% charging efficiency, the regional grid must provide 9,239 kWh. Since this energy is consumed in 8 hours, the additional power of the electric grid during the charging period is 9,239/8 = 1,155 kW (1.155 MW). This is a much lower amount of power than that in part a, and the electricity grid has spare capacity to provide it during the night hours.

Example 8.13 illustrates a major drawback of "fast-charging batteries:" they demand a very high quantity of power, which must be instantaneously provided by the electric grid. If a large number of electric cars simultaneously refuel—and there are thousands of cars refueling in large urban areas at any time of the day—they will cause a strain to the regional electricity grid and potential instability. Alternatively, if the batteries were slowly charged during the night hours, when there is a great deal of excess electric power-producing capacity in all the regions of the world, there is no strain to the electricity production and distribution system. The use of higher numbers of electric vehicles makes sense only when the vehicles are charged during the evening and night hours, when there is adequate spare electric production capacity.

8.6.2 Fuel Cell-Powered Vehicles

Fuel cells operate with a light fuel—hydrogen, alcohol, methane—and directly convert the chemical energy of this fuel to electricity. For all practical purposes, a fuel cell is like a battery that is recharged with its type of fuel. The fuel cell is an open thermodynamic system continuously operating and is charged with its fuel the same way an IC engine car is charged with gasoline or diesel fuel. A vehicle powered by a fuel cell does not suffer from the last three limitations associated with electric cars: The fuel may be purchased in a central station and stored in a tank similar to the gasoline or diesel tanks, which all current vehicles with IC engines use. When the vehicle is in operation, the fuel cell converts the chemical energy of the fuel directly to electricity, which is used in the electric motor that powers the vehicle.

Figure 8.16 shows the energy conversion processes associated with an electric car powered by a hydrogen fuel cell and the well-to-wheels efficiency. Electric power is produced at first in a power plant. Secondly, hydrogen gas is produced by electrolysis. Thirdly, the chemical energy in the hydrogen is directly converted to electricity. Fourthly, the electric energy from the fuel cell is consumed by an electric motor that propels the vehicle. The well-to-wheels efficiency of the fuel cell powered vehicle is equal to the product of the efficiencies of the four processes in the figure. With typical efficiencies for these processes, $\eta_t = 40\%$, $\eta_{el} = 70\%$, $\eta_{tr} = 85\%$, $\eta_{fc} = 70\%$, and $\eta_m = 95\%$, the well-to-wheels efficiency of the fuel cell powered vehicle is $\eta_{wwfc} = 16\%$, a value which is in the low range of vehicle efficiencies with IC engines and certainly lower than that of electric cars.

Table 8.5 offers a summary of the range of the well-to-wheels efficiency, of several automobile technologies and the intermediate processes for some of them. The numbers were adapted from [16] after including the intermediate steps for battery charging and transmission losses.

It is apparent in Table 8.5 that it does not make good sense to produce hydrogen in order to use it with an IC engine and that a hydrogen fuel cell-operated vehicle does not have an energetic advantage relative to efficient IC engine powered vehicles, especially those fitted with hybrid engine technology. However, if the fuel cell were to operate with natural gas or another primary energy source, the first two processes in Figure 8.16 are bypassed;

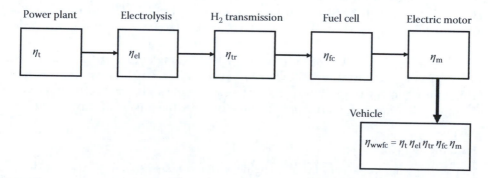

FIGURE 8.16
Energy conversion sequence for fuel cell powered vehicles.

TABLE 8.5

Automobile Technologies and Associated Well-to-Wheels Efficiencies

Fossil Fuel, IC Engine	Fuel → Mechanical	12–30%	
	Transmission	92–99%	
	Well-to-wheels		11–30%
Hybrid, IC engine	Well-to-wheels		30–45%
Electric	Fuel → electricity	40%	
	Electricity transmission	90%	
	Battery charging	88%	
	Battery → mechanical	95%	
	Well-to-wheels		30%
Hydrogen, IC engine	Fuel → electricity	40%	
	Electricity → hydrogen	70%	
	Hydrogen transmission	85%	
	Hydrogen → mechanical	25%	
	Well-to-wheels		6%
Hydrogen, fuel cell	Fuel → electricity	40%	
	Electricity → hydrogen	70%	
	Hydrogen transmission	85%	
	Hydrogen → electricity	70%	
	Electricity → mechanical	95%	
	Well-to-wheels		16%

the primary source is directly fed to the fuel cell; and the well-to-wheels efficiency becomes 63%, which is significantly higher than the efficiencies of hybrid vehicles and electric vehicles. Clearly, the development of fuel cells operating with natural gas, light hydrocarbons, or similar primary energy sources will revolutionize the transportation industry.

Additional advantages of the fuel cell powered cars are (a) the significant reduction of emissions and pollution currently produced by the IC engines; (b) because fuel cell-powered vehicles are charged with a fluid, they will be quickly recharged at any time and the recharging process will not have any effect on the electricity grid, as the electric vehicles may have (Example 8.13). If they are fuelled by hydrogen, the fuel may be produced in the nighttime, when the electric power production corporations have excess production capacity. In the case of electric power produced by wind or solar energy, the needed hydrogen fuel may be produced when there is abundant supply of the renewable energy—spring and autumn for wind and summer for solar. The produced hydrogen is stable and may be stored for a long time in central facilities until it is needed in the vehicles.

In a future hydrogen economy (Section 7.5.2), fuel cell-powered vehicles will become the primary means of transportation. After the depletion of fossil fuels and with the abundance of hydrogen produced from renewable (or nuclear) sources, the fuel cell-powered vehicle would become an economical and, perhaps, the only viable means of transportation.

Example 8.14

The well-to-wheels efficiency of an IC engine powered car is 22%, and the car consumes 2,400 L of gasoline annually. If the car were substituted by any of the types of vehicles listed in Table 8.5 and the usage of the car remained the same, estimate the annual equivalent energy consumption for this car, in liters of gasoline, using the well-to-wheels efficiency values of the table.

Solution: Since the IC engine car has an efficiency of 22%, the annual energy actually used for the propulsion of the car is the equivalent of 0.22 × 2,400 = 528 L of gasoline. This is the equivalent energy at η_{ww} = 100%.

a. If a hybrid car is used at the minimum efficiency (30%), the equivalent energy is 528/0.3 = 1,760 L of gasoline (640 L savings). At the maximum efficiency of 45%, the equivalent energy is 1,173 L (1,227 L savings).
b. If an electric car is used (η_{wwel} = 30%), the equivalent energy is 1,760 L of gasoline (640 L savings).
c. If a hydrogen IC engine car is used (η_{wwfc} = 6%), the equivalent energy is 8,800 L of gasoline (additional 6,400 L).
d. If a hydrogen fuel cell engine is used (η_{wwfc} = 16%), the equivalent energy is 3,300 L of gasoline (additional 900 L).

8.7 Myths and Reality on Conservation and Efficiency

Myth 1: Energy conservation measures worsen our standards of living.

Reality: While some conservation efforts may impose a small inconvenience, several other conservation measures are transparent to the citizens, and some may even improve the standard of living. For example, carpooling allows individuals to socialize; taking a form of public transportation to work allows individuals to spend more time reading a book; and transporting Idaho potatoes to New York City by rail rather than a fleet of trucks is irrelevant to the consumers.

Myth 2: With conservation alone, we can reduce by 60% the energy demand of this nation.

Reality: While the adoption of conservation measures will reduce the national energy consumption, a 60% reduction is unrealistic. Since energy use is directly related to economic activity and the GDP (Figure 2.1), imposing unrealistic goals for energy conservation may hurt the national economy. If improved efficiency is also included in the "conservation" concept of this statement, the technology must become available and the capital must be available for the implementation of the improved efficiency measures.

Myth 3: Electric power plants waste 62% of their energy. We can save this energy and use it to run our national industry.

Reality: While it is correct that on the average, thermal power plants reject 62% of the thermal energy supplied as waste heat, this heat is at very low temperatures. The rest of the national economy needs heat at much higher temperatures—e.g., for metal smelting, manufacturing processes, and IC engines—and the waste heat is useless. A few applications where the low-temperature waste heat may be used are included in Section 8.4. However, these applications alone do not consume enough energy to utilize all the waste heat that is rejected by the thermal power plants.

Myth 4: Car efficiency will improve with the introduction of higher ethanol content in the gasoline mixture.

Reality: The efficiency of the IC engine depends on the cycle pressure and the highest temperature the IC engine cycle attains. As seen in Figure 6.23, the LHV of ethanol is significantly lower than that of gasoline and diesel fuel. If the vehicle fuel pump does not change and the same amount of fuel is supplied to the engine, lower temperatures will be reached with the higher ethanol content fuel than with gasoline alone, and this implies that the efficiency of the IC engine will be lower with the higher ethanol fuel. Of higher importance here is the well-to-wheels efficiency, which includes the energy spent for the production of ethanol. As it becomes apparent from Section 6.5.2 and Table 6.13, more primary energy is spent for the production of ethanol from corn than the energy produced from the combustion of this fuel. As a consequence, the well-to wheels efficiency of corn-produced ethanol would be negative!

Myth 5: "This electric car does not contribute to global warming."

Reality: Several electric cars display this sticker. While the car itself does not emit any CO_2 (or any other pollution for that matter), the associated emissions happen at the point(s) where electricity is produced. Actually, if a high fraction of the electricity used by the electric car is generated by coal units, the operation of the electric car may result in higher CO_2 emission as Example 8.12 illustrates.

Myth 6: The conversion of the IC engine vehicle fleet to electric vehicles is impossible at present because we do not have the capacity to produce the additional electricity.

Reality: The conversion of all passenger cars, light trucks, sport utility vehicles, and vans in the United States would require an additional 980 TWh of electric energy per year, and this is slightly less than 25% of the current electricity consumption in the country [15]. As shown in Figure 7.4, the electric power plants operate at a reduced capacity factor—currently at 44% for the plants in the United States [17]. The country has 1,068.4 GW of installed electric power capacity, which is capable of generating more than two times the current electricity consumption, at 9,360 TWh. A fact that helps here is that most electric vehicles charge in the nighttime (when they are not used), and hence, a very high fraction of the additional electricity for the charging of electric vehicles will occur at nighttime, when the electric demand is at its lowest and there is ample spare capacity (Figures 7.1 and 7.2 and Example 8.13) to do this. Most of the vehicle charging will be met by the spare capacity of the electricity grid; the charging will not strain the current electricity production in the United States and very likely in most other countries. In a market-oriented economy, with appropriate pricing of electricity for car battery charging (e.g., if the price of power for electric vehicle chargers is high during the peak demand hours and low between midnight and 5:00 am), very few electric vehicles will be expected to be charging their batteries in the summertime between 1:00 and 6:00 pm, when the electricity demand is at its highest. Any small additional peak demand will be easily met by a few additional gas turbines.

Myth 7: Electric cars use very low energy in comparison to regular cars

Reality: Since electricity is not a primary energy source and is derived from other sources, one needs to apply the well-to-wheels efficiency for an electric car. This includes the thermal efficiency of the power plant that produces the electric energy for the car batteries and the efficiency for the transmission of this power. Typical well-to-wheels efficiencies of electric cars are in the range of 21–31%, and this is comparable to the range of efficiencies of the newer and more efficient vehicles with IC engines. Where the energy savings may come is that on the average (e.g., the car in Figure 8.14), electric cars are significantly smaller and lighter than the conventional cars with IC engines. Energy savings are chiefly realized because of the reduction of car size and weight, not so much because of the change of the propulsion method.

PROBLEMS

1. For each of the following three desired actions, provide details on the methods and engineering systems they are currently accomplished.

    - Maintain a comfortable temperature for the residents of the households of Houston, Texas, during the summer months.

    - Provide adequate lighting in the classrooms of a high school so that students may learn comfortably.

    - Cook 0.4 kg of pasta.

    For each one of these tasks, provide an alternative method to accomplish it and state if primary energy resource savings would result from the alternative method.

2. Consider the heating and cooling needs of your household. What improvements can be made to reduce the overall energy resource consumption? Separate these improvements into conservation and improved efficiency measures.

3. One of the reasons the transportation sector consumes a great deal more energy in the United States is the commute to work by individuals. The average, one-way commuting trip in Houston, Texas, is 21.5 mi. Because of traffic jams, the average mileage of the cars in the same area is 12 mpg. Consider that four neighbors driving the average commuting daily distance decide to carpool.

    a. What is the amount of gasoline saved weekly and annually?

    b. What would be the annual gasoline savings if 500,000 persons carpooled in a similar manner? Consider that in the latter case, the driving conditions will get better to the point that the average mileage will improve from 12 to 16 mpg.

    c. How many tons of CO_2 are not emitted to the atmosphere annually because of the carpooling activity?

    The week in Houston has 5 working days, and the year, 50 working weeks. Also, you may consider that gasoline is composed solely of octane C_8H_{18}.

4. A tank of 0.7 m³ is to be filled with air at 12 bar. What is the minimum electric work needed? How much electric work is used by a reciprocating compressor with an isentropic efficiency of 72%?

5. One hundred fifty kilograms/hour of steam is needed for an industrial process. The steam is produced by an electric heater. The local power plant uses natural gas for the production of electricity with an overall efficiency of 38% (this includes the transportation losses). What would be the percent reduction of the natural gas consumption if the electric heater were to be substituted with a gas heater?

6. A large oil refinery uses 40 MW of heat, which is supplied by steam at 120°C and 1 bar. The refinery also uses 18 MW of electric power. Design a Rankine cycle and a small power plant that satisfies the heating and electric needs of this refinery.

7. A large food processing plant uses 10 kg/s of steam at 140°C and 1 bar. The plant also needs 5 MW of electric power to run its electric components and sells another 10 MW to the local utility. Design a Rankine cycle to accomplish these tasks.

8. A 5 ton* air-conditioning unit for a household is to be replaced with a GSHP. As a result, the COP of the system is expected to improve from 2.8 to 3.9. The air-conditioning system is used in this household 2,350 hours/year.

   a. What are the annual energy savings due to this improvement?

   b. If all this energy came from a coal power plant of 37% overall efficiency, how many less kilograms of CO_2 are emitted because of this improvement every year?

9. What does the SEER number mean for an air-conditioner? The total rated power for the local utility is 5,000 MW, of which, 95% is used during peak demand hours and the power-generating factor for the utility is 52%. The air-conditioning of the customers accounts for 60% of the peak demand and 22% of the total annual energy demand. As a result of an *energy conservation campaign*, the utility company plans to improve the air-conditioning equipment of its customers from 11.2 to 13.5 SEER. What will be the reduction of the total electricity demand for the utility and the reduction of the peak power that is needed?*

10. A large supermarket, which operates 24 hours every day for 361 days/year, is to substitute 950 kW of fluorescent lamps with LEDs that will provide the same amount of illumination, but will only consume 150 kW. The supermarket is heated for 65 days every year and air-conditioned for 210 days. Using typical efficiency and COP values determine the annual electric energy savings resulting from this energy efficiency measure.

11. Five large hotels in Miami, United States, use their own air-conditioning units, which have a COP of 3.1. It is proposed that the cooling systems of the five hotels be substituted with a larger, modern unit that would have a COP of 3.7. The total installed capacity of the five hotels is 2,600 tons of air-conditioning and the current systems are in use for 3200 hours every year. Determine (a) the annual energy savings from this substitution and (b) the annual avoidance of CO_2 emissions if, currently, 72% of the electric energy in the Miami area is obtained from coal power plants with 36% efficiency and the rest from nuclear power plants.

12. The relative humidity in Tucson, Arizona, frequently hovers near 40%, while the temperature is 38°C. Determine the temperature reduction that would occur in a building if the relative humidity were to adiabatically increase to 60% and to 70%

13. In 2009, the United States consumed 17,910,000 barrels of oil per day, of which 62% was used by small cars and light trucks with an average mileage of 18.5 mpg. If the national standard were to increase to 22 mpg, what would be the annual savings in barrels per year? What are the monetary savings, if the average price of oil is $90/bbl?

14. It is proposed that a percentage of cars in Japan be substituted by electric cars. Japan used 4,680,000 barrels of oil per day in 2016, with 69% going to small cars and trucks. If 20% of these were to be converted to electric cars and trucks, what would be the annual savings in barrels of oil? Assuming that all these oil savings

---

* It appears to be counterintuitive for a utility company to seek demand reduction. However, given the high capital cost and environmental restrictions on the construction of new power plants, many local utilities in areas of growing demand prefer the demand reduction strategy to constructing new and more costly power plants. Thus, they satisfy the new customers with the surplus power that comes from conservation and higher efficiency measures.

come from octane (C_8H_{18}) and that the electricity will be produced from nuclear power plants, what is the annual avoidance of CO_2 emissions?

15. It is proposed that a percentage of cars in Germany be substituted by cars powered by hydrogen fuel cells. Germany used 2,460,000 barrels of oil per day in 2016, with approximately 70% consumed by small cars and trucks. If 20% of these were to be converted to fuel cell cars and trucks, what would be the annual savings in barrels of oil in the country? If all these oil savings come from octane (C_8H_{18}), what is the annual avoidance of CO_2 emissions? All the hydrogen will be produced by electrolysis of water, with electric power provided by wind turbines.

16. "OECD countries have an enormous potential to conserve energy in comparison to developing countries. Therefore, developing countries should be excluded from all international protocols and agreements on CO_2 reduction." Comment in a short essay of 250–300 words.

17. "If we were to only apply energy conservation measures in the United States, we would not have to build another electric power plant until 2045." Comment in a short essay of 250–300 words.

18. A compressor in a Brayton cycle raises the air pressure from 1 atm and 27°C to 25 atm and has an isentropic efficiency 78%. It is proposed to substitute this compressor with two other compressors and an intercooler. Both compressors will have a pressure ratio of 5 (that is, the first produces air at 5 atm, and the second, at $5 \times 5 = 25$ atm) and efficiency of 80%. The intercooler cools the air after the first compressor to 37°C. Determine the following:

   a. The specific work, in kJ/kg required for the operation of the old and the new system of compressors.

   b. If the Brayton cycle admits 2 kg/s of air and is in operation for 25% of the time, the power savings and annual energy savings from this substitution.

19. A commuter drives 15,000 km/year in a midsize car that consumes 10 L of gasoline per 100 km. The commuter substitutes this car with a smaller electric car, operated with a battery, which consumes 14 kWh/100 km. This car has $\eta_M = 90\%$; the batteries charge with efficiency of $\eta_B = 88\%$; the electricity is produced at plants power plants with $\eta_t = 38\%$; and the transmission efficiency of electricity is $\eta_{tr} = 90\%$; 55% of the electricity in the region is produced by carbon sources (LHV = 32,700 kJ/kg), and the rest, by nuclear and renewables. Determine (a) the annual amount of gasoline saved per year, (b) the annual amount of electricity used by the car, (c) the additional amount of carbon needed to produce the energy for this car, and (d) the CO_2 avoidance by the substitution.

References

1. Michaelides, E.E., *Alternative Energy Sources*, Springer, Berlin, 2012.
2. Kestin, J., *A Course in Thermodynamics*, vol. 1, Hemisphere, Washington, DC, 1978.
3. Moran, M.J., Shapiro, H.N., *Fundamentals of Engineering Thermodynamics*, 6th edition, Wiley, New York, 2008.
4. Michaelides, E.E., Exergy and the conversion of energy, *Int. J. Mech. Eng. Educ.*, **12**, 65, 1984.
5. Horlock, J.H., *Cogeneration—Combined Heat and Power*, Krieger, Malabar, FL, 1997.

6. International Energy Agency, *Key World Statistics*, IEA-Chirat, Paris, 2016.

7. Yarosh, M.M., Nichols, B.L., Hirst, E.E., Michel, J.W., Yee, W.C., *Agricultural and Aquacultural Uses of Waste Heat*, Oak Ridge National Laboratory, Oak Ridge, TN, July 1972.

8. Kreith F., Goswami D.Y., eds., *Energy Management and Conservation Handbook*. CRC Press, New York, 2008.

9. Ochsner, K., *Geothermal Heat Pumps—A Guide for Planning and Installing*, Earthscan, London, 2008.

10. Hughes, P.J., Shonder, J.A., *The Evaluation of a 4000-Home Geothermal Heat Pump Retrofit at Fort Polk, Louisiana*, ORNL/CON-640, Oak Ridge National Laboratory, Oak Ridge, TN, 1998.

11. USE EIA (US Energy Information Administration), *Residential Energy Consumption Survey (RECS)*, US EIA, Washington, DC, 2011.

12. ACEEE (American Council for an Energy Efficient Economy), *How Your Refrigerator Has Kept Its Cool Over 40 Years of Efficiency Improvements*, ACEEE, Washington, DC, 2016.

13. US Department of Transportation Bureau of Transportation Statistics, *National Transportation Statistics*, US Department of Transportation Bureau of Transportation Statistics, Washington, DC, 2017.

14. IEA (International Energy Agency), *Global EV Outlook 2016: Beyond One Million Electric Cars*, IEA, Paris, 2016.

15. Smil, V., *Energy—Myths and Realities*, AEI Press, Washington, DC, 2010.

16. Dunlap R.A., A simple and objective carbon footprint analysis for alternative transportation technologies, *Energy Environ. Res.*, **3**, 33–39, 2013.

17. US DOE (Department of Energy) Energy Information Administration, *Electric Power Annual Report—2015*, US DOE Energy Information Administration, Washington, DC, 2017.

9

Energy Economics and Decision-Making Methods

When one asks the question "If renewable energy is a panacea, then why there are so few large renewable energy generation systems in the world?" the simple answer is "Because fossil fuels and nuclear energy are cheaper and more convenient to use." At present, there are no financial costs associated with the emission of greenhouse gases (GHGs) and very little cost is associated with the production of nuclear waste. In addition, these fuels are available—as energy stored for millennia—to be used at any time of day or night and regardless of the weather conditions.

Nuclear and coal power plants, built in the 1970s and 1980s, produce electricity at less than US\$0.025/kWh. Because of this cost structure, it is difficult for an electricity-generating corporation to justify investing in wind turbines that would produce electricity at US\$0.05/kWh or solar farms that would produce at US\$0.12 to US\$0.15 per kWh and only during daylight hours. At the higher prices of renewable energy, it is the good will, governmental subsidies and mandates, and the unencumbered environmental cost that may induce corporations to produce more renewable energy. If the price of electricity produced from renewable sources (with any needed energy storage) becomes less than the price produced from conventional energy sources, simple economic and financial considerations would favor the development of more geothermal units, solar power plants, and wind generators.

A combination of rising fossil fuel prices, the probable initiation of carbon credits or a probable "carbon tax," and a favorable regulatory environment that provides tax credits and accelerated depreciation for renewable energy projects may change the economic and financial circumstances for renewable energy. This chapter offers a succinct exposition of the entire decision-making process that leads to the construction and operation of energy projects. The process includes the following steps: the realization of the need for more power in a community; the enumeration of the alternative choices; and the rational selection of the optimum alternative solution that maximizes profitability or, equivalently, provides the needed amount of electric energy at minimum cost. Central to the financial considerations is the financial concept of the *time-value of money* for the appraisal of energy investments. Several examples are given on the effects of variables such as interest rates, depreciation accounting methods, and governmental measures and regulations that provide incentives and disincentives for renewable energy.

9.1 Introduction

The decision-making process for energy projects is similar to the decision-making process of all other engineering projects and, primarily, involves the following stages:

1. *The need for a project is identified.* For example, the increase of the population of the city of Westheimer necessitates the addition of 100 MW of electric power in the next 10 years.

2. *Alternative solutions to the problem are formulated as projects.* For example, the increased electricity demand in the city may be met by one of the following: an additional nuclear power plant; an additional coal or gas-fired plant; a new solar–thermal plant; a wind energy farm; building transmission lines to buy energy from the city of Northeimer, etc.

3. *The alternative projects are evaluated.* This step includes a feasibility analysis and a more detailed financial analysis of the alternatives, which were identified in step 2. In evaluating the projects, all factors must be examined. At this step, some of the proposed alternative solutions may be excluded for reasons other than economic considerations. For example, the 2014 referendum by the citizens of Westheimer that prohibits the construction of coal power plants or that Westheimer is in a seismic area that increases the environmental risk posed by nuclear power plants.

4. *A decision is made on the best alternative solution.* The criteria used in this step are based on considerations related to financial/economic motives, reliability, quality of the environment, etc.

5. *The decision is formulated as a fully specified engineering project.* For example, the electric corporation that supplies the city of Westheimer will construct from June 2018 to May 2020 a wind energy farm of total rated power 180 MW (to account for the lower capacity factors of wind turbines). The farm will be located at the Windstrong site, and the project will be completed by the end of July 2020.

At the end of the decision-making process, competitive bids for the construction of the project are solicited. It must be noted that the final decision is specific enough to not only allow the planning and construction of the facility, but also allow sufficient flexibility for the next stages of the project. The decision may not specify how many and which type of wind turbines will be bought and installed, how tall the wind turbine towers will be and how they will be built, which towers will be built first, or what will be the spatial arrangement of the towers and the wind turbines. Such decisions will be made during the planning and construction stages of the project and primarily include engineering optimization as well as environmental considerations. This chapter will focus on the economic and financial aspects of the decision-making process, with particular emphasis on the appraisal of the energy projects as financial investments.

It must be emphasized that there is a great deal of uncertainty associated with all the financial parameters of energy projects including the price of fuel, investment costs for the development of a project, labor and operational costs, interest rates, and the time for the completion of the project. Of these, the price of fuels is highly influenced and is correlated with the price of petroleum, which depends on several difficult to predict variables, such as political events in oil-producing countries, agreements and directives of the members of Organization of the Petroleum Exporting Countries (OPEC), new fossil fuel discoveries, secondary and tertiary recovery, and perceived or estimated (not necessarily actual) levels of fossil fuel reserves and resources. The highly variable graph of the petroleum price in Figure 2.10 since 1975 proves the unpredictability of petroleum (and other energy forms) prices. In addition, governmental incentives, regulations, and penalties that enter the financial calculations—tax credits, rebates, and depreciation schedules—are often temporary and may be phased out during the long life cycle of energy projects. It is important to note that because of these uncertainties, any economic or financial appraisal of energy projects is laden with higher uncertainty than scientific and engineering calculations of the energy project, such as the engineering feasibility and design calculations.

9.1.1 Fundamental Concepts of Economics

The definitions of a few important concepts used in the fields of economics and management, which are helpful in the decision-making process for energy projects, are given in this section in alphabetical order [1]:

- *Average cost*: The total of all *fixed* and *variable* costs of a product calculated over a period, usually 1 year, divided by the total number of units produced, e.g., the average cost of electricity is $0.059/kWh, the average cost of production of the drill bits the XYZ factory produces is $2,870 per drill bit.

- *Average revenue*: The total revenue of the product units sold over a period, usually 1 year, divided by the total number of units sold, e.g., the average revenue of electricity is $0.072/kWh; the average revenue of the drill bits the XYZ factory produces is $4,280 per drill bit.

- *Average profit*: The difference between average revenue and average cost. Using the two examples earlier, the average profit of electricity is $0.013/kWh and that of the drill bits is $1,410 per drill bit.

- *Fixed cost*: All costs that are not affected by the level of business activity or production level, such as rents, insurance, property taxes, administrative salaries, and interest on borrowed capital. Fixed costs need to be periodically paid regardless of whether or not the business produces anything or makes any profit from sales.

- *Life cycle cost*: The sum of all costs—fixed and variable—associated with a project from its inception to its conclusion. The life cycle cost includes among others the planning cost; the capital cost for construction and machinery; hiring of employees; and any abandonment, disposal, and storage costs at the end of the project.

- *Marginal or incremental cost*: The cost associated with the production of one additional unit of output.

- *Marginal or incremental revenue*: The revenue accrued from the production of one additional unit of output.

- *Opportunity cost*: Monetary equivalent of what is sacrificed, when a certain course of action is chosen. For example, an opportunity cost to building a new power plant for a corporation is not to build the power plant and, instead, invest their available capital in 7% interest-bearing securities.

- *Salvage value*: The price paid by a willing buyer for a plant or business after all operations have ceased. Typically, this is the value of the used equipment, land, and buildings. If there are cleaning costs associated with the abandonment or disposal of hazardous materials, e.g., in a nuclear site, the salvage value may be negative.

- *Sunk cost*: All costs paid in the past that are associated with past activity that may not be recovered and do not affect any future costs or revenues.

- *Time horizon*: The time from the inception to the end of the project, including any disposal and storage of equipment and products.

- *Variable cost*: Any cost associated with the level of business activity or output level, such as fuel cost, materials cost, labor cost, distribution cost, and sales commissions. The variable cost monotonically increases with the number of units produced. The functional relationship of units produced and variable cost is not necessarily linear. When the variable cost of the units decreases with the number of units produced, *economies of scale* are realized.

9.2 Time-Value of Money

The concept of the time-value of money is based on the premise that $1 today is worth more than $1 a year from now, the latter is worth more than $1 two years from now, and so on. The time-value of monetary funds is intricately related to the following concepts:

1. *Return to capital*, which stipulates that an amount of capital invested must reasonably be expected to yield more capital at the end of the investment period.

2. *Interest rate* or *discount rate r*, which is the percentage of additional funds that is earned for the lending of capital.

3. The current and expected future *inflation*, which increases the cost and value of goods in the future.

4. The *investment risk*. When capital investments, such as energy production and conservation projects, are appraised, there is an inherent risk that the project may not succeed and that all or part of the invested capital may be lost. The investment risk is a justification to charge higher interest for any capital spent and the expectation of a higher return on the invested capital. In general, the higher the investment risk of a project, the higher is the expected return on the capital and the interest rate associated with the project.

A few types of investments are considered *risk free*. The interest rate associated with them r_{rf} is the lowest interest charged by the capital markets. Risk-free investments are usually short-term governmental securities, such as 3- or 6-month governmental obligations (called *treasury paper* in the United States), which typically yield a very low interest rate. Since the government is a very reliable debtor and the investment timeframe is short, the investors are certain that their capital and interest will be paid in full.

The return on the capital expected for other types of debt may be significantly higher: Investors will lend funds to corporations at rates 1–10% above the r_{rf} rate and to individuals at even higher rates. The *premium* over the risk-free rate r_{rf} depends on the financial strength of the corporation, the nature of the investment, and the duration of the loan. Loans and bonds to financially weaker corporations are riskier and carry higher interest. Longer duration loans and bonds (e.g., 30 years versus 2 years) are also riskier and, in general, bear higher interest rates. Certain energy-related activities, such as drilling for oil and gas and constructing solar photovoltaic (PV) energy power plants,* are considered among the riskiest investments. As a result, the interest rates typically charged for these investments are among the highest in the energy production industry.

It is apparent that there is not only one value for the interest rate, but a whole array of rates, ranging from the risk free r_{rf} to significantly higher values. The availability of capital and the market conditions determine the exact values of the interest rates charged for energy projects.† There are two attributes one must consider related to the interest rate: The first is the value and the second is the duration: e.g., 6% per year, 1% per month, and

---

* Because of this, several governments provide "loan guarantees" for photovoltaic and other renewable energy projects. The guarantees help lower the interest rates for renewable energy projects.
† The period of 2008–2017 has historically been a period of very low interest rates in most of the world. Corporations were able to undertake several energy-related projects with relatively low interest for bank loans and publicly traded bonds.

37% per 5 years. Usually, interest rates are quoted per year (*per annum*) or per month, but other durations may also be contracted.

9.2.1 Simple and Compound Interests

The *simple interest* is equal to the product of the initial capital invested, (also called *the principal P*), the interest rate charged *r*, and the number of periods the loan/investment is mad, *N*. The formula for calculating the simple interest *I* at the end of *N* periods is

$$I = PNr. \tag{9.1}$$

With simple interest, the total to be paid at the end of the *N* periods for the entire debt is $T = P + I$. Thus, a sum of $10,000, which is lent at a 6% annual rate simple interest for 5 years, will yield $10,000 × 5 × 0.06 = $3,000 interest. The total to be paid after 5 years is $13,000.

While the simple interest is easy to apply, it is not frequently used in commercial practice. The *compound interest* is used in most commercial transactions. With compound interest, the accrued interest at the end of each period is added to the original capital *P*. Hence, during the subsequent periods, interest is earned not only on the original principal, but also on the previously earned interest, which is added to the principal. The formula for calculating the total amount to be paid with compound interest is

$$T = P(1+r)^N, \tag{9.2}$$

and the total interest paid is $I = T - P$. At compound interest, the total interest on the $10,000 of the previous example invested at 6% per year after 5 years is $3,382. This value is higher than the simple interest of $3,000.

It is apparent that the time period of compounding *N* is as important as the value of interest rate *r* in the calculation of the interest paid. Consequently, a borrower must carefully consider this variable. Several credit card lenders charge an interest rate that is compounded daily (this includes credit cards issued to engineering students!). On an average credit card balance of $1,000, a 15% annual rate compounded annually yields $150 interest paid at the end of 1 year. On the same balance, a daily compounded interest rate of 15/365% (considered by some the same as the 15% annual rate) on the same credit card balance generates $162 of interest to be paid at the end of the year. While the difference of $12 appears small to an individual, it translates to significant profits for a corporation that issues millions of credit cards.

Example 9.1

A contractor is building a new housing development and is given the option to take a $4,000,000 loan for 5 years at a simple interest of 7.0% or at a compound interest of 6.5%. All interest and capital will be paid at the end of the loan period. Which type of interest should the contractor choose?

Solution: With the simple interest, at the end of the 5-year period, the contractor will pay back the amount borrowed ($4,000,000) plus an interest: $0.07 × 5 × 4,000,000 = $1,400,000 for a total payment $5,400,000.

With the compound interest, at the end of the 5 years, the contractor will pay back a total of $4,000,000 × (1.065)^5 = $5,480,347. The contractor should choose to take the simple interest option, even though the annual interest rate appears to be higher.

Example 9.2

A young engineer carries a $5,000 credit card debt at 18% annual rate, compounded daily. A coworker offers to give her a loan of $5,000 at 8% annual rate. How much money does she save in a year? The credit card does not impose any other penalties if the debt is paid within a year.

Solution: The daily rate of the credit card is $18/365 = 0.0493\%$. Therefore, in a year, the young engineer will owe to the bank $5,000 \times (1 + 0.000493)^{365} = \$5,985$. After taking the loan from the coworker, she will have to pay $5,000 \times (1 + 0.08) = \$5,400$. By taking the loan, she will save $585.

9.2.2 Cash Flow

The objective of the decision-making process is the financial evaluation—often called *financial appraisal*—of several alternative investments that are considered as solutions to a specified need. Every alternative solution involves a complex timetable of investments, expenditures, revenues, and tax considerations. The concept of *cash flow* or *net cash flow* is a tool that keeps track of the net influx of funds for each year of the duration of the project and assists in the final decision to undertake a project. The *net cash flow* is the sum of all annual receipts of funds minus all annual expenses associated with the development, operation, and closure or abandonment of a project. Such a project may be as simple as the lending of capital in the first year to obtain the principal and interest in the future, to a complex investment involving the construction and operation of a PV power plant that is expected to produce energy for more than 35 years. In the more complex cases, cash flow diagrams, which depict the net annual cash flow for the lifetime of the project, are used to graphically depict the progress of a project and the yearly cash flows. Net expenditures appear as negative values and net receipts as positive values.

Cash flow is a simple and very convenient method in financial accounting that summarizes all the tangible revenues and expenses associated with a project during a period, typically one calendar year. As the name indicates, the method is based only on the tangible flow of cash items. At does not include any intangible items, such as good will, environmental benefit, and externalities. Depreciation and depletion, which are not cash flows, are taken into account only in the computation of the taxes, which are cash expenditures. The following items are added to make up the annual cash flow of a project, with the positive items denoted by "+" and negative items denoted by "−":

1. *Revenues* (+): All cash receipts from selling products, services, and capital assets related to the project.
2. *Interest earned* (+): Any interest earned by funds directly associated with the project.
3. *Tax credits* (+): Credits on taxes granted by local, regional, and national governments. These are typically a percentage (10–30% for renewable energy projects) of the project investment, during the year when the investment expenditures occur. Because of the global energy challenge and climate change considerations, alternative energy projects are beneficiaries of tax credits in most Organisation for Economic Co-operation and Development (OECD) countries. A necessary condition for a corporation to take advantage of the available tax credits is that it must have taxable income during the year at which tax credits are sought. Depending on the agency that grants the credits and the tax laws, certain tax credits may roll in future years if the corporation does not have enough taxable income in the current year.

4. *Rebates* (+): All monetary receipts that are directly connected to the investment of the project and are received as a result of the investment.

5. *Salvage value* (+): The fair market price of any equipment purchased that may be sold in the market.

6. *Capital expenditures* (–): All expenses of capital equipment and construction associated with the project. Most of the capital expenditures may be depreciated (over a period of several years) to reduce the future taxable revenue.

7. *Other fixed costs* (–): These include rents of space or equipment, management and administration costs, marketing, interest paid, and insurance. The fixed costs do not depend on the units of energy produced and must be paid even if the business or project does not produce anything.

8. *Variable costs* (–): These are monotonically increasing with the level of production. Among the variable costs for energy projects are the cost of fuel, labor costs excluding administration, and distribution costs.

9. *Taxes* (–): These are usually a percentage of the *taxable income* of a corporation. Detailed tax codes in every country specify what the *tax rate* is and how the taxable income is calculated. While a full description of the tax code is beyond this short exposition, an item of importance for energy projects is the *depreciation schedule*. The depreciation schedule, which is different in the tax codes of different countries, specifies the way capital expenditures of a project may be depreciated (subtracted from the taxable income). In general, higher depreciation expenses at the beginning of the project favor the project because they result in lower taxes and higher cash flow earlier in the life of the project.

10. *Closure costs* (–): Any cost associated with the termination of the project. For energy projects, this item includes land cleanup and rehabilitation, the shutting of oil or gas wells, and the disposal of nonmarketable equipment. In the case of nuclear power plants, closure costs include the decontamination of the land and equipment and any expense for the long-term storage of nuclear waste.

The cash flow method considers all the receipts and expenses of a project during the years of the projected life of the project. The method adds all the items denoted by (+) and subtracts the items denoted by (–). The result is the *net cash flow* for the year and is used for the calculation of the present value of the project.

Example 9.3

Consider the construction of a PV power plant where the owner and operator must invest $500,000 in the first year and $1,000,000 in the second year. The plant is completed at the end of the second year and continuously operates for 10 more years. The total revenue of the operation of the plant, including tax credits, is $200,000 during the third year and increases afterward at a rate of 7% annually. The operational expenses associated with the operation of the solar project are $50,000 during the third year and increase at a rate of 5% annually in the subsequent years of operation. At the end of the project, that is the end of the twelfth year, the operations will discontinue and the solar energy project will be sold for $300,000. Determine the annual net cash flows during the lifetime of this project.

Solution: Based on the information provided, the principal streams of cash for this project (revenues, expenses, capital investment, and salvage value) from the inception to the end of the project are depicted in the following table, for every year of the lifetime of the project. The sum of these parameters is the annual net cash flow of the project for the particular year.

Annual Expenditures, Revenues, and Cash Flow for the Small Solar PV Power Plant[a]

Year	Investment	Annual Revenue	Annual Expenses	Salvage Value	Cash Flow
1	−500	0	0	0	−500
2	−1,000	0	0	0	−1000
3	0	200	−50	0	150
4	0	214	−52.5	0	161.5
5	0	229.0	−55.1	0	173.9
6	0	245.0	−57.9	0	187.1
7	0	262.2	−60.8	0	201.4
8	0	280.5	−63.8	0	216.7
9	0	300.1	−67.0	0	233.1
10	0	321.2	−70.4	0	250.8
11	0	343.6	−73.9	0	269.8
12	0	367.7	−77.6	300	590.1

[a] In thousands of US dollars.

The net cash flow diagram for this project is also depicted in Figure 9.1. This diagram is a succinct and convenient depiction of the annual income, from which a more detailed financial analysis of the project may be made. The monetary amounts comprising the cash flow pertain to funds earned or spent during an entire calendar year. For example, if the first year for the PV project of example 9.3 is 2016, the cash flow item 250.8 pertains to the tenth year of the project, the calendar year 2025. For accounting convenience, the first year of a project is oftentimes designated as year 0.

9.2.3 Equivalence of Funds and Present Value

The concept of the time-value of money complicates the appraisal method by inserting time as an additional variable for the financial appraisal of a project. It is axiomatic that

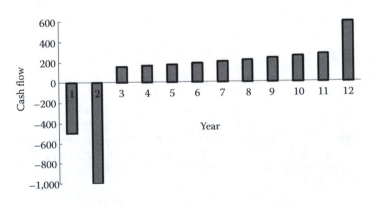

FIGURE 9.1
Cash flow diagram for a small solar PV power plant.

when two or more projects are evaluated or appraised, they must have a common basis of comparison. In the comparison of several engineering and, in particular, energy-related potential projects, one quickly discovers that the projects involve a multitude of monetary expenses and receipts over the entire lives of the projects, which span several years or decades. The lives of the projects may significantly vary. For the evaluation or appraisal of the projects, one must reduce all the monetary expenses and benefits of each project to an *equivalent basis* that transcends the time frames of the individual projects.

Equivalency, in the monetary sense, implies that an informed investor is indifferent between two sums of funds to be received at two different times in the future. For example, if an informed investor is indifferent between \$100,000 to be paid in 2020 and \$120,000 to be paid in 2022, then these two sums are *equivalent* to the investor. A consequence of the equivalency principle is that if the same investor is offered \$100,000 to be paid in 2020 or \$110,000 to be paid in 2022, the investor will unequivocally choose the first option.

The *present value* P_V of future funds or future returns is a method to establish the equivalent basis for the comparison of several projects. The P_V establishes the currently equivalent monetary value of future funds for an informed investor or a corporation. The present value P_V of a principal amount of funds M to be obtained after N time periods at a *discount rate* r_d is defined as

$$P_V = \frac{M}{\left(1+r_d\right)^N}.\tag{9.3}$$

This implies that an amount equal to P_V today is equivalent to a greater amount M after N periods. A glance at Equations 9.2 and 9.3 proves that the present value P_V is directly related to the compound interest concept, with the discount rate being equivalent to the interest rate. The P_V is equivalent to the principal invested today, and the future amount M is equivalent to the total sum T, which is expected after N periods of the investment.

Another way to look at Equation 9.3 is that an informed investor is indifferent between having a sum equal to P_V today and a sum M, to be received after N periods. By evaluating the present value of all the cash flows for all the alternative projects, one establishes the P_V of the total receipts and expenditures of each project. This provides an equivalent basis for the financial appraisal of alternative projects that have different durations, complex cash flows, and different life cycles. Informed investors establish a common basis for the rational economic comparison of alternative projects using the present value concept. This rational comparison assists in the final stages of the decision-making process for all capital projects, including energy-related investments.

In the calculation of the P_V of a project, there are several parameters and mathematical tools used that facilitate the calculations. Oftentimes in financial calculations, there is a constant amount of funds, expenditures, or incomes, which is repeated for several years. This is called an *annuity*, and since it is earned or expended over several years, it has a present value. The present value of the annuity, the constant amount A earned or spent in every year of the duration of the project, starting at present, in year 0, and ending in year N is

$$P_{VA} = A + \frac{A}{\left(1+r_d\right)} + \frac{A}{\left(1+r_d\right)^2} + \ldots + \frac{A}{\left(1+r_d\right)^N} = A\left[\frac{\left(1+r_d\right)^N - 1}{\left(1+r_d\right)^N r_d}\right].\tag{9.4}$$

Associated to the annuity concept is the borrowing of a principal sum of funds at present with the promise to repay both the principal and the interest in equal amounts over a fixed number of time periods. An individual or a corporation borrows an amount P today with the promise to repay an amount M during each of the subsequent N time periods. This is the concept most of the housing *mortgages* are based on, as well as most bank loans to small corporations. In consideration of the funds received at present P, the owner promises to make a number of equal and regular payments in the future. In the case of house mortgages, the time period N is defined in months, and the agreed upon annual interest rate i takes the place of the discount rate in Equation 9.4. The periodic monthly payment M for a house mortgage is obtained from the substitution of the parameters in the last form of the preceding equation with $P = P_{VA}$, $i = r_d$, and $M = A$:

$$M = \frac{iP(1+i)^N}{(1+i)^{N+1}-1}. \tag{9.5}$$

For a typical house mortgage of 30-year duration, the number of monthly periods is $N = 30 \times 12 = 360$, and the interest rate i is the monthly interest to be charged. The latter is equal to 1/12th of the quoted annual rate. For example, if the quoted annual rate by the lender is 6%, $i = 0.06/12 = 0.005$.

Example 9.4

A young professional is to buy her first house, which costs $235,000. She will put a 10% down payment ($23,500) and plans to borrow the remaining $211,500. Two options are given to her: a 30-year mortgage at annual rate 6% and a 15-year mortgage at 5.4% annual rate. (a) How much will her monthly payment be for each of the options? (b) How much is the total interest she will pay for each option?

Solution:

a. The mortgage is paid monthly. The monthly interest of the 30-year mortgage is 6/12 = 0.5% ($i_{30} = 0.005$) with $N = 12 \times 30 = 360$ payments. The interest rate of the 15-year mortgage is 5.4/12 = 0.45% ($i_{15} = 0.0045$) with $N = 12 \times 15 = 180$ payments.

Using Equation 9.5, the monthly payment for the 30-year mortgage is $1,260.5, and for the 15-year mortgage, the monthly payment is $1,703.1.

b. The total payment for the 30-year (360 months) mortgage is 1,260.5 × 360 = $453,780. After subtracting the amount borrowed ($211,500), the total interest paid to the bank is $242,280.

The total payment for the 15-year (180 months) mortgage is $1,703.10 × 180 = $306,558. After subtracting the amount borrowed ($211,500), the total interest paid to the bank is $95,058.

9.2.4 Note on the Discount Rate and Interest Rates

It is apparent from Equation 9.4 that the discount rate r_d is an important variable in the calculation of the present value of any future sum of cash M. This variable is not universal, but specifically pertains to the entity that will perform the project and to the type of the project. It is common for corporations to have a range of discount rates for the different types of projects they pursue and to have a range of discount rates for the several periods of the life cycle of the projects. These ranges takes into account the type of the project (known technology or development of new technology), the risk of the project, the availability of capital (internal funds), and the ease to borrow additional funds. The choice of the correct value for the discount rate is of paramount importance in any engineering project. As with all other interest rates, there is not a single discount rate used by corporations, governmental agencies, and individuals, but a range of discount rates that depend upon the following parameters:

1. *The type of entity/corporation that performs the project, public or private.* In general, public corporations and nonprofit organizations do not aim for monetary profits and may borrow funds at lower rates than private, for-profit corporations. Because their cost of capital is lower, their discount rates are lower too.

2. *The inherent risk of the project.* The more risky the project (e.g., because it uses new and unproven technology), the higher is the return sought and the higher would be the discount rate. Building an engine to harvest the waves off the coast of Scotland is a more risky investment than substituting the existing incandescent bulbs with light-emitting diodes (LEDs) in an office building in the heart of Edinburgh.

3. *The overall economy and ease of borrowing capital.* In general, when the overall economy is recovering or robust, it is easier to borrow funds. Interest rates are lower and so are the discount rates for the corporations.

4. *Inflation.* Inflation, or expected inflation in the future, is directly and monotonically correlated with the discount rates. The discount rates significantly rise in years when high inflation occurs, or is expected.

5. *The length of the project.* Because the future is mostly unpredictable, longer projects are associated with higher economic uncertainty and higher risk. The discount rates for longer projects may be significantly higher. Projects involving larger coal and nuclear power plants, which are expected to be in operation after 40–50 years, are currently in this category because they are associated with future risks such as the banning of coal as a fuel because of international treaties related to global environmental change and future nuclear moratoria and decisions to reduce or phase-out nuclear reactors. The nuclear moratorium that was announced in Germany in 2011 has put a hold on all ongoing projects and all investment related to nuclear energy in the country.

The several discount rates r_d that are used by corporations and public utilities are significantly higher than the interest rates r charged by financial institutions. The discount rates are usually tied to an interest rate by a simple formula such as $r_d = r + \delta r$, where δr is in the range of 2–15%. Corporations have a range of discount rates, which depends on the perceived risk and the duration of the project (short, intermediate, or long term). The interest rate, to which the discount rate may be tied, is usually one of the following:

1. The *interbank discount rate*, which is the rate banks charge each other for borrowing funds. In the United States, the interbank discount rate is set by the Federal Reserve Board and is frequently referred to as *the discount rate.*

2. The *short-term interest rate* pertains to government securities. This is typically the interest rate on the 6-month or 12-month treasury bonds/paper.

3. The *intermediate term interest rate* is the interest rate of 10-year treasury bonds.

4. The *long-term interest rate* of government securities is typically the interest rate on 20-year or 30-year treasury bonds. The long-term interest rate is pertinent to most energy projects that typically last more than 30 years.

5. The *prime rate* charged by financial institutions. This rate is usually offered to the more affluent customers, institutions, and corporations for almost risk-free investments.

6. The *cost of borrowed* capital represents the interest rate paid by corporations and public utilities when they issue long-term *bonds.** Because corporations, public utilities, and nonprofit organizations have different financial ratings and ability to attract capital by issuing bonds, the cost of borrowed capital significantly varies between the several borrowing entities. Entities with a robust financial situation and good reputation have high ratings (e.g., AAA, AA+, and AA) and lower cost of borrowed capital.

7. The *expected return on equity* rate is set by the management and used for internally financed projects. It represents the return expected by the shareholders of the corporation. The expected return on equity significantly varies among the different corporations. It is typically lower with public utilities and very low for nonprofit organizations.

8. The minimum acceptable rate of return (MARR) is a rate set by the management of several corporations and public utilities for capital projects. It represents the minimum rate of return that will make a project acceptable to undertake.

Large energy projects are commonly financed by a combination of bonds and available capital (equity). The pertinent discount rate is the MARR of the entity that undertakes the projects. It must be noted that the interest rates are not constant and vary daily, usually by a very small amount. The daily values of the first five interest rates in the preceding list are publicized in the financial pages of most newspapers. Because the last two rates pertain to individual corporations, they are not publicized. Typically, these rates are set by the management and depend on the nature of the project, its duration, and the associated risk.

9.3 Decision-Making Process

The decision-making process for engineering projects follows a well-defined need and is accomplished by multistep, structured procedures, often followed by computer simulations and mathematical modeling. Alternative solutions to what is needed are formulated,

---

* The bond is a promissory note through which—in exchange for the capital borrowed—the borrower will make defined quarterly, semiannual, or annual interest payments to the lender and will repay the entire borrowed amount after a number of years. For example, corporation XYZ that issues today a $10,000, 15-year bond at 8%, paid quarterly, agrees to pay to its lender $200 every 3 months ($800 annually) for 15 years. At the end of the 15-year period, the bond obligates XYZ to pay $10,000 to the lender. Bonds are invariably used to finance large energy projects, such as nuclear power plants, gas turbines, and wind farms.

and a detailed economic analysis is made for several of the alternatives. The final decision takes into consideration the economic and financial analyses as well as previous experience, environmental factors, social factors, and input from the engineers who will be in charge of the project. The decision-making process of engineering projects includes the following steps [2,3]:

1. Identify and describe precisely what is needed (the need).
2. Develop a comprehensive list of alternative projects that will satisfy the need individually or in groups.
3. Eliminate several of the proposed projects, based on noneconomic nonengineering criteria (environmental regulations, political considerations, etc.).
4. Perform an investment appraisal method for the remaining projects.
5. Prioritize the list of the remaining projects, based on their economic impact, and decide to develop a single project or a group of projects that will satisfy the need.

9.3.1 Developing a List of Alternatives

The first stage of this process is the reformulation of the problem or need in several ways, which may stimulate creative thinking and inspire at least one alternative solution. For example, the problem that the city of Westheimer will need another 100 MW of electric power by 2028 may be expressed in several ways including the following [1]:

1. "By 2028, we need to install another 100 MW of electric capacity." Alternative: Build a 100 MW power plant (or two 50 MW power plants or any other combination of power plants that will add 100 MW capacity).
2. "By 2028, we will be short of 100 MW of electric power." Alternative: Start conservation efforts that would save 100 MW.
3. "By 2028, we need to increase the power produced by our plants by 100 MW." Alternative: Build additional reheaters and feed-water heaters in the three existing coal power plants and increase their rated capacity by a total of 100 MW.
4. "We need 100 MW of green power by 2028." Alternative: Build a wind farm or a solar power plant that will provide the needed 100 MW of additional power.
5. "Unless we have another 100 MW of power by 2028, we will not be able to sustain the city's growth." Alternative: Restrict housing building permits and delay the growth of the city.
6. "Can we buy 100 MW electric capacity by 2028?" Alternative: Buy the additional energy from the nearby city of Eastheimer, which is expected to have a surplus of electric production capacity between 2028 and 2040.

The second stage is the development of a long list of alternative solutions to the need. Several of these alternatives may be mutually exclusive, while others may be combined to yield another, usually superior, alternative. Two management processes are used for the development of the longer list of alternatives, *conventional brainstorming* and *the nominal group technique* [2,3].

1. The *conventional brainstorming process* is based on the principles that "a quantity of ideas and many possible solutions evolves into a single quality solution" and "deferment of judgment and selection process until all ideas are heard." The process involves the following three phases:

   a. Preparation
   b. The brainstorming session
   c. The evaluation session

   During the preparation phase, the participants are selected and a loosely defined statement of the problem/need is circulated. The number of participants is limited to individuals who have the ability and experience to contribute or have demonstrated creative abilities to tackle the problem that is discussed. Typically, 5–12 individuals constitute a good group for brainstorming. During the brainstorming session, ideas are generated and duly recorded. Participants are encouraged to pose different questions to the problem at hand, which would elicit alternative solutions, as in items 1–6 earlier. For the unhindered expression of the participating individuals and the smooth flow of ideas, it is important that

   a. All participants perceive that they have equal standing during the session. More importantly, participants must be certain that they have nothing to lose by proposing their ideas. Unconventional seating arrangements (round table or theater seating), casual dressing, and use of first names without titles facilitate the smooth exchange of ideas at this level.
   b. Quantity of solutions/ideas is encouraged from the beginning. The determination of their quality and evaluations come at a later stage.
   c. Critique or list of disadvantages is ruled out at this phase. Evaluation and critique are parts of the process to follow.
   d. Improvement of contributed solutions and combination of solutions is actively encouraged.
   e. All solutions/ideas are succinctly recorded in a way that all the participants can see and, if appropriate, propose corrections, modifications, and combinations with other solutions.

   The evaluation stage follows at the end of the brainstorming session if all participants are expected to attend. If this is not desirable by the management, the evaluation stage will become a separate session with a smaller number of participants. During this session, the outcomes of each solution are critically evaluated relative to the problem at hand, some of the solutions are eliminated, and a smaller list of the most desired alternatives is formulated for further evaluation.

2. The *nominal group session* is a more formal meeting with the objective to achieve consensus among the participants. The list of participants and the loosely defined statement of the problem/need are circulated to all the participants before the

meeting, and well-thought solutions are sought. The actual meeting has a more formal setting and includes the following stages [23]:

a. All participants are expected to present and discuss their ideas in an orderly fashion. Formal presentations are given. Questions, comments, and critique of the ideas are allowed at this stage of the nominal group session.

b. Clarification, further study, and modification by the proposer follows, based on the comments, criticisms, and suggestions from part 1. The expectation at this stage is that a better, more optimal solution will be formulated.

c. Group modification and clarification of each idea follows. The ideas/solutions to the original need are formulated as projects.

d. The group may ask for the input of a "design team" with specialists who will later undertake the completion of the project.

e. Ranking or prioritization of all the solutions to the problem by voting or consensus. Any rules for the voting process must have been agreed before the meeting.

f. Elimination of solutions/projects that are placed at the bottom of the list.

g. Final discussion and clarification of any outstanding issues; finalization of the short list of solutions/projects.

It must be noted that in both processes, the alternative projects to be pursued may be *mutually exclusive*, that is only one of the projects may be pursued at the exclusion of all others, or *complementary*, when a group of projects may be pursued simultaneously or in series.

After the short list of ideas/alternatives/solutions has been selected, a detailed economic evaluation of the pertinent projects in the short list is undertaken. This evaluation is typically conducted by a smaller group of experts, takes into account only the economic aspects of the chosen alternative projects, and determines the *profitability* of the projects. Other issues, which may not be quantified and do not materially affect the cash flow and the overall profitability of the project, are called *intangibles* or *intangible items* and are left out at this stage. Such intangible items may be "the public good," "a greener planet," or "national security." These items must have been discussed and resolved in previous stages of the decision-making method. In the final decision for the completion of the project, the management/leadership may opt to consider one or more intangible items, provided that it does not affect the profitability of the investment. In a market-oriented system of for-profit corporations and public utilities, projects that lead to financial losses are not initiated and are not developed. If an entity (for-profit or nonprofit corporation or public utility) undertakes several money-losing projects, it sooner or later will find itself out of business. The investment appraisal analyses that follow treat the energy-related projects strictly as investments with the final decision based on whether or not a project is profitable to be undertaken.

9.3.2 Externalities

The use of all energy sources for the production of energy inevitably has environmental impacts that may be associated with some type of monetary cost for the entire society. These costs are often called *externalities*, are not well defined, and are rarely included in the economic analyses for energy production projects. Externality costs are linked to environmental and ecological degradations and include several factors that have not been priced by the national or international markets. Because they are not tangible and well-defined market costs and, by customary practice, energy producers do not pay for them, externalities widely vary among different studies. Factors that affect the environment and the ecosystems and their costs are difficult to calculate include the following:

1. Long-term effects of nuclear energy storage
2. Effect of coal–dust pollution on human health
3. Effects of GHGs that cause global climate change
4. Effects of wind turbines on the local and migratory bird populations
5. Ecological disruptions created by the building of a dam on a large river for a hydroelectric power plant
6. Ecosystem disruptions created by the construction of a barrage for a tidal energy system
7. Urban air pollution caused by fossil fuel combustion

Figure 9.2 depicts the estimates of a study on the externalities of seven sources for the production of electricity, including fossil fuels, nuclear, and renewable sources [4]. It is observed that depending on the assumptions made for the societal cost of energy production, the added cost of electric power from all sources ranges from a few fractions of a cent ($0.01) to several dollars. The author of these numbers comments on the high uncertainty of the externalities: "All approaches to the comparison of disparate environmental effects suffer in common from several intractable problems. Difficulties are posed by the obscurity, complexity, uncertainty and unquantifiability of many effects. Incommensurate indices and discrepant frames of reference exacerbate the analyst's predicament …" [4].

Because they do not represent tangible costs to an enterprise, externalities do not come into the calculations of the economic appraisal of energy projects. Externalities that have a clear and quantifiable impact in the city or the region are often included in a qualitative

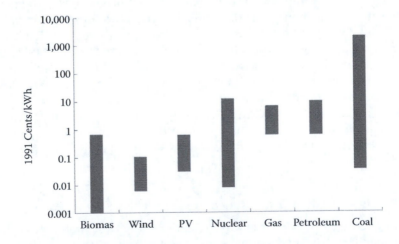

FIGURE 9.2
Estimates of the externalities of seven energy sources.

manner to eliminate one or more alternative solutions. Regarding the 100 MW need for the city of Westheimer, if the city desires to be "nuclear free," the options of building a nuclear power plant and buying energy from a nuclear power plant are *a priori* excluded from the list of alternative solutions. Similarly, if there is a desire to protect the migratory birds in the region, building a wind farm would be erased from the list of alternatives.

9.4 Investment Appraisal Methods

When the short list of the suggested alternative projects has been established and an acceptable method has been determined for the equivalency of the future cash flow, one must use an appraisal method to evaluate the alternatives and to make a final decision, for which projects to develop. An optimum appraisal method of project evaluation must have the following characteristics:

1. It takes into account the entire time horizon (life cycle) of the project.
2. It takes into account all the cash flows pertinent to the project for the entire time horizon.
3. It encompasses a suitable method to discount future cash flows and uses an *equivalent basis* of the future cash flows.

Several methods that are commonly used for the evaluation of investments in energy projects are presented in the next subsections and are critically evaluated.

9.4.1 Net Present Value

Variations of this method are called *present worth* (PW) and *net present worth* (NPW). The net present value (NPV) of a project is the sum of the discounted annual cash flows from the inception of the project to its final disposal. Consider a project that commences at present—the current year is usually denoted as year 0—and is completely disposed at year n, which is $(n + 1)$ years from the present. The pertinent cash flow streams are CF_0, CF_1, CF_2, ..., CF_n, and the discount rate is r_d. The equation that defines the NPV of this project is

$$NPV = CF_0 + \frac{CF_1}{(1+r_d)} + \frac{CF_2}{(1+r_d)^2} + \dots + \frac{CF_n}{(1+r_d)^n} = \sum_{i=0}^{n} \frac{CF_i}{(1+r_d)^i}. \tag{9.6}$$

If the NPV of a project is positive, the entity that considers it (a public utility, a nonprofit, or a for-profit corporation) incurs a net financial benefit (profit) from the development, maintenance, and operation of the project. If the NPV is negative, there is no financial advantage to the entity from pursuing the project, because it is equivalent to a financial loss at present. Unless there are other compelling considerations, the unprofitable project will not be undertaken. Profitable projects with positive NPV are undertaken, provided the capital for their development becomes available.

It is apparent that the discount rate r_d plays a very important role in the calculation of the NPV and the evaluation of the project. The discount rate is also one of the most difficult variables to precisely determine. Usually, the exact r_d values for classes of projects is established by the management, after taking into account the factors that are listed in Section 9.2.4.

During the decision-making process for energy systems, NPVs of all alternatives are calculated. For revenue-producing projects, the most profitable project is the one with the maximum NPV. Alternative projects with negative NPVs are rejected because they are equivalent to present-time financial losses. For energy conservation projects that involve costs only and are necessary to be undertaken (as in example 9.5 for the lighting of a building), the project with the minimum NPV minimizes the costs and points to the best alternative.*

If the alternative revenue-producing projects are exclusive—that is, only one must be selected—the alternative project with the highest positive NPV is selected to be developed. If the projects are not exclusive, they are ranked in order of decreasing NPV, with the first few projects having the priority for further development. When several nonexclusive projects are feasible, the decision of how many projects to pursue is dictated by the availability of critical resources, such as

1. The availability of capital (do we have enough funds available for all the investments needed?).
2. The capacity of the entity to simultaneously pursue these projects (can the engineering department undertake the design of all the projects? Do we have enough engineers and managers to supervise all the projects?).

Example 9.5

The owner of a new building is considering three alternatives for lighting in the building: incandescent bulbs (IBs), fluorescent bulbs (FBs), LEDs. The cost of purchasing and installing the three types of lighting are $640, $1,400 and $7,000, respectively. The current annual cost of electric energy (including its contribution to air-conditioning) by using IBs, FBs, and LEDs are $4,560, $2,380, and $785, respectively, and are expected to increase by 3% annually. IBs need to be replaced every 2 years; FBs, every 3 years; and LEDs, every 6 years. The cost of replacement of all types of lighting increases at 2.5% per year. Which type of lighting the owner should choose? The discount rate r_d for the owner is 7%.

Solution: We are going to calculate the NPV of all costs associated with the operation of the three lighting systems for 6 years. The latter is the minimum common multiple of the lifetimes of the three alternative projects (2, 3, and 6 years). Whatever the decision is, when the project ends in 6 years, it can be repeated for another 6 years. The following table shows all the cash flows (fixed and variable costs) associated with the three projects as well as the discounted cash flow. The current year is designated as year 0, and the sixth year is designated as year 5.

---

* Because costs are negative numbers, the so-called minimum cost actually represents the maximum negative number, e.g., of the numbers in the list (−125, −398, −15, and −28), the number −15 is the maximum. Effectively, one always chooses the maximum NPV.

	0	1	2	3	4	5	NPV
IB installation	640		672		742		
IB energy	4,560	4,697	4,838	4,983	5,132	5,286	
IB total	5,200	4,697	5,510	4,983	5,875	5,286	
IB discounted	5,200	4,390	4,813	4,067	4,482	3,769	26,720
FB installation	1,400			1,508			
FB energy	2,380	2,451	2,525	2,601	2,679	2,759	
FB total	3,780	2,451	2,525	4,108	2,679	2,759	
FB discounted	3,780	2,291	2,205	3,354	2,044	1,967	15,641
LED installation	7,000						
LED energy	785	809	833	858	884	910	
LED total	7,785	809	833	858	884	910	
LED discounted	7,785	756	727	700	674	649	11,291

The NPV of the three projects represents the present costs of using each type of lighting for the next 6 years. From the NPV of the three projects, the decision for the owner of building is apparent: The LED option is the best (i.e., less costly or most profitable) lighting alternative for this building.

The NPV is a simple and straightforward method to calculate the present monetary equivalent of benefits and costs associated with all energy projects. It takes into account the three criteria listed at the beginning of this section and provides an excellent method to assess the financial benefits of all energy production and conservation projects. It also helps to prioritize and rank different energy projects. This method is recommended for the appraisal of projects pertinent to energy generation, energy conservation, etc. Several additional examples on the use of the NPV method are given in Section 9.5.

It must be noted that the NPV method may be used for all types of investments including stocks and bonds, as example 9.6 demonstrates.

Example 9.6

XZY Corporation has issued a number of $10,000, 15-year bonds at 6%, with interest paid annually at the end of the year. Ten years after their issuance, the market price of the bonds is $9,400. Your discount rate for such investments is 6.8%. Would you buy this bond at its current market price?

Solution: At 6% rate on the nominal price, the bond pays $600 annually. With 5 years remaining in the life of the bond, you will derive a cash flow stream of $600 at the end of years 1–5 and $10,000 (the nominal price of the bond) at the end of year 5. The NPV of this cash flow stream to you is

$$NPV = -9400 + \frac{600}{1.068} + \frac{600}{1.068^2} + \frac{600}{1.068^3} + \frac{600}{1.068^4} + \frac{600}{1.068^5} + \frac{10,000}{1.068^5} = 270.$$

The NPV of the bond is positive. Buy the bond!

9.4.2 Annual Worth Method

The annual worth method (AWM) is based on the same principles as the NPV method to discount future cash inflows and outflows. Instead of determining a single value for the NPV of the project, the AWM determines the equivalent annual amount of the NPV, the annual worth (AW) of all revenues or losses, for the life cycle of the project. If the AW is positive, the project will be undertaken. Effectively, the AWM spreads the NPV of a project in each of the years of the development and operation of the project as an annuity. This method is frequently used with energy-related projects that may be indefinitely repeated (or for very long periods) in the future, as in the following example.

Example 9.7

The *efficient electron* electricity transmission corporation (EEC) is considering updating its medium-range transformers every 5 years to ensure low transformation losses. A new transformer costs $25,000 and may be sold in 6 years for $5,000. The increase in revenue, due to the higher transformer efficiency, is $8,000 per year. If the discount rate for EEC is 20%, calculate the AW of the project.

Solution: The AW of the initial investment (outlay) of the project for the 6-year duration may be calculated from Equation 9.4, using $P_{VA} = 25,000$, $N = 5$ (because the project starts in year 0, not in year 1), and $r_d = 0.2$:

$$A = PV_A \left[\frac{r_d(1+r_d)^N}{(1+r_d)^N - 1} \right] = \$8,360.$$

Similarly, the annuity for the $5,000 salvage value of the transformers is $672.
The AW of the project is AW = $8,000 − $8,360 + $674 = $313.
Since the AW is positive, the project will benefit the EEC, and the recommendation is that the project be developed. Note that if the relative magnitudes of the project parameters do not change, the project of replacing the transformers will be repeated after 6 years and may be repeated *ad infinitum*.

The AWM may be thought of as the annuity that corresponds to the NPV for the life cycle of the project. It is apparent that the AWM also adheres to the three criteria for investment appraisal methods listed at the beginning of this section.

9.4.3 Average Return on Book

The "book" in the average return on book (ARB) method refers to the *book value* of an investment, which is an accounting concept, defined as the initial value of the investment minus the accumulated depreciation. According to the ARB method, a project is undertaken if the ARB is greater than an acceptable discount rate, which is usually determined by management. Let us consider a project that starts in year 0 with a $12,000 investment. Straight-line depreciation* is allowed during years 1–3. The yearly net income of this project is $2,700. Table 9.1 shows the book value of this project during the 4 years of the life of the project:

---

* Straight-line depreciation means that the investment is depreciated in equal amounts over the time period allowed by the tax code. In this example, the depreciation is $12,000/3 = $4,000 in each of the first 3 years of the project. The sum of the depreciation over the allowed time period is equal to the initial investment.

TABLE 9.1

Calculation of Book Value

	Year 0	Year 1	Year 2	Year 3
Investment	12,000			
Net income	2,700	2,700	2,700	2,700
Accumulated depreciation	0	4,000	8,000	12,000
Book value	12,000	8,000	4,000	0

The average book value is $(12,000 + 8,000 + 4,000 + 0)/4 = 6,000$, and the ARB is $2,700/6,000 = 0.45$ or 45%. This project would be undertaken if the acceptable ARB rate to the management is less than 45%.

The ARB method is a very simple method, but suffers from several deficiencies for the appraisal of more complex energy related projects: At first, there is no differentiation between immediate income and future income. Income and depreciation in year 1 are treated the same way as income and depreciation in year 12. The *equivalent basis* of revenues and costs is not an inherent attribute of this method. Secondly, it does not use the cash flows, which are real and tangible revenues and expenses, but the accounting income, an accounting concept, which does not consider all the capital expenses for a project during the year they occur, but only considers the allowed depreciation of these expenses. Under most taxation systems, the capital expenses and depreciation during a year are entirely different. For example, under the US tax code, the depreciation of the structure that houses the turbines of a power plant occurs in 27.5 years, but the expense to the corporation for the building of this structure occurs at the beginning of the project (and immediately available cash is needed for its construction). Thirdly, as with the revenues, the method uses the average book value, another accounting concept, which does not differentiate between recent and distant capital expenditures. And, fourthly, the acceptable ARB is usually based on historical data for a corporation and is more difficult to determine based on the current economic environment and the current interest rates. The ARB is not an investment appraisal method that is recommended to be used with energy-related projects.

9.4.4 Payback Period

According to the payback period (PBP) method, the initial investment on the project must be recovered from net receipts/income within a specified period, which ranges from 2 to 8 years. This is the *PBP* and is calculated by counting the number of years it takes for the cumulative cash flows to be equal to the initial investment. For example, a project that involves an initial investment of $1,000,000 and pays net cash flows of $100,000 in the first 2 years and $210,000 thereafter would have a PBP of 6 years: During the first 5 years, the cumulative cash flow is $830,000, which is less than the initial investment, while during the first 6 years, the cumulative cash flow is $1,040,000, and this surpasses the initial investment of $1,000,000. Sometimes the PBP will be given more precisely in years and months: Assuming that all the payments occur monthly, the PBP of this project is 5 years and 10 months.

The PBP method makes use of the cash flow, which is the correct parameter for the calculation of the annual costs and benefits of the project, but does not differentiate between recent and distant cash flows. Similarly, the method does not differentiate between capital

TABLE 9.2

PBP and NPV for Three Hypothetical Projects

Project	CF_0	CF_1	CF_2	CF_3	PBP, Years	NPV
A	−2,000	1,000	1,000	5,000	2	3,492
B	−2,000	0	2,000	5,000	2	3,409
C	−2,000	1,000	1,000	100,000	2	74,857
D	−2,000	0	2,010	0	2	−339

Note: Numbers are in $1,000s.

expenditures that occur at the beginning of the project or at later time during the life of the project. Therefore, the *equivalence basis* is not inherent in the PBP method.

Because it does not encompass the equivalence basis of future cash flows, the application of the PBP method may lead to erroneous conclusions for projects. Table 9.2 shows the cash flows, the PBP, and the NPV (calculated at a discount rate of 10%) for four projects, which have the same initial investment of $2,000,000. It is apparent that although all projects have the same PBP of 2 years, one of the projects (C) is superior to the others by a wide margin, and one of the projects (D) is actually unprofitable. Even when one examines only projects A and B, the first project is superior to the second because it provides positive cash flow during the first year and its NPV, which is listed in the last column, is higher. It is apparent in this case that the NPV is a much better method to make a rational choice among the four projects.

The *discounted PBP* method has been proposed, where the cash flows are discounted by a rate r_d as in the NPV method. While this modification takes into account the time-value of funds, it still does not take into account any funds that are obtained after the cut-off period, which may be significant and may sway the investment decision among several projects. In addition, the value of the PBP (years) is arbitrary, it does not consider the current economic environment and the ability of the corporation or utility to borrow and use funds. In the four projects of Table 9.2, the large payoff in year 3 of project C does not receive the emphasis it deserves. Projects C and D not only have the same PBP but also have entirely different outcomes for the entity that undertakes the investment.

The use of the PBP as an investment appraisal method for energy-related projects should be avoided, except as a measure of how quickly all investment funds for a project are recovered. Because of inherent uncertainty of future revenues, the longer it takes to recover the investment funds, the higher is the associated risk of the project.

9.4.5 Internal Rate of Return

The internal rate of return (IRR) method is directly connected, and in most cases, it is equivalent to the NPV method: The IRR of a project is the value of the rate of return r_{ir}, which makes the NPV of the project equal to zero. Using Equation 9.6, the expression that defines the rate r_{ir} is

$$CF_0 + \frac{CF_1}{\left(1+r_{ir}\right)} + \frac{CF_2}{\left(1+r_{ir}\right)^2} + \ldots + \frac{CF_n}{\left(1+r_{ir}\right)^n} = \sum_{i=0}^{n} \frac{CF_i}{\left(1+r_{ir}\right)^i} = 0. \tag{9.7}$$

The IRR method involves the solution of the nonlinear Equation 9.7. A value of r_{ir} is computed for every alternative project. For the decision-making process of exclusive projects,

TABLE 9.3

IRR and NPV for Three Mutually Exclusive Projects

	CF_0	CF_1	CF_2	CF_3	CF_4	CF_5	CF_6-CF_∞	IRR, %	NPV
A	−9,000	6,000	5,000	4,000	0	0	0	33.4	3,592
B	−9,000	1,800	1,800	1,800	1,800	1,800	11,000	20.2	4,033
C	0	−6,000	1,200	1,200	1,200	1,200	7,290	20.2	2,119

Note: Numbers are in $1000s.

the project with the maximum value of r_{ir} is chosen, provided that this rate is higher than the MARR for the entity that will develop and operate the project. Because Equation 9.7 is highly nonlinear, the calculation of the IRR is more cumbersome than calculating the NPV, and it is typically accomplished by iteration (trial-and-error) methods.

The IRR method is based on the NPV method. In most cases, the two methods are equivalent and may be used interchangeably. However, there are projects when the IRR method is not equivalent to the NPV. This happens because of the nonlinearity in the computation of the IRR and applies to mutually exclusive projects and projects that involve investments with different durations. Table 9.3 shows three mutually exclusive projects where the NPV (calculated with 10% discount rate) and IRR methods yield different results for the investments.

While projects A and B entail the same initial investment of $9,000,000 for each one, project A has the highest IRR because the returns occur early in the life of the project. However, because this project breaks even only after year 3, its NPV is lower. For the appraisal of the projects A and B, the NPV and the IRR will generate different rankings for the two projects as long as the value of r_d used in the NPV method is lower than 15.6%. If $r_d > 15.6$, the two methods yield the same rankings. Comparing projects B and C, the two have the same IRR (20.2%), but the NPV of project B is higher. This case demonstrates that while the NPV and the IRR methods are usually equivalent, there are cases where they may differ and the application of the IRR may lead to a nonoptimum decision. As a rule, the NPV is simpler to perform and always leads to financially sound decisions, especially when one deals with mutually exclusive projects with complex payoff schedules.

9.4.6 External Rate of Return

The external rate of return (ERR) method, also known as the *modified internal rate of return* [5], is a variance of the previous method. According to the ERR, all the cash outflows of a project are discounted to the present, year 0, at a discount rate equal to an *external reinvestment rate* i_{er}, which is defined by the management. All the cash inflows of the project are discounted to the last year of the project, year N, using the same rate i_{er}. The ERR is the interest rate i_{err} that makes the two sums of discounted cash inflows and cash outflows to be equivalent. A project is accepted to be developed if i_{err} is greater than the MARR.

The EER method, in general, yields results similar to the NPV. As with the IRR method, the ERR will also fail to identify the best project if the highest returns of a project occur at an early stage. In addition, because the method uses two discount rates i_{er} and i_{err} as well as the MARR, it suffers from the uncertainties involved in the correct computation and interpretation of the three rates.

9.4.7 Profitability Index

The *profitability index* (PI) is sometimes referred to as *benefit/cost ratio*. It is defined as the sum of all discounted future positive cash flows divided by the initial investment. If the entire initial investment occurs in year 0 and is denoted by I, and the cash flows in years 1–n are denoted as CF_1, CF_2, CF_3, ..., and CF_n, respectively, the expression for the PI is as follows:

$$PI = \left(\frac{CF_1}{\left(1+r_d\right)} + \frac{CF_2}{\left(1+r_d\right)^2} + ... + \frac{CF_n}{\left(1+r_d\right)^n} \right) \Big/ I = \frac{1}{I}\sum_{i=1}^{n} \frac{CF_i}{\left(1+r_d\right)^i}. \qquad (9.8)$$

It must be noted that the cash flows do not include any part of the project investment, which is denoted by I. For the decision-making process, if $PI > 1.0$, the project is profitable and should be undertaken.

The PI method has several common features with the NPV method: The time-value of funds is inherent in both methods; the present value of all positive cash flows is used in the PI method, and both methods may use the same discount rate. However, when the investment occurs over a period of several years and when the investment and revenue streams follow a complex pattern and are mixed during several years, as it often occurs in the early stages of an energy project, the PI method becomes more cumbersome to use and may lead to erroneous decisions, especially when complex and mutually exclusive projects are considered.

From the presentation and discussion of the seven most commonly used investment appraisal methods, it is apparent that the NPV method is a relatively simple method to use, especially with long-term projects and complex projects that involve several streams of revenues and expenses. The NPV method is recommended to be used for energy-related projects because it is well defined; it is relatively simpler to use than the other methods; and it is widely accepted by the business, financial, and engineering communities.

There are several quantitative project development and financial tools that have been created and are available to assist with feasibility studies and the financial appraisal of energy projects. Most corporations have developed their own proprietary tools for the appraisal of energy projects. There are also several financial tools that are free and available online to the general public. One such tool is the *RETScreen Clean Energy Project Analysis Software*, which was developed with the assistance of the government of Canada and is available through the website http://en.openei.org/wiki/RETScreen_Clean_Energy_Project _Analysis_Software (last visited on July 1, 2017). According to the website, this software "... is used by over 490,000 people in every country and territory of the world."

9.5 Case Studies: Financial Analysis of a Wind Farm Project

Having established the advantages of using the NPV method as a tool for the appraisal of energy related projects, this section offers a few examples on the use of the method in complex, practical situations. The objective of the case studies is to underscore the effects of several variables that come into the decision-making process of an energy project. These variables include the following:

1. The effect of energy prices on the investment
2. The effect of the allowable depreciation for the project

3. The effect of the cost of borrowed capital and of the discount rate

4. The effect of tax credits and the combination of tax credits and accelerated depreciation, two variables that are related to the governmental/regulatory settings of energy projects

5. The effect of time delay in the construction and operation of the project

Let us consider the development of a wind farm for the production of electricity at a location in western Texas. The nominal power rating of this wind farm is 50 MW. The investment for this project is estimated to be $115,000,000 ($115M) to be spent as follows: $30M in year 0 for the preparation of the field, the permits, the engineering study of the farm, and the construction of the towers; $70M in year 1 to be spent mostly on the wind turbines and generators; and $15M in year 2 for the installation of the turbines/generators and connections to the electricity grid. The corporation plans to invest $50M of their own funds into the project and finance the other $65M in year 1 of the project by issuing 12-year bonds at 6% interest rate. Based on the wind conditions in the area, it is estimated that the power plant will produce 215×10^6 kWh/year. Zero electricity is produced in years 0 and 1, while the wind turbines are in the construction stages; 80% of the expected energy will be produced in year 2 and 100% thereafter. It is estimated that the useful life of this power plant to the corporation will be 12 years after the completion of the installation. At that point, the year 14 of the project, the wind farm will be sold to another corporation for an estimated $66M (salvage value). The fixed costs of the plant (including the leases on the land) are estimated to be $350,000 per year, starting in year 0 and increasing at an annual rate of 3%. The operating (variable) costs are estimated at $400,000/year, starting in year 2 and increasing at 5% annually for the duration of the project. The average price of electricity from wind is currently $0.04/kWh and is expected to increase at an annual rate of 2%. The corporation is taxed at 28% rate, and a straight-line 10-year depreciation schedule is allowed for the entire investment. The corporate discount rate r_d for this type of investment is 15%.

It is common practice for energy-producing corporations to issue interest-bearing bonds to finance part of their projects, in this case $65 M of the total investment. The corporation puts up $50M of the capital cost, and the remaining $65M is financed with bonds. For simplicity, it will be assumed that the entire revenue of the bonds will be collected at the end of year 0; interest for the bonds will be paid during years 1–12. The entire principal of $65M for the bonds will be repaid to the bondholders in year 13. Thus, in year 0, the corporation finances the $30M investment out of their own funds; in year 1, the actual investment by the corporation is $70M with $5M of their own funds and the other $65M from the issuing of bonds; and in year 2, the entire $15M invested comes from the own funds of the corporation.

The revenues of this project principally come from the production and sale of electricity, which is 0 during years 0 and 1; 172×10^6 kWh during year 2; and 215×10^6 kWh during years 3–14. At the end of the 14-year period, the corporation receives $66M as "salvage value" from the sale of the wind farm. The costs of the project consist of investment of capital, annual fixed and variable costs, annual interest expense on the bonds, and taxes paid by the corporation. The taxes are 28% of the *taxable income* of the corporation, which for simplicity is defined here as the total annual revenue (excluding bond revenue) minus annual fixed and variable expenses, minus interest, minus the allowable depreciation.*

---

* The tax code of each country defines what the taxable income is and what kind of depreciation schedule is allowed. Tax codes are complex, and typically, all corporations employ several accountants to determine the accounting and financial parameters of their projects.

Given the schedule of the investment for this project, the depreciation of all the expenses is allowed over a period of 10 years. The depreciation schedule in this case is as follows:

1. For year 0, the allowable depreciation is $3M (=$30M/10).
2. For year 1, the allowable depreciation is $10M ($30M/10 + $70M/10).
3. For years 2–9, the allowable depreciation is $11.5M ($30M + $70M + $15M/10).
4. For year 10, the allowable depreciation is $8.5M ($70M/10 + $15M/10).
5. For year 11, the allowable depreciation is $1.5M ($15M/10).

At the end of year 11, the entire investment has been depreciated, and there is no depreciation allowed in the subsequent years.

It will be assumed that the corporation has other profitable activities to take advantage of the accrued depreciation during years 0 and 1, when there is no revenue from this project. In these two years, the project will incur a "negative tax," which is equivalent to a tax benefit for the corporation by offsetting taxes from other operations. Actually, and because of the allowable depreciation, the corporation incurs a "negative tax" until the end of year 10 of the project, as shown in Table 9.4.

The cash flow diagram for this project is shown in Figure 9.3. It is apparent that the corporation has negative cash flow in years 1 and 2, when most of the capital investment occurs, and very small revenue appears from the sale of electricity. In years 3–12, the corporation has positive cash flow from the sale of electricity. The negative cash flow in year 13 is due to the repayment of the principal of the bonds. The large cash flow in year 14 is primarily due to the sale of the project for $66M.

The accounting formulas used for the computation of the yearly cash flow from the several variables are given for reference purposes. These formulas and the tables that follow are for illustration purposes only. Accounting principles and the tax laws differ from country to country, and hence, the *taxable income* for corporations and individuals are differently defined in different countries. The formulas used for the several parameters that appear in the tables of this and the following cases are as follows:

- Total revenue = (revenue from electricity) + (bond revenue) + (salvage value).
- Total costs = (capital investment) + (fixed costs) + (variable costs) + (interest on bonds) + (bond repayment).
- Pretax income = (total revenue excluding bond revenue) − (closing costs) − (fixed costs) − (variable costs) − (interest on bonds).
- Taxable income = (pretax income) − (depreciation).
- Tax = (taxable income) × (tax rate) − (tax credit). If the tax is negative, this implies that the corporation receives a benefit by having to pay less taxes in their other projects.
- Cash flow = (total revenue) − (total costs) − (tax).
- Discount factor for year i: $1/(1 + r_d)^i$.
- Discounted cash flow = (cash flow of year i) × (discount factor for year i).
- NPV = sum of all discounted cash flows from the beginning to the end of the project.

Table 9.4 shows the actual calculations for the NPV of this project. With the preceding parameters for costs and revenues, the NPV of the project is negative $10,395,515. This indicates that the project is unprofitable, and therefore, the corporation should not undertake the development of this wind farm. It is worth noting in Table 9.4 that the sale of the wind

TABLE 9.4

NPV Calculation of the Wind Farm: Basic Scenario

	Year														
	0	1	2	3	4	5	6	7	8	9	10	11	12	13	14
Revenue															
Price, per kWh	0.0400	0.0408	0.0416	0.0424	0.0433	0.0442	0.0450	0.0459	0.0469	0.0478	0.0488	0.0497	0.0507	0.0517	0.0528
kWh produced	0	0	172,000,000	215,000,000	215,000,000	215,000,000	215,000,000	215,000,000	215,000,000	215,000,000	215,000,000	215,000,000	215,000,000	215,000,000	215,000,000
Revenue from electricity	0	0	7,157,952	9,126,389	9,308,917	9,495,095	9,684,997	9,878,697	10,076,271	10,277,796	10,483,352	10,693,019	10,906,879	11,125,017	11,347,517
From bonds	65,000,000	0	0	0	0	0	0	0	0	0	0	0	0	0	0
Salvage value	0	0	0	0	0	0	0	0	0	0	0	0	0	0	66,000,000
Total revenue	65,000,000	0	7,157,952	9,1263,89	9,308,917	9,495,095	9,684,997	9,878,697	10,076,271	10,277,796	10,483,352	10,693,019	10,906,879	11,125,017	77,347,517
Costs															
Capital investment	30,000,000	70,000,000	15,000,000	0	0	0	0	0	0	0	0	0	0	0	0
Closing costs	0	0	0	0	0	0	0	0	0	0	0	0	0	0	0
Fixed costs	350,000	360,500	371,315	382,454	393,928	405,746	417,918	430,456	443,370	456,671	470,371	484,482	499,016	513,987	529,406
Variable costs	0	0	400,000	420,000	441,000	463,050	486,203	510,513	536,038	562,840	590,982	620,531	651,558	684,136	718,343
Interest on bonds	0	0	3,900,000	3,900,000	3,900,000	3,900,000	3,900,000	3,900,000	3,900,000	3,900,000	3,900,000	3,900,000	3,900,000	0	0
Bond repayment	0	0	0	0	0	0	0	0	0	0	0	0	0	65,000,000	0
Total costs	30,350,000	74,260,500	19,671,315	4,702,454	4,734,928	4,768,796	4,804,121	4,840,968	4,879,408	4,919,511	4,961,353	5,005,013	5,050,574	66,198,123	1,247,749
Tax Calculation															
Pretax income	−350,000	−4,260,500	2,486,637	4,423,934	4,573,988	4,726,299	4,880,876	5,037,728	5,196,863	5,358,285	5,521,999	5,688,006	5,856,305	9,926,894	76,099,768
Depreciation	3,000,000	10,000,000	11,500,000	11,500,000	11,500,000	11,500,000	11,500,000	11,500,000	11,500,000	11,500,000	8,500,000	1,500,000	0	0	0
Tax credit	0	0	0	0	0	0	0	0	0	0	0	0	0	0	0
Taxable income	−3,350,000	−14,260,500	−9,013,363	−7,076,066	−6,926,012	−6,773,701	−6,619,124	−6,462,272	−6,303,137	−6,141,715	−2,978,001	4,188,006	5,856,305	9,926,894	76,099,768
Tax	−938,000	−3,992,940	−2,523,742	−1,981,298	−1,939,283	−1,896,636	−1,853,355	−1,809,436	−1,764,878	−1,719,680	−833,840	1,172,642	1,639,765	2,779,530	21,307,935
Cash flow	35,588,000	−7,026,756	−9,989,621	6,405,233	6,513,272	6,622,935	6,734,231	6,847,164	6,961,741	7,077,965	6,355,839	4,515,364	4,216,540	−57,852,636	54,791,833
Discount factor	1.0000	0.8696	0.7561	0.6575	0.5718	0.4972	0.4323	0.3759	0.3269	0.2843	0.2472	0.2149	0.1869	0.1625	0.1413
Discounted cash flow	35,588,000	−6,110,226	−7,553,589	4,211,544	3,723,984	3,292,769	2,911,394	2,574,103	2,275,806	2,012,000	1,571,066	970,547	788,101	−9,402,671	7,743,656
NPV	−10,395,515														

Note: Production 215,000,000 kWh/year; price growth = 2% annually; discount rate = 15%; fixed cost growth rate = 3%; variable cost growth rate = 5%; corporate tax rate = 28%.

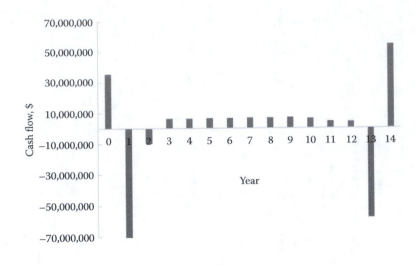

FIGURE 9.3
Cash flow diagram for the wind farm example. The numbers correspond to Table 9.4.

farm for $66 M contributes little to the NPV (slightly less than $9.5 M) because the sale occurs far in the future and the discount factor for that year is 0.1413.

The question arises, "under what conditions would this *green energy* project be profitable for the corporation?" or "what can make this project profitable to be undertaken by the corporation?" The obvious answer is "if some of the important parameters that affect the NPV changed to favor a higher NPV." For example, if the current price of electricity were $0.053/kWh instead of $0.040/kWh and everything else remained the same, the NPV would be a positive $348,099, and the project would be profitable for the corporation to develop. This implies that if consumers are willing to pay approximately 1.3 cents/kWh more, the renewable energy project would materialize in a market-oriented economy. This practice of higher prices for "green energy" has been applied in several states, where electricity consumers have the option to buy "green energy" for 1–2 cents higher price per kWh.

Alternatively, if the project were determined to be less risky and the discount rate for the corporation were 5.3% instead of 15%, the NPV for the lifetime of the project would be a positive $155,706, and again, this project would be worth undertaking.* Issuing the bonds at a significantly lower rate, or receiving a zero-interest loan from a state bank, has the same effect. The low-interest loan method for "green energy" projects has also been applied in several countries. In the case of the project considered here, the NPV of the project becomes positive if the corporation can issue bonds at 1.9% annually (or equivalently if it receives a $65M loan from a central bank at 1.9% or lower interest rate). All other parameters remain the same as in Table 9.4.

9.5.1 NPV and Governmental Incentives or Disincentives

The calculations of the last section show that there are several additional variables that play an important role in the calculation of the NPV, which determines if projects are profitable and worth undertaking by a corporation. It is also apparent that several of these

---

* Projects are considered less risky when corporations accumulate a great deal of expertise in developing and operating them and when financial guarantees are received from local or state governments. Governmental loan guarantees, as in the development of four new nuclear reactors in the United States in 2011, result in the issuance of bonds at significantly lower interest rates.

parameters are not within the control of the corporation that will undertake and develop the project. Estimates of the future values of most parameters represent mere assumptions that are made by professionals (managers, accountants, and engineers) using past experience on similar projects. All recent experience dictates that many of these parameters (e.g., the cost of fuel in the period of 2005–2017) are unpredictable and fluctuate a great deal. This adds to the risk of an energy project. To counteract the higher risk, corporations commonly use higher discount rates, especially for energy projects with long time horizons.

Several of the parameters that determine the NPV of a project are controlled by regional and national governments. Through regulations and tax policy, governments influence these parameters in order to encourage or discourage categories of energy projects. Renewable energy projects, energy conservation projects, and higher-efficiency projects are currently favored to be developed in most countries. Among the parameters that governments influence and regulate are as follows:

1. The tax rate on the taxable income (28% in the example of Table 9.4).

2. The allowable schedule of depreciation. Shorter depreciation schedules favor the profitability of projects.

3. Offering tax credits for desirable categories of projects. In 2017, almost all national governments are offering tax credits for *green energy* projects.

4. Imposing taxes on undesirable energy projects. The *carbon tax*, which has been proposed in several countries, has the effect of increasing the price of electric energy produced by power plants that use fossil fuels. The carbon tax is "passed" to the consumers through increased electricity prices, and this makes renewable energy units more attractive to investors. In the case of the project in Table 9.4, a mere increase of the electricity price from $0.040/kWh to $0.053/kWh makes the project attractive for the corporation.

5. Guaranteeing a favorable price for the energy produced. This is particularly important for solar energy projects that have high capital cost. This practice has been applied in several European Union (EU) countries and regions.

6. Enacting environmental regulations, e.g., on the reduction or elimination of sulfur or mercury emissions, which would ultimately discourage the use of coal and petroleum and would encourage renewable energy projects.

Combinations of such incentives are offered in 2017 for renewable energy projects in most countries. Since 2004, renewable energy and improved energy efficiency projects have attracted significant tax credits, from 5–30%, in most OECD countries. Such incentives play a major role in reducing the risk of renewable energy projects and making the NPV positive and the projects worth developing. The momentous annual rates of growth of renewable energy produced globally, shown in Table 6.1, are partly due to a combination of governmental incentives. We will consider here three such cases that affect the taxable income of the corporation, which undertakes the wind farm project:

1. *Allow the capital costs of the project to be depreciated faster*: Instead of a 10-year depreciation schedule, let us consider that the capital investment may be depreciated, again with the straight-line method, but in 5 years instead of the 10 years of the basic scenario: The allowable depreciation in year 0 is $6M; in year 1, it is $20M; in years 2–4, it is $23M; in year 5, it is $17M; and in year 6, it is $3M. The calculations for this case are shown in Table 9.5. The effect of the faster depreciation is

TABLE 9.5
NPV Calculation of the Wind Farm: Basic Scenario with 5-Year Depreciation

	Year														
	0	1	2	3	4	5	6	7	8	9	10	11	12	13	14
Revenue															
Price, per kWh	0.0400	0.0408	0.0416	0.0424	0.0433	0.0442	0.0450	0.0459	0.0469	0.0478	0.0488	0.0497	0.0507	0.0517	0.0528
kWh produced			172,000,000	215,000,000	215,000,000	215,000,000	215,000,000	215,000,000	215,000,000	215,000,000	215,000,000	215,000,000	215,000,000	215,000,000	215,000,000
Revenue from electricity			7,157,952	9,126,389	9,308,917	9,495,095	9,684,997	9,878,697	10,076,271	10,277,796	10,483,352	10,693,019	10,906,879	11,125,017	11,347,517
From bonds	65,000,000	0	0	0	0	0	0	0	0	0	0	0	0	0	0
Salvage value	0	0	0	0	0	0	0	0	0	0	0	0	0	0	66,000,000
Total revenue	65,000,000	0	7,157,952	9,126,389	9,308,917	9,495,095	9,684,997	9,878,697	10,076,271	10,277,796	10,483,352	10,693,019	10,906,879	11,125,017	77,347,517
Costs															
Capital investment	30,000,000	70,000,000	15,000,000	0	0	0	0	0	0	0	0	0	0	0	0
Closing costs	0	0	0	0	0	0	0	0	0	0	0	0	0	0	0
Fixed costs	350,000	360,500	371,315	382,454	393,928	405,746	417,918	430,456	443,370	456,671	470,371	484,482	499,016	513,987	529,406
Variable costs	0	0	400,000	420,000	441,000	463,050	486,203	510,513	536,038	562,840	590,982	620,531	651,558	684,136	718,343
Interest on bonds	0	3,900,000	3,900,000	3,900,000	3,900,000	3,900,000	3,900,000	3,900,000	3,900,000	3,900,000	3,900,000	3,900,000	3,900,000	0	0
Bond repayment	0	0	0	0	0	0	0	0	0	0	0	0	0	65,000,000	0
Total costs	30,350,000	74,260,500	19,671,315	4,702,454	4,734,928	4,768,796	4,804,121	4,840,968	4,879,408	4,919,511	4,961,353	5,005,013	5,050,574	66,198,123	1,247,749
Tax Calculation															
Pretax income	−350,000	−4,260,500	2,486,637	4,423,934	4,573,988	4,726,299	4,880,876	5,037,728	5,196,863	5,358,285	5,521,999	5,688,006	5,856,305	9,926,894	76,099,768
Depreciation	6,000,000	20,000,000	23,000,000	23,000,000	23,000,000	17,000,000	3,000,000	0	0	0	0	0	0	0	0
Tax credit	0	0	0	0	0	0	0	0	0	0	0	0	0	0	0
Taxable income	−6,350,000	−24,260,500	−20,513,363	−18,576,066	−18,426,012	−12,273,701	1,880,876	5,037,728	5,196,863	5,358,285	5,521,999	5,688,006	5,856,305	9,926,894	76,099,768
Tax	−1,778,000	−6,792,940	−5,743,742	−5,201,298	−5,159,283	−3,436,636	526,645	1,410,564	1,455,122	1,500,320	154,610	1,592,642	1,639,765	2,779,530	21,307,935
Cash flow	36,428,000	−67,467,560	−6,769,621	9,625,233	9,733,272	8,162,935	4,354,231	3,627,164	3,741,741	3,857,965	3,975,839	4,095,364	4,216,540	−57,852,636	54,791,833
Discount factor	1.0000	0.8696	0.7561	0.6575	0.5718	0.4972	0.4323	0.3759	0.3269	0.2843	0.2472	0.2149	0.1869	0.1625	0.1413
Discounted cash flow	36,428,000	−58,667,443	−5,118,806	6,328,747	5,565,030	4,058,422	1,882,454	1,363,585	1,223,182	1,096,675	982,767	880,271	788,101	−9,402,671	7,743,656
NPV	−4,848,031														

Note: Price growth rate = 2% annually; discount rate = 15%; fixed cost growth rate = 3%; variable cost growth rate = 5%; corporate tax rate = 28%.

to change the NPV of the project from a negative $10,395,515 to a less negative $4,848,031.

The project is still unprofitable in a market-oriented economy, but the shorter depreciation schedule has a positive effect on the bottom line. A shorter depreciation schedule allows a corporation to receive the tax benefits of the investment earlier rather than later. Because the early cash flows of the project affect the NPV the most, a shorter depreciation schedule always has a positive effect on the NPV and on the profitability of projects. In the US tax code, the *accelerated depreciation* is allowed for several energy-related projects. This is a depreciation method, whereby an asset is allowed to be depreciated at faster rates than the straight-line method, e.g., 30% in the first year, 25% in the second year, 20% in the third year, etc. The accelerated depreciation schedule allows higher deductions in the earlier years of a project and, in general, has a positive impact on the NPV of a project, provided that the corporation has sufficient other taxable income to take advantage of the tax benefits from the accelerated depreciation.

2. *Tax credits on the investment*: A tax credit is a partial relief from the taxes the corporation has to pay otherwise and is equal to a fraction of the investment. Since the current project does not produce profits to be taxed in the first 2 years, it is important for the corporation to have other profitable operations/projects to realize the allowable tax credit. In 2017, tax credits offered by national and regional governments for renewable energy projects were in the range of 5–35%.

Let us consider that there is a governmental 10% tax credit on the investment for renewable energy projects, such as the wind farm under consideration. With the 10% tax credit for renewable energy investments, the corporation saves from other taxable activities $3M during year 0, $7M in year 1, and $1.5M in year 2. In this case, the tax credits are sufficient to change the negative NPV of the wind farm project to a positive NPV of $5,373,141, which strongly favors the development of the project. The calculations for this scenario are shown in Table 9.6. The parameters for this table are the same as in the base scenario of Table 9.4 (which was unprofitable with −$10,395,515 NPV) with the only modifications the accelerated depreciation and the 10% tax credit.

It must be emphasized that for a corporation to receive the tax benefits of accelerated depreciation and the tax credits, it must have other profitable taxable operations, whose taxes would be offset by the benefits. If a corporation does not have other profitable operations, claiming the depreciation and credits may be delayed for later years, depending on the applicable tax codes. However, delays to fully realize the tax benefits as early as possible have a negative impact on the NPV because they are discounted in the later years of the project. Individuals, smaller corporations, and small start-up companies that do not have enough taxable income may not benefit from accelerated depreciation schedules and tax credits. Direct monetary subsidies, rebates, and guaranteed loans with low interest are more effective incentives for individuals and corporations with lower taxable incomes.

3. *Guaranteed price or annual price increases*: Consumers in several countries have voluntarily accepted to pay higher prices (between $0.01 and $0.02 in the United States) for renewable "green" energy. The higher than market "green electricity" prices always improve the NPV of a project and make it attractive to investors.

TABLE 9.6

NPV Calculation of the Wind Farm: Basic Scenario with 5-Year Depreciation and 10% Tax Credit

	Year														
	0	1	2	3	4	5	6	7	8	9	10	11	12	13	14
Revenue															
Price, per kWh	0.0400	0.0408	0.0416	0.0424	0.0433	0.0442	0.0450	0.0459	0.0469	0.0478	0.0488	0.0497	0.0507	0.0517	0.0528
kWh produced	0	0	172,000,000	215,000,000	215,000,000	215,000,000	215,000,000	215,000,000	215,000,000	215,000,000	215,000,000	215,000,000	215,000,000	215,000,000	215,000,000
Revenue from electricity	0	0	7,157,952	9,126,389	9,308,917	9,495,095	9,684,997	9,878,697	10,076,271	10,277,796	10,483,352	10,693,019	10,906,879	11,125,017	11,347,517
From bonds	65,000,000	0	0	0	0	0	0	0	0	0	0	0	0	0	0
Salvage value	0	0	0	0	0	0	0	0	0	0	0	0	0	0	66,000,000
Total revenue	65,000,000	0	7,157,952	9,126,389	9,308,917	9,495,095	9,684,997	9,878,697	10,076,271	10,277,796	10,483,352	10,693,019	10,906,879	11,125,017	77,347,517
Costs															
Capital investment	30,000,000	70,000,000	15,000,000	0	0	0	0	0	0	0	0	0	0	0	0
Closing costs	0	0	0	0	0	0	0	0	0	0	0	0	0	0	0
Fixed costs	350,000	360,500	371,315	382,454	393,928	405,746	417,918	430,456	443,370	456,671	470,371	484,482	499,016	513,987	529,406
Variable costs	0	0	400,000	420,000	441,000	463,050	486,203	510,513	536,038	562,840	590,982	620,531	651,558	684,136	718,343
Interest on bonds	0	3,900,000	3,900,000	3,900,000	3,900,000	3,900,000	3,900,000	3,900,000	3,900,000	3,900,000	3,900,000	3,900,000	3,900,000	0	0
Bond repayment	0	0	0	0	0	0	0	0	0	0	0	0	0	65,000,000	0
Total costs	30,350,000	74,260,500	19,671,315	4,702,454	4,734,928	4,768,796	4,804,121	4,840,968	4,879,408	4,919,511	4,961,353	5,005,013	5,050,574	66,198,123	1,247,749
Tax Calculation															
Pretax income	-350,000	-4,260,500	2,486,637	4,423,934	4,573,988	4,726,299	4,880,876	5,037,728	5,196,863	5,358,285	5,521,999	5,688,006	5,856,305	9,926,894	76,099,768
Depreciation	6,000,000	20,000,000	23,000,000	23,000,000	23,000,000	17,000,000	3,000,000	0	0	0	0	0	0	0	0
Tax credit	3,000,000	7,000,000	1,500,000	0	0	0	0	0	0	0	0	0	0	0	0
Taxable income	-6,350,000	-24,260,500	-20,513,363	-18,576,066	-18,426,012	-12,273,701	1,880,876	5,037,728	5,196,863	5,358,285	5,521,999	5,688,006	5,856,305	9,926,894	76,099,768
Tax	-4,778,000	-13,792,940	-7,243,742	-5,201,298	-5,159,283	-3,436,636	526,645	1,410,564	1,455,122	1,500,320	1,546,160	1,592,642	1,639,765	2,779,530	21,307,935
Cash flow	39,428,000	-60,467,560	-5,269,621	9,625,233	9,733,272	8,162,935	4,354,231	3,627,164	3,741,741	3,857,965	3,975,839	4,095,364	4,216,540	-57,852,636	54,791,833
Discount factor	1.0000	0.8696	0.7561	0.6575	0.5718	0.4972	0.4323	0.3759	0.3269	0.2843	0.2472	0.2149	0.1869	0.1625	0.1413
Discounted cash flow	39,428,000	-52,580,487	-3,984,591	6,328,747	5,565,030	4,058,422	1,882,454	1,363,585	1,223,182	1,096,675	982,767	880,271	788,101	-9,402,671	7,743,656
NPV	5,373,141														

Note: Price growth rate = 2% annually; discount rate = 15%; fixed cost growth rate = 3%; variable cost growth rate = 5%; corporate tax rate = 28%.

The mere increase of the electric energy price from $0.040/kWh to $0.053/kWh changes the negative $10,395,515 NPV of the original project in Table 9.4 to a positive NPV of $348,099.

Another way to offer higher prices for renewable energy is to ensure (or guarantee) that electricity prices grow at a faster rate in the future. This can be one of the effects of a carbon tax, which may be gradually imposed in the future. The carbon tax (or any other kind of tax on nonrenewable energy sources) is always passed to the consumers as an increased average price of electricity. Table 9.7 shows the effect to the entire project, if the annual electricity price increase is guaranteed to be 6.5%, because of some type of regulatory action. The numbers in the table consider 10-year linear depreciation and zero tax credit. It is observed that the increased annual price growth from 2% to 6.5% has a positive impact on this project because it transforms the negative $10,395,515 NPV of the project to a positive NPV of $854,828.

It must be noted, however, that while the high growth rates of the electricity prices favor alternative energy projects, promote energy efficiency and savings, and may minimize environmental pollution, such programs make electricity very expensive and have negative social impacts for the poorer segments of the population that will not be able to afford electricity at the fast increasing prices. Incentives to promote "green energy" should take into account the energy needs of the entire population.

4. *A regulatory disincentive*: The regulatory environment, which is largely controlled by central, regional, and local governments, may also impose incentives and disincentives to energy projects. An obvious disincentive is the taxation of alternative energy activities or by-products, such as the imposition of a disposal fee on nuclear and biomass waste products. A disincentive that seldom comes to the attention of the public is a prolonged delay in the commencement of an energy project. This delays the development of the project and the realization of the project revenue. A common cause of such delays is a lawsuit related to real or perceived environmental or ecological effects and local judicial decisions that delay the energy projects.

Let us consider again the case of the wind farm with the parameters depicted in Table 9.6, which results in a positive NPV of $5,373,141. Given the positive NPV with the tax credit and the accelerated depreciation and the other financial considerations shown in Table 9.6, XYZ Corporation has decided to undertake the project, has obtained all the local permits, has finished the construction of the towers, installed the wind turbines, and is in the final stage of connecting to the grid and commencing power production. However, a local environmental group alleges that wind turbines are harmful to the migratory geese that happen to pass twice a year near the wind farm site. The environmental group persuades a local judge to issue an injunction for the construction and operation of the wind farm, pending a "… complete and thorough environmental and ecological impact of the proposed project." XYZ Corporation appeals the decision to a higher court and, eventually, prevails in the court system. However, the effect of the judicial process is to delay the commencement of operation of the plant for 15 months (a very short time for most judicial systems). Because the wind turbines are idle for 15 months, there is no revenue for the corporation in year 2 of the project, and the revenue in year 3 is only 50% of the expected revenue. Electricity is produced, and the full expected revenue is generated in years 4–14 as reflected in Table 9.8, which is the same as Table 9.6, except that the revenue

TABLE 9.7

NPV of the Wind Farm: Basic Scenario

								Year							
	0	1	2	3	4	5	6	7	8	9	10	11	12	13	14
Revenue															
Price, per kWh	0.0400	0.0426	0.0454	0.0483	0.0515	0.0548	0.0584	0.0622	0.0662	0.0705	0.0751	0.0800	0.0852	0.0907	0.0966
kWh produced	0	0	172,000,000	215,000,000	215,000,000	215,000,000	215,000,000	215,000,000	215,000,000	215,000,000	215,000,000	215,000,000	215,000,000	215,000,000	215,000,000
Revenue from electricity	0	0	7,803,468	10,388,367	11,063,611	11,782,745	12,548,624	13,364,284	14,232,963	15,158,105	16,143,382	17,192,702	18,310,228	19,500,392	20,767,918
From bonds	65,000,000	0	0	0	0	0	0	0	0	0	0	0	0	0	0
Salvage value	0	0	0	0	0	0	0	0	0	0	0	0	0	0	66,000,000
Total revenue	65,000,000	0	7,803,468	10,388,367	11,063,611	11,782,745	12,548,624	13,364,284	14,232,963	15,158,105	16,143,382	17,192,702	18,310,228	19,500,392	86,767,918
Costs															
Capital investment	30,000,000	70,000,000	15,000,000	0	0	0	0	0	0	0	0	0	0	0	0
Closing costs	350,000	0	0	0	0	0	0	0	0	0	0	0	0	0	0
Fixed costs	0	360,500	371,315	382,454	393,928	405,746	417,918	430,456	443,370	456,671	470,371	484,482	499,016	513,987	529,406
Variable costs	0	0	400,000	420,000	441,000	463,050	486,203	510,513	536,038	562,840	590,982	620,531	651,558	684,136	718,343
Interest on bonds	0	3,900,000	3,900,000	3,900,000	3,900,000	3,900,000	3,900,000	3,900,000	3,900,000	3,900,000	3,900,000	3,900,000	3,900,000	0	0
Bond repayment	0	0	0	0	0	0	0	0	0	0	0	0	0	65,000,000	0
Total costs	30,350,000	74,260,500	19,671,315	4,702,454	4,734,928	4,768,796	4,804,121	4,840,968	4,879,408	4,919,511	4,961,353	5,005,013	5,050,574	66,198,123	1,247,749
Tax Calculation															
Pretax income	-350,000	-4,260,500	3,132,153	5,685,912	6,328,683	7,013,949	7,744,503	8,523,316	9,353,555	10,238,595	11,182,029	12,187,689	13,259,654	18,302,270	85,520,169
Depreciation	3,000,000	10,000,000	11,500,000	11,500,000	11,500,000	11,500,000	11,500,000	11,500,000	11,500,000	11,500,000	8,500,000	1,500,000	0	0	0
Tax credit	0	0	0	0	0	0	0	0	0	0	0	0	0	0	0
Taxable income	-3,350,000	-14,260,500	-8,367,847	-5,814,088	-5,171,317	-4,486,051	-3,755,497	-2,976,684	-2,146,445	-1,261,405	2,682,029	10,687,689	13,259,654	18,302,270	85,520,169
Tax	-938,000	-3,992,940	-2,342,997	-1,627,945	-1,447,969	-1,256,094	-1,051,539	-833,472	-601,005	-353,194	750,968	2,992,553	3,712,703	5,124,636	23,945,647
Cash flow	35,588,000	-70,267,560	-9,524,850	7,313,857	7,776,651	8,270,044	8,796,042	9,356,787	9,954,560	10,591,788	10,431,061	9,195,136	9,546,951	-51,822,366	61,574,522
Discount factor	1.0000	0.8696	0.7561	0.6575	0.5718	0.4972	0.4323	0.3759	0.3269	0.2843	0.2472	0.2149	0.1869	0.1625	0.1413
Discounted cash flow	35,588,000	-61,102,226	-7,202,155	4,808,980	4,446,326	4,111,673	3,802,772	3,517,563	3,254,163	3,010,847	2,578,399	1,976,432	1,784,393	-8,422,583	8,702,245
NPV	854,828														

Note: As in Table 9.4, but with 6.5% annual electricity price increase. Price growth rate = 6.5% annually; discount rate = 15%; fixed cost growth rate = 3%; variable cost growth rate = 3%; corporate tax rate = 28%.

TABLE 9.8

NPV of the Wind Farm: Basic Scenario with 5-Year Depreciation and 20% Tax Credit

	Year														
	0	1	2	3	4	5	6	7	8	9	10	11	12	13	14
Revenue															
Price, per kWh	0.0400	0.0408	0.0416	0.0424	0.0433	0.0442	0.0450	0.0459	0.0469	0.0478	0.0488	0.0497	0.0507	0.0517	0.0528
kWh produced	0	0	0	107,500,000	215,000,000	215,000,000	215,000,000	215,000,000	215,000,000	215,000,000	215,000,000	215,000,000	215,000,000	215,000,000	215,000,000
Revenue from electricity	0	0	0	4,563,194	9,308,917	9,495,095	9,684,997	9,878,697	10,076,271	10,277,796	10,483,352	10,693,019	10,906,879	11,125,017	11,347,517
From bonds	65,000,000	0	0	0	0	0	0	0	0	0	0	0	0	0	0
Salvage value	0	0	0	0	0	0	0	0	0	0	0	0	0	0	66,000,000
Total revenue	65,000,000	0	0	4,563,194	9,308,917	9,495,095	9,684,997	9,878,697	10,076,271	10,277,796	10,483,352	10,693,019	10,906,879	11,125,017	77,347,517
Costs															
Capital investment	30,000,000	70,000,000	15,000,000	0	0	0	0	0	0	0	0	0	0	0	0
Closing costs	0	0	0	0	0	0	0	0	0	0	0	0	0	0	0
Fixed costs	350,000	360,500	371,315	382,454	393,928	405,746	417,918	430,456	443,370	456,671	470,371	484,482	499,016	513,987	529,406
Variable costs	0	0	400,000	420,000	441,000	463,050	486,203	510,513	536,038	562,840	590,982	620,531	651,558	684,136	718,343
Interest on bonds	0	3,900,000	3,900,000	3,900,000	3,900,000	3,900,000	3,900,000	3,900,000	3,900,000	3,900,000	3,900,000	3,900,000	3,900,000	0	0
Bond repayment	0	0	0	0	0	0	0	0	0	0	0	0	0	65,000,000	0
Total costs	30,350,000	74,260,500	19,671,315	4,702,454	4,734,928	4,768,796	4,804,121	4,840,968	4,879,408	4,919,511	4,961,353	5,005,013	5,050,574	66,198,123	1,247,749
Tax Calculation															
Pretax income	-350,000	-4,260,500	-4,671,315	-139,260	4,573,988	4,726,299	4,880,876	5,037,728	5,196,863	5,358,285	5,521,999	5,688,006	5,856,305	9,926,894	76,099,768
Depreciation	6,000,000	20,000,000	23,000,000	23,000,000	23,000,000	17,000,000	3,000,000	0	0	0	0	0	0	0	0
Tax credit	3,000,000	7,000,000	1,500,000	0	0	0	0	0	0	0	0	0	0	0	0
Taxable income	-6,350,000	-24,260,500	-27,671,315	-23,139,260	-18,426,012	-12,273,701	1,880,876	5,037,728	5,196,863	5,358,285	5,521,999	5,688,006	5,856,305	9,926,894	76,099,768
Tax	-4,778,000	-13,792,940	-9,247,968	-6,478,993	-5,159,283	-3,436,636	526,645	1,410,564	1,455,122	1,500,320	1,546,160	1,592,642	1,639,765	2,779,530	21,307,935
Cash flow	39,428,000	-60,467,560	-10,423,347	6,339,733	9,733,272	8,162,935	4,354,231	3,627,164	3,741,741	3,857,965	3,975,839	4,095,364	4,216,540	-57,852,636	54,791,833
Discount factor	1.0000	0.8696	0.7561	0.6575	0.5718	0.4972	0.4323	0.3759	0.3269	0.2843	0.2472	0.2149	0.1869	0.1625	0.1413
Discounted cash flow	3,9428,000	-52,580,487	-7,881,548	4,168,477	5,565,030	4,058,422	1,882,454	1,363,585	1,223,182	1,096,675	982,767	880,271	788,101	-9,402,671	7,743,656
NPV	-684,086														

Note: As in Table 9.6, but with a 15-month delay in the production of power. Price growth rate = 2% annually; discount rate = 15%; fixed cost growth rate = 3%; variable cost growth rate = 5%; corporate tax rate = 28%.

stream is delayed by 15 months. It is observed that the effect of this delay is to reduce the NPV of the project from a positive $5,371,141 to a negative $684,086. The negative NPV simply makes the project unprofitable for XYZ Corporation. Because the investment for this project has already been made, the lack of planned revenue will also have a negative impact on the financial health of the corporation and will increase the cost of borrowing for future projects.

Longer delays, which are more common in the judicial systems of most OECD countries, in the commencement of the power generation projects, may become economically disastrous and even threaten the viability of energy corporations. Such risks significantly add to the risk of energy projects and influence the internal discount rate r_d.*

9.5.2 Use of the NPV Method for Improved Efficiency Projects

Because the NPV is a general investment appraisal method, it is used in all energy projects, including projects of energy conservation and improved efficiency. Such projects are very similar to the electricity generation projects but, usually, of shorter duration. The main difference between the electricity generation and the conservation/efficiency projects is that the "revenue" is actually generated from energy savings, which need to be quantified.

A very important source of revenue for conservation and energy efficiency projects are tax credits from central and local governments. In the wake of the realization of global and regional environmental changes, the governments of several countries, as well as of regions within countries, offer generous incentives for energy conservation/efficiency projects in the form of tax credits, grants, and accelerated depreciation. Oftentimes, these incentives are augmented by the local electricity production corporations, who perceive such projects as a means to avoid (or at least delay) new electricity generation projects that are costly and more capital intensive. The electricity production corporation CPS in San Antonio, Texas, offers rebates for all energy efficiency improvements including $225 for every ton of air-conditioning (12,000 Btu/hour) for the installation of ground source heat pumps (GSHPs). Because energy efficiency projects are smaller engineering projects, which may be completed in a few weeks, the savings (revenues) usually start accruing immediately.

Let us study the substitution of a simple, conventional air-conditioning unit in a large commercial building with a GSHP. The building is occupied and used throughout the year. It is proposed that the older air-conditioning and heating systems of the building be replaced with a GSHP, which is significantly more efficient. While the mechanical systems will be replaced, the ductwork and air distribution systems in the building will be kept intact to reduce the overall cost and the inconvenience of significant construction work inside the building. The entire replacement of the system will cost $320,000. It is estimated that the more efficient GSHP will result in savings of 270,000 kWh/year from the air-conditioning load of the building. At $0.093/kWh, this amounts to savings of $25,110/year for the owner of the building. The GSHP system will also save the equivalent of $4,450/ year from the reduced space heating cost during the winter and from the hot water supply during the entire year. Such savings are described in more detail in Section 8.3.3, where a complete analysis of the GSHP system is presented. The total savings for the project

---

* A series of lawsuits filed during the 1980s in the United States opposing the construction of nuclear power plants resulted in nuclear reactor construction delays of 2–7 years. The lawsuits and the delays rendered several nuclear energy projects unprofitable, and severely distressed the nuclear energy industry in the United States. Ironically, the building of an additional 56 nuclear reactors in the 1990s would have made the United States compliant with the Kyoto protocol.

amount to \$29,560/year; they start accruing in year zero and are expected to increase at an annual rate of 3%, reflecting higher energy prices in the future. In addition, the federal US tax code allows for 30% tax credit for the owner of the building and depreciation of the investment in 5 years. The GSHP system is expected to have a lifetime of 15 years and does not need significant maintenance. The discount rate for the investment is 15%, and the tax rate for the owner of the building is 28%.

Table 9.9 shows the monetary benefits for the owner of the building for the 15 years of the operation of this GSHP system. Figure 9.4 is a schematic of the cash flow for this project. The cash flows in the first 5 years, shown as years 0–4, include the tax benefits from the depreciation of the investment. The NPV for this project is positive (\$74,130), and this implies that it is profitable for the owner of the building to pursue this energy efficiency project.

TABLE 9.9

NPV Calculation of the GSHP Project

Year	Investment	Savings	Tax Benefits	Cash Flow	Discounted Cash Flow
0	−320	29.560	113.92	−176.52	−176.52
1	0	30.447	17.92	48.37	42.06
2	0	31.360	17.92	49.28	37.26
3	0	32.301	17.92	50.22	33.02
4	0	33.270	17.92	51.19	29.27
5	0	34.268		34.27	17.04
6	0	35.296		35.30	15.26
7	0	36.355		36.36	13.67
8	0	37.446		37.45	12.24
9	0	38.569		38.57	10.96
10	0	39.726		39.73	9.82
11	0	40.918		40.92	8.80
12	0	42.145		42.15	7.88
13	0	43.410		43.41	7.06
14	0	44.712		44.71	6.32
					NPV = 74.13

Note: The numbers are in thousands.

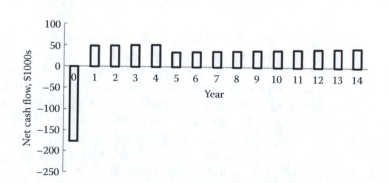

FIGURE 9.4
Cash flow diagram for the installation of a GSHP system.

The tax benefits column includes the initial 30% tax credit, which amounts to $96,000, and the tax benefits due to the depreciation. The positive NPV in this case justifies the investment for the energy efficiency project. It is evident, however, that the tax credit of 30% ($96,000 in year 0) plays a very important role in the positive NPV and the profitability of this project. The fast depreciation over 5 years also helps in the financial picture of the project. Without the tax credit, the NPV of the project would have been negative, and the project, unprofitable to undertake. This is a common characteristic of several alternative energy and improved efficiency projects. Because of the low price of fossil fuels, the maturity of fossil fuel technology, and the long-term engineering experience with fossil fuel projects, switching to alternative energy sources is a riskier investment and becomes unprofitable without economic and financial "externalities," such as favorable regulations and governmental incentives. These incentives have been justified in most countries by the following reasons:

1. The potential adverse effects of atmospheric carbon dioxide accumulation for the global climate
2. The declining supply of fossil fuels and the certainty of their depletion in the future, which necessitates the replacement of fossil fuels in the near or far future by renewable energy sources
3. The creation of new professions and new jobs that may support the national economies in the future
4. Energy security for the countries that import fossil fuels

It has been observed that several electricity generation corporations (utilities in the United States) are willing to subsidize with additional rebates all the energy efficiency projects, such as the one described in this section. This trend is more evident with the state- or city-owned energy production entities that include in their mission concern for the environment and sustainability.

For the case exposed in Table 9.9, the underlining reason for the cash rebate is that the more efficient air-conditioning units in the region served by the electricity generation corporation will cause lesser peak power demand during the summer months, and this includes three beneficial effects for the power producing corporations:

1. The lesser peak power demand implies that the most inefficient and most expensive power producing units will not need to be operated for the production of power.
2. Any growth in the demand for electricity, e.g., because of population growth, will not need to be met immediately, thus deferring the high capital costs associated with the building of new power units further in the future. Adding new electricity generation capacity, e.g., by the construction of a new nuclear power plant, may be significantly more expensive and capital intensive for a corporation than encouraging energy conservation and higher efficiency projects for their customers.
3. There are lesser atmospheric emissions of pollutants from the power plants of the corporation. Emission reductions are often mandated by governmental regulations and are expected by the citizens in the area served by electricity generating corporations.

9.5.3 Financing Energy Efficiency Projects as Mortgages

A convenient method to finance energy efficiency projects by homeowners and owners of commercial buildings is by taking a loan connected to the mortgage of the building. Such a loan is repaid in equal installments in 5–30 years. Because a mortgage uses the present market value of the building as collateral, it bears significantly lower interest rate than a commercial loan. As with the common mortgage loans, the installments include all the interest as part of the borrowed capital. The following example illustrates the financing of such a project by a mortgage.

Example 9.8

A homeowner is considering financing a PV system with a 6-year mortgage. The PV system costs $45,000; it is estimated to last for 30 years and will be financed in its entirety with a 6% loan payable within 6 years. The tax credit for the investment is 30%; the monetary savings from the electric power produced are estimated at $1,000 in the first year (year 0 of the project) and $2,500 per year in years 1–29. The insurance and maintenance costs of the system are estimated at $400 per year in all the years of its operation. The marginal tax rate for the homeowner is 30%, and he/she may deduct the interest paid for this loan from his/her taxes.

Determine (a) the annual amount of the mortgage payment and (b) the NPV of this project if the discount rate for the homeowner is 5%.

Solution: The mortgage on this loan may be calculated from Equation 9.4 or with "mortgage calculators," which are available in several Internet sites. For the specified loan, the annual payment is $8,949.36. Assuming that the project starts at the beginning of year 0, the following table gives the principal and interest paid for the 6 years of the duration of the loan:

	Year					
	0	**1**	**2**	**3**	**4**	**5**
Total payment ($)	8,949.36	8,949.36	8,949.36	8,949.36	8,949.36	8,913.36
Principal ($)	6,424.11	6,820.34	7,241.00	7,687.61	8,161.77	248.19
Interest ($)	2,525.25	2,129.02	1,708.36	1,261.75	787.59	8,665.17
Loan balance ($)	38,575.89	31,755.55	24,514.55	16,826.93	8,665.17	0.00

The NPV of the energy revenue/savings, projected to year 0, is

$$\text{NPV: Revenue} = 1,000 + \sum_{n=1}^{29} \frac{2,500}{(1.05)^n} = 1,000 + 2,500 \times 14.9 = 38,245.$$

Similarly, the NPV for the costs, projected to year 0 is

$$\text{NPV: Costs} = \sum_{n=0}^{29} \frac{400}{(1.05)^n} = 400 \times 15.9 = 6,359.$$

The following table gives the cash flow for the life cycle of this project. Because in years 6–29, the revenue and expense are constant, these items have been projected to year 0. In this case, with the revenues and costs projected to year 0, we only need to perform the calculations in the years 0–5, when this loan is active.

	Year					
	0	1	2	3	4	5
Revenue ($)	38,245.00					
Cost ($)	6,359.00					
Loan payment ($)	8,949.36	8,949.36	8,949.36	8,949.36	8,949.36	8,913.36
Tax credit ($)	13,500.00					
Interest deduction from taxable income ($)	757.58	638.71	512.51	378.53	236.28	85.26
Cash flow ($)	37,194.22	−8,310.65	−8,436.85	−8,570.84	−8,713.08	−8,828.10
Discounted cash flow ($)	37,194.22	−7,914.91	−7,652.47	−7,403.81	−7,168.27	−6,917.05
NPV	137.70					

This example introduces a useful shortcut in the application of the NPV method, when a long stream of costs and revenues is projected to a single year. The spreadsheet with the pertinent computations is considerably shortened. One may note in this example that the building owner does not risk any of his/her own capital because the entire investment is financed by the mortgage loan. For this reason, she/he could have decided on another, perhaps lower discount rate, even $r_d = 0.0$. A low discount rate in the range of 0–5% is justified because the owner takes a small risk by taking the mortgage on his/her house and because the revenue/savings from the production of electric energy is not guaranteed. As with the other examples in this chapter, one notes that the tax credit of $13,500 is a significant factor for a positive NPV and the decision of the home owner to undertake the project. The interest deduction from taxes for this project is also helpful in rendering the NPV positive and the project worth developing.

9.6 Project Financing for Alternative Energy Technology

It is apparent from the examples of the previous sections that the building of new alternative energy power plants and the completion of energy efficiency projects require significant injection of capital. Energy projects are capital intensive, and especially, new technology for the production of electric power from renewable energy sources requires significant capital investment. Renewable energy projects are characterized by relatively high capital cost and very low variable cost, because the cost of their "fuel" is zero. In addition, all energy generation and efficiency projects are long-term projects with timescales that extend far into the future: wind turbines are expected to operate for 25–40 years; PV panels for 30–40 years; geothermal and fossil fuel plants have been in operation for more than 40 years; most of the nuclear reactors that were built in the 1960s operate 55 years later and have received extensions to operate for 60–70 years; hydroelectric power plants are expected to be in operation for 100–150 years from the commencement of their operations*; and new GSHPs for buildings are expected to operate for more than 35 years.

The long timescale of energy projects, and especially of renewable energy projects, makes investments in them qualitatively different from other market investments. In the early twenty-first century, the world economy and the international financial markets are very much influenced by the 1985–2015 "high-tech" economic era when new inventions and technological developments in electronics, computers, and the Internet created the "new world economy" and produced new investment paradigms with several success-ful, high-technology corporations—Microsoft, Apple, Google, Facebook, Alibaba, Netflix, Amazon, etc. The development and establishment of these, now gigantic, corporations involved an excellent original concept and a relatively small amount of initial capital that helped to develop the concept. The initial capital was provided by teams of investors—in the United States, they are called "angels" and typically invest a few hundred thousand dollars in the early stages of a startup corporation. After the success of the initial stages "venture capitalists" invest a few million dollars in the second stage of development of the startup corporation. While several projects undertaken by angels and venture capitalists have failed, the financial rewards from successful projects were enormous, and the inves-tors realized 10–1000 times return on the capital they invested. The rewards were also very quick to materialize, typically within 3–6 years, usually at the time when the original corporation made its first initial public offering (IPO) of stock to be publically traded.

Investments in energy production and energy efficiency projects differ from the type of investments and time frames of the "high technology" projects for the following reasons:

1. Oftentimes, the technology involved in alternative energy projects is not new and has been applied elsewhere. Energy projects are not as risky. As a consequence, they do not warrant the high internal discount rates r_d and the very high profit-ability expected by the successful high-technology projects.

2. Energy production and conservation projects are easier to be reproduced. Patents that may ensure exclusive rights to all similar projects and high profitability to the investors are more difficult or impossible to obtain.

3. All energy production projects are long-term projects with payoffs that extend far into the future. The original investors that risk their capital, typically, do not recoup their investment within 3–6 years in the IPO of the corporation's stock.

4. The initial investment required for even a small alternative energy power plant is very high in comparison to investments in internet-related corporations. A new 60 MW wind farm costs approximately $100 million, and a 1000 MW nuclear power plant in the United States requires an investment of approximately $9 bil-lion. Energy hardware is by far more expensive than Internet software.

5. Because of the high energy price uncertainty, the revenue of energy projects is also uncertain. The uncertainty of energy projects is by far lower than the uncer-tainty of high-technology projects, where the investors do not know if the entire concept/idea is going to work and produce any revenue.

---

* Maintenance and replacement of parts of power plants is periodically required. These periodic expenses are by far lower than the construction of new power plants.

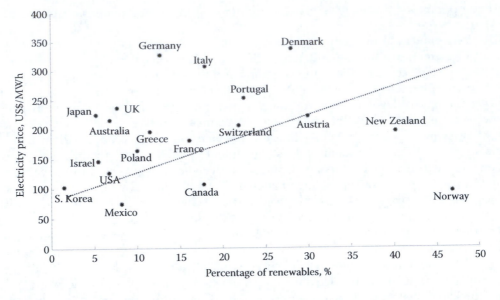

FIGURE 9.5
Percentage of renewables in the TPES and electricity price for households.

The high-tech financial models are not applicable to energy production investment models. It is not rational to expect that the free markets will provide the means to finance all the alternative energy projects that have reasonable potential to become successful and profitable. However, energy production and energy projects are necessary for the functioning of the contemporary human society, and they are vital for the national economy. It is a matter of national interest that ways must be found for energy projects to be developed so that the energy supply is not interrupted.

Several governments—both regional and national—have instituted incentives for the development of new energy projects, especially renewable energy projects. The continuation of regulatory intervention, governmental subsidies, financial guarantees, and financial incentives may be necessary for the development and the long-term success of alternative energy and energy efficiency projects, at least until the expected or perceived depletion of the fossil fuel resources ushers a more favorable and realistic price regime for solar and wind energy. Oftentimes, the governmental incentives for renewable energy are paid by taxes on all other energy forms or taxes on the sale of electric power. The taxes collected, effectively subsidize renewable energy, especially solar and wind projects. Figure 9.5 depicts the fraction of the total primary energy source (TPES) supplied by renewable energy in several nations, versus the price paid for electric energy by the households [6,7]. It is observed in this figure that there is a strong, positive correlation between the percentage of renewables used and the price of electricity paid by the households.* The two outlier countries in the figure—Norway and New Zealand—have small populations and extensive renewable energy resources, hydroelectric for Norway and geothermal and hydroelectric for New Zealand. Canada, which also sells electricity at very low price, generates 60% of its electric energy by hydroelectric power plants. On the other side of the trend line, Germany, Denmark, several other EU countries, and Japan impose taxes on the sale of electric energy

---

* Industrial users of electric energy are often excluded from such taxation. In all countries that appear in the figure, the price of electric energy for industrial users is significantly lower [6].

and use the proceeds to subsidize renewable energy, energy conservation, and higher efficiency projects. The trend in the figure shows the higher cost of renewables (except for hydroelectric energy) at present for the production of electricity. It also implies that more technological advances are needed in the future to reduce the price of renewables and make the renewable energy sources competitive with other sources of electricity. Care for the environment, the mitigation of global environmental change, and the reduction of atmospheric pollution—a significant public health benefit in all countries—are good reasons for the continuation of governmental incentives and fully justify governmental interventions in the energy sector even in the free-market economies.

PROBLEMS

1. You are considering the construction and operation of a 5 MW solar PV power plant. Enumerate three of your expected fixed costs and three of your expected variable costs.

2. List all the costs you will expect to incur from the construction and operation of a geothermal power plant, which is to produce energy for 40 years. Differentiate them in fixed and variable costs.

3. Because of population growth, it is expected that the city you live in will need an additional 300 MW of electric power in the next 10 years. Enumerate 6–8 alternatives that would satisfy the increased demand.

4. The expected average rate of return in the next decade is 4%. What is the value of $60,000 10 years from now?

5. A financial company uses a 5% rate of return today and expects that this rate of return will increase by one percentage point in each of the next 9 years. How much is today's sum of $100,000 worth to this company 10 years from now?

6. You borrow $250,000 for a business venture to be repaid 3 years from now. You have the option to take a simple interest rate of 40% or a 1% rate compounded monthly interest. Which option will you choose?

7. What is the present value of $1,000,000 20 years from now? The discount rate is 7%.

8. A small geothermal power plant is expected to generate a net income of $300,000 in each of the next 30 years. What is the present value of this monetary stream if the discount rate is 6.5%?

9. From the point of view of an investment company, explain the difference between the following:
   a. The interbank discount rate
   b. The prime rate
   c. The cost of borrowed capital for a corporation
   d. The return on equity for a corporation

10. For the small PV power plant of example 9.3, determine the following:
   a. The NPV for a discount rate 7%
   b. The PBP
   c. The ARB if the straight-line depreciation is taken
   d. The IRR
   e. The PI

11. For the small PV power plant of example 9.3, what would be the NPV if there were a 25% investment credit in addition to the cash flow shown in the years when the investment is made?

12. *The initial investment for a geothermal power plant is $60,000,000 and is to be paid in 2 years (year 0 and year 1) in equal installments. The electricity generated by the plant is expected to bring revenue of 5,000,000 per year, starting in year 2 and increasing at a rate of 3% annually. The total annual expenses of the plant will be 1,500,000 increasing at a rate 4% annually. The corporation that owns the plant will issue 15-year bonds in year 1 for 50% of the invested capital at a rate 6%. The discount rate for this company is 12% and its taxation rate is 35%. A 10-year depreciation period is allowed for all investments in this power plant. The plant is expected to generate electricity until year 30, after which it will have zero value. What is the present value of this investment? Should the company undertake the investment?

13. What would be the answer to the plant of problem 12 if the company were to borrow 80% of the cost of the geothermal power plant with 15-year bonds at an annual rate of 6%?

14. In order to promote the development of alternative energy sources the government allows 5-year depreciation for the geothermal project of problem 12 and the company borrows 80% of the invested capital as in problem 13. Should the company undertake the investment?

15. In order to accelerate the use of geothermal power, the government offers a 20% tax credit for investments in geothermal energy (in the years the investment is made) as well as 5-year depreciation for the project of problem 12. How does this affect the NPV of the investment?

16. "The government does not need to subsidize energy projects. In a free market economy we should allow the market forces to determine if a project is going to succeed or fail." Comment.

References

1. Michaelides, E.E., *Alternative Energy Sources*, Springer, Berlin, 2012.
2. Park, C.S., *Contemporary Engineering Economics*, 4th edition, Pearson, Old Tappan, NJ, 2007.
3. Sullivan, W.G., Wicks, E.M., Luxhoj, J.T., *Engineering Economy*, 12th edition, Pearson, Old Tappan, NJ, 2003.
4. Stirling, A., Regulating the electricity supply industry by valuating environmental effects: How much is the emperor wearing?, *Futures*, **24**, 1024–1047, 1992.
5. Park, C.S., Sharp-Bette, G.P., *Advanced Engineering Economy*, Wiley, New York, 1990.
6. International Energy Agency, *Key World Statistics—Prices*, IEA-Chirat, Paris, 2016. (The 2015 edition was consulted for the price data in Italy).
7. International Energy Agency, *Energy Statistics of OECD Countries 2015*, OECD, Paris, 2015.

---

* Construct a spreadsheet to facilitate the solution of the next four problems.

Answers to Selected Problems

Chapter 1:

2. (a) 10 MJ = 9,479 Btu; (b) 45 kWh = 153,540 Btu; (c) 215 MWh = 734 × 10^6 Btu.

4. 2,000 MW.

6. 65.3%

8. In United States, savings are $444/year; in Germany, savings are €673/year.

10. 2,658 Wh.

11. 16.2 kW.

Chapter 2:

2. (a) 136.8 × 10^9 MJ; (b) 153 × 10^9 MJ; (c) 439.5 × 10^3 MJ.

6. In 2014, from renewables: 5,266 TWh; this is approximately 60% of the TWh produced from coal.

9. $T \approx 109$ years.

10. $T = 34$ years; with the higher amount of reserves, $T = 57$ years.

Chapter 3:

2. 352,500 lb.

4. (a) 33 × 10^9 MJ annually; (b) 3.43 × 10^6 kg coal daily; 8.8 × 10^6 kg CO_2, 138 × 10^3 kg SO_2 annually.

6. 5.62 × 10^9 kg of CO_2 annually (58% emission reduction).

8. (a) $T_E \downarrow$; (b) $T_E \downarrow$; (c) $T_E \uparrow$; (d) $T_E \uparrow$; (e) no difference; (f) $T_E \downarrow$.

10. (a) 45.1 × 10^9 MJ/year; (b) 1.9 × 10^9 kg coal; (c) 99.8 × 10^8 kg Ca(OH) annually.

12. (a) 18,922 TJ; (b) 64,028 TJ; (c) 1,437 TJ; (d) 2,962 TJ.

14. 243°C.

Chapter 4:

2. 47,190 kJ/kg.

4. (a) 26,853 kJ/kg; (b) 12.2 kg of air per kg of coal; (c) 40.3×10^6 kg air per day.

6. (a) 4.7×10^3 t of peat daily; (b) 1,043 t of ash daily; 377,410 t ash yearly.

8. (a) 47,028 kJ/kg; (b) 640 kg, 539 m^3.

10. 3 mm → 14.9 m/s; 5 mm → 19.2 m/s; 10 mm → 27.1 m/s.

12. Issues to be discussed in essay: Environmental effects of fracking, benefits of fracking, property rights (for the community and for the owners of mineral wealth).

Chapter 5:

2. $X \to {}_{93}Np^{239}$; $Y \to {}_{94}Pu^{239}$; $Z \to {}_{54}Xe^{143}$; $W \to {}_{-1}e^0$.

4. 643.2×10^{12} J.

6. 15 billion years.

8. 455,700 MWh/t.

9. (a) 76.9×10^9 MJ; (b) 8.61×10^9 kg CO_2; 149,383 kg.

10. Use the material in Section 5.2.

Chapter 6:

2. 2,743,000 m^3/s.

4. 147–1840 kW.

6. (a) 10,512 kWh; (b) 11,388 kWh; (c) 21,900 kWh; (d) 20,148 kWh; (e) 15,768 kWh.

8. 318.2 m^2.

10. 4.8 MW.

12. 12 m^2.

13. 64,350 m^2.

18. 15.9 W/m^2.

20. 281 kW.

24. 316,200 kWh/year.

26. Approximately 401 kWh.

28. (a) 900 kWh; (b) 900 kWh (the air velocity does not exceed 12 m/s).

32. 4,704 kW.

34. (a) 1.04 kg/s; (b) 553 kW; (c) 4.844×10^6 kWh/year; (d) $421,000/year.

36. (Use exergy) 64,691 kWh.

38. 5.9 MW.

40. max power 4,748 kW; optimum $T \approx 100°C$; optimum power $\approx 1,580$ kW.

42. (a) 277,200 kW; (b) 147.3×10^6 kg/year; (c) 441.8×10^6 kg/year for CO_2.

46. 4.8 million trees.

49. Approximately 0.59 million gal of diesel annually.

50. 178×10^6 kg coal saved; 543×10^6 kg of CO_2 avoidance per year.

52. 4.06×10^9 kg bagasse.

56. 2.88 million mi.

58. 104 km².

60. Approximately 47,000 turbines needed.

62. 13.0 MW.

64. 16.4 kW.

66. Max power of 10.6 kW.

Chapter 7:

2. AF = 96.4%; POF = 90.7%.

4. (a) 3.68×10^{12} J; 204.5 MW; 49.4%.

6. 65.4 kJ; 305×10^6 N.

8. (a) 1.42×10^9 J; (b) 45.2°C; (c) 1.36×10^9 kJ.

10. 1.48 V.

12. 50.4 kg.

14. Air: 23,016 kJ; hydrogen: 347,743 kJ.

Chapter 8:

3. (a) 3,585 gal/year; (b) 597 million gal/year; (c) CO_2 avoidance: 4.9 million t/year.

4. −2,087 kJ.

5. 62% reduction.

8. (a) Electricity savings approximately 14,200,000 Btu; (b) 4,538 less kg of CO_2 emitted per year.

10. With COP = 3 for air conditioning (a/c) and 4 for heat pump (hp) savings: 7,963,200 kWh.

12. T_2 = 33°C and 30.5°C.

14. 211,816,800 bbl; 73 × 10^9 kg CO_2.

18. (a) 467 kJ/kg; (b) 2,505,360 kWh.

Chapter 9:

2. Land, construction, equipment, maintenance, management, operations, sales, taxes, closing.

4. $88,800.

6. Simple interest: $350k; compound interest: $357.7k.

8. $4,218,000.

10. (a) NPV = $49,596; (b) PBP = 10 years; ARB = 82%; IRR = 7.62%; PI = 92.7%.

12. −$10,269 (do not undertake).

14. $2,952 (undertake).

Index